Lecture Notes in Networks and Systems

Volume 1149

The series "Lecture Notes in Networks and Systems" publishes the latest developments in Networks and Systems—quickly, informally and with high quality. Original research reported in proceedings and post-proceedings represents the core of LNNS.

Volumes published in LNNS embrace all aspects and subfields of, as well as new challenges in, Networks and Systems.

The series contains proceedings and edited volumes in systems and networks, spanning the areas of Cyber-Physical Systems, Autonomous Systems, Sensor Networks, Control Systems, Energy Systems, Automotive Systems, Biological Systems, Vehicular Networking and Connected Vehicles, Aerospace Systems, Automation, Manufacturing, Smart Grids, Nonlinear Systems, Power Systems, Robotics, Social Systems, Economic Systems and other. Of particular value to both the contributors and the readership are the short publication timeframe and the world-wide distribution and exposure which enable both a wide and rapid dissemination of research output.

The series covers the theory, applications, and perspectives on the state of the art and future developments relevant to systems and networks, decision making, control, complex processes and related areas, as embedded in the fields of interdisciplinary and applied sciences, engineering, computer science, physics, economics, social, and life sciences, as well as the paradigms and methodologies behind them.

Indexed by SCOPUS, EI Compendex, INSPEC, WTI Frankfurt eG, zbMATH, SCImago.

All books published in the series are submitted for consideration in Web of Science.

For proposals from Asia please contact Aninda Bose (aninda.bose@springer.com).

Siba Kumar Udgata · Srinivas Sethi ·
George Ghinea · Sanjay Kumar Kuanar
Editors

Intelligent Systems

Proceedings of 4th International Conference on Machine Learning, IoT and Big Data (ICMIB 2024), Volume 2

Springer

Editors
Siba Kumar Udgata
University of Hyderabad
Hyderabad, Telangana, India

George Ghinea
Department of Computer Science
College of Engineering Design
and Physical Sciences
Brunel University London
Uxbridge, UK

Srinivas Sethi
Department of Computer Science
Engineering and Applications
Indira Gandhi Institute of Technology
Dhenkanal, Odisha, India

Sanjay Kumar Kuanar
School of Engineering and Technology
GIET University
Gunupur, Odisha, India

ISSN 2367-3370 ISSN 2367-3389 (electronic)
Lecture Notes in Networks and Systems
ISBN 978-981-97-8159-1 ISBN 978-981-97-8160-7 (eBook)
https://doi.org/10.1007/978-981-97-8160-7

This Springer imprint is published by the registered company Springer Nature Singapore Pte Ltd.
The registered company address is: 152 Beach Road, #21-01/04 Gateway East, Singapore 189721, Singapore

Committees

Chief Patrons

Prof (Dr.) Satya Prakash Panda, President, GIET University, India
Prof (Dr.) Chandra Dhwaj Panda, Vice President, GIET University, India
Prof (Dr.) Jagadish Panda, Director General, GIET University, India

Patrons

Dr. A. V. N. L. Sharma, Vice Chancellor, GIET University, India
Dr. N. V. Jagannadha Rao, Registrar, GIET University, India

General Chair

Prof. Lalit Mohan Patnaik, National Institute of Advanced Studies and Indian Institute of Science, Bengaluru

Program Chair

Prof. Siba K. Udgata, University of Hyderabad, India

Program Co-chairs

Prof. Srinivas Sethi, IGIT Sarang
Prof. Gheorghita (George) Ghinea, Computer Science, Brunel University
Prof. Sanjay Kumar Kuanar, GIET University, India

Organising Chairs

Prof. S. N. Mishra, IGIT Sarang
Dr. Raghvendra Kumar, GIET University, India

Convenor

Prof. Srinivas Sethi, IGIT Sarang
Dr. Sanjay Kumar Kuanar, GIET University, India

Publicity Chairs

Dr. K. Murali Gopal, GIET University, India
Dr. Neelamadhab Padhy, GIET University, India
Dr. A. V. S. Pavan Kumar, GIET University, India
Dr. Sachikanta Dash, GIET University, India
Prof. B. P. Panigrahi, IGIT Sarang
Prof. Ashima Rout, IGIT Sarang

Finance Chairs

Dr. Nilambar Sethi, GIET University, India
Dr. Bidush Kumar Sahoo, GIET University, India

Hospitality Chairs

Dr. Premansu Sekhara Rath, GIET University, India
Dr. Jyotirmaya Mishra, GIET University, India

Mrs. Rasmita Panigrahi, GIET University, India
Mr. K. Siva Krishna, GIET University, India

Logistic Chairs

Prof. Satya Narayan Das, GIET University, India
Mr. Ranjit Patnaik, GIET University, India
Mr. Ranjeet Panigrahi, GIET University, India
Mr. Santosh Kumar Panda, GIET University, India

Accommodation Chairs

Mr. Sibo Prasad Patro, GIET University, India
Mr. P. Sudheer Babu, GIET University, India
Mr. Chandrakanta Mahanty, GIET University, India

Organising Committee

Mr. Bhavani Sankar Panda, GIET University, India
Ms. Jemarani Jaypuria, GIET University, India
Mr. G. V. S. Narayana, GIET University, India
Mr. Jagadish Sahoo, GIET University, India

Technical Committee Members

Gheorghita (George) Ghinea, Computer Science, Brunel University
Kumar Padmanabh, EBTIC (A Research Lab of British Telecom)
Deepak Tosh, Computer Science, University of Texas at El Paso
Haesik Kim, VTT Technical Research Centre, Finland
Soroj Meher, ISI, Bengaluru
Manas Ranjan Patra, Berhampur University
Siba Kumar Udgata, University of Hyderabad
O. B. V. Ramanaiah, OBV, JNTU Hyderabad
G. Suvarna Kumar, MVGRCE, Vijayanagaram
G. Sandhya MVGRCE, Vijayanagaram
R. Hemalatha University College of Engineering, Osmania University
Amit Kumar Mishra University of Cape Town South Africa

R. Thangarajan Kongu Engineering College Tamilnadu
Birendra Biswal Gayatri Vidya Parishad College of Engineering, Vishakhapatnam
Somanath Tripathy, IIT Patna
A. K. Turuk, NIT Rourkela
B. D. Sahoo, NIT Rourkela
D. P. Mohapatra, NIT Rourkela
P. M. Khilar, NIT Rourkela
P. G. Sapna, CIT, Coimbatore
R. K. Dash, NIC, Mizoram
B. K. Tripathy, VIT, Vellore
Moumita Patra IIT, Guwahati
S. N. Das, GIET University, Gunupur, Odisha
Ram Kumar Dhurkari IIM, Sirmaur, Himachal Pradesh
Chitta Ranja Hota, BITS Pilani, Hyderabad,
A. Kavitha, JNTU, Hyderabad
Subasish Mohapatra, CET Bhubaneswar
Sanjaya Kumar Panda, NIT Warangal
S. Mini, NIT, Goa
S. Nagender Kumar, University of Hyderabad
Lalit Garg, University of Malta, Malta
Lalitha Krishna, Kongu Engineering College, Tamilnadu
C. Poongodi, Kongu Engineering College, Tamilnadu
Sumanth Yenduri, Kennesaw University, USA
Shaik Shakeel Ahamad, Majmaah University, Saudi Arabia
K. Srujan Raju, CMR Technical Campus, Hyderabad
R. Hemalatha, University College of Engineering, Osmania University, Hyderabad
P. Sakthivel, Anna University
Pavan Kumar Mishra, NIT, Raipur
Tapan Kumar Gandhi, Department of Electrical Engineering, IIT, Delhi
Annappa B., NIT Surathkal
Prafulla Kumar Behera, Utkal University, Bhubaneswar, Odisha
Nekuri Naveen, School of Computer and Information Sciences, University of Hyderabad
Ch. Venkaiah, School of Computer and Information Sciences, University of Hyderabad
Rajendra Lal, School of Computer and Information Sciences, University of Hyderabad
Dillip Singh Sisodia, Department of Computer Science and Engineering, NIT, Raipur
Pradeep Singh, Department of Computer Science and Engineering, NIT, Raipur
Jay Bagga, Ball State University, USA
Sumagna Patnaik, JB Institute of Engineering and Technology, Hyderabad
Ajit K. Sahoo, University of Hyderabad
Atluri Rahul, Neurolus Systems, Hyderabad
Samrat L. Sabat, Center for Advanced Studies in Electronic Science and Technology (CASEST), University of Hyderabad

Nihar Satapathy, Sambalpur University
Susil Kumar Mohanty, Computer Science and Engineering, IIT, Patna
Kagita Venkat, NIT, Warangal
Sanjay Kuanar, GIET University, Gunupur, Orissa
Bhabendra Biswal, College of Engineering, Bhubaneswar
Padmalaya Nayak, GR Institute of Engineering and Technology, Hyderabad
Bhibudendu Pati, R. D. Womens University, Bhubaneswar
Chabi Rani Panigrahi, R. D. Womens University, Bhubaneswar
Rajesh Verma, Infosys Ltd., Hyderabad
Arun Avinash Chauhan, School of Computer and Information Sciences, University of Hyderabad
Khusbu Pahwa, Delhi Technological University, New Delhi
Soumen Roy, DRDL, Hyderabad
Satyajit Acharya, Tech Mahindra, Hyderabad
Subhrakanta Panda, BITs Pilani, Hyderabad
Vineet P. Nair, School of Computer and Information Sciences, University of Hyderabad
Subash Yadav, Department of Computer Science, Central University of Jharkhand, Ranchi
Layak Ali, Central University Karnataka, Gulbarga
Deepak Kumar, NIT Meghalaya
Bunil Balabantaray, NIT, Meghalaya
Sumanta Pyne, NIT Rourkela
Asis Tripathy, VIT, Vellore
Mousumi Saha, NIT Durgapur
Abhijit Sharma, NIT, Durgapur
Mayukh Sarkar, MNNIT, Allahabad
Oishila Bandyopadhyay, IIIT, Kalayani
Subrat Kumar Mohanty, IIIT, Bhubaneswar
Ramesh Chandra Mishra, IIIT Manipur
Hirak Maity, Kolaghat Engineering College
Sandeep Kumar Panda, ICFAI, Hyderabad
Ashim Rout, IGIT Sarang
Srinivas Sethi, IGIT Sarang
S. N. Mishra, IGIT Sarang
Sangita Pal, IGIT Sarang
Sanjaya Patra, IGIT Sarang
Biswanath Sethi, IGIT Sarang
Niroj Pani, IGIT Sarang
Dillip Kumar Swain, IGIT Sarang
B. P. Panigrahy, IGIT Sarang
Rabindra Behera, IGIT Sarang
L. N. Tripathy, OUTR Bhubaneswar
B. B. Choudhary, IGIT Sarang
Dhiren Behera, IGIT Sarang

Anand Gupta, IGIT Sarang
Ayaskanta Swain, NIT Rourkela
Gayadhr Panda, NIT Meghalaya
S. K.Tripathy, IGIT Sarang
Devi Acharya, VIT, Vellore
Saurov Bhoi, PMEC Berhampur
Kalyan Kumar Jena, PMEC Berhampur
Alok Ranjan Prusty, Skill Development, Delhi
Babita Majhi, GGU, Chhattisgarh
Prof. Subhrashu Das, GCE, Keonjhar
Puspalata Pujahari, GGU, Chhattisgarh
Tirimula Rao, JNTU Kakinada
Kshirsagar Sahoo, VIT, AP
Niranjan Panigrahy, PMEC Berhampur
Trilochon Rout, PMEC Berhampur
P. K. Panigrahy, GIET, India
Mihir Kumar Sutar, UCE BURLA
Gopal Behera, GCEK, Bhawanipatna
Rajendra Prasad Nayak, GCEK, Bhawanipatna
Kaliprasan Sethi, GCEK, Bhawanipatna
Jitendra Kumar Rout, NIT Raipur

Reviewers

Alekha Mishra
Alok Prusty
Alok Tripathy
Anitha A.
Anup Maharana
Arunima Pattanayak
Ashima Rout
Ashis Das
Atish Nanda
Bhargavi B.
Bibhudatta Sahoo
Bibudhendu Pati
Bichitra Mandal
Bichitra Sethi
Bunil Balabantaray
Chhabi Rani Panigrahi
Chinmayee Rout
Chiranji Lal Chowdhary
Debabrata Dansana
Debasis Mohapatra
Debendra Muduli
Deepak Gupta
Deepak Tosh
Devesh Bandil
Diptendu Singha Roy
P. K. Panigrahi
Rabindra Behera
Srinivas Sethi
Abhinav Tomar
Bikramaditya Das
Krishnaveni

Sangharatna Godboley
Santosh Majhi
Suvendu Narayan Mishra
Gopal Behera
Hemant Apat
IGIT Sarang
Indrajeet Gupta
J. Chandrkanta Badajena
Janmenjoy Nayak
Jitendra Kumar
Jitendra Rout
K. Hemant Reddy
Kali Rath
Kalyan Kumar Jena
Kaushik Mishra
Krishna Kumar
Kshira Sahoo
Lakhmi Das
Lalatendu Muduli
Lalitha Krishnasamy
Madhu Panda
Malti Nagle
Manoj Kumar Patra
Niranjan Panigrahi
Om Prakash Jena
Pavan Mishra
Pradumn Kumar
Prahallad Sahu
Prajna Paramita Nanda
Prasant Dhal
Pratap Puhan
Preeti Chandrakar
Pritee Parwekar
Priyadarsan Parida
Priyanka Sen
Punyaban Patel
Purna Sethi
Pushkar Kishore
Raghunandan Swain
Raghvendra Kumar
Rajendra Nayak
Rajiv Senapati
Rakesh Chandra Balabantaray
Rakesh Swain
Ramesh Sahoo

Ranumayee Sing
Rashmi Sahoo
Reviewer Name
Rohit Kumar Bondugula
Rohit Sharma
S. Gopal Krishna Patro
Saifulla Md
Salman Abdul Moiz
Sambit Kumar Mishra
Sampa Sahoo
Sandhya Sahoo
Sangita Pal
Sanjay Kuanr
Sanjaya Panda
Sanjib Nayak
Santanu Dash
Saroj Panigrahy
Sashi Nayak
Sasmita Acharya
Sasmita Behera
Satyapada Das
Satyasundara Mahapatra
Sibarama Panigrahi
Sipra Swain
Sohan Pande
Sourav Bhoi
Sreyasee Rout
Srichandan Sobhanayak
Subasish Mohapatra
Subhashish Nayak
Subhasish Pani
Sudhansu Patra
Suman Das
Sumit Kar
Sunita Dhalbisoi
Suraj Sharma
Surya Das
Susil Mohanty
Suvendra Jayasingh

Swarupananda Bissoyi
Tirimula Rao Benala
Umakanta Samantsinghar
Umashankar Ghugar
Umesh Sahu
Yatindra Gopal

Preface

It is really a pleasure to pen down a message on the eve of the International Conference on Machine learning, Internet of Things and Big data (ICMIB-2024) being organized by the prestigious GIET University, Gunupur, Odisha, India, during 29–31 March 2024. The conference draws some excellent technical keynote talks and papers. We received 368 full paper submissions and we accepted only 75 papers. The contributing authors are from different parts of the globe that includes India. All the papers are reviewed by at least three independent reviewers and in some cases by as many as four reviewers. All the papers are also checked for plagiarism and similarity score. It was really a tough job for us to select the best papers out of so many good papers for presentation in the conference. We had to do this unpleasant task, keeping the Springer guidelines and approval conditions in view. We take this opportunity to thank all the authors for their excellent work and contributions and also the reviewers who have done an excellent job.

On behalf of the program chair, we are indebted to Prof. L. M. Patnaik, General Chair of the Conference, for his timely and valuable advice. We cannot imagine the conference without his active support at all the crossroads of decision-making process. We would also like to place on record our thanks to all the keynote speakers, tutorial speakers, reviewers, session chairs, authors, technical program committee members, various chairs to handle finance, accommodation, hospitality, publication and publicity and above all to several volunteers.

We are also thankful to Springer Nature publication house for agreeing to publish the accepted papers in their *Lecture Notes in Networks and Systems* (LNNS) series.

The organizing committee made all arrangements to make the conference enjoyable and a good learning experience for everybody. Hope all the delegates attending the virtual sessions also enjoy the presentations and also use the opportunity to exchange their ideas.

Gunupur, India

Prof. Siba Kumar Udgata
Prof. Srinivas Sethi
Prof. Sanjay Kumar Kuanar
Prof. George Ghinea
Program Chairs, ICMIB-2024

Contents

Editors and Contributors

About the Editors

Prof. Siba Kumar Udgata is a Professor in Computer and Information Sciences at University of Hyderabad, India. He has a Ph.D. in Computer Science in the area of mobile computing and wireless communications and also worked as United Nations Fellow and worked in the UNU/IIST, Macau. His research focus is on wireless communication, Wi-Fi sensing, mobile computing, intelligent sensors, sensor network algorithms, Internet of Things, and applications. He has published more than 150 research papers in reputed international journals and conference proceedings. He has worked as Principal Investigator in many Government of India funded research projects mainly for the development of wireless sensor network applications, Wi-Fi sensing for assistive healthcare, and Wi-Fi sensing-enabled smart-home applications.

Prof. Srinivas Sethi is a Professor in Computer Science Engineering and Application, Indira Gandhi Institute of Technology, Sarang (IGIT, Sarang), India, and has been actively involved in teaching and research since 1997. He did his Ph.D., in the area of routing algorithms in mobile ad hoc network and is continuing his research work in the wireless sensor network, cognitive radio network, cloud computing, BCI, and Cognitive Science. He is Member of Editorial Board for different journal and Program Committee Member for different international conferences/workshop. He is Book Editor of four international conference proceedings published in Springer, and Taylor & Francis. He has published more than 90 research papers in international journals, edited book chapters, and conference proceedings. He also published two number of Books. He completed eight numbers of research and consultancy project funded by different funding agencies such as DRDO, DST, AICTE, NPIU, and local Government office.

George Ghinea (Member, IEEE) is currently Professor of mulsemedia computing with the Computer Science Department, Brunel University London, UK His work focuses on building ubiquitous and adaptable cross-layer end-to-end communication systems incorporating user multisensorial and perceptual requirements. He has applied his expertise in areas such as eye-tracking, telemedicine, multi-modal interaction, and ubiquitous and mobile computing, leading a team of eight researchers in these areas. He has over 350 publications in his research field and his work has been funded by both national and international fora. He consults regularly for both public and private institutions in his areas of expertise.

Dr. Sanjay Kumar Kuanar is currently serving as Professor and Head of Computer Science and Engineering Department of GIET University. He has earned his Ph.D. degree from Jadavpur University, Kolkata in November, 2015, obtained his M.E. (Computer Engineering) degree from Jadavpur University in 2007, and B.Tech. Degree in Computer Science and Engineering from Utkal University in 1998. His research interests include computer vision, pattern recognition, multimedia computing, and software metrics. He has authored and co-authored several publications in refereed SCI journals including IEEE Transactions and has many international IEEE peer-reviewed conference publications like ICPR, ICAPR, etc., to his credit. His papers got many citations with h-index of 7 and i-10 index of 5. He is Reviewer for many national and international peer-reviewed journals and Member of technical program committee for many national and international conferences.

Contributors

Nibedita Adhikari Post Graduate Department of Computer Science and Applications, Utkal University, Bhubaneswar, India

Mazin R. AL-Hameed National University of Science and Technology, Dhi Qar, Iraq

Mohammed Ayad Alkhafaji College of Engineering, Medical Instruments Technology Engineering, National University of Science and Technology, Dhi Qar, Nasiriyah, Iraq

Hassanain K. Alrammahi General Directorate of Education in Najaf, Najaf, Iraq

Khulood Abdel-Khaliq Al-Salim Dejla University College, Baghdad, Iraq

Pratham Ankalwar Electronics and Telecommunication Department, Vishwakarma Institute of Information Technology, Pune, India

Kachi Anvesh Department of IT, Vardhaman College of Engineering, Hyderabad, India;
Department of CSE, Visvesveraya Technological University, Belagavi, India

Smita Bakshi GIET University, Gunupur, India

Sudarshan Basu National Institute of Technology Rourkela, Odisha, India

Smita Bhagwat Department of E&TC, Vishwakarma Institute of Information Technology, Pune, Maharashtra, India

Rhugved Bhojane Department of Electronics Engineering, Shri Ramdeobaba College of Engineering and Management, Nagpur, Maharashtra, India

Mahesh Bhutada Electronics and Telecommunication Department, Vishwakarma Institute of Information Technology, Pune, India

Swarupananda Bissoyi Department of Computer Application, Maharaja Sriram Chandra Bhanja Deo University, Baripada, Odisha, India

Mohammed Brayyich College of Engineering, Medical Instruments Technology Engineering, National University of Science and Technology, Dhi Qar, Iraq

Siddhi Chourasia Department of Electronics Engineering, Shri Ramdeobaba College of Engineering and Management, Nagpur, Maharashtra, India

A. Christy Jeba Malar Department of Information Technology, Sri Krishna College of Technology, Coimbatore, Tamilnadu, India

Santosh Kumar Das Department of Computer Application, Maharaja Sriram Chandra Bhanja Deo University, Baripada, Odisha, India

Subhalaxmi Das Odisha University of Technology and Research, Bhubaneswar, India

Suman Kumar Das Department of Mechanical Engineering, University of Brescia, Brescia, Italy

Amlan Dash Sterlite Technology Ltd. Gurugram, Haryana, India

Baishnobi Dash Parala Maharaja Engineering College, Berhampur, Odisha, India

Ranjan Kumar Dash Odisha University of Technology and Research, Bhubaneswar, India

Sourabh Debnath National Institute of Technology Rourkela, Odisha, India

M. Deva Priya Department of Computer Science and Engineering, Karpagam College of Engineering, Coimbatore, Tamilnadu, India

Sunita Dhalbisoi Department of ETC Engineering, VSSUT, Burla, Odisha, India

Atharva Divekar Department Electronics and Telecommunications Engineering, Vishwakarma Institute of Information Technology, Pune, India

Raj Dumbre Electronics and Telecommunication Department, Vishwakarma Institute of Information Technology, Pune, India

P. Ezhilarasi Department of Electronics and Communication Engineering, St. Joseph's College of Engineering, Chennai, India

Mohd Sayeed Farhaan Department of Electronics and Communication Engineering, Kakatiya Institute of Technology and Science, Warangal, India

Najlaa Nsrulaah Faris College of Engineering, Medical Instruments Technology Engineering, National University of Science and Technology, Dhi Qar, Iraq

Deepak Garg Department of Electronics and Communication Engineering, Delhi-NCR Campus, SRM Institute of Science and Technology, Ghaziabad, India

Nilesh Gawali Nitya Innovations LLP, Turbhe, Navi Mumbai, Maharashtra, India

Pujari Arjun Goud Department of Electronics and Communication Engineering, Kakatiya Institute of Technology and Science, Warangal, India

Shirish Gund Electronics and Telecommunication Department, Vishwakarma Institute of Information Technology, Pune, India

Muhammet Tahir Güneşer Electronics and Communication Engineering Department, Faculty of Electrical and Electronics, Istanbul Technical University, Istanbul, Turkey

Mohammed Ridha Hammoodi Faculty of Computer Science and Mathematics, University of Kufa, Kufa, Iraq

Ipsita Ankita Hota Odisha University of Technology and Research, Bhubaneswar, India

Ali Ashoor Issa College of Engineering, Medical Instruments Technology Engineering, National University of Science and Technology, Dhi Qar, Nasiriyah, Iraq

Olga Ivanova South Ural State University, Chelyabinsk, Russia

Kadim A. Jabbar College of Engineering, Medical Instruments Technology Engineering, National University of Science and Technology, Dhi Qar, Nasiriyah, Iraq

Anupama Jawale Department of Computer Science, SNDT Women's University, Mumbai, India;
Department of Information Technology, Narsee Monjee College of Commerce and Economics, Mumbai, India

S. Rajesh Kannan Department of Electronics and Communication Engineering, St. Joseph's College of Engineering, Chennai, India

Amit Kar Sterlite Technology Ltd. Gurugram, Haryana, India

Siksha Swaroopa Kar CSIR-CRRI, New Delhi, India

Sanjay Kumar Kuanar GIET University, Gunupur, India

Koppula Koushik Kumar Department of Electronics and Communication Engineering, Kakatiya Institute of Technology and Science, Warangal, India

Raghvendra Kumar GIET University, Gunupur, Odisha, India

Sachin Kumar South Ural State University, Chelyabinsk, Russia

Sunny Kumar CSIR-CRRI, New Delhi, India

Snehal V. Laddha Department of Electronics Engineering, Shri Ramdeobaba College of Engineering and Management, Nagpur, Maharashtra, India

Ganesh Magar Department of Computer Science, SNDT Women's University, Mumbai, India

Soumya Ranjan Mahanta Utkal University, Bhubaneswar, Odisha, India

Bichitra Mandal OUTR, Bhubaneswar, India

Sujogya Mishra Odisha University of Technology and Research, Bhubaneswar, India

Arunanshu Mohapatra Department of ETC Engineering, VSSUT, Burla, Odisha, India

Debasis Mohapatra Parala Maharaja Engineering College, Berhampur, Odisha, India

Ramesh Kumar Mohapatra National Institute of Technology Rourkela, Odisha, India

Mayur More Department of E&TC, Vishwakarma Institute of Information Technology, Pune, Maharashtra, India

M. Mufti Nottingham University, Nottingham, UK

Md Abdul Muqueem Department of Electronics and Communication Engineering, Kakatiya Institute of Technology and Science, Warangal, India

Israa Falih Muslm Department of Performance Evaluation, University of Babylon, Babil Govermorate, Iraq

Subrata Kumar Nayak GVHSS, Balipatna, Odisha, India

Ahmed J. Obaid Faculty of Computer Science and Mathematics, University of Kufa, Kufa, Iraq

Huda Qasim Owaied College of Engineering, Medical Instruments Technology Engineering, National University of Science and Technology, Dhi Quar, Iraq

Anwesh Padhi Department of Electrical Engineering, IGIT Sarang, Dhenkanal, Odisha, India

Rashmita Padhi Integrated M.C.A, Utkal University, Bhubaneswar, India

Amey Panchbhayye Department of E&TC, Vishwakarma Institute of Information Technology, Pune, Maharashtra, India

Mrutyunjaya Panda Utkal University, Bhubaneswar, Odisha, India

Dipti Pandit Electronics and Telecommunication Department, Vishwakarma Institute of Information Technology, Pune, India

Dheeman Panigrahi Department of Electrical Engineering, IGIT Sarang, Dhenkanal, Odisha, India

Satya Sobhan Panigrahi Inovaare Clouds Pvt. Ltd, Bhubaneswar, Odisha, India

Punyaban Patel Department of Computer Science and Engineering-Cyber Security, CMR CET, Hyderabad, India

Riyam Patel Department of Computational Intelligence, College of Engineering and Technology, SRM Institute of Science and Technology, Chennai, India

Geetanjali Pradhan Odisha University of Technology and Research, Bhubaneswar, India

Nissan Kumar Pradhan Department of Electrical Engineering, IGIT Sarang, Dhenkanal, Odisha, India

Sabyasachi Pradhan CPP, NALCO, Angul, Odisha, India

Sipali Pradhan Department of Computer Science, RBVRR Women's College, Hyderabad, India

Adesh Jothi Prasanna Department of Computational Intelligence, College of Engineering and Technology, SRM Institute of Science and Technology, Chennai, India

D. Beaulah Princiba Department of Electronics and Communication Engineering, St. Joseph's College of Engineering, Chennai, India

V. Rajeshwari Department of Electronics and Communication Engineering, St. Joseph's College of Engineering, Chennai, India

Masna Rashmitha Department of Electronics and Communication Engineering, Kakatiya Institute of Technology and Science, Warangal, India

Reddyvari Venkateswara Reddy Department of Computer Science and Engineering-Cyber Security, CMR CET, Hyderabad, India

Bharati M. Reshmi Department of CSE, Visvesveraya Technological University, Belagavi, India;
Department of CSE, Basaveswar Engineering College, Bagalkote, Karnataka, India

Ashima Rout Department of ETC Engineering, IGIT Sarang, Sarang, Odisha, India

Saroj Kumar Rout Department of IT, Vardhaman College of Engineering, Hyderabad, India

Nicola Saccani Department of Mechanical Engineering, University of Brescia, Brescia, Italy

Bibhudatta Sahoo Department of CSE, NIT Rourkela, Rourkela, Odisha, India

Ramesh K. Sahoo IGIT Sarang, Dhenkanal, Odisha, India

Bandita Sahu Birla Global University, Bhubaneswar, India

Jayesh Sarode Department of E&TC, Vishwakarma Institute of Information Technology, Pune, Maharashtra, India

Cihat Şeker Electric Electronic Department, Collage of Engineering, Karabuk University, Karabuk, Turkey

J. Sengathir Department of Information Technology, CVR College of Engineering, Vastunagar, Mangalpally, Hyderabad, Telangana, India

Bijaya Kumar Sethi Department of CSE (Data Science), Vardhaman College of Engineering, Hyderabad, India

Padmini Sethi RDW University, Bhubaneswar, India

Devendra Kumar Sharma Department of Electronics and Communication Engineering, Delhi-NCR Campus, SRM Institute of Science and Technology, Ghaziabad, India

Govind Sharma Department of E&TC, Vishwakarma Institute of Information Technology, Pune, Maharashtra, India

Shaymaa Shnain College of Basic Education, University of Babylon, Babil Govermorate, Iraq

Borra Sivaiah Department of Computer Science and Engineering, CMR CET, Hyderabad, India

Atharva Sonwane Department Electronics and Telecommunications Engineering, Vishwakarma Institute of Information Technology, Pune, India

Arkay Swami Department Electronics and Telecommunications Engineering, Vishwakarma Institute of Information Technology, Pune, India

Olga Vorfolomeeva West Kazakhstan State University, Uralsk, Kazakhstan

Liver Segmentation from MR T1 In-Phase and Out-Phase Fused Images Using U-Net and Its Modified Variants

Siddhi Chourasia, Rhugved Bhojane, and Snehal V. Laddha

Abstract In medical imaging, MR T1 in-phase and out-phase images, when combined, produce a clear representation of pure water, aiding in tissue analysis and diagnosing conditions like fatty liver. However, accurately segmenting the liver from these complex images poses challenges due to fuzzy boundaries and variable appearances. To address this, deep learning techniques, particularly the U-Net architecture, have shown promise. Different variants of U-Net, such as Residual U-Net, Attention U-Net, and RA-UNet, have been developed to improve liver segmentation accuracy. In our study, we applied these U-Net variants to segment the liver from MR T1 fused images. Among them, the Attention U-Net model demonstrated superior performance, achieving a Mean Intersection over Union (IoU) value of 0.9513. This indicates its effectiveness in accurately delineating liver boundaries and its potential for enhancing clinical tasks like surgical planning and postoperative evaluation.

Keywords Liver segmentation · Deep learning · U-Net · Residual U-Net · Attention U-Net · RA-UNet

1 Introduction

The liver functions as both an organ and a gland within the human body. Located at the right part of one's abdomen, just below the ribs, it plays a crucial role in various physiological processes. The liver is a vital organ that carries out numerous tasks required for life support such as detoxification, metabolism, and the production

S. Chourasia · R. Bhojane · S. V. Laddha (✉)
Department of Electronics Engineering, Shri Ramdeobaba College of Engineering and Management, Nagpur, Maharashtra, India
e-mail: laddhasv2@rknec.edu

S. Chourasia
e-mail: chourasiass_3@rknec.edu

R. Bhojane
e-mail: bhojanerv@rknec.edu

S. Kumar Udgata et al. (eds.), *Intelligent Systems*, Lecture Notes in Networks and Systems 1149, https://doi.org/10.1007/978-981-97-8160-7_1

of bile. As it produces proteins and hormones that are required by other bodily components, it is also a gland. Hepatitis, cirrhosis, liver cancer, fatty liver disease, and autoimmune liver diseases are a few examples of liver conditions that can harm the liver's health and function [1]. The biggest abdominal organ, the liver, is also vulnerable to physical harm due to its location and function.

Segmentation, often referred to as partitioning, involves dividing a targeted organ for volumetric or morphological analysis, commonly conducted on magnetic resonance imaging (MRI) or multiplanar computed tomography (CT). It's critical to accurately partition the liver from medical pictures in various medical scenarios, including diagnosing hepatic diseases, planning surgeries, and evaluating postoperative outcomes. To better comprehend the anatomy of the liver and its surrounding tissues, surgeons frequently use correct segmentation of the liver to aid in surgical planning. Accurate segmentation of the liver also aids clinicians in the detection and diagnosis of a variety of liver illnesses, including liver cancer, cirrhosis, and fatty liver disease. Liver segmentation remains challenging in medical imaging because of the intricate background, indistinct boundaries, and diverse appearances of the liver [2]. Manual segmentation of MRI images in clinical settings is labor-and time-intensive because lesions have uneven forms, limited contrast between lesions and surrounding tissue, variations in liver size and shape among patients, and similar intensities between the liver and other organs [3]. The density, signal intensity, or architectural structure of the liver may also change as a result of a range of diseases that affect it. Examples include liver fat, fibrosis, iron build-up, and tumors [4]. Larger data volumes are the result of recent performance improvements in various imaging techniques. Therefore, clinicians require computer assistance to analyze this enormous amount of obtained data [5]. Considering these challenges, researchers have developed diverse computer-aided diagnostic methods for liver segmentation [3]. Many strategies for liver segmentation have been put forth to this point. Machine learning (ML) is a new discipline that includes a class of algorithms called deep learning (DL), which is appropriate for methods like big data analysis and image segmentation [6]. Deep learning has lately shown itself to be extremely effective at tackling computer vision challenges by automatically recognizing hierarchies of pertinent characteristics from the input data [5]. New technologies have also shown promise in numerous other fields including object identification, image categorization, barrier evasion, facial identification, linguistic analysis, and material examination [7–15]. For many computer vision issues, the segmentation approach based on deep learning may offer the most recent accuracy, and it has been successful in a variety of ways in the image segmentation sector [16].

2 Related Work

The U-Net architecture has seen significant developments and adaptations aimed at improving its effectiveness in medical image segmentation tasks. Notable variations such as the Residual U-Net, Attention U-Net, and RA-UNet have emerged, each addressing specific challenges encountered in segmentation tasks. Several studies propose novel methodologies, contributing to the broader landscape of medical image analysis [6, 17, 18].

One study introduces an autonomous liver segmentation technique utilizing CT scans, CNNs, and a basic UNet architecture [2, 16]. Trained on CT volumes, the model achieves an average IOU of 92.8%. Another proposes H-DenseUNet, a hybrid approach combining 2D and 3D DenseUNet architectures, achieving hierarchical feature extraction for liver and tumor segmentation. The model integrates intra- and inter-slice representations through a hybrid feature fusion (HFF) layer, optimizing feature extraction across image slices [19].

Furthermore, a UNet model integrates full-scale connection blocks and compression extraction modules to extract spatial and contextual information from images. By connecting full-scale blocks via skip links, the model enhances its ability to capture tumor edge information, achieving an F1 score of 94.52% [20, 21]. Additionally, the UNet model with the Grey Wolf Class Topper Optimization (GW-CTO) algorithm maximizes encoder depth and epoch count during segmentation, enhancing feature selection.

Another approach suggests using a Modified-UNet for autonomous liver and tumor segmentation, featuring a pair of tracks for encoding and decoding, achieving high dice scores of 96.15% and 89.38% for liver and tumor segmentation, respectively. Additionally, the Attention UNet model utilizes multiscale residual blocks and attention mechanisms to reinforce positive features and reduce semantic differences between encoder and decoder [22, 23].

Moreover, a modified ResUNet model employs fewer filters in convolutional blocks for efficient training, achieving high accuracy in liver and tumor segmentation tasks, with dice scores of 0.96 and 0.74, respectively [24, 25]. The model demonstrates remarkable accuracy, with an accuracy of 99.71 and 99.72% for liver and tumor segmentation, respectively [26].

After understanding the nature of the UNet and some of its tweaked versions [27, 28], we decided to generalize all the models under UNet, Attention UNet, Residual UNet, and Residual-Attention UNet, and study and test them.

3 Methodology

In this study, we used a private dataset of 282 abdominal MR T1 fused (in-phase + out-phase) images to examine the performance of U-Net, Attention U-Net, Residual U-Net, and Residual Attention U-Net for liver segmentation. 197 images are used for training and 85 images are used for testing in the dataset (Fig. 1).

We compare segmentation accuracy, IOU, and loss of each architecture to determine the optimal U-Net-based model for liver segmentation, aiming to enhance clinical precision and automated segmentation technology development.

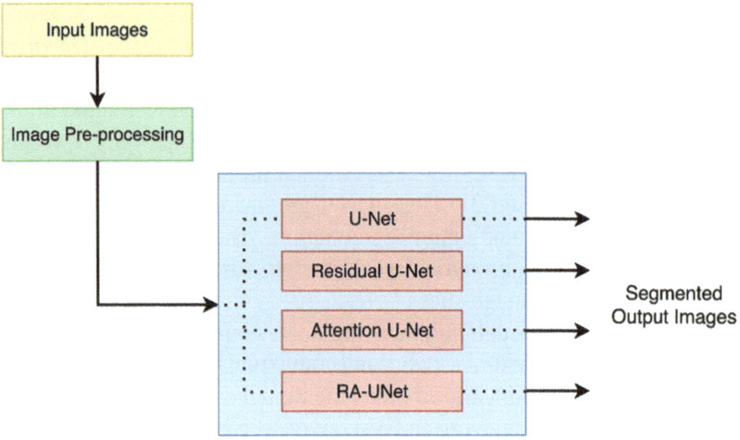

Fig. 1 Image segmentation processing blocks

3.1 U-Net

U-Net, an end-to-end network, segments target features by extracting layer characteristics via up-sampling and down-sampling. It enhances FCN, working effectively with smaller datasets for more accurate segmentation. Its design comprises two pathways: the contracting (encoder) and expansion (decoder) paths (Fig. 2).

U-Net, resembling a U shape, connects encoder and decoder networks with skip connections for detail preservation. Encoder reduces resolution with max-pooling and convolution, maintaining context. Decoder upsamples for segmentation. Skip connections retain detail. Features are concatenated, followed by convolutions and ReLU activation. A 1 × 1 convolution adjusts channels for segmentation. This

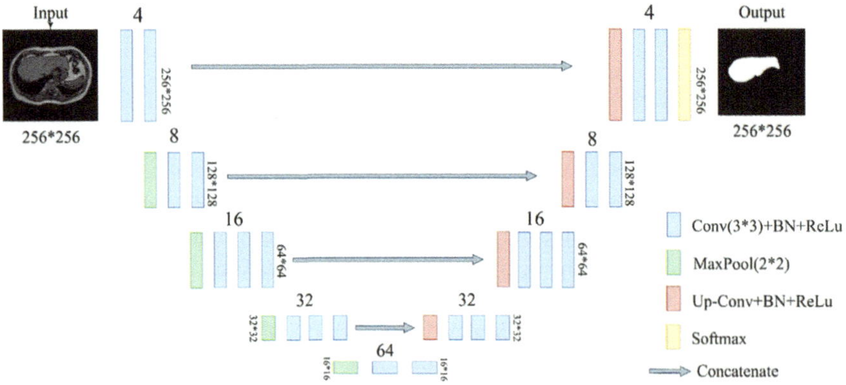

Fig. 2 U-NET architecture

architecture balances spatial information preservation with parameter efficiency for accurate segmentation.

The network's energy function is provided by:

$$E = \sum w(x)(p_{k(x)}(x)) \tag{1}$$

where p_k is the final feature map's applied pixel-wise SoftMax function, which is defined as:

$$p_k = (a_k(x))/\sum_{k=1}^{k} \exp(a_k(x)). \tag{2}$$

3.2 *Residual U-Net (Res-UNet)*

ResUNet merges UNet with Deep Residual Learning for efficient deep neural network training. It comprises encoding, bridging, and decoding stages for pixel-level categorization in semantic segmentation. Bridging aids information flow, boosting segmentation accuracy. Leveraging both architectures enhances ResUNet's performance [29] (Fig. 3).

ResUNet, similar to U-Net, includes encoding, decoding, and bridging parts. It uses residual units with two 3 × 3 convolution blocks and identity mapping. Unlike U-Net, ResUNet uses pre-activated residual blocks instead of convolutions and ReLU. Encoder blocks create skip connections for decoder blocks, enhancing abstract representation learning. Each residual unit can be represented by a generic

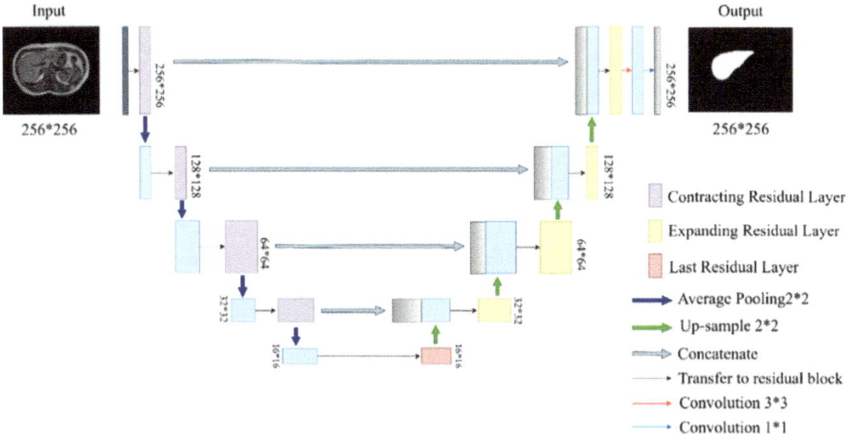

Fig. 3 Residual U-Net architecture

form, as follows:

$$y_1 = (x_1) + F(x_1, W_1) \tag{3}$$

$$x_{l+1} = f(y_1) \tag{4}$$

where x_1 and x_{l+1} stand for the residual unit's input and output, $F(.)$ stands for the residual function, $f(.)$ denotes the activation function, and $h(.)$ denotes the identity mapping function [17].

In the second and third encoder blocks, stride 2 reduces feature map dimensions from 256 to 128. The bridge includes a pre-activated residual block with the same stride. Each decoder block starts with 2 × 2 upsampling, then concatenates with encoder block's skip connection, facilitating access to learned features. After final decoding level, multi-channel feature maps are projected onto segmentation using 1 × 1 convolution and sigmoid activation [29].

3.3 Attention U-Net

The Attention U-Net, an advancement of the U-Net architecture, incorporates attention mechanisms for semantic segmentation tasks. It enhances the decoder with channel-wise and spatial attention, improving segmentation accuracy [30] (Fig. 4).

The following formula describes additive attention:

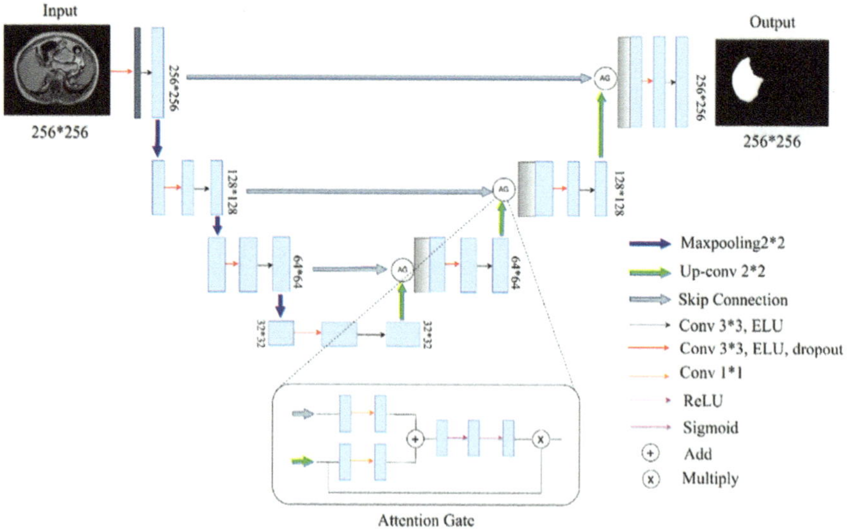

Fig. 4 Attention U-Net architecture

$$q_{att}^{1} = \psi^{T}(\sigma_1\left(W_x^T\left(x_i^l + W_g^T g_i + b_i\right)\right) + b_\psi \tag{5}$$

$$\alpha_i^l = \sigma_2(q_{att}^l(x_i^l, g_i; \Theta_{att})) \tag{6}$$

where g is the gating signal and x^l is the characteristics of the contracting route. The sigmoid function is denoted by the phrase $\sigma_2(x_{i,c})$:

$$\sigma_2\left(x_{i,c}\right) = \frac{1}{1 + \exp(-x_{i,c})} \tag{7}$$

The Attention U-Net design has been employed in many applications, including medical image segmentation, satellite image analysis, and more, and is quite successful for semantic segmentation tasks.

3.4 Residual Attention U-Net (RA-UNet)

The Residual Attention U-Net merges U-Net with attention and residual blocks for image segmentation. Attention gates highlight important elements by collecting spatial information, while residual blocks aid detailed feature extraction and gradient flow, utilizing shortcut links for learning residual mappings across levels. The residual block is described as:

$$OR_{i,c}(x) = x + f_{i,c}(x) \tag{8}$$

where $c \in \{1, \ldots..C\}$ indicates the index of channels, C is the total number of channels, and f represents the residual mapping to be learned. x marks the initial input of a residual block. OR symbolizes the output of a residual block [31] (Fig. 5).

RA-UNet combines UNet's contraction and expansion paths with a bridging component. Residual Attention U-Net enhances UNet with attention gates and RAMs in each layer. RAMs selectively attend to crucial aspects, improving segmentation accuracy. RA-UNet's success lies in its efficient integration of residual attention mechanisms.

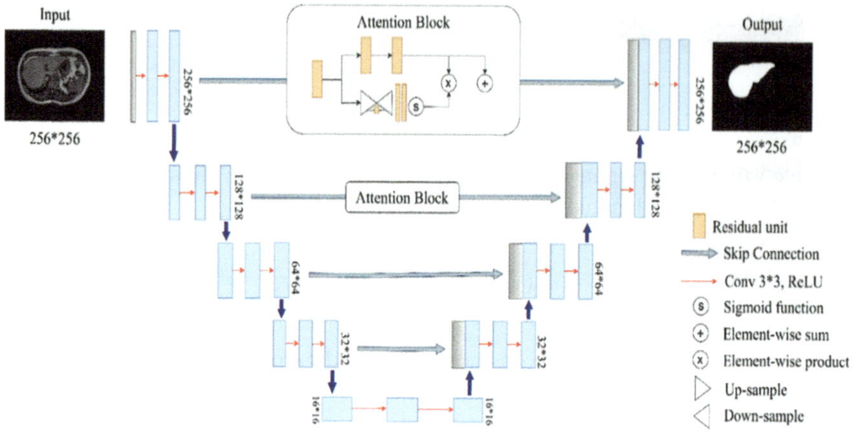

Fig. 5 Residual attention U-Net architecture

4 Experiments and Results

4.1 Experimental System Details

The experiment utilized TensorFlow 2.10, a deep learning framework by Google, on a system with an Intel Core i5 10210U CPU, 8 GB RAM, and an NVIDIA GeForce MX230 GPU. Python was the programming language, with GPU acceleration via cuDNN and NVIDIA CUDA.

4.2 Dataset

The dataset comprised 282 MRI samples from the Chaos Grand Challenge, employing T1-DUAL sequences with in and out phases. It was later divided into 70% training and 30% testing sets (Fig. 6).

4.3 Evaluation Metrics

Assessment metrics including Accuracy, Loss, and IoU were used to appropriately analyze the models' experimental findings.

Accuracy: It is the percentage of correctly classified pixels in your image.

$$\text{Accuracy} = \frac{\text{TP} + \text{TN}}{\text{TP} + \text{TN} + \text{FP} + \text{FN}} \tag{9}$$

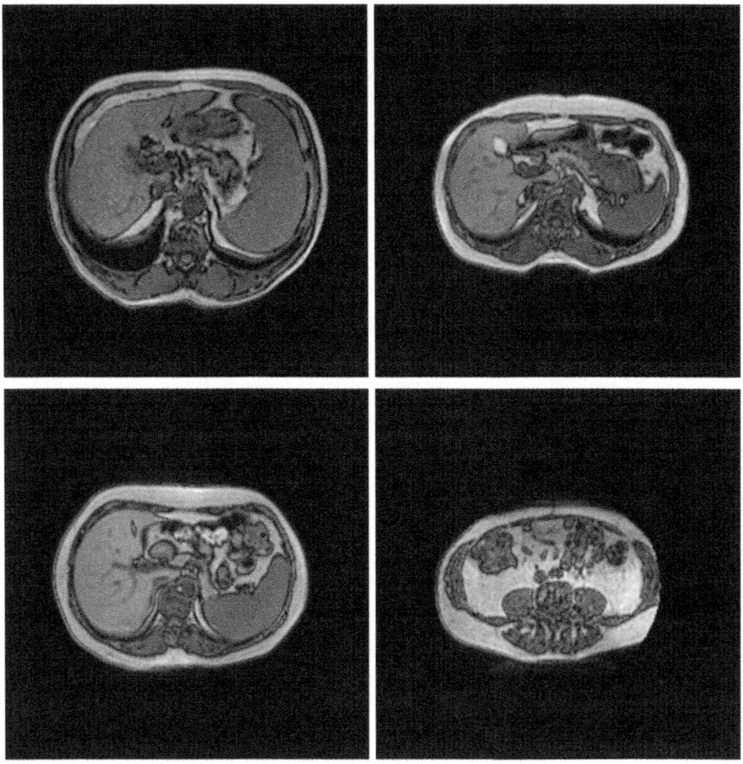

Fig. 6 Sample MR fused images from the dataset

The CNNs in this study employ the weighted cross-entropy (WCE) loss function, aiding in distinguishing between background and liver pixels, particularly useful in class imbalances [18]. WCE penalizes each class based on its median frequency.

$$WCE = -\frac{1}{n}\sum_{i=1}^{n} W_{c,i}[T_i \log P_i + (1 - T_i)\log(1 - P_i)] \tag{10}$$

IoU: The intersection over union (IoU), which is determined by dividing the overlap area between the two, is the area where the predicted segmentation and the ground truth cross.

$$\text{IoU} = \frac{\text{TP}}{\text{TP} + \text{FP} + \text{FN}} \tag{11}$$

where the terms True Negative, True Positive, False Negative, and False Positive, respectively, stand for TN, TP, FN, and FP.

Table 1 The evaluation parameters and respective values for various models

Model	Number of trainable parameters	Accuracy (%)	Loss	IOU (%)
UNET	1,965,283	99.39	0.2401	94.23
Res-UNET	2,076,243	99.51	0.2361	93.60
Attention-UNET	2,337,609	99.43	0.0408	95.13
RA-UNET	2,448,631	99.50	0.2362	92.88

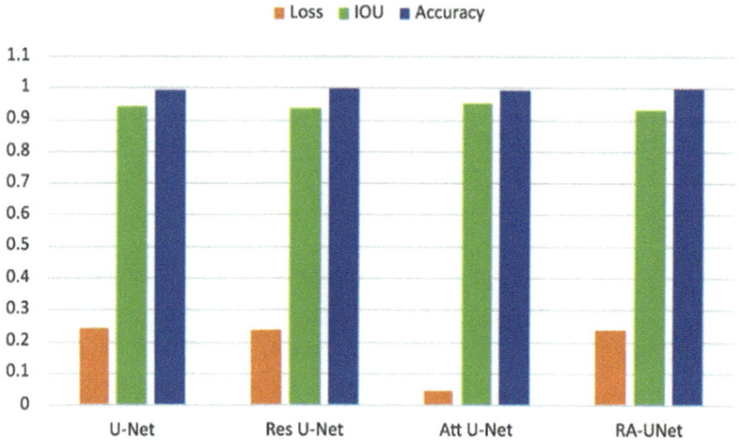

Fig. 7 Performance of different models for liver segmentation

4.4 Results

One of the simplest and most popular methods for examining the effectiveness of the models is the IoU. Table 1 provides an overview of the study's findings. Accuracy and loss are two additional performance indicators that are used, but for this experiment, the IoU is our main focus (Fig. 7).

The results showed that U-Net and Attention U-Net achieved Mean IoU scores of 0.9424 and 0.9513, respectively. Res-UNet and Residual Attention U-Net attained scores of 0.9360 and 0.9292. Residual Attention U-Net outperformed with the highest accuracy, while Attention U-Net excelled in Mean IoU performance (Figs. 8 and 9).

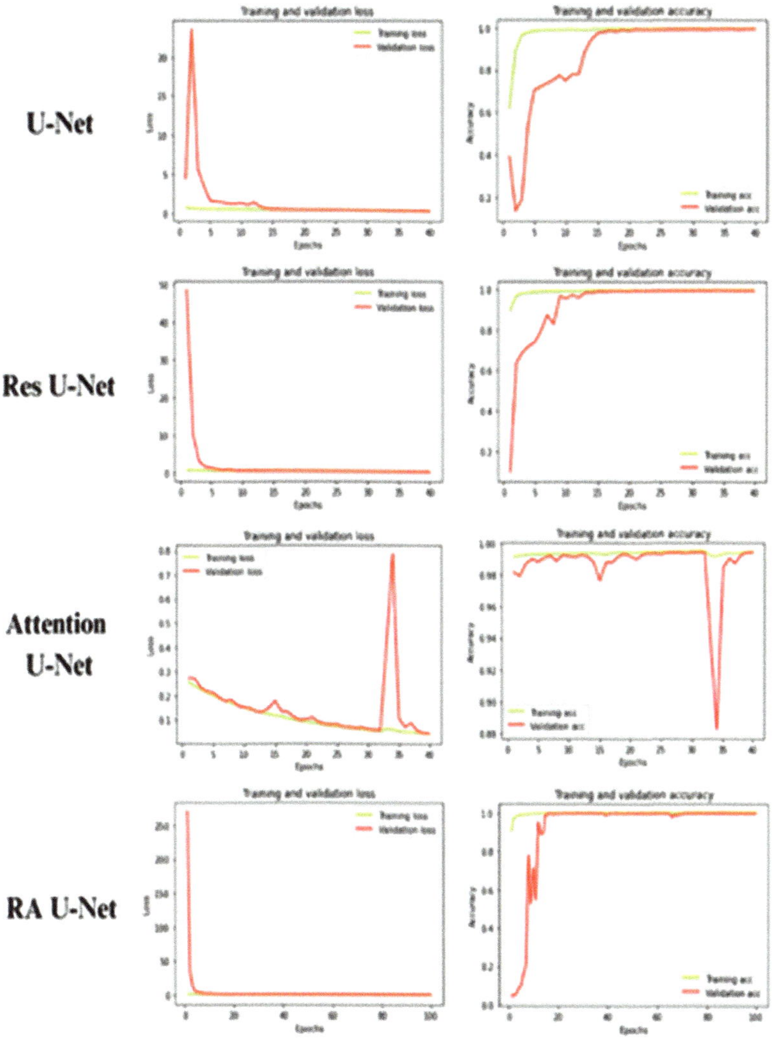

Fig. 8 Accuracy and loss graphs of various models

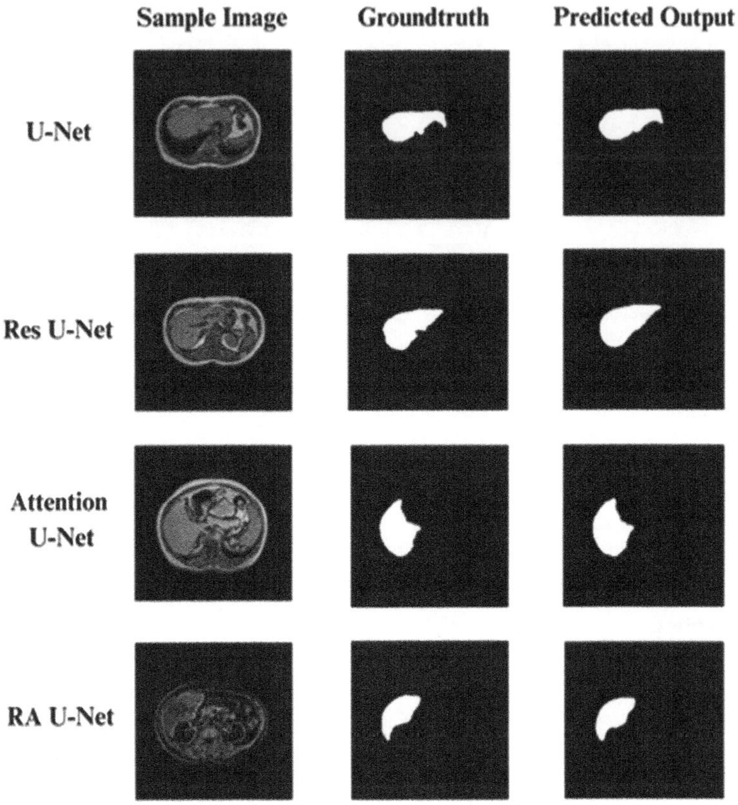

Fig. 9 Result of liver segmentation from MRI T1 fused images using various models

5 Conclusion

All models demonstrated high accuracy (0.9939–0.9960) in segmenting the liver from MR T1 fused images. Attention U-Net achieved the highest Mean IoU (0.95135725), indicating superior segmentation performance. Res-UNet and RA-U-Net had slightly higher loss values compared to U-Net and Attention U-Net, suggesting potential for improvement. Future research could focus on refining deep learning architectures for enhanced liver segmentation accuracy.

References

1. Shantan Cheemerla, M.D., Maya Balakrishnan, M.D.: Global epidemiology of chronic liver disease. Clin. Liver Dis. (Hoboken) **17**(5), 365–370. Published online 2021 June 4. https://doi.org/10.1002/cld.1061 https://www.ncbi.nlm.nih.gov/pmc/articles/PMC8177826/
2. Yang, D., Xu, D., Kevin Zhou, S., Georgescu, B., Chen, M., Grbic, S., Metaxas, D., Comaniciu, D.: Automatic liver segmentation using an adversarial image-to-image network. In: International Conference on Medical Image Computing and Computer-Assisted Intervention MICCAI 2017: Medical Image Computing and Computer Assisted Intervention—MICCAI 2017 (2017)
3. Abebe Ayalew, Y., Fante, K.A., Mohammed, M.A.: Modified U-Net for liver cancer segmentation from computed tomography images with a new class balancing method. BMC Biomed. Eng., Article number 4 (2021)
4. Gotra, A., Sivakumaran, L., Chartrand, G., Vu, K.-N., Vandenbroucke-Menu, F., Kauffmann, C., Kadoury, S., Gallix, B., de Guise, J.A., Tang, A.: Liver segmentation: indications, techniques and future directions. Published: 14 June 2017
5. Enokiya, Y., Iwamoto, Y., Chen, Y.-W., Han, X.-H.: Automatic liver segmentation using U-Net with Wasserstein GANs. J. Image Graph **6**(2), December 2018
6. Du, G., Cao, X., Liang, J., Chen, X., Zhan, Y.: Medical image segmentation based on U-Net: a review. J. Imaging Sci. Technol. Soc. (2020)
7. Rahman, H., Fatima Naik Bukht, T., Imran, A., Tariq, J., Tu, S., Alzahrani, A.: A deep learning approach for liver and tumor segmentation in CT images using ResU-Net. Bioengineering
8. Yin, X.-X., Sun, L., Fu, Y., Lu, R., Zhang, Y.: U-net-based medical image segmentation, Article ID 4189781 (**2022**). https://doi.org/10.1155/2022/4189781
9. Shirvani, Z., Abdi, O., Goodman, R.C.: High-resolution semantic segmentation of woodland fires using residual attention U-Net and time series of Sentinel-2. Remote Sens. **15**(5), 1342 (2023). https://doi.org/10.3390/rs15051342
10. Chen, X., Yao, L., Member, IEEE, Zhang, Y., Senior Member, IEEE: Residual attention U-Net for automated multi-class segmentation of COVID-19 chest CT images, arXiv:2004.05645v1 [eess.IV], 12 Apr. 2020
11. Kavur, A.E., Selver, M.A., Dicle, O., Barış, M., Gezer, N.S.: CHAOS—Combined (CT-MR) healthy abdominal organ segmentation challenge data (version v1.03), April 2019. Zenodo. https://doi.org/10.5281/zenodo.3362844
12. Chen, Y., Wang, K., Liao, X., Qian, Y., Wang, Q., Yuan, Z., Heng, P.-A.: Channel-unet: a spatial channel-wise convolutional neural network for liver and tumors segmentation. Front. Genet. **10**, 1110. https://doi.org/10.3389/fgene.2019.01110
13. Liu, Y., Yang, W.: Automatic liver segmentation using U-net in the assistance of CNN. In: 2020 International Conference on Intelligent (2020)
14. Jiang, L., Ou, J., Ruihua, L.,, Zou, Y., Xie, T., Xiao, H., Bai, T.: RMAU-Net: residual multi-scale attention U-Net for liver and tumor segmentation in CT images. Comput. Biol. Med. **158**, May 2023:106838 https://www.sciencedirect.com/science/article/abs/pii/S0010482523003037
15. Liu, H., Wang, X.: Liver segmentation in CT based on ResUNet with 3D probabilistic and geometric post process. https://doi.org/10.1109/SIPROCESS.2019.8868690. Conference: IEEE international conference on signal and image processing
16. Li, X, Qian, W., Xu, D., Liu, C.: School of Information, Yunnan University, Kunming, 650504, China. Image segmentation based on improved U-Net. J. Phys.: Conf. Ser. **1815**, 012018. IOP Publishing. https://doi.org/10.1088/1742-6596/1815/1/012018
17. Siddique, N., (Graduate Student Member, IEEE), Paheding, S., Elkin, C.P. (Member, IEEE), Devabhaktuni, V (Senior Member, IEEE): U-Net and its variants for medical image segmentation: a review of theory and applications (2021). Received May 4, 2021, accepted May 31, 2021, date of publication June 3, 2021, date of current version June 14, 2021
18. Sabir, M.W., Khan, Z., Saad, N.M., Khan, D.M., Al-Khasawneh, M.A., Perveen, K., Qayyum, A., Ali, S.S.A.: Segmentation of liver tumor in CT scan using ResU-Net". Appl. Sci. **12**, 8650. https://doi.org/10.3390/app12178650

19. Li, X., Chen, H., Member, IEEE, Qi, X., Dou, Q., Fu, C.-W., Member, IEEE, Heng, P.-A.: H-DenseUNet: hybrid densely connected UNet for liver and tumor segmentation from CT volumes. arXiv:1709.07330v3 [cs.CV]
20. Rela, M., Suryakari, N.R., Patil, R.R.: A diagnosis system by U-net and deep neural network enabled with optimal feature selection for liver tumor detection using CT images. Multimedia Tools Appl. **82**, 3185–3227 (2023). https://doi.org/10.1007/s11042-022-13381-2
21. Manjunath, R.V., Kwadiki, K.: Modified U-NET on CT images for automatic segmentation of liver and its tumor. Biomed. Eng. Adv. **4,** 100043 (2022)
22. Wu, J., Zhou, S., Zuo, S., Chen, Y., Sun, W., Luo, J., Duan, J., Wang, H.: Wang, D.: U-Net combined with multi-scale attention mechanism for liver segmentation in CT images. Wu et al. BMC Med Inform Decis Mak **21**, 283 (2021) https://doi.org/10.1186/s12911-021-01649-w
23. Ayalew, YA., Fante, K.A., Mohammed, M.A. (2021) Modified U-Net for liver cancer segmentation from computed tomography images with a new class balancing method. Ayalew et al. BMC Biomed Eng 3, 4 (2021). https://doi.org/10.1186/s42490-021-00050-y
24. Sun, J., Hui, Z., Tang, C., X. Wu, C.: Liver segmentation based on complementary features U-Net. Visl Comput (2022) https://link.springer.com/article/10.1007/s00371-022-02617-9
25. Wang, J., Zhang, X., Lv, P., Zhou, L., Wang, H.: EAR-U-Net: EfficientNet and attention-based residual U-Net for automatic liver segmentation in CT
26. Manjunath, R.V., Kwadiki, K.: Automatic liver and tumour segmentation from CT images using deep learning algorithm. Results Control Optimization **6**, 100087
27. Li, C., Yusong, T., Chen, W., Luo, X., Gao, Y., Jia, X, Wang, Z.: Attention UNet++: a nested attention-aware U-Net for liver CT image segmentation. In: 2020 IEEE International Conference on Image Processing, ICIP 2020. https://doi.org/10.1109/ICIP40778.2020.9190761
28. Jin, Q., Meng, Z., Sun, C., Wei, L., Su, R.: RA-UNet: a hybrid deep attention-aware network to extract liver and tumor in CT scans. arXiv:1811.01328v1
29. Zhang, Z., Liu, Q., Member, IEEE, Wang, Y., Senior Member, IEEE: Road extraction by deep residual U-Net. IEEE Geosci. Remote Sens. Lett. (2017). arXiv:1711.10684v1 [cs.CV], 29 Nov. 2017
30. Oktay, O., Schlemper, J., Folgoc, L.L., Lee, M., Heinrich, M., Misawa, K., Mori, K., McDonagh, S., Hammerla, N.Y., Kainz, B., Glocker, B., Rueckert, D.: Attention U-Net: learning where to look for the Pancreas. arXiv:1804.03999v3 [cs.CV]
31. Jin, Q., Meng, Z., Sun, C., Cui, H., Su, R.: RA-UNet: a hybrid deep attention-aware network to extract liver and tumor in CT scans. Front. Bioeng. Biotechnol., 23 December 2020 Sec. Preclinical Cell and Gene Therapy, vol. 8 (2020). https://doi.org/10.3389/fbioe.2020.605132

Performance Benchmarking of Iterative Dichotomiser 3 (ID3) Using Clustering-Based Discretization for Sensor Data

Anupama Jawale and Ganesh Magar

Abstract Rise in sensor-based devices has made sensor data cheap and easily available from various sensors. In most cases, sensor data are in the form of converted amplitude/voltage data that essentially become continuous data. Iterative Dichotomiser 3 (ID3) is the basis algorithm used for decision trees. Decision tree algorithms are known to be best suited for discrete data. The accuracy of ID3 algorithm can be improved by converting continuous data to discrete form using some sort of binning/quantization technique. This research study uses clustering technique for quantization of continuous data obtained from accelerometer and gyroscope sensor. The continuous sensor data are divided into clusters, generating discrete data in the form of centroids of each cluster. The obtained discrete data are used for performance benchmarking of ID3 algorithm against statistical binning technique of discretization as well as original continuous data. Our research study shows that the accuracy of ID3 with clustering-based discretization overperforms other techniques by 2–3% improvement in accuracy.

Keywords Clustering · Continuous · Discrete · Features · ID3 · Statistical

1 Introduction

Sensor data stream is in the form of continuous numeric values, indicating a numeric form of information. For data mining or decision tree applications, most of the methods use item-set processes and operations and hence the continuous numeric

A. Jawale (✉) · G. Magar
Department of Computer Science, SNDT Women's University, Mumbai, India
e-mail: anupama.jawale26@gmail.com

G. Magar
e-mail: drgmmagar@gmail.com

A. Jawale
Department of Information Technology, Narsee Monjee College of Commerce and Economics, Mumbai, India

S. Kumar Udgata et al. (eds.), *Intelligent Systems*, Lecture Notes in Networks and Systems 1149, https://doi.org/10.1007/978-981-97-8160-7_2

values fail to generate desired solution. Crisp sensor values are divided into discrete bins or groups for quantization of data. Increase in cheap sensor-based technology has produced tremendous real-time data as a continuous flow. The unexplored research area of processing this continuous data into discrete data has given rise to motivation of this research study. As decision tree algorithm works well on discrete data compared to continuous data, we define the main objective of this work to investigate and compare methods for improving decision tree performance using clustering techniques of discretization. The objective of this study is to compare and study cluster-based discretization with statistical discretization techniques. Firstly, the below section briefs about the decision tree algorithm Iterative Dichotomiser 3 (ID3) and measures used for performance measurement. Further, this study is organized into Related Work (Sect. 2), Methodology (Sect. 3), Experimental Setup (Sect. 4), Result and Discussion (Sect. 5). Conclusion of this study is presented in Sect. 6.

1.1 Iterative Dichotomiser 3

The algorithm works in the steps below:

1. Entropy calculation for entire dataset.
2. Calculate entropy for all categorical values.
3. Calculate information gain for the feature.
4. Select feature with maximum information gain.

For continuous variables, typically decision tree splits on the basis of threshold value (Cut point T). The attribute value $A \leq T$ results in left branch of tree and $A > T$ results in right branch of the tree. ID3 and other decision tree algorithms use this cut-point technique [1]. Entropy calculation in decision tree is given by the mathematical Eq. (1),

$$\text{Entropy} = \sum_{i=1}^{c} -p_i \log_2 p_I \tag{1}$$

where P_i indicates the highest number of probabilities of an element of i-th class in the dataset. As seen in Eq. (1), if the formula is applied to continuous numerical variable, we found, number of observations (N) is equivalent to number of classes and this results in N-1 evaluations (Assuming we do not have identical values). Thus, we must obtain discrete values from these continuous variables [2]. Discretization can be obtained by partitioning the continuous variable into subranges, based on equal windowing techniques, frequency binning technique, or statistical features binning [3].

In this research study, we have proposed a discretization method based on clustering technique. Clustering using k-means is used to obtain cluster bin. Each cluster bin represents a group of observations closer to cluster centroid. The closeness is computed using Euclidian Distance measured from centroid to individual observation. K-means is a well-known algorithm for unsupervised learning, based on distance computation. In general, cluster assignment problem can be formulated as shown in Eq. (2)

$$Cluster_i = \left\{ x_p : \|x_p - m_i{}^t\|^2 \leq \|x_p - m_j{}^t\|^2 \forall j, 1 \leq j \leq k \right\} \quad (2)$$

where each $x(p)$ belongs to exactly one cluster. Once the clustering process is completed, each observation is assigned to a cluster number that performs a discretization of the value.

2 Related Work

Decision Trees are a widely used Data Mining technique that employs a tree-like structure to offer outcomes based on input decisions. The fact that decision trees can be utilized for both regression and classification is a significant feature. Decision tree algorithms produce decision trees from training data to generate solutions for classification and regression problems. ID3, C4.5, C4.5 are a few commonly used algorithms for building a decision tree in data mining. Entropy and Gini index are two measures used in ID3 and similar algorithms [4]. This kind of classification approach can handle both heterogeneous and missing data. Decision Trees can also generate rules that are easy to understand. Furthermore, classifications do not necessitate a large number of computations.

Among all the decision tree algorithms, C4.5 and ID3 are classical algorithms with several applications. There are several suggestions for improvement of these algorithms. In research study [5], researchers have worked on simplification of information gain formula to improve its computation time. They have presented the improved version of C4.5 algorithm by selecting the optimized split based on information gain. Another improvement presented by researchers for ID3 algorithm is to add attribute weights, use of conditional probability, and calculation of distance between attribute and decision attribute [6].

Further, few researchers have studied and worked on the type of data to be input to decision trees. Decision trees work well with discrete data without any missing values. Continuous variables for decision tree result in high information gain, making the tree dwarf and without generalization. Results obtained from testing data are inaccurate and less relevant for real-life situations.

Various applications require discretization of continuous variables, viz, Biomedical Application [7], Applications in Bayesian Network [8]. Researchers have suggested many techniques for discretization. For example, Sum of Continuous Variables [9], Estimated Probability Distribution and Parzon Window [10]. Probabilistic Rounding Mechanism, Approximation of Continuous Random Variable [11].

Different discretization techniques are suggested for conversion of continuous data to discrete data, applicable to areas like fault detection, driver behavior recognition, and EEG signal classification, wherever sensor data are used for decision tree. The existing survey paper [3] has presented taxonomy of discretization techniques along with their classification. Researchers have classified discretization methods as Static vs Dynamic, Univariate vs Multivariate, Global vs Local, Supervised s Unsupervised and Direct vs Incremental. They have also presented criteria to compare discretization like number of intervals, inconsistency, and predictive classification rate.

There are some recent developments in discretization methods, depending upon the type of data. In motion mechanism, the time-variant performance function is discretized into timestamped discrete performance functions using correlation coefficient of normal variables [12]. The discretization method based on multiple second-order generalized integrator, using backward Euler method, and parallel processing of output signal and orthogonal output signal is proposed in the research study for frequencies generated from experimental setup, useful for harmonic compensation, motor control, and sequence component detection [13]. Another research study has proposed the method of discretization using clustering, to overcome errors in common methods like equal width binning, Equal Frequency Binning, and Mean and Standard Deviation Binning [14].

3 Methodology

To address the issue of discretization, we propose a cluster-based discretization method. For each feature in the dataset, we apply clustering and compute centroids. We assign these centroids to each observation that falls under that cluster in dataset. The decision tree algorithm is applied to this dataset to classify the data into three classes for a road condition, smooth, bumpy, or washed.

We also apply statistical binning techniques on the same dataset by dividing data into groups based on statistical features of observation values. The statistical features considered for binning are Minimum, First Quantile, Median, Mean, Third Quantile, and Maximum. Figure 1 shows box plot of summary statistics of accelerometer reading continuous values and statistical binning therein, whereas Fig. 2 shows the summary statistics of accelerometer features with cluster centroids binning.

The proposed methodology is described in Fig. 3. The raw dataset collected by the accelerometer sensor has noise and missing values due to atmospheric conditions like temperature, humidity, etc. In this research study, exploratory data analysis indicated that the percentage of missing values is very small ($\approx$0.12%). Hence, the observations containing missing values are deleted. Since the précised numeric value of feature vectors AccX, AccY, and Accz (x, y, and z reading of accelerometer) determines the classification of the road ride, imputing missing values with mean and median would have affected covariance between variables.

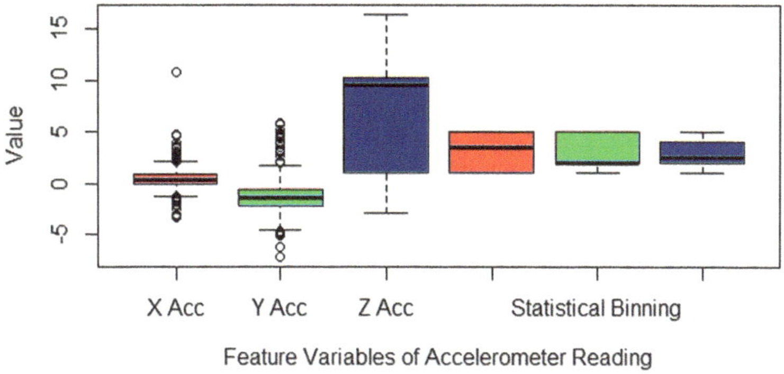

Fig. 1 Statistical features (min, max, 1st, 3rd quantile, and median)

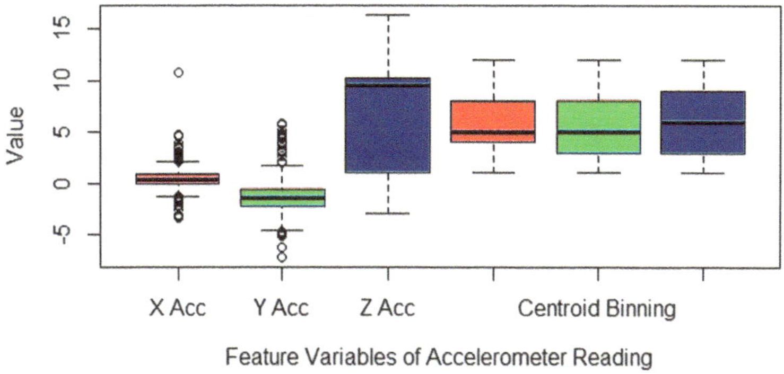

Fig. 2 Clusters features (centroid-based bins)

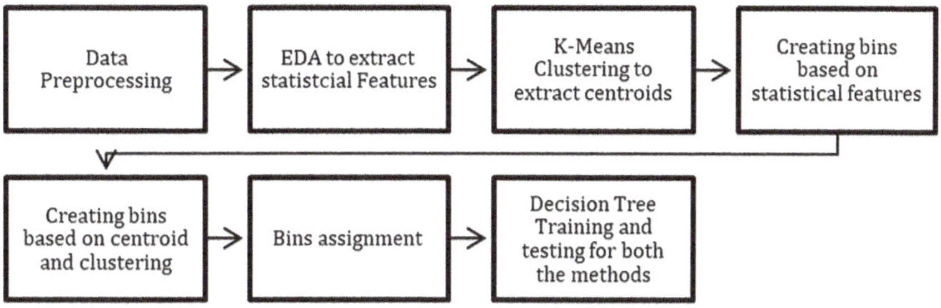

Fig. 3 Flow of proposed methodology

4 Experimental Setup

The data set collected for this research study is smartphone accelerometer data for three different types of roads, viz; Smooth, Bumpy, and Washed. The data are collected from a scooter ride on these roads. A total of 2043 observations with four feature variables are considered for experimental setup, after the initial preprocessing step.

For computation of clustering, we have used K-means MacQueen algorithm [15]. MacQueen starts centroid initialization randomly but recomputes the centroid values after each assignment. In this research study, we have tried computing multiple centroids at the initial stage and choose the best one. This results in a stable cauterization solution. We have experimented on different number of centroids initialization (6, 8, 10, 12, and 24). The results obtained are presented in the section below.

5 Results and Discussion

To evaluate and compare effectiveness of discretization using clustering, we have trained a dataset on decision tree using both the methods, with statistical binning and with cluster-based binning. The continuous sensor data are divided into clusters, generating discrete data in the form of centroids of each cluster. Figure 4 a, b, and c shows decision tree splits for continuous data, statistical binned data, and cluster-based binned data.

Results of these experiments are shown in Table 1.

Performance metrics used for ID3 benchmarking are described below.

i. Sensitivity: In a diagnostic test, sensitivity refers to the probability of a positive test result given that the individual truly has the disease. It reflects the test's ability to correctly identify true positives. Sensitivity is often expressed as a percentage.
ii. Specificity: Specificity, in contrast to sensitivity, refers to the probability of a negative test result given that the individual truly does not have the disease. It reflects the test's ability to correctly identify true negatives. Similar to sensitivity, specificity is often expressed as a percentage.
iii. Positive Predictive Value (PPV): PPV, also known as precision, indicates the probability that an individual with a positive test result actually has the disease. It takes into account both sensitivity and disease prevalence. A high PPV suggests a test is good at identifying true positives and not generating many false positives.
iv. Negative Predictive Value (NPV): NPV refers to the probability that an individual with a negative test result truly does not have the disease. Similar to PPV, it considers both specificity and disease prevalence. A high NPV suggests the test is effective at identifying true negatives and avoiding false negatives.

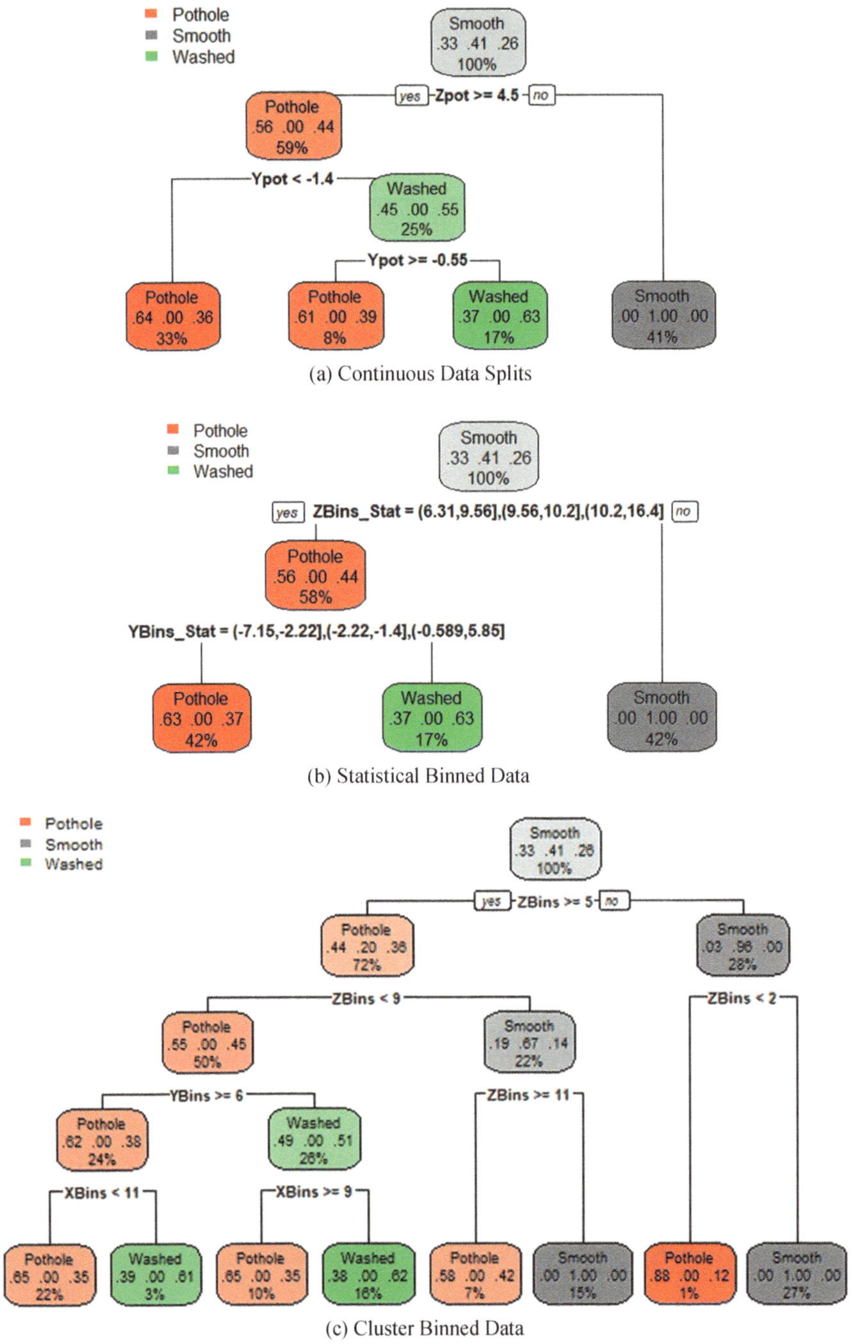

Fig. 4 Decision tree splits (**a**–**c**)

Table 1 Accuracy comparison of continuous data, statistical binned discrete data, and cluster-based binned data

Continuous data-no discretization	Pothole	Smooth	Washed	Initial centroids 06	Pothole	Smooth	Washed
Accuracy 73.11%				Accuracy 77.26%			
Sensitivity	0.59	1	0.5	Sensitivity	0.64	1	0.51
Specificity	0.78	1	0.83	Specificity	0.83	1	0.85
Positive predictive value	0.52	1	0.57	Positive predictive value	0.65	1	0.51
Negative predictive value	0.83	1	0.78	Negative predictive value	0.83	1	0.85
Prevalence	0.28	0.4	0.3	Prevalence	0.32	0.44	0.23
Detection rate	0.16	0.4	0.15	Detection rate	0.21	0.44	0.11
Detection prevalence	0.32	0.4	0.26	Detection prevalence	0.32	0.44	0.23
Balanced accuracy	0.68	1	0.66	Balanced accuracy	0.74	1	0.68
Statistical binning	Pothole	Smooth	Washed	Initial centroids 08	Pothole	Smooth	Washed
Accuracy 72.08%				Accuracy 75.79%			
Sensitivity	0.62	1	0.48	Sensitivity	0.59	1	0.43
Specificity	0.79	0.96	0.84	Specificity	0.84	1	0.83
Positive predictive value	0.53	0.94	0.61	Positive predictive value	0.65	1	0.37
Negative predictive value	0.84	1	0.76	Negative predictive value	0.8	1	0.86
Prevalence	0.27	0.38	0.34	Prevalence	0.33	0.47	0.18
Detection rate	0.17	0.38	0.16	Detection rate	0.2	0.47	0.08
Detection prevalence	0.32	0.4	0.27	Detection prevalence	0.3	0.47	0.21
Balanced accuracy	0.7	0.98	0.66	Balanced accuracy	0.71	1	0.63
Initial centroids 10	Pothole	Smooth	Washed	Initial centroids 12	Pothole	Smooth	Washed
Accuracy 77.26%				Accuracy 76.77%			
Sensitivity	0.64	1	0.51	Sensitivity	0.6	1	0.55

(continued)

Table 1 (continued)

Continuous data-no discretization	Pothole	Smooth	Washed	Initial centroids 06	Pothole	Smooth	Washed
Accuracy 73.11%				Accuracy 77.26%			
Specificity	0.83	1	0.85	Specificity	0.84	1	0.82
Positive predictive value	0.65	1	0.51	Positive predictive value	0.66	1	0.49
Negative predictive value	0.83	1	0.85	Negative predictive value	0.81	1	0.86
Prevalence	0.32	0.44	0.23	Prevalence	0.33	0.44	0.22
Detection rate	0.21	0.44	0.11	Detection rate	0.2	0.44	0.12
Detection prevalence	0.32	0.44	0.23	Detection prevalence	0.3	0.44	0.25
Balanced accuracy	0.74	1	0.68	Balanced accuracy	0.72	1	0.69
Initial centroids 24	Pothole	Smooth	Washed				
Accuracy 77.75%							
Sensitivity	0.6	1	0.72				
Specificity	0.97	1	0.78				
Positive predictive value	0.96	1	0.131				
Negative predictive value	0.68	1	0.98				
Prevalence	0.53	0.42	0.04				
Detection rate	0.32	0.42	0.03				
Detection prevalence	0.33	0.42	0.24				
Balanced accuracy	0.78	1	0.75				

v. Prevalence: Prevalence refers to the proportion of individuals in a population who have a specific disease or condition at a given point in time. It can be expressed as a percentage or a proportion.
vi. Detection Rate: Detection Rate is the proportion of individuals with the disease that are correctly identified by the test, regardless of the test result for those

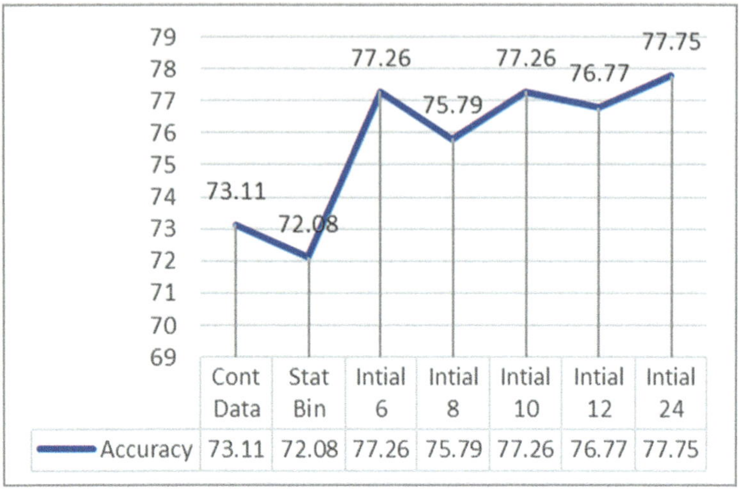

Fig. 5 Accuracy comparison chart

without the disease. It can be calculated as the sum of sensitivity and (1-specificity).

vii. Detection Prevalence: Detection Prevalence refers to the proportion of positive test results in a population, regardless of whether the individuals truly have the disease. It can be helpful for understanding the overall positivity rate of a test.

viii. Balanced Accuracy: Balanced Accuracy is a metric used to address potential biases in classification tasks, particularly when dealing with imbalanced datasets. It takes the average of sensitivity and specificity, providing a more balanced view of the test's performance compared to solely using accuracy.

The obtained discrete data using cluster-based binning are used for performance benchmarking of ID3 algorithm against statistical binning technique of discretization, original continuous data. From the results obtained we can state that the accuracy of ID3 with clustering-based discretization overperforms other techniques by 2–3% improvement in accuracy. In future, we wish to continue this research study for different types of classification problems for datasets of larger size. Figure 5 shows the comparison of accuracy values obtained from the experiments.

The contribution of this work is to bring to light the potential mechanism effect of clustering-based discretization on decision tree algorithm.

6 Conclusion and Future Scope

The contribution of this work is to bring to light the potential mechanism effect of clustering-based discretization on decision tree algorithm. As described in the Results section, it is observed that cluster-based discretization outperforms the statistical binning method. The outcome of this study reveals the potential use of clustering-based discretization and decision tree entropy computations. In future, this study may open up causal inference between clustering mechanism and decision tree split techniques.

Acknowledgements The authors are thankful for the support provided by Visvesvaraya Ph.D., Scheme. This publication is an outcome of the R&D work undertaken project under the Visvesvaraya Ph.D., Scheme of Ministry of Electronics & Information Technology, Government of India, being implemented by Digital India Corporation.

References

1. Quinlan, J.R.: Induction of decision trees. Mach. Learn. **1**(1), 81–106 (1986). https://doi.org/10.1007/BF00116251
2. Fayyad, U.M., Irani, K.B.: On the handling of continuous-valued attributes in decision tree generation. Mach. Learn. **8**(1), 87–102 (1992). https://doi.org/10.1007/BF00994007
3. Garcia, S., Luengo, J., Sáez, J.A., López, V., Herrera, F.: A survey of discretization techniques: taxonomy and empirical analysis in supervised learning. IEEE Trans. Knowl. Data Eng. **25**(4), 734–750 (2013). https://doi.org/10.1109/TKDE.2012.35
4. Behera, G.: Privacy preserving C4.5 using Gini index. In: 2011 2nd National Conference on Emerging Trends and Applications in Computer Science, Shillong, India, March 2011, pp. 1–4. IEEE (2011). https://doi.org/10.1109/NCETACS.2011.5751385
5. Chen, P.: The application of an improved C4.5 decision tree. In: 2021 7th Annual International Conference on Network and Information Systems for Computers (ICNISC), Guiyang, China, July 2021, pp. 392–396 (IEEE). https://doi.org/10.1109/ICNISC54316.2021.00078
6. Liang, X., Qu, F., Yang, Y., Cai, H.: An Improved ID3 Decision Tree Algorithm Based on Attribute Weighted, Guilin, China (2015). https://doi.org/10.2991/cmes-15.2015.167
7. Alexandre, L., Costa, R.S., Henriques, R.: DI2: prior-free and multi-item discretization of biomedical data and its applications, March 2021. arXiv:2103.04356 [cs, q-bio]. http://arxiv.org/abs/2103.04356. Accessed 4 Mar. 2022
8. Farnaz Nojavan, A., Qian, S.S., Stow, C.A.: Comparative analysis of discretization methods in Bayesian networks. Environ. Model. Softw. **87**, 64–71, January 2017. https://doi.org/10.1016/j.envsoft.2016.10.007
9. Zheng, H.: NAT-Modeled Dynamic Discretization for Inference with Sum of Continuous Variables, p. 102
10. Beynon, M.J.: Stability of continuous value discretisation: an application within rough set theory. Int. J. Approx. Reason. **35**(1), 29–53 (2004). https://doi.org/10.1016/S0888-613X(03)00057-4
11. Barbiero, A., Hitaj, A.: Approximation of continuous random variables for the evaluation of the reliability parameter of complex stress–strength models. Ann. Oper. Res. (2021). https://doi.org/10.1007/s10479-021-04010-6
12. Yao, Q., Zhang, Q., Tang, J., Wang, X., Hu, M.: Time-variant reliability analysis method for uncertain motion mechanisms based on stochastic process discretization. IEEE Access 1–1. https://doi.org/10.1109/ACCESS.2022.3153524
13. Wu, S., Wang, C., Zhao, R.: An improved integrator discretization method for multiple second-order generalized integrators. In: 2021 IEEE 4th International Electrical and Energy Conference (CIEEC), Wuhan, China, May 2021, pp. 1–6. IEEE (2021). https://doi.org/10.1109/CIEEC50170.2021.9510844

14. Tan, S.C.: Improving association rule mining using clustering-based discretization of numerical data. In: 2018 International Conference on Intelligent and Innovative Computing Applications (ICONIC), Plaine Magnien, December 2018, pp. 1–5. IEEE (2018). https://doi.org/10.1109/ICONIC.2018.8601291
15. Macqueen, J.: Some methods for classification and analysis of multivariate observations. In: Multivariate Observations, p. 17

Studies on Effect of Load Variation on Terminal Voltage and Frequency of Three Phase Synchronous Generator

Dheeman Panigrahi, Anwesh Padhi, Nissan Kumar Pradhan, and Sabyasachi Pradhan

Abstract Synchronous generators are run at a speed in synchronism with grid frequency and are used to convert mechanical energy to AC electrical energy. The generator is synchronised onto a large interconnected grid thus operating in unison or in parallel with all other generators of the overall system supplying energy to all the loads connected to it. It is essential to study the effect of the load variations on the voltage magnitude and the frequency. This paper studies the effect of load variations on the voltage magnitude and the frequency of an isolated three-phase synchronous generator. The three-phase synchronous machine has been modelled in the MATLAB Simulink environment. Separate PI controllers have been developed for keeping the magnitude and frequency of the terminal voltage constant. The simulation has been done in MATLAB Simulink environment. It can be concluded from the simulation results obtained that the magnitude and frequency of the terminal voltage of the synchronous generator remain constant at their rated values irrespective of variation in connected load (both real power and reactive power). These simulation results show the proper functioning of the proposed PI speed regulator and PI voltage regulator in this paper.

Keywords Three-phase synchronous generator · PI controllers · MATLAB Simulink environment

D. Panigrahi (✉) · A. Padhi · N. K. Pradhan
Department of Electrical Engineering, IGIT Sarang, Dhenkanal, Odisha 759146, India
e-mail: panigrahidheeman@gmail.com

S. Pradhan
CPP, NALCO, Angul, Odisha 759145, India

S. Kumar Udgata et al. (eds.), *Intelligent Systems*, Lecture Notes in Networks and Systems 1149, https://doi.org/10.1007/978-981-97-8160-7_3

1 Introduction

Synchronous generators are run at a speed in synchronism with grid frequency and are used to get electrical energy from mechanical energy. In these machines, a DC current is applied to the field winding mounted on the rotor to produce a magnetic field. The rotor of the generator is rotated with the help of a prime-mover, thus producing a rotating magnetic field in the air-gap of the synchronous generator. This rotating magnetic field cuts the stationary conductors of the stator winding and induces a balanced set of three-phase AC electro-motive-forces (emfs) in them [1–4]. In reality, the mode of operation of alternator is different. The generator would be synchronised onto a large interconnected grid thus operating in unison or in parallel with all other generators of the overall system supplying energy to all the loads connected to it. In the first mode of operation, the frequency and the magnitude of the terminal voltage must be controlled by means of the prime-mover torque and the field current respectively [5–8].

2 Real Power Mismatch and Frequency Relationship

The generator powers, real and reactive, are essentially determined by the load itself. Although this mode of operation is comparatively uncommon, we presently study it for the purpose of comparing it with the more common parallel operation. Let us understand this by taking a suitable example. Consider a synchronous machine initially running at rated speed with no load and a full terminal voltage. Its terminals are connected to a load having three star-connected resistors. As the load is being applied, a stator current wave is immediately created by the resulting stator currents, hence leading to an air-gap torque. This torque will tend to decelerate the rotor. In order to maintain a constant speed, it is thus necessary to increase the prime-mover torque until torque balance is restored. The air-gap torque thus creates the mechanism by means of which mechanical power from the prime-mover is transformed into electrical power to the load impedances. Primarily, the real power output depends upon the load and is controlled by regulating the prime-mover torque so as to achieve torque balance and, as a result, constant speed and frequency [1–8].

3 Reactive Power Mismatch and Voltage Relationship

The stator current while flowing through the synchronous reactance causes a voltage drop across it. Therefore, the terminal voltage is expected to drop below its full rated value. To restore constant terminal voltage, the excitation level must be boosted, thereby increasing the emf magnitude. Or we can say that the terminal voltage is controlled by regulating the excitation level. However, there is an interaction or cross-coupling between the two control channels; an increase in excitation level causes an increase in the real power output which, in turn, requires a prime mover adjustment. A single generator, while operating alone, supplies both the real power (P) and reactive power (Q) as demanded by the load attached to its terminals. The governor controls the mechanical power input to the generator. The set-points of the governor control the operating frequency of the generator for a particular real power demand. Similarly, the DC field current flowing through the rotor field winding controls the magnitude of the terminal voltage of the generator for a particular reactive power demand [1–7].

4 Synchronous Generator Modelling Using MATLAB Simulink

The synchronous generator operating alone can be modelled by using MATLAB Simulink (version R2023a). The Simulink blocks used to model this operation of the synchronous machine are taken from the existing library of MATLAB. The detailed model of the three-phase synchronous machine has been shown in Fig. 1.

Figure 1a represents the Simulink model of the synchronous generator working alone and feeding a three-phase RLC balanced load. And, Fig. 1b shows the block diagram of the Simulink model of the synchronous generator. Two controllers are used here. One is PI voltage regulator shown in Fig. 2 and the other is the PI speed regulator shown in Fig. 3. In the PI voltage regulator, the error between the actual pu terminal voltage and 1 pu rated voltage is fed to the PI regulator and the output obtained is fed as an excitation voltage to the synchronous machine. The proportional and integral constants are tuned at values 0.00045 and 0.1 respectively. The purpose of this regulator is to ensure availability of rated terminal voltage irrespective of variation in reactive power load on the generator. Similarly, in the PI speed regulator shown in Fig. 3, the error between the actual pu rotor speed and 1 pu rated speed is fed to the PI regulator and the output is fed as the mechanical power input to the synchronous generator from the prime-mover. The proportional and integral constants are tuned at values 10 and 100 respectively. The purpose of this regulator is to ensure the frequency of the terminal voltage to be equal to the rated frequency irrespective of variation in real power load on the generator.

5 Simulation and Results

The above model of synchronous generator shown in Fig. 1 was simulated in MATLAB Simulink environment by assigning values to the parameters as given below.

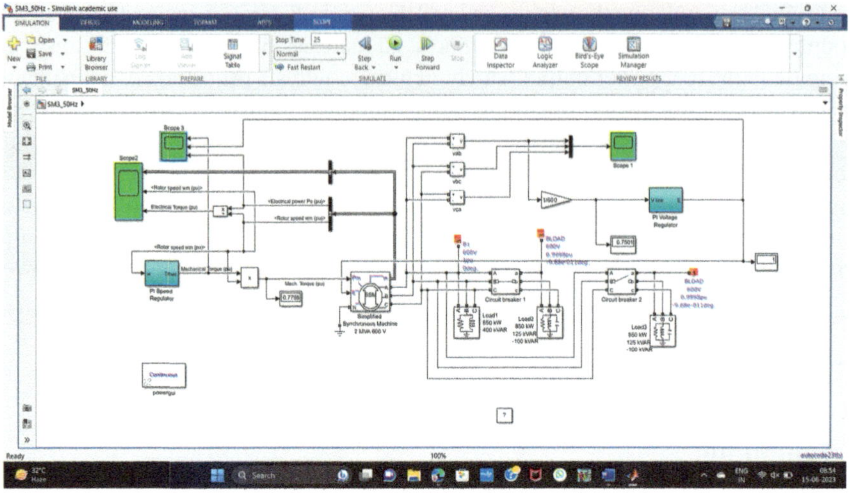

(a) Simulink model of the Synchronous Generator

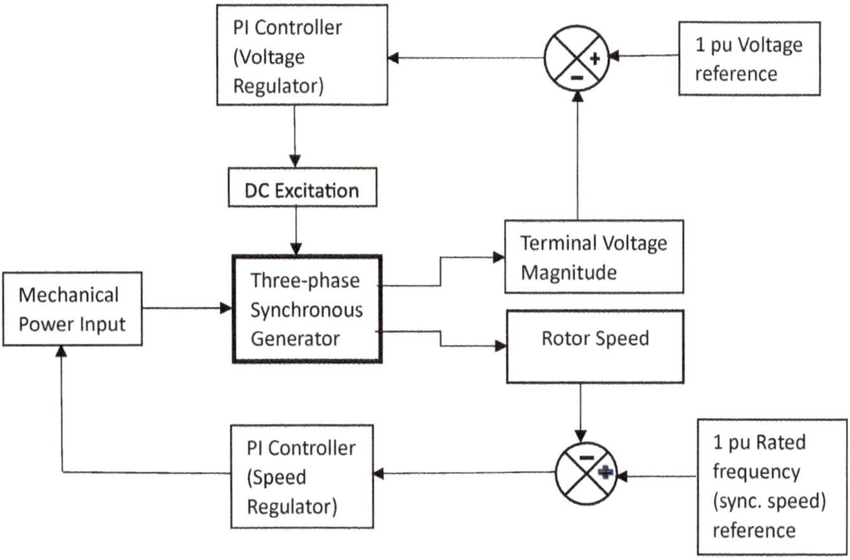

(b) Block diagram of the Simulink model of the synchronous generator

Fig. 1 **a** Simulink model of the synchronous generator. **b** Block diagram of the Simulink model of the synchronous generator

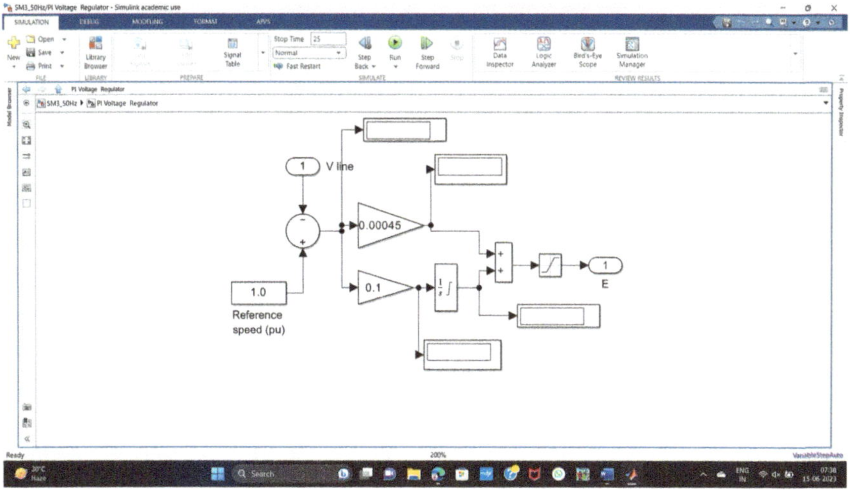

Fig. 2 PI voltage regulator

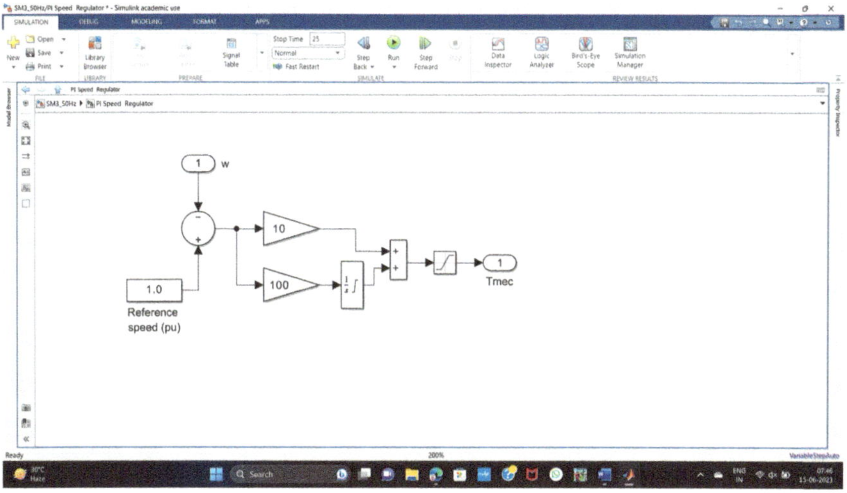

Fig. 3 PI speed regulator

Three-phase synchronous generator:

Nominal power, line-to-line voltage and frequency [Pn(VA) Vn(Vrms) fn(Hz)]: [2000e3, 600, 50].
Inertia, damping factor and pairs of poles [H(sec) Kd(pu_T/pu_w) p()]: [0.6, 0.0, 2].
Internal impedance [R(pu) X(pu)]: [0.0036 0.16]

Initial conditions [dw(%) th(deg) ia,ib,ic(pu) pha,phb,phc(deg)]: [0 7.01083 0.824535 0.824535 0.824535 −14.0377 −134.038 105.962].

Loads:

Load1—600 V, 50 Hz, 850 kW, 400 kVAR
Load2—600 V, 50 Hz, 850 kW, 125 kVAR, −100 kVAR
Load3—600 V, 50 Hz, 850 kW, 125 kVAR, −100 kVAR

The load 'Load1' was connected permanently to the terminals of the synchronous generator. The load 'Load2' was switched off at time t = 10 s. The load 'Load3' was switched on at time t = 15 s. The entire simulation was done for a time t = 25 s. The results and waveforms as obtained in the scopes are described below in Figs. 4, 5, 6, 7, 8, 9, 10, 11, 12 and 13.

Figure 4 shows the terminal voltage waveform of the synchronous generator. The time period of the voltage waveform is 0.02 s before switching off the load 'Load2' at t = 10 s. This time period of the voltage waveform remains same at 0.02 s after switching off the load 'Load2' at t = 10 s. Thus, the frequency of the terminal voltage remains same at its rated value 50 Hz irrespective of the variation in the load. This shows that the PI speed regulator is working properly ensuring constant frequency against load variation. However, the amplitude of the terminal voltage shows a variation of 0.0172 pu increase while switching off the load 'Load2' as t = 10 s. Figure 5 shows the terminal voltage waveform of the synchronous generator. The time period of the voltage waveform is 0.02 s before switching on the load 'Load3' at t = 15 s. This time period of the voltage waveform remains same at 0.02 s after switching on the load 'Load3' at t = 15 s. Thus, the frequency of the terminal voltage remains same at its rated value 50 Hz irrespective of the variation in the load. This shows that the PI speed regulator is working properly ensuring constant

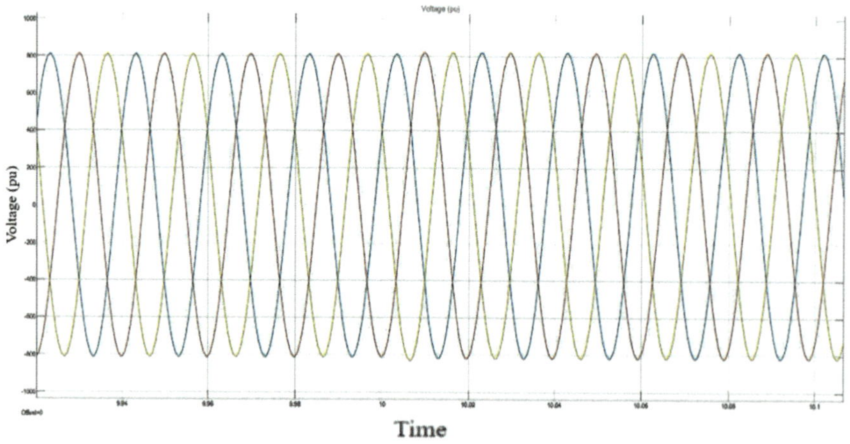

Fig. 4 Voltages 50 Hz around time t = 10s

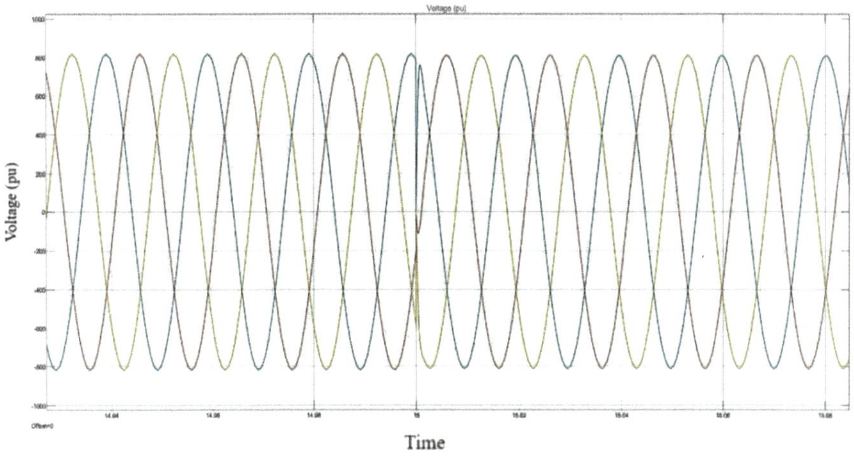

Fig. 5 Voltages 50 Hz around time t = 15s

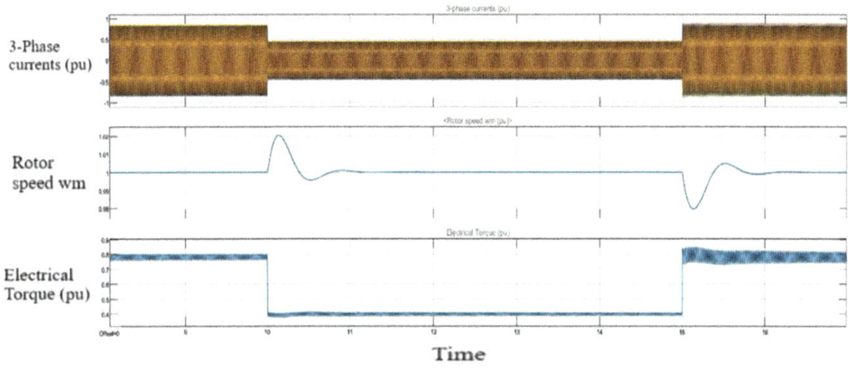

Fig. 6 Currents, rotor speed & electrical torque developed

frequency against load variation. However, the amplitude of the terminal voltage shows a variation of 0.0172 pu decrease while switching on the load 'Load3' as t = 15 s.

Figure 6 shows the variation in stator three-phase currents, rotor speed and electrical torque developed in the synchronous generator while switching off the load 'Load2' at t = 10 s and switching on the load 'Load3' at t = 15 s. It can be observed that the magnitude of the currents reduces while switching off the load 'Load2' and increases while switching on the load 'Load3'. But the frequency of the currents remains same at 50 Hz throughout the load variation. The rotor speed remains constant throughout the load variation except for showing little undershoots and overshoots of 0.05 pu and 0.2 pu respectively during transient conditions of switching off and switching on the loads. This can be accepted with the present

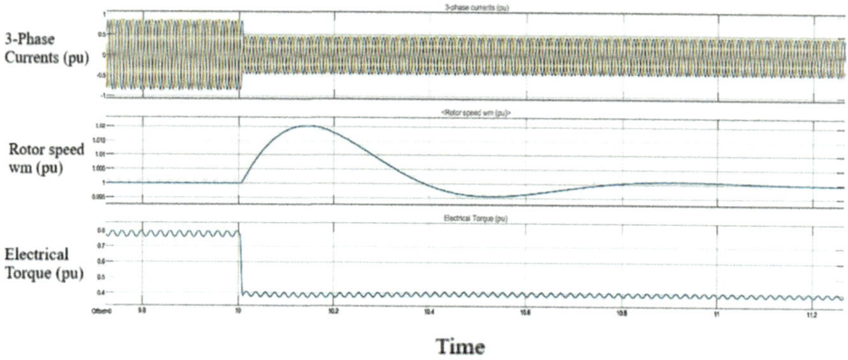

Fig. 7 Currents, rotor speed & electrical torque developed ('Load2' switched off at t = 10s)

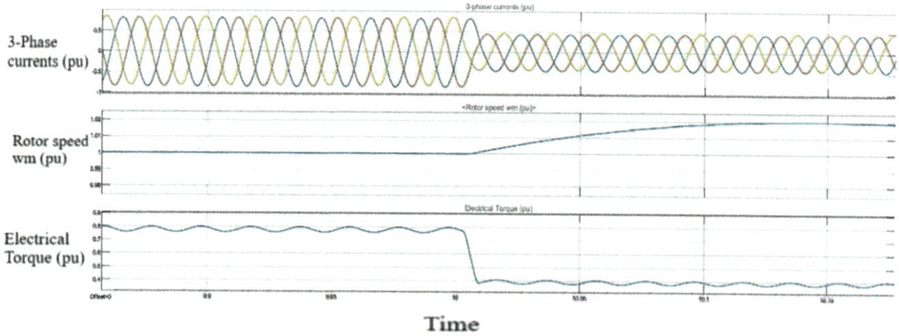

Fig. 8 Currents, rotor speed & electrical torque developed ('Load2' switched off at t = 10s, magnified view)

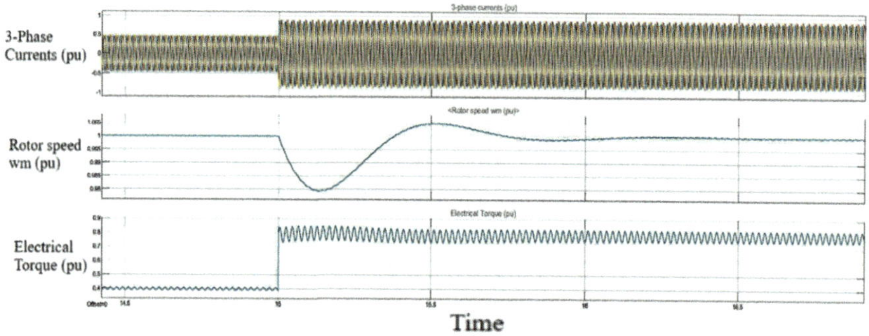

Fig. 9 Currents, rotor speed & electrical torque developed ('Load3' switched on at t = 15s)

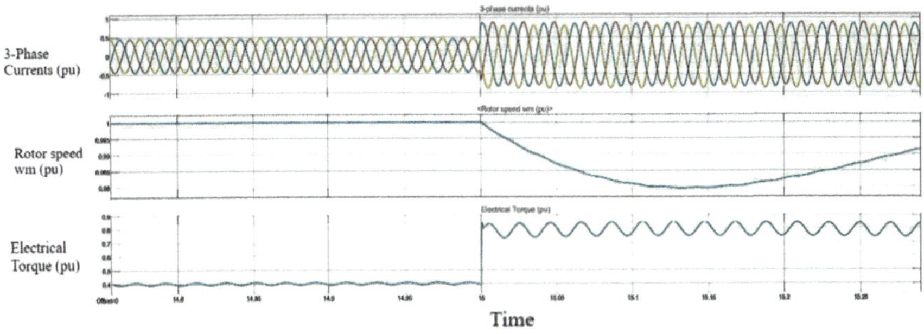

Fig. 10 Currents, rotor speed & electrical torque developed ('Load3' switched on at t = 15s, magnified view)

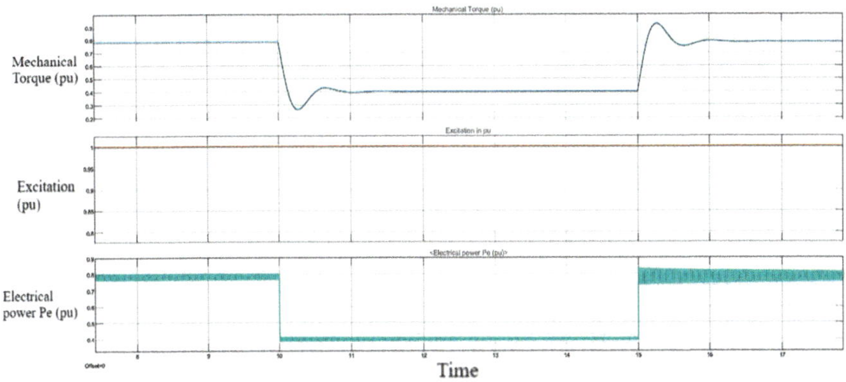

Fig. 11 Variation in mechanical torque input, the excitation and the electrical power developed

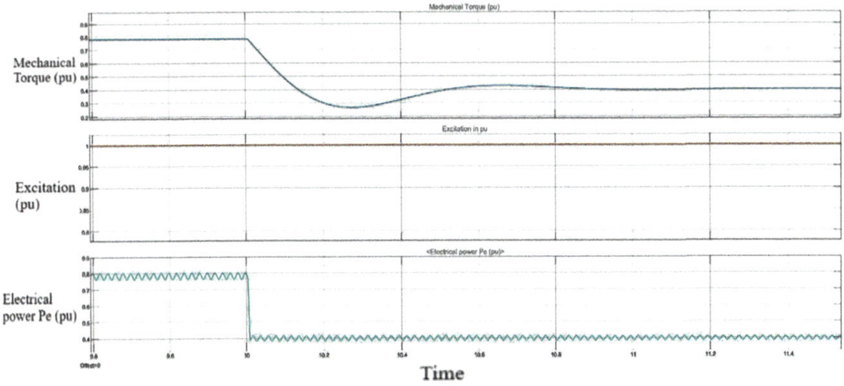

Fig. 12 Variation in mechanical torque input, the excitation and the electrical power developed while switching off 'Load2' at t = 10s

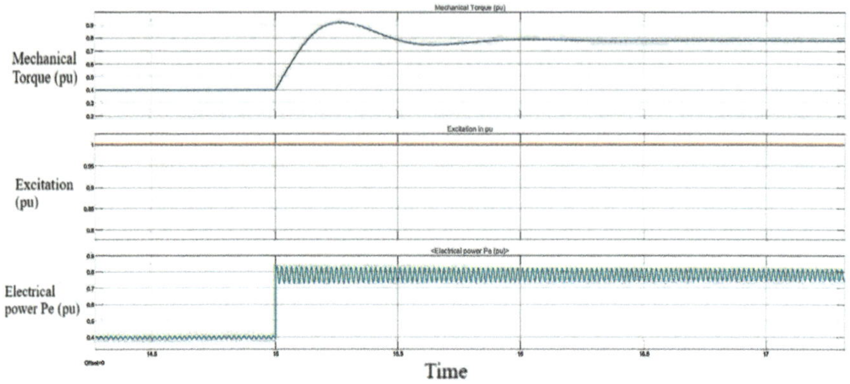

Fig. 13 Variation in mechanical torque input, the excitation and the electrical power developed while switching on 'Load3' at t = 15s

tuning of the PI speed regulator. The electrical torque developed varies in tune with the load variation.

The results of Fig. 6 have further been magnified in Figs. 7, 8, 9 and 10 during switching off the load 'Load2' and switching on the load 'Load3' respectively. Figure 11 shows the variation in mechanical torque input, the excitation and the electrical power developed in the synchronous machine. It can be seen that when the load 'Load2' is switched off at t = 10 s, the mechanical torque input to the synchronous machine reduces from 0.78 pu to 0.4 pu. Similarly, when the load 'Load3' is switched on at t = 15 s, the mechanical torque input to the synchronous machine increases from 0.4 pu to 0.78 pu. Thus, the variation in the mechanical input to the torque is in tune with the load variation thereby showing the proper working of the PI speed regulator. The undershoots and overshoots during the transient switching condition are lying between 0.12 pu & 0.07 pu which are acceptable.

The electrical power developed in the synchronous machine changes in tune with the load variation. The results of Fig. 11 have further been magnified in Figs. 12 and 13 during switching off the load 'Load2' and switching on the load 'Load3' respectively.

6 Conclusion

It can be concluded from the results obtained in this work that the magnitude and frequency of the terminal voltage of the synchronous generator remain constant at their rated values irrespective of variation in connected load (both real power and reactive power). This shows the proper functioning of the PI speed regulator and PI voltage regulator proposed in this work.

References

1. Chapman, S.J.: Electric Machinery Fundamentals, 4th edn. McGraw Hill Publication, Singapore (2005)
2. Say, M.G.: Alternating Current Machines, English Language Book Society/Longman, England, 5th edn. (1990 reprint)
3. Fitzgerald, A.E., Kingsley Jr., C., Umans, S.D.: Electric Machinery, 6th edn. Tata McGraw Hill Publishing Company Ltd, New Delhi (2003 reprint)
4. Elgerd, O.I.: Electric Energy Systems Theory. Tata McGraw Hill Publishing Company Ltd, New Delhi, TMH Edition (1989 reprint)
5. Sanjenbam, C.D., Singh, B.: Universal active filter for standalone hydro-electric system with SPV and battery support. In: 2022 IEEE 1st Industrial Electronics Society Annual On-Line Conference (ONCON), pp.1–6 (2022)
6. Sanjenbam, C.D., Singh, B.: Modified adaptive filter based UPQC for battery supported hydro driven PMSG system. IEEE Trans. Industr. Inf. **19**(7), 8018–8028 (2023). https://doi.org/10.1109/TII.2022.3215950
7. Sanjenbam, C.D., Singh, B.: PMSG based standalone distributed generation integrated universal active filtering system. In: 2023 IEEE IAS Global Conference on Emerging Technologies (GlobConET), pp. 1–6 (2023)
8. Sarker, K.: FC-PV-battery-Z source-BBO integrated unified power quality conditioner for sensitive load & EV charging station. J. Energy Storage **75**, 109671 (2024)

Latent Challenges of Multimodal Deep Learning Models: Taxonomy and Survey

Sachin Kumar, Olga Ivanova, Olga Vorfolomeeva, and Raghvendra Kumar

Abstract In the evolution of artificial intelligence technology, multimodal deep learning models are a crucial and evolutionary step. Algorithms will soon be widely employed to process multimodal information in every domain. However, researchers often require in-depth exploration of methods and models when venturing into a new subdomain of studied approaches and technologies. Various classification features support prior research on the categorization of multimodal deep learning models. Typically, such taxonomies classify these models as early, intermediate, and late fusion models. This article presents an in-depth review of multimodal deep learning models. It proposes a classification of obstacles that are rarely addressed in literature, but which hold significant implications for future advancements in the field. These challenges comprise computational complexity, generalizability, scalability, security, and a new data processing philosophy. Various aspects of each obstacle are taken into account. This survey can aid researchers in anticipating possible future issues and generating ideas for extended experiments.

Keywords Multimodal data · Deep learning · Taxonomy · Computational complexity · Generalizability

S. Kumar · O. Ivanova
South Ural State University, Chelyabinsk, Russia
e-mail: kumars@susu.ru

O. Ivanova
e-mail: onivanova@susu.ru

O. Vorfolomeeva
West Kazakhstan State University, Uralsk, Kazakhstan
e-mail: olga.vorfolomeyeva@wku.edu.kz

R. Kumar (✉)
GIET University, Gunupur, Odisha, India
e-mail: raghvendraagrawal7@gmail.com; raghvendra@giet.edu

S. Kumar Udgata et al. (eds.), *Intelligent Systems*, Lecture Notes in Networks and Systems 1149, https://doi.org/10.1007/978-981-97-8160-7_4

1 Introduction

Nowadays, analyzing multimodal data sources in various real-world domains is a dominant trend in machine learning research. In [1], a multimodal classification is described that employs several feature extraction methods, including manual, CNN, WE, AE, LTSM, DNN, GGA, DT, and others, as well as a concatenation method that includes feature concatenation, deep concatenation, score merge, deep merge, and tensor fusion. The authors also present a comprehensive list of unresolved issues in multimodal models, which are widely acknowledged but under-investigated in literature. Ideally, it is possible to categorize the challenges into two groups: those arising from unbalanced data (such as the absence of data in datasets, lack of multimodal datasets, and various learning complexities for each modality) and those originating from inadequate computational resources (such as unpredictable results from the amalgamation of diverse parts of multimodal models and the failure to create universal models).

In [2], a taxonomy is presented that describes approaches for processing multimodal data based on their features. The taxonomy identifies two main categories: task-specific methods, including encoder-decoder-based models, attention-based models, and reinforcement learning models, and pre-training approaches, encompassing pre-training types, pre-training tasks, transformer-based architectures, and unifying architectures. More focused surveys of multimodal deep learning models have been conducted, concentrating on individual task types. In [3], the authors provide a current summary of self-supervised multimodal deep learning models. Their classification system employs a three-dimensional categorization of objective functions, data alignment techniques, and deep learning model architectures. In their study [4], the authors examine multimodal fusion models in the smart healthcare field. They categorize reviewed literature into five types of multimodal approaches, providing examples and defining classes for feature selection, rule-based systems, machine learning, deep learning, and natural language processing. Traditionally, surveys and taxonomies aim to forecast the evolution of multimodal techniques and approaches.

The aim of this article is to systematically analyze pertinent data in order to pinpoint current tasks and inefficiencies. As time goes by, novel challenges arise, making it necessary to periodically update, re-evaluate, and even overhaul our research domain. The primary objective of this study is to constantly enhance and perfect existing solutions by administering more comprehensive surveys.

2 Methodology

In this section, we present the methodologies used to identify challenges, and the steps taken to gather and analyze relevant literature.

2.1 Search Strategy

We searched extensively through various peer-reviewed journals in the field of multimodal techniques, including Scopus, ResearchGate, ScienceDirect, IEEE Xplore, Google Scholar, and arXiv. Our search query followed the structure of established protocols:

<name of the database> + ″multimodal models″ + <challenge>
+ <year of publication>.

The <challenge> keyword in the search query was initially categorized based on predetermined challenges. However, our research led to a shift in our understanding of the less-discussed challenges of multimodal models. As a result, we updated our taxonomy and conducted search queries based on the new categories. At the initial stage, we developed a descriptive framework to identify issues, list methods, present main findings, and extract peer-reviewer notes (if available) for each paper. This approach helped us to organize the field and identify obscure issues and gaps. The investigation ran from July 10, 2023, through September 10, 2023.

2.2 Process and Statistics of the Search

The steps of the search and reviewing process are presented in Fig. 1.

Statistical data of the revised papers are presented in Table 1.

The majority of the papers were indexed in multiple databases, while only 7% of the reviewed papers were found in a single source. This trend was especially prominent in works published in 2023, a period when journal indexing was still in progress, even if the journals had already been included in a database index.

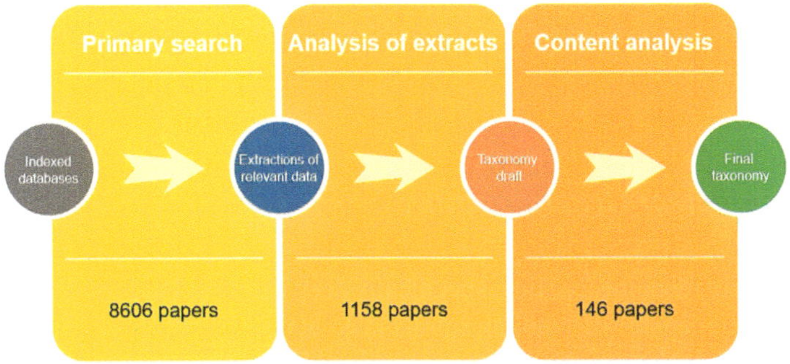

Fig. 1 Steps of the reviewing process

Table 1 Requirements implied to a paper for being reviewed

Resource	Weblink	Number of papers considered
Scopuse	scopus.com	1795
Science Direct	sciencedirect.com	148
IEEE Xplore	ieeexplore.ieee.org	338
ResearchGate	researchgate.net	2256
Google Scholar	scholar.google.com	1899
ArXiv	arxiv.org	515

3 Taxonomy of Rarely Discussed Challenges of Multimodal Deep Learning Models

In our survey, we investigate four sets of challenges confronted by modern multimodal models, which have not garnered ample attention from researchers in contrast to other challenges in the development of multimodal models.

3.1 Computational Complexity

One of the primary difficulties in contemporary deep learning models is the inherent complexity of algorithms. Multimodal models require the management of sizable amounts of varied data and the use of intricate, iterative, and resource-intensive algorithms that often strain both hardware and software resources. New software and hardware architectures enable a partial reduction of the gap between the demand and the actual availability of resources for computational experiments based on models and practical problem solutions. However, implementing new, resource-saving approaches and algorithms for processing multimodal data, which are detailed in this section, can bring significant success in bridging the gap. We review research that examines various approaches for addressing the temporal and spatial intricacies of multimodal models.

3.2 Insufficient Generalizability

The progress of machine learning, specifically deep learning, has brought forth multitasking models capable of handling similar tasks across a wide range of subject areas. However, the problem of insufficient generalizability remains important despite the considerable advancements made in the creation of transformers with self-awareness and mapping capabilities. Researchers have access to various pre-trained models that can be adapted or adjusted for alternate datasets. However, it is not enough to simply combine the features identified through the processing of distinct modalities in a coplanar manner for multimodal models. This section includes articles on the abundance of current architectures, the intricacies of concatenation and transfer of results for additional generalization or processing on higher levels.

3.3 *Insufficient Scalability*

Although interconnected, the concepts of generalization and scalability differ in subject matter and research methodologies. Generalization involves identifying commonalities among research subjects to transfer traits to other environments and unfamiliar data. Meanwhile, scalability refers to the ability of multimodal models to function effectively despite fluctuations in available data or resources. Therefore, this section presents research that aims to expand the applicability of current models and algorithms across different environments.

3.4 *Data Security and New Philosophy of Data Processing and Generation*

Modern society is witnessing an unprecedented surge in the integration of AI models in everyday objects, ranging from smart irons to fertilizing agricultural land. Therefore, the responsible and secure application of emerging technologies, along with their associated data storage and processing protocols, have become critical concerns of our era. In this section, authors from diverse fields investigate various aspects related to the security of multimodal computing technologies. A matrix in Figure 2 displays the common challenges analyzed and examined in both single and

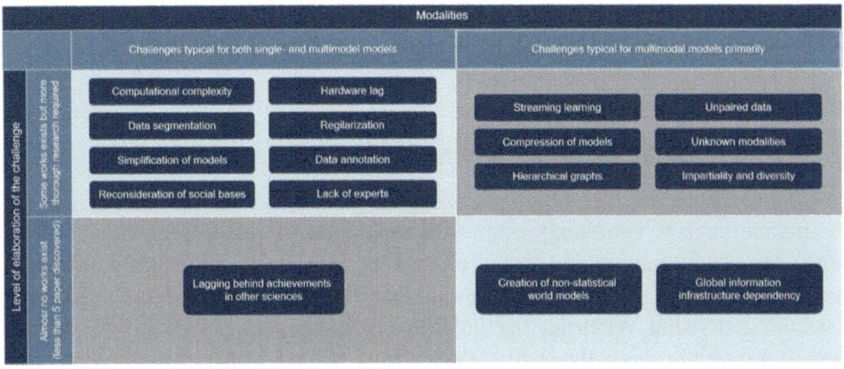

Fig. 2 Matrix of challenges

multimodal models. For each of the four challenge categories, we establish subcategories of problems and ideas to sort current research and identify accomplished outcomes and gaps in applicable fields.

4 Multimodal Computational Complexity

Deep neural networks processing multimodal data can receive billions and trillions of data segments as input. The size of neural networks is often measured in billions and trillions of parameters, as seen in [5]. Modern hardware and software enable the training of huge neural networks. Nevertheless, the updates for software and hardware components to cope up with researchers' increasing demands for neural network efficiency are slow. Currently, there are numerous approaches to address the challenges posed by temporal and spatial complexity.

4.1 Multimodal Ensembling

One potential strategy involves integrating multimodal learning principles with other solution approaches. Ensemble models have been extensively tested in practical applications. The classical Ensembling method combines outputs from multiple models that analyze data of the same modality using a rule, similar to the late merging approach in multimodal learning. The notion of multilevel Ensembling is suitable for models that function with multimodal data. Nevertheless, the currently available methods for concatenating the outcomes of processing heterogeneous modalities with distinct levels of significance, informativity, and non-uniformity of mistakes may not be adequate to achieve the desired level of precision. Multilevel Ensembling is a suitable concept for models that work with multimodal data. However, current methods for integrating processed data from diverse modalities with varying degrees of importance, informativeness, and error heterogeneity may be inadequate for achieving the desired level of precision.

4.2 Multimodal Data Segmentation

Segmentation is a commonly employed technique for identifying boundaries of objects and materials within visual representations. It is a task that is performed unconsciously and automatically by individuals but presents considerable challenges for computer vision. When dealing with low-performance devices, such as smart gadgets and mobile robots, the segmentation process must utilize minimal computing resources and avoid cloud connectivity. Numerous studies have investigated data segmentation, encompassing both automatic and semi-automatic methods [6].

4.3 Multimodal Streaming Data Learning

Researchers analyzed several AI multimodal techniques, such as PaLM 2 and ChatGPT-4, to assess their effectiveness [7]. The study discovered that allowing the models access to elusive data greatly reduces their performance. Server-less and distributed technologies are recommended for data storage, management, and preprocessing [8]. The authors suggest that incorporating serverless computing frameworks, with all computational and processing units located outside the laboratory, will optimize both cost and calculation speed. Currently, streaming multimodal learning is highly sought after and continues to attract many researchers. There is currently insufficient research concerning the optimization of computational complexity and acquisition of necessary technical resources for the proposed models in the existing literature.

4.4 Multimodal Compression

Multimodal models require considerable computing resources. Researchers are addressing this issue by using cloud frameworks to handle large datasets and complex transformers commonly used in multimodal models. Additionally, pre-trained models are often located in the cloud as well, making it another popular method. These models typically have multiple layers dedicated to each modality within their networks. Several compression techniques have been developed to address the issue of effectiveness, although their frequency of use may be limited. Another instance of quantifying compression is available in the prediction of agent behavior [9], and in analyzing compression of multimodal models [10]. While there have been no new methods proposed for compressing multimodal models, utilizing existing approaches has significantly improved necessary resources, such as processors, memory, and time.

4.5 Simplification

Simplification of multimodal models is based on two approaches:

- Usage of multistage pre-training of models
- Usage of surrogate models

Pre-trained models have accelerated the creation of multimodal models despite their significant computational expense. In their research [11], the authors present an iterative training method that starts with a small amount of in-domain data, only a few percent, combined with out-of-domain data. In subsequent stages, a multimodal model is pre-trained using a portion of in-domain data as well as out-of-domain data from other tasks previously mentioned. This approach improves the model's effectiveness for various future problems. The authors utilized this technique to reduce the number of multimodal model parameters by over 50% and minimize the size of the training dataset by 11.76%.

4.6 Hardware Lag

Moore's Law has hit its theoretical limit due to technological constraints. The utilization of GPUs and tensor processors provides merely momentary respite from the issue of computational complexity that contemporary multimodal models present. Currently, multiple manufacturers are developing complete lines of tensor processors using 3-nm technology. These processors are a groundbreaking achievement in machine learning, especially for mobile and server applications. Their potential will be fully realized as they tackle more and more challenging research tasks. To compare GPUs and CPUs in machine learning tasks, single-modal datasets were utilized [12]. Different models have shown the superiority of both graphic and tensor processors across multiple measures. In analyzing machine learning models trained on different processor architectures, researchers measured precision, training and inference duration, memory usage, required bandwidth, availability, and power efficiency. To date, no studies have investigated the performance of multimodal models across different processor architectures.

5 Insufficient Generalizability

The development stages of a multimodal machine learning model have been extensively laid out. However, the diversity of data and tasks prevents the creation of universal models. Obstacles regarding data include inadequate diversity, an imbalanced preference for certain types, counterfeit patterns, an unequal distribution of features, under-filled feature space, data becoming outdated over time, and other potential hindrances. Enhancing the generalizability of multimodal models is an arduous task that necessitates the vast amount of time and the expertise of highly qualified researchers. Generalizability describes a model's ability to function accurately with previously presented data [13]. Researchers aim to achieve a balance between under-training and over-training models to minimize the difference between learning error (during training) and generalizability error (during testing). Low-capacity models are inadequate in approximating a diverse range of functions and characteristics, whereas high-capacity models are prone to overfitting.

5.1 Regularization

In fact, regularization is the most commonly employed set of techniques to increase generalizability in multimodal models. This widely used method has been incorporated in works such as [14], utilizing L1 regularization, and works such as [15], utilizing L2 regularization. In several studies, the aim of minimizing a function with constraints has been attained by imposing a penalty determined by the weight norm of a neural network. Achieving this requires the formulation of a generalized Lagrange function [16].

To prevent overfitting of large datasets, researchers commonly employ the early stopping method. This technique returns the results with the lowest error, enabling model control without altering the core training process or target function [17]. Frequently, this approach is combined with other regularization techniques to optimize model performance. The abundance of regularization methods indicates that contemporary architectures of deep neural networks, such as transformers, are task-specific and contingent on the model's capacity and dataset size. This is equally applicable to multimodal models.

5.2 Data Annotation

Active learning is a prevalent method for annotating multimodal data. The pool-based sampling method is a prominent technique for this purpose. This method involves evaluating unmarked data with a pre-trained model on labeled data to determine its usability in subsequent training. If the assessment is satisfactory, the model forwards the data to an expert for annotation. Technical terms and abbreviations must be accurately explained when used for the first time. The newly annotated data are then utilized in the active learning phase. Numerous human annotators should be utilized to randomly annotate multimodal data, according to [18]. The weight of each individual's annotations is minimal. However, through the proposed procedure's deduction of accurate labels from scattered and noisy estimates, the required training accuracy can be achieved, and the necessary number of annotations is minimized.

5.3 Hierarchical Graphs

Generalizing multimodal models is essential for business tasks. Companies' data centers house substantial volumes of multimodal data that require quick processing, often in mere milliseconds. As a result, hierarchical ranking techniques have been developed for real-time systems. Typically, such a system has multiple levels of hierarchies to classify millions of tokens within fractions of seconds [19]. In [20], the authors present a generalized model that mimics the natural strategy of human thinking in information retrieval by progressively filtering out non-essential data and narrowing the focus to more relevant topics. The authors use document-by-document search and comparison as a means of implementing their model.

6 Insufficient Scalability

Scalability presents a considerable challenge for researchers utilizing multimodal models. They are required to process large datasets within tight time constraints and conduct multiple experiments using varying model architectures and computing resources. While scalability and computational complexity are closely related, we have chosen to address scalability challenges in multimodal models separately, given the many obstacles involved in solving them.

6.1 Unpaired Data

The tasks addressed by multimodal models allow for the addition and modification of data during the operation of the systems. To address gaps and align data, diverse forms of artificial data generation are often utilized. Processing a large amount of unpaired data created during continuous model training requires a thorough evaluation of methodologies to align original, newly generated, and synthetic data. Currently, one of the most frequently used methods is cross-modal contrast training [21]. Compared to other popular multimodal models, the presented approach scales the data size, model size, and batch size during training without any labeled data. The model, proposed in [22], separates model learning, control, and output functions. Model training occurs on the main computing node while control and output functions are implemented in a distributed environment with efficient data collection, management, and optimization of computing resources.

6.2 Hidden Modalities

Researchers have extensively investigated the utilization of pre-trained deep neural networks, the hierarchical structure of multimodal models, algorithms for retraining and generalization, and the transfer of models across various subject areas. However, there is currently inadequate research on building models that can learn from these modalities in completely disparate subject areas. Amid current methodologies for constructing multimodal models, only incorporating federated multimodal training can provide a significant resolution to this issue. The referenced research, including [23], implemented well-known datasets with familiar properties, constraints, and uncertainties to the scientific community to help reduce the potential impact of the aforementioned challenge. Nevertheless, it is undeniable that the challenge will become more significant when utilizing models in specific business sectors.

6.3 Creation of Non-Statistical World Models

The most significant multimodal models currently available are called Large Language Models (LLMs). Many of them have been continuously trained for years. LLM deep neural networks operate using not only vast input data but also generate a significant amount of data on intermediate layers. However, comprehending how a deep neural network reaches a decision is insufficient. Recently, studies have shown that Language Models (LMs) can create concepts that were not in the original algorithms [24]. These studies demonstrate that LMs are more than just statistical models and can generate intricate conceptualizations on their own.

7 Security and New Philosophy of Data Processing and Generation

One of the foremost challenges that humanity will soon encounter pertains to forming innovative social relations in all areas of activity. The implementation of artificial intelligence (AI) technologies for solving multiple tasks is an irrevocable process, thereby necessitating our adaptation to their amalgamation. Currently, there is an exceptional opportunity for humankind to promptly establish common values and guidelines that will expedite the advancement of artificial intelligence technologies while determining the allowable boundaries of such technologies' intrusion into critical human life and society aspects.

7.1 Impartiality and Diversity

In the context of multimodality, data preprocessing creates additional complexity. Although deep neural networks may reveal previously undetected limitations in data preprocessing, they do not explicitly demonstrate this. The authors in [25] noted the challenge of selecting an appropriate metric to evaluate the quality of multimodal models. The authors chose to utilize the KR20 metric to evaluate the reliability of dichotomous variables, rather than the regularly utilized metrics of AUC–ROC, F1, and specificity. This decision was influenced by the specific features of the problem under examination. Transformers and cross-modal learning, while utilizing self-attention mechanisms, do not always provide reliable solutions to the problem at hand.

Table 2 International standards in the field of artificial intelligence

International standardization institution	Website	Number of standards
ETSI	etsi.org	22
ISO	iso.org	17
ITU	itu.int	4
IEEE	ieee.org	22

7.2 Global Information Infrastructure and Standardization Dependency

The complexity of model architecture, together with the increase in data volume, the shift to multimodal models, distributed and parallel computing, all lead to a dependence on the global information infrastructure. There is a need to formulate foolproof network connectivity protocols and practices to ensure the integrity, security, and accessibility of multimodal data, even in isolated collaborations within the business community. It is imperative to make intensive efforts in developing and implementing national, regional, and international standards in this area to avoid future hindrance. Currently, numerous standards have been established in the field of artificial intelligence (see Table 2).

Specific standards have not yet been established for the procedures involved in developing and operating multimodal deep learning models. Therefore, these models must adhere to the requirements of other artificial intelligence models and algorithms.

7.3 Reconsideration of the Social Basis Fundamentals

The potentially infinite complexity and uncontrollable growth of large multimodal models is of concern to many researchers. As a result, issues that arise when dealing with a single unit modality are amplified across all modalities.

The concern regarding content originality has gained relevance with the widespread availability of LLM. It is possible that journals and publishers may demand authors to confirm that no AI assistants trained on multimodal data were used in generating their article content. Unless thoroughly validated by human oversight, relying solely on results generated by a multimodal model is improbable. Several investigations [26, 27] highlighted the presence of erroneous correlations and a lack of comprehension concerning the data's associations.

7.4 *Lack of Experts at Data Pre-Processing*

Research using multimodal models is an interdisciplinary endeavor with the primary objective of resolving specific issues. These models are tested on subject-specific datasets to ensure their relevance. The involvement of specialists in the respective fields is critical to the effective operation of deep machine learning. In the past, attempts were made to function without experts by developing different models and techniques. Therefore, it is essential to investigate creative channels of communication between experts and models to guarantee continuous surveillance of data quality and effectiveness of pre-processing. However, many studies have shown that incorrect interpretation of input data presents substantial risks [28]. Therefore, it is crucial to investigate new channels of communication between experts and models to maintain continuous monitoring of data quality and pre-processing efficiency. The development of multidimensional models requires specialized training for professionals and the implementation of interaction protocols among experts and data analysts.

7.5 *Lagging Behind Achievements of Other Sciences*

Multimodal models have the potential to aid in problem-solving regarding the management of waste disposal and air and water pollution, optimization of logistics processes, and the development of autonomous vehicles for various purposes. Although digitalization has played a significant role in these areas, the application of multimodal models is still at an early stage. To ensure the high-quality, intensive, and safe development of multimodal deep learning models, it is necessary to conduct a scientific study on the potential risks associated with developing, implementing, and using such models in human activities. This will enable us to predict potential consequences and scenarios for advancing new technologies.

8 Conclusions

This paper intends to present scientific discoveries by exploring multimodal model challenges that researchers confront in different domains. It is important to note that multimodal models are usually specific to their domains. The utilization of matching architectures, fusion methods, data pre-processing techniques, and challenge solutions for other datasets is limited, and a thorough feasibility analysis is required. The survey findings may aid scholars and professionals in pinpointing deficiencies in their approach to creating and evaluating multimodal models. This, in turn, can enhance the quality of their solutions and scientific outcomes. Furthermore, the suggested taxonomy can expand comprehension of potential vulnerabilities and risks in the research process, ultimately advancing scientific development within the relevant field.

Acknowledgements The study is supported by Russian Science Foundation regional grant number: 23-21-10009.

References

1. Sleeman, W.C., Kapoor, R., Ghosh, P.: Multimodal classification: current landscape, taxonomy and future directions. ACM Comput. Surv. **55**(7), December 2022 (2022)
2. Manzoor, M.A., Albarri, S., Xian, Z., Meng, Z., Nakov, P., Liang, S.: A survey on evolution, pretraining and its applications, multimodality representation learning (2023)
3. Zong, Y., Aodha, O.M., Hospedales, T.: A survey, self-supervised multimodal learning (2023)
4. Shaik, T., Tao, X., Li, L., Xie, H., Velásquez, J.D.: A survey of multimodal information fusion for smart healthcare: mapping the journey from data to wisdom. Inform. Fusion **102**, 102040 (2024)
5. Xia, H., Chu, L., Pei, J., Liu, W., Bian, J.: Model complexity of deep learning: a survey. Knowl. Inform. Syst. **63**, 2585–2619 (2021)
6. Wen, X., Famouri, M., Hryniowski, A., Wong, A.: Attendseg: a tiny attention condenser neural network for semantic segmentation on the edge (2021). arXiv preprint arXiv:2104.14623
7. Gonzalez, D.T., Djulbegovic, M., Kim, C., Antonietti, M., Gameiro, G.R., Uversky, V.N.: A comparative study of large language models in explaining intrinsically disordered proteins (2023)
8. Nesen, A., Bhargava, B.: Towards situational awareness with multimodal streaming data fusion: serverless computing approach. In: Proceedings of the International Workshop on Big Data in Emergent Distributed Environments, pp. 1–6 (2021)
9. Sedlmeier, A., Kölle, M., Müller, R., Baudrexel, L., Linnhoff-Popien, C.: Quantifying multimodality in world models (2021). arXiv preprint arXiv:2112.07263
10. Msuya, H., Maiseli, B.J., et al.: Deep learning model compression techniques: advances, opportunities, and perspective. Tanzania J. Eng. Technol. **42**(2), 65–83 (2023)
11. Liu, T., Feng, F., Wang, X.: Multi-stage pre-training over simplified multimodal pre-training models (2021). arXiv preprint arXiv:2107.14596
12. Kochura, Y., Gordienko, Y., Taran, V., Gordienko, N., Rokovyi, A., Alienin, O., Stirenko, S.: Batch size influence on performance of graphic and tensor processing units during training and inference phases. In: Advances in Computer Science for Engineering and Education II, pp. 658–668. Springer (2020)
13. Heaton, J., Goodfellow, I., Bengio, Y., Courville, A.: Deep Learning, 800 pp. The MIT Press (2016). ISBN: 0262035618. Genet. Program. Evolv. Mach. **19**(1–2), 305–307 (2018)
14. Magazzù, G., Zampieri, G., Angione, C.: Multimodal regularized linear models with flux balance analysis for mechanistic integration of omics data. Bioinformatics **37**(20), 3546–3552 (2021)
15. Wang, P., Qiu, C., Wang, J., Wang, Y., Tang, J., Huang, B., Jian, S., Zhang, Y.: Multimodal data fusion using non-sparse multi-kernel learning with regularized label softening. IEEE J. Sel. Top. Appl. Earth Observat. Remote Sens. **14**, 6244–6252 (2021)
16. Theljani, A., Chen, K. An augmented Lagrangian method for solving a new variational model based on gradients similarity measures and high order regularization for multimodality registration. In: Inverse Problems and Imaging (2019)

17. He, L., Li, H., Chen, M., Wang, J., Altaye, M., Dillman, J.R., Parikh, N.A.: Deep multimodal learning from MRI and clinical data for early prediction of neurodevelopmental deficits in very preterm infants. Front. Neurosci. **15**, 753033 (2021)
18. Liu, Y., Wei, Y.-S., Yan, H., Li, G.-B., Lin, L.: Causal reasoning meets visual representation learning: a prospective study. Mach. Intell. Res. **19**(6), 485–511 (2022)
19. Sangamithra, B., Manjunath Swamy, B.E., Sunil Kumar, M.: Personalized ranking mechanism using Yandex dataset on machine learning approaches. In: Proceedings of the International Conference on Cognitive and Intelligent Computing: ICCIC 2021, vol. 1, pp. 629–639. Springer (2022)
20. Ianina, A., Vorontsov, K.: Regularized multimodal hierarchical topic model for document-by-document exploratory search. In: 2019 25th Conference of Open Innovations Association (FRUCT), pp. 131–138. IEEE (2019)
21. Nakada, R., Gulluk, H.I., Deng, Z., Ji, W., Zou, J., Zhang, L.: Understanding multimodal contrastive learning and incorporating unpaired data. In: International Conference on Artificial Intelligence and Statistics, pp. 4348–4380. PMLR (2023)
22. bibitemch4127soldati2023design Soldati, P., Ghadimi, E., Demirel, B., Wang, Y., Gaigalas, R., Sintorn, M.: Design principles for generalization and scalability of AI in communication systems (2023). arXiv preprint arXiv:2306.06251
23. Xu, Z., Shen, Y., Huang, L.: Multiinstruct: improving multi-modal zero-shot learning via instruction tuning (2022). arXiv preprint arXiv:2212.10773
24. Piantadosi, S.T., Hill, F.: Meaning without reference in large language models (2022). arXiv preprint arXiv:2208.02957
25. Yusuf, A., Noor, N. Md., Bello, S.: Using multimodal learning analytics to model students' learning behavior in animated programming classroom. Educ. Inform. Technol. 1–44 (2023)
26. Yuan, C., Zhang, M., Huang, X., Xie, W., Lin, X., Zhao, W., Li, B., Qian, D.: Diffuse large b-cell lymphoma segmentation in PET-CT images via hybrid learning for feature fusion. Med. Phys. **48**(7), 3665–3678 (2021)
27. Wang, W., Gao, K., Luo, Y., Jiang, T., Gao, F., Shao, J., Sun, J., Xiao, J.: Triple correlations-guided label supplementation for unbiased video scene graph generation. In: Proceedings of the 31st ACM International Conference on Multimedia, pp. 5153–5163 (2023)
28. Rahate, A., Walambe, R., Ramanna, S., Kotecha, K.: Multimodal co-learning: challenges, applications with datasets, recent advances and future directions. Inform. Fusion **81**, 203–239 (2022)

SpaCy and NLTK NLP Techniques for Text Summarization: A Comprehensive Comparison

Raj Dumbre, Pratham Ankalwar, Smita Bhagwat, Dipti Pandit, Mahesh Bhutada, and Shirish Gund

Abstract In the era of information overload, efficient text summarization is crucial for quick comprehension. This research focuses on abstractive summarization using Spacy and NLTK, comparing their effectiveness on the Kaggle English Article dataset. The model is a natural language processing-based model. It demonstrates superior performance, as evidenced by higher BLEU scores. The study gives insights about the strengths of Spacy's deep learning architecture and NLTK's versatility. The research introduces a novel framework, emphasizing data pre-processing, model training, and evaluation using BLEU scores, laying the groundwork for future hybrid approaches.

Keywords Summarization · Natural language processing · Spacy · NLTK · Deep learning

1 Introduction

On a daily basis, people grapple with an overwhelming influx of information, primarily in textual format. The need to decipher lengthy articles for essential insights can be laborious. Comparing articles on the same subject often reveals overlapping details, necessitating prolonged analysis. Extracting vital information from each article becomes a time-consuming task. This underscores the importance of a robust text processing tool, facilitating a comprehensive understanding of lengthy documents by presenting concise summaries of crucial information.

R. Dumbre (✉) · P. Ankalwar · S. Bhagwat · D. Pandit · M. Bhutada · S. Gund
Electronics and Telecommunication Department, Vishwakarma Institute of Information Technology, Pune, India
e-mail: rajdumbrev@gmail.com

S. Bhagwat
e-mail: smita.bhagwat@viit.ac.in

D. Pandit
e-mail: dipti.pandit@viit.ac.in

S. Kumar Udgata et al. (eds.), *Intelligent Systems*, Lecture Notes in Networks and Systems 1149, https://doi.org/10.1007/978-981-97-8160-7_5

The primary objective of text summarization is leveraging computer tools to streamline information extraction from articles. This methodology reduces the time required to comprehend extensive articles, enabling a swift review of multiple reports. Text summarization is a sophisticated procedure that encompasses intricate stages, including the identification of the central topic, analysis of content, generation of a summary, and assessment of the quality of the summary. While many summarization models predominantly utilize extractive methods, which involve selecting existing sentences, there is a critical demand for automatic abstractive text summarization models.

Two models stand out in addressing the overwhelming volume of online data, aiding in the rapid identification of pertinent information within articles. The ultimate aim of these models is to produce precise and succinct summaries. Our strategy involves training and evaluating a pre-trained model on the Kaggle English Articles dataset, seeking to gauge the efficacy of our approach in generating accurate summaries across five specific classes. There are two main approaches to summarizing textual content:

Extractive Summarization: This method scans the document and selects sentences that closely resemble the key points in the original text. Instead of generating a completely new summary, it ranks each sentence based on how well it conveys the content compared to others.

Abstractive Summarization: In the context of abstractive summarization, the objective is to make a more precise summary by extracting crucial elements from the original text. The resultant summary is generated through this distinctive and not merely a replication of sentences from the original content.

The study consists of 2 extractive approaches.

Spacy: A Brief Overview: Spacy stands as a robust natural language processing (NLP) library, renowned for its efficient architecture tailored for speed. With features encompassing tokenization, part-of-speech tagging, and named entity recognition, Spacy is a preferred choice for diverse NLP tasks. Specifically, relevant to text summarization, it facilitates both extractive and abstractive methods. For extractive summarization, it identifies crucial sentences, while for abstractive summarization, it generates concise summaries by leveraging deep learning models.

NLTK: Core Capabilities: The Natural Language Toolkit (NLTK) emerges as a well-established NLP library, offering a rich set of tools for text processing. In line with summarization needs, NLTK provides functionalities such as tokenization, lemmatization, stop word removal, named entity recognition, and part-of-speech tagging. Its versatility extends to implementing both extractive and abstractive text summarization techniques, enabling the extraction of key sentences or the generation of concise, new summaries.

2 Literature Review

In various contexts, Spacy and NLTK prove invaluable for text summarization. In the realm of news articles, both tools efficiently condense lengthy content, providing concise versions for readers seeking quick updates on current events. Spacy and NLTK make scientific papers easier to understand, helping students and researchers quickly grasp the main findings without getting lost in the details. By summarizing product descriptions, shoppers can quickly compare key information and make informed decisions online [2]. Similarly, businesses benefit from Spacy and NLTK in summarizing customer reviews, enabling efficient identification of prevalent concerns and suggestions [6]. This adaptability showcases the broad applicability of Spacy and NLTK in enhancing information accessibility across diverse domains. Taking a new approach than prevailing literature, this research introduces a distinctive methodology by incorporating five varied datasets. The approach involves training on both NLTK and Spacy models, contributing a unique perspective to existing studies.

This distinction from conventional methods enables an exploration of the efficacy and adaptability of these models across diverse datasets [8]. The utilization of multiple datasets enhances the breadth of these findings, offering a comprehensive understanding of the performance of NLTK and Spacy in different contexts [4, 13]. This work aims to improve natural language processing by showing how robust and versatile these models are.

3 Methodology

The research utilizes the Kaggle English Test Articles dataset, consisting of 3101 articles spanning five distinct topical areas, covering the period from 2002 to 2006. The dataset encompasses politics (1265 articles), entertainment (400 articles), sports (455 articles), technology (954 articles), and business (1027 articles). Each article in the news articles column is paired with a corresponding summary in the summaries column. The dataset exhibits variations in the lengths of both the text and the summaries across categories, as illustrated in Fig. 1. The graph in Fig. 2 visually represents the distribution of Text in the files within each category, providing a comprehensive overview of the dataset's composition.

3.1 Data Pre-processing

In addressing NLP-related challenges, an indispensable preliminary stage involves data pre-processing. The detailed methodology and workflow of our research are delineated in Fig. 3. The initial scrutiny of the dataset was directed towards the identification and resolution of missing or null values. Consequently, all rows containing

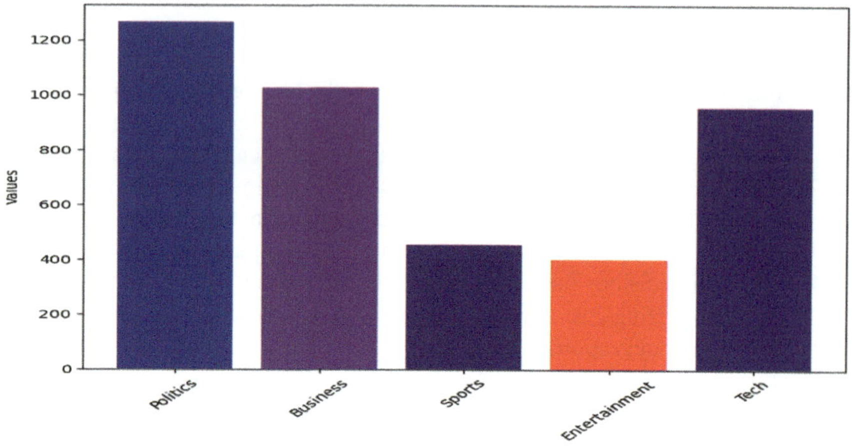

Fig. 1 Dataset diversity

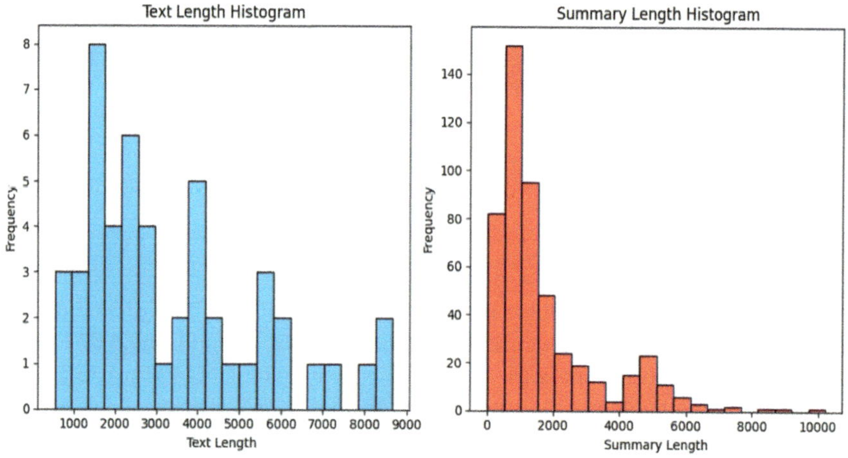

Fig. 2 Contrast between text length and summary length

null values were systematically removed. Tokenization, a crucial step in our methodology, entailed the conversion of input sentences and words into tokens. To determine input length parameters, both the maximum length (max-length) and minimum length (min-length) of Articles and Summaries were computed. The mean article length and summary length for each category were calculated, and the findings are detailed in Table 1.

To define input length parameters, the maximum length (max-length) and minimum length (min-length) of both Articles and Summaries were calculated. The mean article length and summary length for each category were then computed, and the outcomes are outlined in Table 1. To accommodate inputs shorter than the

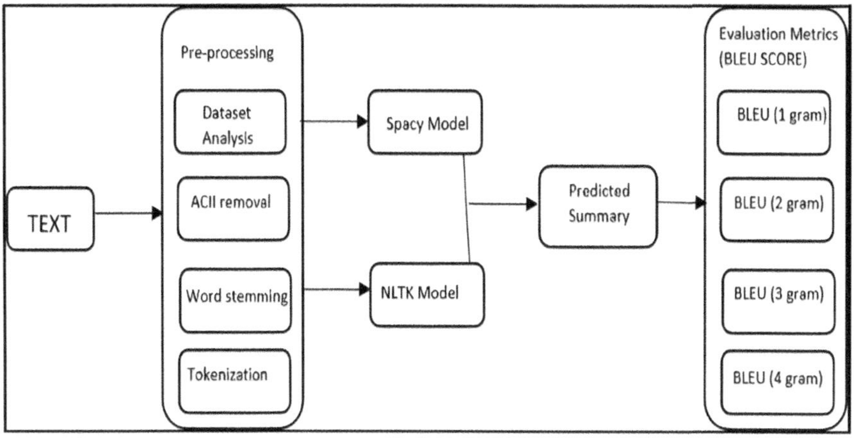

Fig. 3 Methodology and its workflow

Table 1 Mean of article and summary length

Categories	Articles-length	Summaries-length
Business	901.880392	297.929412
Entertainment	502.621762	169.049223
Politics	780.973621	246.709832
Sport	359.262231	153.187867
Tech	892.695761	280.837905

required length. Figure 4 illustrates the token count distribution. Furthermore, it was ensured that the inclusion of paddings did not contribute to the calculation of loss.

Spacy

In the Spacy-based text summarization process, the initial step involves pre-processing the input document for summarization. This includes text tokenization to break the input into manageable units, followed by the removal of stop words and punctuation to reduce noise in the text. Additionally, part-of-speech tagging and dependency parsing are performed using Spacy to understand the grammatical structure and relationships between words and phrases.

Sentence scoring involves TF-IDF calculation, determining Term Frequency (TF) and Inverse Document Frequency (IDF) for each term in a sentence. The sentence's importance score is the sum of these TF-IDF scores for its terms. Threshold selection is a crucial step in deciding which sentences should be included in the summary. This can be done through methods like percentile-based selection, fixed numbers of top sentences, or dynamic thresholds based on sentence scores, adapting to the document's content.

Following the threshold, the summary is created by selecting sentences above the determined threshold and reconstructing them into a coherent and concise

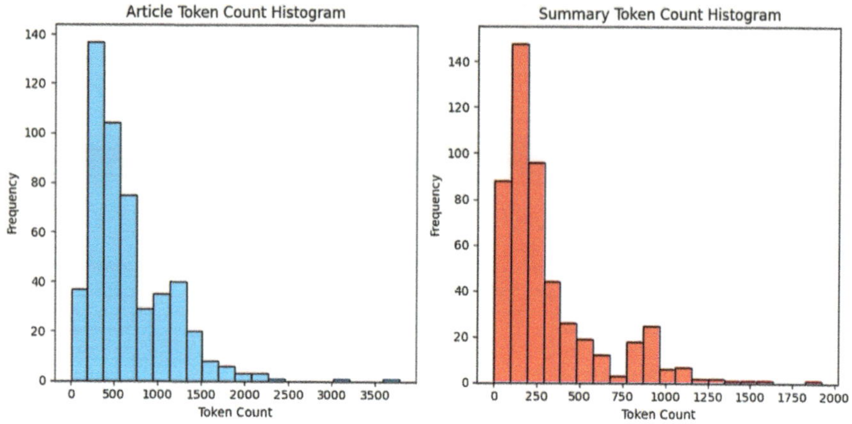

Fig. 4 Token count comparison between the articles and their summaries

form. Spacy's sentence segmentation and dependency parsing capabilities ensure the natural flow of the summary.

To evaluate the extractive summary's quality, metrics like BLEU can be employed. This comprehensive Spacy-based summarization method, which includes pre-processing, scoring, thresholding, summary generation, and evaluation, provides a systematic and effective approach to extracting key information from the input document.

NLTK

The proposed text summarization approach involves a systematic six-step process. Initially, the textbook undergoes tokenization, breaking it into words and rulings. Subsequently, lemmatization reduces words to their root form, contributing to a more cohesive understanding. The removal of stop words follows, eliminating common words that lack substantive meaning.

Identifying important rulings involves various ranking methods, including word frequency, TF-IDF, and position in the textbook. Keyword extraction is a key step, focusing on identifying keywords and linking them to summarization techniques. These methods are simple and effective for generating concise summaries of shorter textbooks, but may have limitations for longer or more complex texts. This comprehensive approach, rooted in systematic ranking and keyword extraction, provides a structured framework for effective text summarization.

Evaluation Metrics: In the realm of in-text summarization, the evaluation of summaries is a crucial task. The manual and semi-automatic assessment of largescale summarization models is not only expensive but also inconvenient. To tackle these challenges, various efforts have been undertaken to develop automatic metrics, enabling the prompt and cost-effective evaluation of summarization models.

BLEU (Bilingual Evaluation Understudy): BLEU is a metric employed for assessing machine-generated translations by comparing them to human reference translations. Extensively utilized in tasks such as machine translation and text summarization, BLEU gauges the similarity between machine-generated text and one or more human references. A higher BLEU score signifies superior quality, indicating a closer approximation to human-generated content.

Equations for BLEU Score Calculation: BLEU computes a score by assessing the precision of n-grams (contiguous sequences of *n* words) in the machine-generated translation in comparison to the reference translations. The fundamental equations for BLEU are as follows:

Modified Precision (Pn):
Multiple datasets enhance the breadth of these findings, offering a comprehensive understanding of the performance of NLTK and Spacy in different contexts [4, 13]. This work aims to improve natural language processing by showing how robust and versatile these models are.

(r). If the machine-generated translation is shorter, BP is less than 1, and if it's longer, BP is greater than 1. In simpler terms, if the machine translation is shorter than the human translation, it will be penalized. The penalty is more severe the shorter the machine translation is. This is because shorter translations are more likely to omit important information (Table 2).

BLEU Score

The BLEU score is calculated as the geometric mean of the modified precision scores (Pn) with a brevity penalty (BP) term. Particularly the modified precision and brevity penalty, help assess the quality of machine-generated text by taking into account both the precision of matched n-grams and the length of the translation. A higher BLEU score indicates a higher level of similarity and quality between the machine-generated text and human references.

Table 2 Modified precision calculation table

Symbol	Description
Pn	Modified precision score for n-grams
Count (n-grams in reference and candidate)	Number of n-grams that occur in both the reference and candidate translations
Count (n-grams in candidate)	Number of n-grams in the candidate translation
Max_Ref_Coun t	Maximum number of n-grams in any of the reference translations

Table 3 Results on the dataset test using spacy and NLTK model (statistical)

Paper Ref.	Method	Dataset	Evaluation method	Result
(II)	Extractive text summarization using neural networks	DUC 2002	ROUGE-1 ROUGE-2	55.1 22.6
(XVI I)	Deep learning with feed forward approach	English and Malayalam of CNN	Validation Score	0.59
(X)	Graph based approach	CNN daily-mail	ROUGE-1 ROUGE-2	47.58 45.42
(I)	Spacy	Unknown	F- Score	0.69
This paper	Spacy	Kaggle English	BLEU	0.659
	NLTK	Test articles	BLEU	0.592

4 Result

The BLEU score ranges between 0 and 1, with 1 indicating a perfect match between the machine-generated translation and the reference translations. The BLEU score results are presented in Table 3.

This paper presents a rigorous comparison between NLTK and SPACY models applied to a diverse dataset of over 500 randomly selected articles, employing BLEU scores for evaluation. The uniqueness of our work lies in the inclusion of five distinct article datasets, giving us a complete picture of how well our models perform. Importantly, our models performed consistently well across all five datasets, showing that they are robust. This study, following the IEEE format, fills a crucial gap by providing a concise yet thorough comparative analysis contributing significantly to the understanding of natural language processing techniques.

Key Findings:

Spacy achieved a significantly higher BLEU score compared to NLTK, indicating that the summaries generated by Spacy were more similar to the references. Spacy's summaries were generally more concise and informative, capturing the key points of the original text while maintaining fluency and coherence. The Graphical results and scores have been mentioned in Figs. 5 and 6. Spacy's ability to understand the context and semantics of the text contributed to its superior performance in generating accurate and faithful summaries.

5 Conclusion and Future Work

In the course of this research, our focus was directed towards the exploration of efficient text summarization techniques, with a specific emphasis on abstractive methods employing Spacy and NLTK. Evaluating the Kaggle English Articles

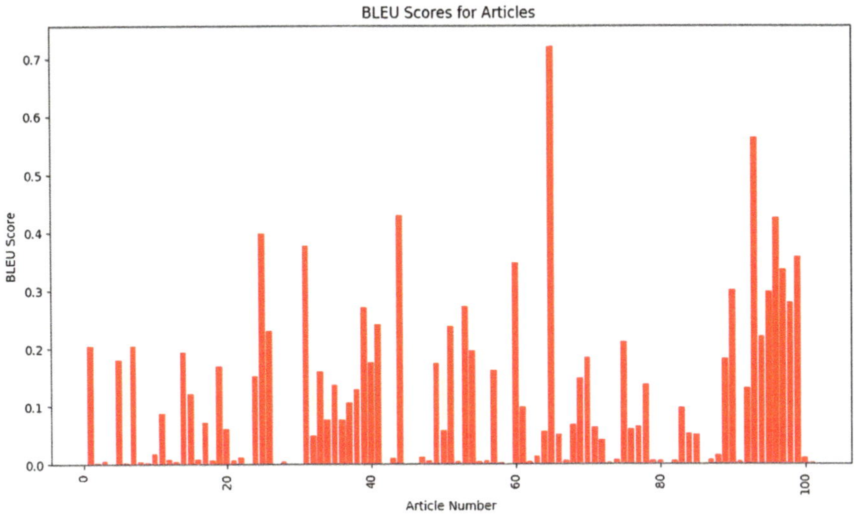

Fig. 5 BLEU scores using spacy model

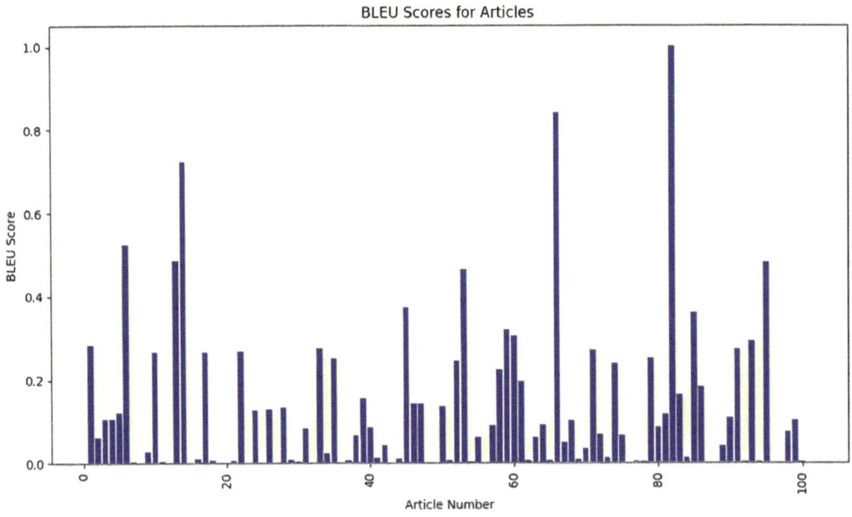

Fig. 6 BLEU scores using NLTK model

dataset revealed that neural network-based abstractive models performed better than traditional extractive methods, as indicated by significantly higher BLEU scores. Highlighting Spacy's powerful deep learning architecture and NLTK's versatility, this study offers valuable insights for practitioners in the field of natural language processing. The proposed framework, comprising comprehensive stages of data pre-processing, model training, and evaluation utilizing BLEU scores, establishes a robust foundation for subsequent research and development endeavors. This study not only enhances understanding of summarization techniques but also provides practical guidance for choosing the right NLP libraries for specific needs. Given the dynamic nature of text summarization, further exploration and innovation are warranted. Future research endeavors may consider hybrid approaches, leveraging the pre-processing efficiency of Spacy and the summarization flexibility of NLTK. Additionally, the exploration of larger and more diverse datasets spanning various domains holds the potential for enhancing the generalizability of proposed models. As the field progresses, the integration of emerging technologies, such as advanced deep learning architectures, stands to further refine the accuracy and efficiency of abstractive summarization methods. This research, therefore, lays a foundational framework for such potential advancements, offering a launching pad for future investigations in the ever-evolving field of natural language processing.

References

1. i. Extractive automatic text summarization using SpaCy in Python & NLP. In: 2021 International Conference on Advance Computing and Innovative Technologies in Engineering (ICACITE), pp. 582–585. IEEE. ii. Zhang, Y., Yao, M., Zhang, Y., Liu, Y.: A hybrid text summarization model based on deep learning and spaCy (2018). arXiv preprint arXiv:1804.10586. iii. Rahnama, M., Ghadiri, A.: A novel text summarization model based on spaCy and transformers (2021). arXiv preprint arXiv:2104.13084
2. Kumar, A., Sharma, R.: An extractive text summarization system based on NLTK and spaCy. Int. J. Eng. Technol. **7**(2), 132–137 (2018)
3. Gupta, P., Kumar, A.: An extractive text summarization system for Hindi language using NLTK. Int. J. Eng. Adv. Technol. **8**(3), 124–131 (2019)
4. Jain, A., Jain, N.: A novel extractive text summarization model based on NLTK and transformers. Int. J. Comput. Appl. **185**(18), 31–35 (2020). vii. Singh, P., Jain, N.: A novel extractive text summarization model based on NLTK and deep learning. Int. J. Comput. Intell. Data Min. **12**(3), 165–172. viii. Singh, A., Sharma, R.: Extractive text summarization using NLTK and spaCy: a comparative study on different sentence ranking methods. Int. J. Emerg. Technol. Comput. Sci. Eng. **10**(3), 135–142 (2021)
5. Mishra, R., Gupta, A.: Extractive text summarization using NLTK and spaCy: a comparative study on different summary lengths. Int. J. Adv. Res. Comput. Commun. Eng. **11**(18), 123–128 (2022)
6. Routray, J.K., Routray, S.: Extractive text summarization using NLTK (2016)
7. Jadhav, S.M., Chaudhari, D.S.: Extractive text summarization using NLTK and machine learning (2014)
8. Jadhav, A.M., Chaudhari, D.S.: Extractive text summarization using NLTK and Deep Learning (2015)
9. xiii. Dhoke, A.V., Shah, D.B.: Automatic text summarization using NLTK (2012)
10. Christian, H., Agus, M., Suhartono, D.: Single document automatic text summarization using term frequency-inverse document frequency (TF–IDF). ComTech: Comput. Math. Eng. Appl. **7**, 285 (2016)
11. Devlin, J., Chang, M.-W., Lee, K., Toutanova, K.: BERT: "Pre-training of deep bidirectional transformers for language understanding" (2018) CoRR abs/1810.04805

EV Lithium-Ion Battery SOH Estimation Based on LightGBM Model

Mayur More, Jayesh Sarode, Amey Panchbhayye, Govind Sharma, Smita Bhagwat, and Nilesh Gawali

Abstract This paper introduces a robust methodology for predicting the State of Health (SOH) of battery utilizing the LightGBM gradient-boosting algorithm. Through an extensive analysis of battery data collected from diverse sources, it showcases the effectiveness of the approach used. The methodology encompasses critical steps such as data pre-processing, feature selection, hyperparameter tuning, model evaluation, and visualization. In this paper, the results affirm 90.69 high accuracy of LightGBM in predicting battery SOH, establishing it as a valuable tool for battery management and diagnostics.

Keywords EV Lithium ion battery · Battery State of Health (SOH) · LightGBM · SOH prediction

M. More (✉) · J. Sarode · A. Panchbhayye · G. Sharma · S. Bhagwat
Department of E&TC, Vishwakarma Institute of Information Technology, Pune, Maharashtra, India
e-mail: mayur.22120124@viit.ac.in

J. Sarode
e-mail: jayesh.22011156@viit.ac.in

A. Panchbhayye
e-mail: amey.22011155@viit.ac.in

G. Sharma
e-mail: govind.22120211@viit.ac.in

S. Bhagwat
e-mail: smita.bhagwat@vit.edu

N. Gawali
Nitya Innovations LLP, Turbhe, Navi Mumbai, Maharashtra, India
e-mail: nilesh@nityainnovations.com

S. Kumar Udgata et al. (eds.), *Intelligent Systems*, Lecture Notes in Networks and Systems 1149, https://doi.org/10.1007/978-981-97-8160-7_6

1 Introduction

The escalating global demand for electric vehicles (EVs) and the widespread integration of renewable energy storage systems underscore the pivotal role of effective battery management in today's dynamic energy landscape. As the backbone of these systems, batteries are subject to continuous stress, necessitating precise monitoring and predictive maintenance strategies to ensure their optimal performance, longevity, and safety. At the heart of these strategies, lies the accurate prediction of a battery's State of Health (SOH), a critical parameter that serves as a barometer of its overall health, remaining capacity, and potential lifespan.

As we navigate a future increasingly reliant on electric mobility and sustainable energy solutions, the development of advanced methodologies for predicting and managing battery SOH becomes imperative. This research addresses this pressing need by introducing a robust and sophisticated methodology centered around the utilization of LightGBM, a powerful gradient-boosting algorithm known for its efficiency in handling complex and extensive datasets [2, 3, 8].

2 Methodology

Figure 1 represents the flow of the methodology implemented.

2.1 Dataset

The battery raw dataset comprises data from 20 files, collectively containing 16,100,728 observations. Each observation includes parameters such as record time, state of charge (SOC), pack voltage (V), charge current (A), maximum and minimum cell voltages (V), maximum and minimum temperatures (°C), available energy (kW), and available capacity (Ah). The dataset provides detailed insights into battery performance and health across various electric vehicles.

Individual batteries within the dataset have varying numbers of observations, ranging from 704,958 to 856,192. On average, each file contains approximately 772,362 readings. This comprehensive dataset offers a rich source of information for analyzing battery behavior and trends over time. With such extensive data, researchers and analysts can conduct in-depth studies on battery management, diagnostics, and

Fig. 1 Methodology flow diagram

predictive modeling [4, 5]. By leveraging this dataset, insights can be gained into factors influencing battery health, performance degradation patterns, and strategies for optimizing battery lifespan and efficiency in electric vehicles and renewable energy systems. (Dataset used of: Prognostics of battery capacity based on charging data and data-driven methods for on-road vehicles [1]).

2.2 Data Preprocessing

During the initial data processing stage, a meticulous approach was undertaken to prepare the dataset for subsequent analysis and modeling. This involved several key steps.

Energy Calculation. Energy consumption during charging is computed by multiplying pack voltage (V) and charging current (amp-hour) to form energy (watt-hour) of each time instance taken by battery-management-system (BMS), given in Eq. (1).

$$\text{Energy (Wh)} = \text{Voltage (V)} \times \text{Current (Ah)} \tag{1}$$

This measure provides insight into the actual power consumed during every recorded instance from BMS while charging the battery pack.

Charge Cycle Identification. Charge cycles were identified within the dataset based on relevant criteria, defining a charge cycle as a sequence of charging events from initiation to completion. Across all 20 batteries, there is an average of 1200–1800 charge cycles in each battery.

Initial Capacity Determination. Benchmark capacities corresponding to specific state of charge (SOC) levels in the top charge cycle were extracted to serve as reference points for subsequent SOH calculations.

State of Health (SOH) Calculation. The SOH of the battery is calculated using the general formula for the calculation of health, given in Eq. (2).

$$\text{SOH} = \left(\frac{\text{current capacity}}{\text{initial capacity}}\right) \times 100 \tag{2}$$

Considering additional parameters such as battery voltage, charging current, energy, charge cycles, and time.

2.3 LightGBM Methodology

Figure 2 represents the flow of the LightGBM methodology implemented.

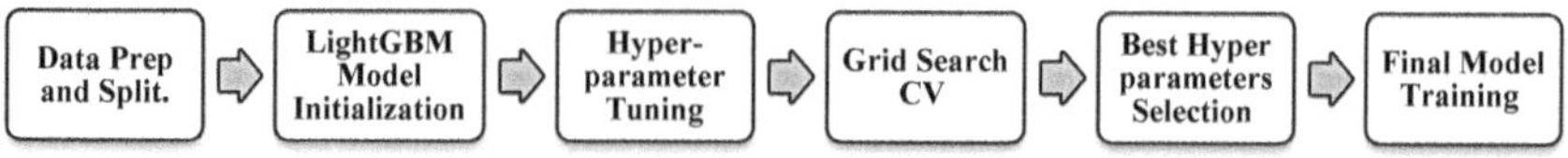

Fig. 2 LightGBM methodology flow diagram

Data Preparation and Train-Test Split. Battery data from various files (Total 10 of 20 files were selected) is loaded into a data frame. Important characteristics concerning battery health, including state of charge (SOC), pack voltage, charging current, and energy, are chosen for examination. The focal variable is the State of Health (SOH) of the battery. The dataset is divided into training and testing subsets utilizing an 80–20 distribution ratio. This ensures enough data for model training and an independent dataset for assessing the model's generalization.

LightGBM Model Initialization and Hyperparameter Tuning with Grid Search. A LightGBM regressor model is initialized with default hyperparameters. The default settings are used as a starting point for subsequent hyperparameter tuning. A grid search is employed to find the optimal hyperparameters for the LightGBM model. The hyperparameters considered include:

Number of Estimators (Trees). The n_estimators parameter determines the number of boosting rounds. Different values, such as 100, 200, and 300, are explored in the grid search.

Learning Rate. The learning rate determines the magnitude of each step taken during the optimization procedure. Various values, such as 0.01, 0.1, and 0.2, are tested to find the optimal learning rate.

Maximum Tree Depth. The max_depth parameter governs the maximum depth of each tree. Different values, such as 3, 5, and 7, are considered to strike a balance between model complexity and generalization.

Grid Search Cross Validation. Grid search, utilizing cross-validation with a fold count of 3 (CV = 3), is conducted to optimize the model. This approach facilitates a comprehensive evaluation of the model's performance across various subsets of the training data, thereby enhancing its robustness. Throughout each iteration of cross-validation, the LightGBM model undergoes training with distinct combinations of hyperparameters, enabling an exhaustive exploration of the parameter space.

Best Hyperparameters Selection. The hyperparameters corresponding to the best-performing model are determined based on the results of the grid search. This includes the optimal number of estimators, learning rate, and maximum tree depth. Table 1 displays best hyperparameters which are selected.

The Light GBM model was trained with a learning rate of 0.2, chosen for a slower convergence that helps prevent overshooting and improves generalization. The maximum depth was limited to 3 to avoid overfitting and maintain model simplicity. Utilizing 200 estimators, the model gains advantages from an increased

Table 1 Best hyperparameters selection

Hyperparameters	Selected
Number of estimators (trees)	200
Learning rate	0.2
Maximum tree depth	3

number of trees within the ensemble, enabling it to effectively capture intricate data relationships. This configuration, featuring a low learning rate alongside a higher number of estimators, enhances the model's predictive performance through optimized parameter settings.

Final Model Training and Understanding Tree Formation. The conclusive training of the LightGBM model involves utilizing the complete training dataset along with the finest hyperparameters identified through grid search. This approach guarantees the refinement of the model with an optimal blend of parameters, encompassing the number of trees, learning rate, and tree depth. Operating as a gradient-boosting algorithm, LightGBM constructs decision trees in a sequential manner. Each tree is designed to rectify the errors of its predecessors. The learning rate dictates the extent of contribution from each tree towards the ultimate prediction. Higher learning rates make the model learn faster but may lead to overshooting. The maximum tree depth controls the complexity of individual trees, preventing overfitting by limiting their depth.

3 Results and Discussion

3.1 Model Evaluation Metrics

The performance of the model was assessed using various regression metrics.

Mean Squared Error (MSE). The Mean Squared Error is computed by averaging the squared variances between the predicted SOH values (pi) and the actual SOH values (yi) across all observations (n) in our dataset as given in Eq. (3).

$$\mathrm{MSE} = \frac{1}{n}\sum_{i=1}^{n}(pi - yi)^2 \tag{3}$$

The model attained an MSE of 1.894, implying an average squared deviation between the actual and predicted values. This suggests a satisfactory level of accuracy in the predictions.

Mean Absolute Error (MAE). The Mean Absolute Error is calculated by averaging the absolute variances between the predicted SOH (pi) and actual SOH values (yi) across all observations (n) as given in Eq. (4).

$$\text{MAE} = \frac{1}{n}\sum_{i=1}^{n}|pi - yi| \tag{4}$$

With an MAE of 1.016, the model exhibits an average absolute deviation of approximately 1.016 units from actual values.

R-squared. R-squared quantifies the proportion of variance in the SOH values explained by our model, relative to the total variance in the data. It is calculated as the ratio of the explained sum of squared error (SSE) to the total sum of square (SST). Where, SSE is residuals between predicted SOH values (pi) and actual SOH values (yi). SST represents the variability in the actual SOH values as given in Eq. (5).

$$R^2 = 1 - \frac{\text{SSE}}{\text{SST}} \tag{5}$$

R-squared, ranging from 0 to 1, signifies the degree of model fit, with higher values denoting better fit. A perfect fit is indicated by a value of 1, while values closer to 0 indicate poorer model performance. The high R-squared value of 0.9069 suggests that the model explains 90.69% of the variance in the dependent variable, demonstrating strong predictive abilities.

3.2 *Correlation Analysis*

The correlation matrix displayed in Table 2 offers valuable insights into the relationships among different parameters in the dataset. The correlation coefficient between two variables, which ranges from −1 to 1, is contained in each cell of the matrix. A correlation coefficient approaching 1 indicates a strong positive correlation, meaning that as one variable increases, the other tends to increase as well. Conversely, a coefficient nearing −1 suggests a strong negative correlation, indicating that as one variable increases, the other tends to decrease. A correlation coefficient close to 0 indicates a weak or negligible linear relationship between the variables [7].

In this correlation matrix: There is a strong positive correlation (0.99) between parameters such as State of Charge (SOC), Pack Voltage, Max Cell Voltage, Min Cell Voltage, and Capacity. This suggests a direct relationship between these variables, implying that as one variable increases, the others also tend to increase.

The Charge Cycle variable shows a negative correlation (−0.96) with Actual SOH, suggesting a strong inverse relationship. This indicates that as the charge cycle increases, the actual state of health tends to decrease.

Table 2 Correlation matrix

	SOC	Pack voltage	Charge vurrent	Max cell voltage	Min cell voltage	Max temp	Min temp	Capacity	Charge cycle	Actual SOH	Predicted SOH
SOC	1	0.99	0.25	0.99	0.99	0.12	0.12	0.99	0	−0.02	−0.02
Pack Voltage	0.99	1	0.21	1	1	0.08	0.08	0.97	−0.05	0.04	0.03
Charge current	0.25	0.21	1	0.2	0.22	−0.13	−0.13	0.24	−0.19	0.09	0.07
Max cell voltage	0.99	1	0.2	1	1	0.09	0.06	0.98	−0.04	0.03	0.03
Min cell voltage	0.99	1	0.22	1	1	0.08	0.08	0.98	−0.05	0.04	0.04
Max temp	0.12	0.08	−0.13	0.09	0.08	1	0.99	0.14	0.15	−0.1	−0.11
Min temp	0.12	0.08	−0.13	0.06	0.08	0.99	1	0.14	0.15	−0.1	−0.1
Capacity	0.99	0.97	0.24	0.98	0.98	0.14	0.14	1	−0.03	0.02	0.03
Charge cycle	0	−0.05	−0.19	−0.04	−0.05	0.15	0.15	−0.03	1	−0.96	−0.98
Actual SOH	−0.02	0.04	0.09	0.03	0.04	−0.1	−0.1	0.02	−0.96	1	0.96
Predicted SOH	−0.02	0.03	0.07	0.03	0.04	−0.11	−0.1	0.03	−0.98	0.96	1

Table 3 Actual SOH versus predicted SOH values

SOC (%)	Pack voltage (V)	Charging current (A)	Charge cycle	SOH (%)	Predicted SOH (%)
84	371.3	73.89	286	89.56427	90.28677
82.8	369.1	70	307	89.91487	90.33964
82	368.2	69.7	343	87.93871	89.21373
80.4	365.3	50	389	89.79721	88.63702
80.4	366	70	836	86.83332	86.39193
80.4	364.9	70.1	1038	85.84674	85.35269
80.4	363	69.9	1092	84.39312	84.56241
80.4	358.6	43.2	1184	80.03401	83.16585
82.4	361.8	63.6	1278	81.24785	82.5806
84	362.8	60	1379	79.7744	81.57901

The Predicted SOH variable exhibits moderate to strong positive correlations with parameters such as Pack Voltage (0.03), Charge Current (0.07), Max Cell Voltage (0.03), Min Cell Voltage (0.04), and Capacity (0.03). This suggests that the predicted state of health tends to increase as these parameters increase, albeit to varying degrees.

3.3 Predictive Analysis

The model's predictions were evaluated on a test dataset containing over 800,000 samples, demonstrating its accuracy in predicting battery SOH. The alignment between actual and predicted SOH values visually underscores the methodology's effectiveness in achieving the research objectives [6]. Table 3 displays multiple instances of SOC around 80 during the recorded charging data of the battery.

3.4 Visualization

Visualizations, including scatter plots and polynomial regression curves, provide intuitive insights into the model's precision and enhance understanding of battery health dynamics, contributing to the fulfillment of the research objectives.

Actual SOH Versus Predicted SOH. In Fig. 3, the alignment between actual SOH values and the predictions made by this model is striking. The scatter plot visually demonstrates the model's accuracy, with each point representing a battery instance. The proximity of points to the diagonal line signifies precise predictions, showcasing the model's ability to capture nuanced SOH variations.

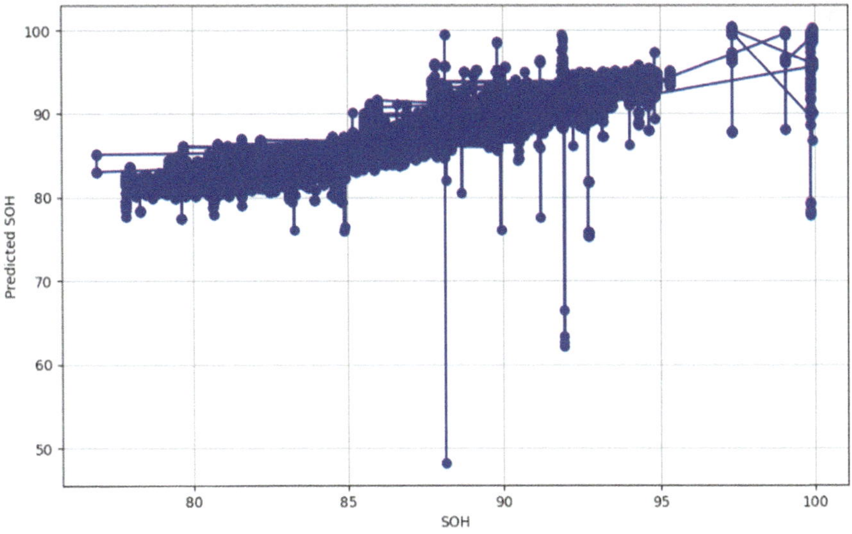

Fig. 3 Actual SOH versus predicted SOH

Charge Cycle Versus Actual SOH and Predicted SOH. Moving beyond point-wise evaluation, Fig. 4 presents a series of graphs depicting the intricate relationship between charge cycle and SOH. These plots illustrate how SOH evolves over successive charge cycles. The polynomial trend lines, visually capturing the trends, facilitate a nuanced understanding of battery health dynamics. Peaks and valleys in these graphs provide insights into critical points in a battery's life.

Polynomial Regression Curve Graphs. Complementing the Charge Cycle versus Actual SOH and Predicted SOH graph, Fig. 5 introduces a polynomial regression curve graph. The graph provides a mathematical representation of actual and predicted SOH values, aiding in understanding underlying patterns. The flexibility to adjust the polynomial degree enables the exploration of non-linear dependencies within the dataset, enhancing its interpretative capabilities.

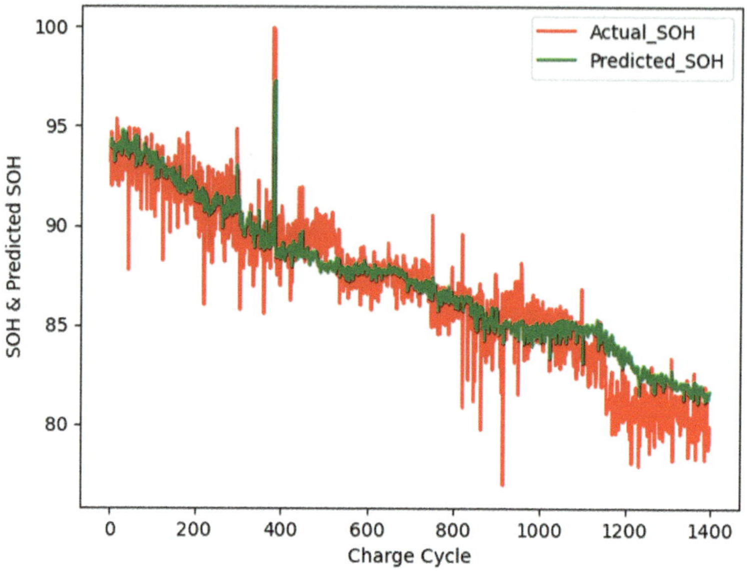

Fig. 4 Charge cycle versus actual SOH and predicted SOH

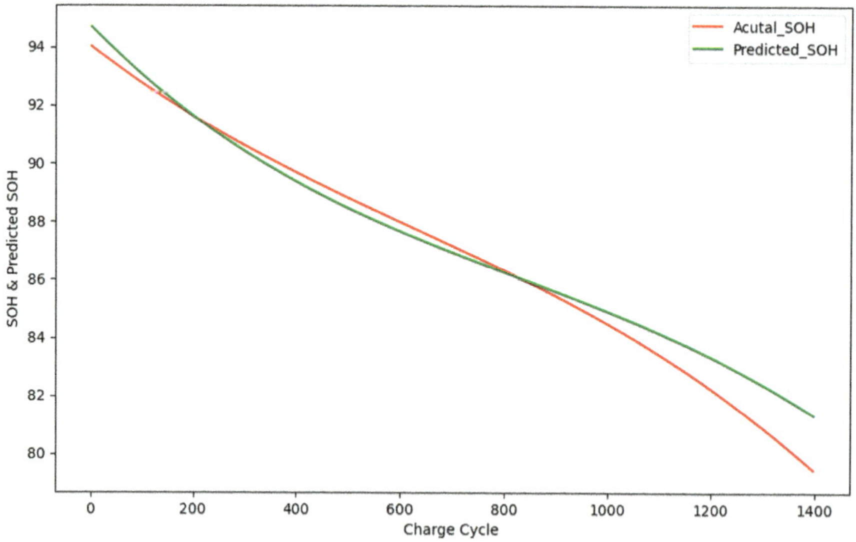

Fig. 5 Polynomial regression curve (Degree 3) of Charge Cycle versus Actual SOH and predicted SOH

4 Conclusion

This research introduces a robust methodology for predicting state of health (SOH) of battery using the LightGBM algorithm, leveraging a dataset comprised of charging data from 20 commercial electric vehicles over a span of 29 months. Through meticulous data pre-processing, feature selection, and hyperparameter tuning, we have developed a model with high predictive accuracy and reliability. This methodology encompasses critical steps from data collection to model evaluation, addressing the pressing need for advanced battery management strategies in electric vehicles and renewable energy systems. By accurately predicting battery SOH, our approach contributes to enhancing the reliability, longevity, and efficiency of energy storage systems.

Accurate prediction of the State of Health (SOH) is crucial for addressing the challenges in the rapidly evolving energy landscape. The model demonstrated an impressive accuracy of 90.69%, highlighting its effectiveness in predicting battery health accurately. As the world transitions towards electrification and sustainable energy solutions, our research offers essential insights and approaches for enhancing battery management strategies and maintaining the reliability of energy storage systems in this dynamic environment.

References

1. Deng, Z., Xu, L., Liu, H., Hu, X., Duan, Z., Xu, Y.: Prognostics of battery capacity based on charging data and data-driven methods for on-road vehicles. Appl. Energy **1**(339), 120954 (2023)
2. Zhang, M., Yin, J., Feng, T.: Lithium battery SOH estimation based on manifold learning and LightGBM. Appl. Sci. **13**(11), 6540 (2023)
3. Liu, H., Xiao, Q., Jin, Y., Mu, Y., Meng, J., Zhang, T., Jia, H., Teodorescu, R.: Improved LightGBM-based framework for electric vehicle lithium-ion battery remaining useful life prediction using multi health indicators. Symmetry **14**(8), 1584 (2022)
4. Wang, G., Cui, N., Li, C., Cui, Z., Yuan, H.: A state-of-health estimation method based on incremental capacity analysis for Li-ion battery considering charging/discharging rate. J. Energy Storage **10**(73), 109010 (2023)
5. Dai, Q., Kelly, J.C., Gaines, L., Wang, M.: Life cycle analysis of lithium-ion batteries for automotive applications. Batteries **5**(2), 48 (2019)
6. Shu, X., Shen, S., Shen, J., Zhang, Y., Li, G., Chen, Z., Liu, Y.: State of health prediction of lithium-ion batteries based on machine learning: advances and perspectives. iScience **24**(11) (2021)
7. Garche, J., Jossen, A.: Monitoring and safety tests of batteries: from state of charge (SOC) and health (SOH) to misuse, abuse and crash. In: AIP Conference Proceedings 2016 Aug 25, vol. 1765, no. 1. AIP Publishing (2016)
8. Weicker, P.: A Systems Approach to Lithium-Ion Battery Management. Artech House (2013)

A Precision Tessellated Fundus Detection: Leveraging Color and Texture Features with SVM Classification

Kachi Anvesh, Bharati M. Reshmi, Saroj Kumar Rout, Bijaya Kumar Sethi, and Satya Sobhan Panigrahi

Abstract The automatic tessellated fundus detection approach we provide in this paper makes use of texture and color information. The color fundus image is described using a combination of color moments, Local Binary Patterns (LBP), and Histograms of Oriented Gradients (HOG). The tessellated fundus is identified using an SVM classifier that has been trained after feature extraction. This paper employs and contrasts both linear and RBF kernels. To assess the suggested strategy, an 836-fundus image collection is created. The mean accuracy for the linear SVM is 98%, with sensitivity and specificity values of 0.99 and 0.98, respectively. The average RBF kernel accuracy is 97%, with 0.99 sensitivity and 0.95 specificity. According to the detection results, color and texture attributes can accurately represent the fundus images as normal or tessellated.

Keywords Tessellated fundus · Histograms of oriented gradients · Image classification · Myopia · Neovascular membranes · Local binary patterns

K. Anvesh (✉) · S. K. Rout
Department of IT, Vardhaman College of Engineering, Hyderabad 501218, India
e-mail: anveshitse@gmail.com

K. Anvesh · B. M. Reshmi
Department of CSE, Visvesveraya Technological University, Belagavi 590018, India

B. M. Reshmi
Department of CSE, Basaveswar Engineering College, Bagalkote, Karnataka, India

B. K. Sethi
Department of CSE (Data Science), Vardhaman College of Engineering, Hyderabad, India

S. S. Panigrahi
Inovaare Clouds Pvt. Ltd, Bhubaneswar, Odisha, India

S. Kumar Udgata et al. (eds.), *Intelligent Systems*, Lecture Notes in Networks and Systems 1149, https://doi.org/10.1007/978-981-97-8160-7_7

1 Introduction

Eye is the most vital organ in human body, and too a sensitive part as well. High myopia, also known as Pathologic myopia, resulting from posterior myopia-specific diseases like axial elongation [1], can profoundly impact the shape of the eye. Among the consequences of pathologic myopia is myopic maculopathy, a term used to characterize these changes in eyes afflicted with this condition. Legal blindness is primarily caused by myopic maculopathy [2]. Several studies have been conducted in the field of pathologic myopia, including META-PM (Meta-Analysis for Pathologic Myopia), conducted by Matsui et al. Myopic maculopathy is categorized into five types, which are defined and researched in their study. A variety of all myopic lesions are shown in Table 1.

Fundus tessellation, indicating prominent choroidal vessels beyond the peripapillary area, offers insights into choroidal vascular structure. Linked to age, myopia, and ocular conditions like glaucoma, traditional imaging struggles with its complexity [3]. This study employs AI and ROI extraction to quantitatively analyze fundus tessellated density distribution, aiming to establish a novel index for assessing choroidal vessel visibility.

In clinical practice, fundus photos offer convenient assessment. Researchers studied factors related to fundus tessellation in older adults, considering potential influences like age-related choroidal thinning and retinopathy in pathological myopia [2]. Choroidal veins are clearly visible in tessellated fundi, prompting the need for automated detection methods due to manual identification's time-consuming nature. While automated detection of tessellated fundi remains unexplored, ongoing research focuses on automating the detection of myopia-related symptoms like peripapillary atrophy (PPA). In [4], Cheng introduced a self-configured and automated method, which uses biological stimulated features (BSF) for PPA detection, achieving over 90% accuracy (Fig. 1).

According to ophthalmoscopic visibility of large choroidal vessels in the macular region, subregions and feature images of macular fundus tessellation are defined. In this study, we provide LBP, color moments, and HOG feature-based automatic detection approaches on tessellated fundus images. To measure the colored features, we calculate the R, G, and B means and standard deviations. Edge information is captured by Histograms of Oriented Gradients (HOG) [5], while the texture is

Table 1 Various categories of myopic maculopathy

Serial number	Types of myopic synthesis	Category
1	Non-existent myopic retinal lesions	0
2	Geometric fundus patterns	1
3	Atrophic macular	2
4	Scattered chorioretinal lesions	3
5	Widespread chorioretinal degeneration	4

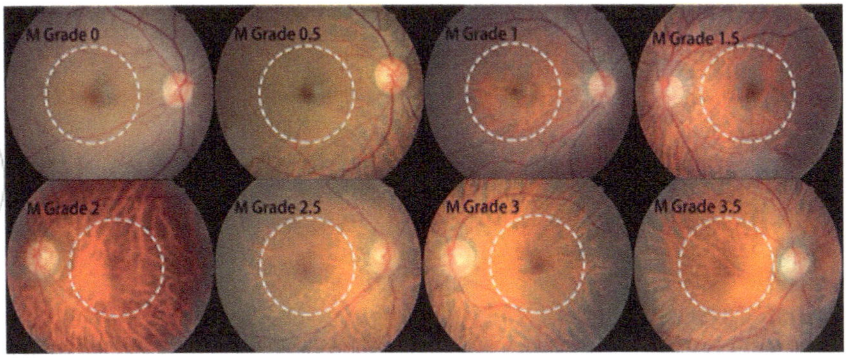

Fig. 1 Images of macular fundus tessellation

described by Local Binary Patterns (LBP) descriptor [6]. Face detection, picture categorization, and other applications employ these two attributes frequently.

1.1 Tessellated Down Syndrome

Down syndrome, the most prevalent chromosomal abnormality globally, presents a spectrum of multi-systemic irregularities [7]. Individuals with Down syndrome possess three copies of chromosome 21, which carries fewer functional genes compared to other autosomes [8]. Despite this, the prognosis is relatively manageable due to the extra chromosome 21. Not all chromosome 21 genes contribute to the Down syndrome phenotype; other genomic segments also play a role. Understanding the genotype–phenotype relationship and epigenetic influences is crucial [9, 10] for comprehending Down syndrome's variability and range in clinical manifestations.

Down syndrome children often exhibit ocular abnormalities, though posterior segment issues are less documented than other linked anomalies [11]. Retinal abnormalities like hemorrhage, coloboma, and retinal detachment have been noted, albeit with prevalence under 5% [12]. While retinal detachment may relate to increased ocular trauma risk in those with intellectual disabilities, other anomalies seem sporadic. Reports also mention choroidal and retinal pigment anomalies. The research structure includes sections on methodology, experimental findings, validity threats, and conclusions [11, 12].

2 Related Work and Methodology

In Adatia et al., choroidal neovascular membranes (CNVM) in pathological myopia (PM) can cause severe vision loss. Anti-VEGF treatment, like ranibizumab or bevacizumab, shows promise in improving visual outcomes, especially when considering patient age, previous treatments, and initial visual acuity.

In Kyoko Ohno-Matsui et al., experts defined myopic macular lesions via literature review, developed a classification using fundus photographs, and evaluated reproducibility among specialists.

In [13], the study examined 577 7-year-old children, assessing various ocular parameters using advanced imaging techniques and AI for defining fundus features.

In [4], utilizing sparse transfer learning, label information from ophthalmologists is transferred to sample distribution knowledge, enhancing PPA detection accuracy over 90%.

In [18], tessellated fundus appearance may serve as a marker for visual performance, myopia severity, and risk of myopia progression. A systematic review found associations with visual acuity, refractive error, axial length, and choroidal thickness.

The suggested automated tessellated fundus detection method is introduced in this section's technical details. The main steps of the suggested method will follow the Color Fundus Images with Automated Tessellated Fundus Detection, such as extraction of features to provide color, texture information, and detection of tessellation. The proposed method's pipeline is depicted in Fig. 2.

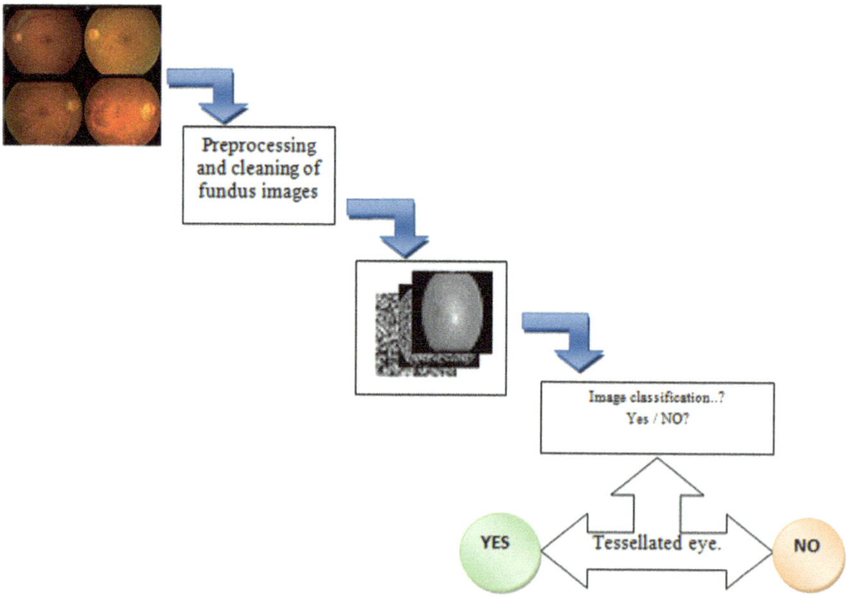

Fig. 2 Architecture flow of the proposed algorithm

2.1 Feature Extraction

Our analysis of color fundus images makes use of various techniques, including color moment analysis, oriented gradient histograms (HOGs), local binary patterns (LBPs), and bag-of-words (BOWs) [14]. Figure 3 illustrates how features are extracted. Color moment features are utilized to capture characteristic information. In this context, an image of the color fundus is calculated by summing the R, G, and B channels. It produces a six-dimensional,

$$\text{vector C} = \left[\sum \text{mean (R, G, B)}, \ \text{standard (R, G, B)}\right],$$

which describes the image. Local Binary Patterns (LBPs) constitute a descriptor formed by creating a histogram of quantized Local Binary Patterns. These LBPs are extracted from local image neighborhoods, as detailed in reference [6]. We extract the

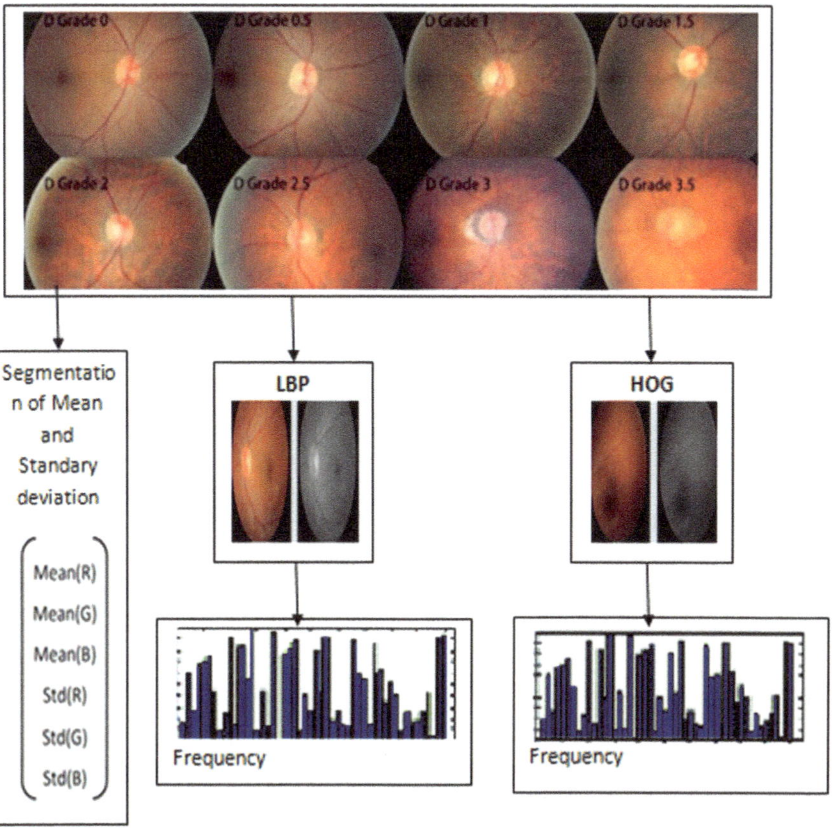

Fig. 3 Exercise scheme of feature extraction

LBP descriptor and use the LBP, bag of words that will represent each fundus picture. By utilizing the rate of recurrence of the words, BOW attempts to convey the retinal image. In this study, codebooks (words) are created using the training images. To determine the codebook centers, the LBP descriptor is determined by quantized Local Binary Patterns extracted from nearby picture neighborhoods, forming histograms. Utilizing the Bag-of-Words technique, we represent fundus images with LBP descriptors. Codebooks are generated from training images by computing LBP descriptors and clustering using K-means. HOG characteristics are commonly employed in object detection research [5, 6]. The steps below can be used to extract the HOG features from an input image:

- Each of the squared patches in the image has an oriented gradient histogram produced for it. Block-wise pattern normalization improves histogram consistency for extracting a reliable feature(s).

Incorporating HOG Bag-of-Words: To incorporate the HOG (Histograms of Oriented Gradients) bag-of-words, we quantize the extracted HOG feature. Similar to the process for the feature of LBP BOW. Here, we tend to calculate the feature of HOG for every individual training fundus image. Using K-means clustering technique, we determine the codebook centers, with K being set to 128 in this particular instance.

2.2 Detection

In this study, photos are classified into two groups—with and without tessellated fundus—utilizing Support Vector Machine (SVM) [14]. Investigation focuses on radial basis function (RBF) and linear kernels, using the LIBSVM package [15, 16] for image categorization. LIBSVM is a library for Support Vector Machines (SVMs) [17]. A hybrid feature selection method, merging Principal Component Analysis and Artificial Intelligence Neural Network, is introduced for data preprocessing and feature extraction. The effectiveness of the proposed algorithm is evaluated on the Wisconsin Breast Cancer Dataset using tenfold cross-validation [18].

The study aims at early sepsis detection using physiological data, employing Long Short-Term Memory (LSTM) and SVM models [19], towards developing an automated system leveraging deep learning and recurrent neural network (RNN) models [20].

Novel Finding

This research introduces novel approaches including hybrid feature selection, Long Short-Term Memory (LSTM), and Convolutional Neural Network (CNN) for various tasks.

3 Proposed Methods and Experimentation

The dataset comprises with 836 photographs of fundus images, which are collected from a local clinics and hospitals. Among these, 418 photographs serve as positive samples, representing images displaying fundus images with tessellation, and remaining 418 retinal images are negative samples which are Normal fundus images. Fig 4 showcases examples from our dataset, including fundus images without tesselated features. In Fig. 4, sample (a) and sample (b) are indicative of Normal cases, representing fundus images without tessellation. Conversely, sample (c) and sample (d) represent abnormal cases, featuring fundus images with tessellation. These images distinctly exhibit discernible choroidal vessels, as depicted in Fig. 4.

RBF Kernel

The Radial Basis Function (RBF) kernel, also referred to as the Gaussian kernel, is a fundamental mathematical function used in various machine learning tasks, notably

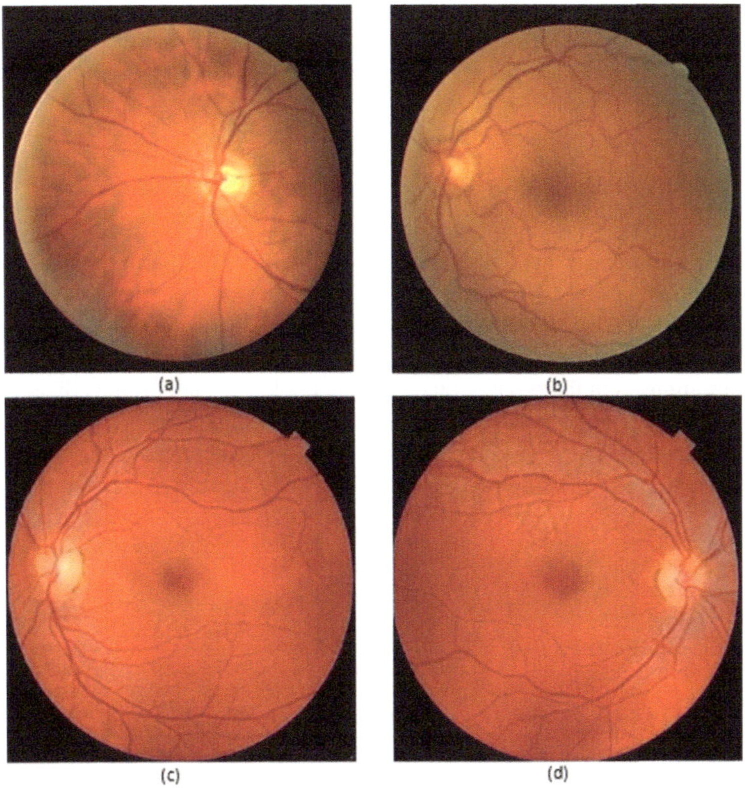

Fig. 4 Fundus dataset Images

in support vector machines (SVMs). It is highly valuable for solving problems that involve nonlinear regression and classification.

The RBF Kernel can be expressed mathematically as:

$$k(a, a') = \exp\left(-\frac{||a - a1||^2}{2\sigma^2}\right) \quad (1)$$

The RBF kernel assigns greater importance to points situated closer to the center (i.e., having a smaller Euclidean distance between a and a'. These weights diminish progressively as the distance between the points increases, owing to the exponential calculation within the formula. In the realm of SVM, the RBF kernel demonstrates significant value by enabling the mapping of data into a feature space of higher dimensions. This transformation simplifies the segregation of data using a linear hyperplane, especially in scenarios where the initial feature space does not permit a straightforward linear division of the data.

Images from the fundus dataset are presented in this figure. Samples (a) and (b) represent the negative category, denoting fundus images without tessellated features, while samples (c) and (d) belong to the positive category, showcasing fundus images with tessellated fundus.

3.1 Detection Results

We divide the dataset at random into two equal-sized subgroups to assess the suggested methodology. The dataset consists of two subsets: the training set and the test sample set. Every individual subset contains 209 abnormal positive samples and 209 normal negative samples. As illustrated previously in the section, training examples are used in the production of all codebooks. To support the efficacy of the separate features, we publish the recital on both individual aspects additionally with their combination. The performance in Eq. (2) is then calculated using sensitivity P $\oplus$, specificity P, and their average.

$$\mathcal{P} = \frac{(\mathrm{P} \oplus + \mathrm{P})}{2} \quad (2)$$

$$\text{True Positives } (+) = \frac{\text{Number of True Positive Values}}{\text{Total Number of Actual Positive Values}} \quad (3)$$

$$\text{False Positives } (-) = \frac{\text{Number of True Negative Values}}{\text{Total Number of Actual Negative Values}} \quad (4)$$

As shown in Table 1, True Positives indicate—total number of normal positive samples correctly classified, Total Positives indicate—a total number of positive

Table 2 Autodetection experimental results

Linear			
	Sensitivity P+	Specificity P_	Accuracy P
Color	0.65	0.88	0.85
LBP	0.99	0.87	0.88
HOG	0.93	0.95	0.92
RBF			
	Sensitivity P+	Specificity P_	Accuracy P
Color	0.45	0.66	0.7
LBP	0.9	0.86	0.86
HOG	0.9	0.85	0.89

samples correctly identified during the test set, and True Negatives indicate how many negative samples were correctly classified.

Table 2 shows that, when utilizing a single feature or a mixture of features, there is no discernible difference in the results between linear SVM and RBF SVM. Texture characteristics outperform color features for a single feature, indicating that texture difference is more significant. The feature combination performs best when both texture and color attributes are considered. Figure 5 displays sample images that have been correctly and incorrectly classified. As can be seen in the first row, negative samples have been correctly categorized (those without tessellated fundi), while positive samples have been properly classified (those with tessellated fundi). An example of misclassification is shown in the third row. In the first two images, which are mistakenly categorized as positive samples, they are negative samples. In the latter two images, incorrectly labeled as negative samples, they are indeed positive samples. Notably, these misclassifications do not exhibit extreme deviations and somewhat resemble the characteristics of the other class.

		Predicted image values	
Actual Values		Positive	Negative
	Positive	**True Positives**	False Negatives
	Negative	False Positives	**True Negatives**

Fig. 5 Displays the predicted image value

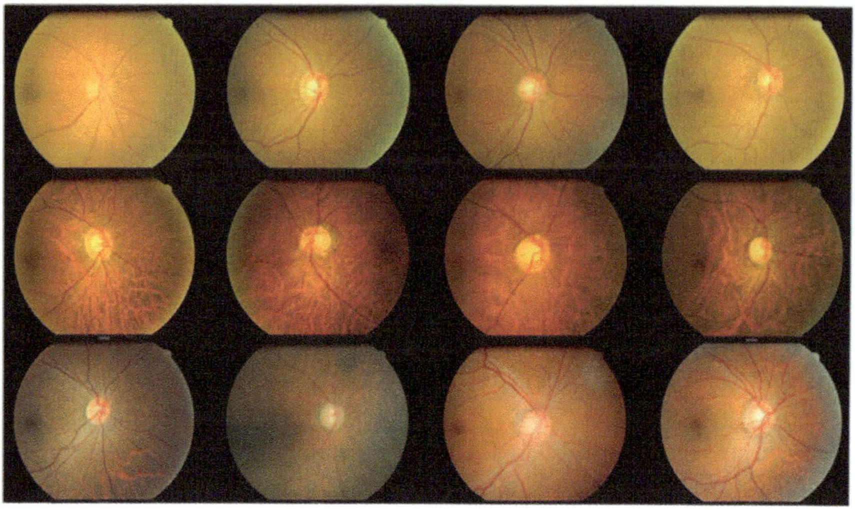

Fig. 6 The detection accuracy rises in fundus photographs

3.2 Size of Codebook

The detection outcomes for various codebook sizes are displayed in Fig. 6. The detection accuracy rises to about 96% for the features residing in HOG and LBP as the codebook size gradually raises from 8 to 126. Proportionally as the codebook size grows further, the performance does not noticeably improve. We chose a codebook size of 126 for mutual features in LBP and HOG, taking efficiency and efficacy into account.

Figure 6 Experimentation findings. Negative samples that were successfully categorized are displayed in the top row (Normal fundus images). The second row of images in Fig. 6. represents correctly identified positive samples (abnormal fundus images with tessellation), while the third row displays instances of misclassification. Among the misclassified cases, the first two images represent the negative sampled images incorrectly categorized as positive, while the end row which consists of two are positive samples erroneously depicted as negative (which is not true).

4 Results and Discussion

In this paper, we have suggested a computerized method to detect a retinal fundus image to depict a tessellation. Our approach involves a comprehensive analysis and comparison of features, including Color Moments, LBP, and HOG. Additionally, we have evaluated the detection performance using Support Vector Machine RBF kernels, including linear mechanism.

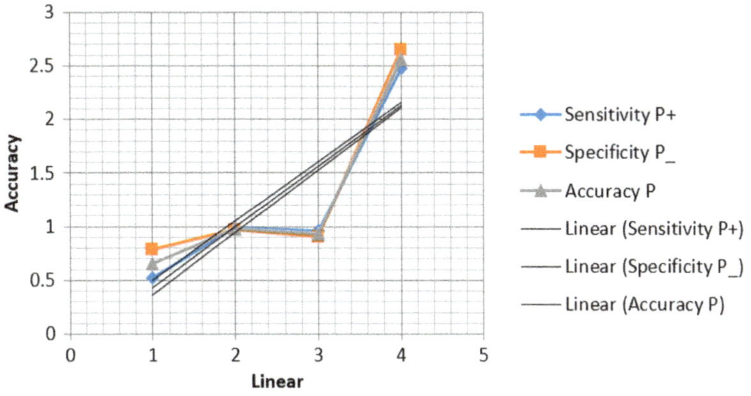

Fig. 7 Linear accuracy tessellated fundus detection

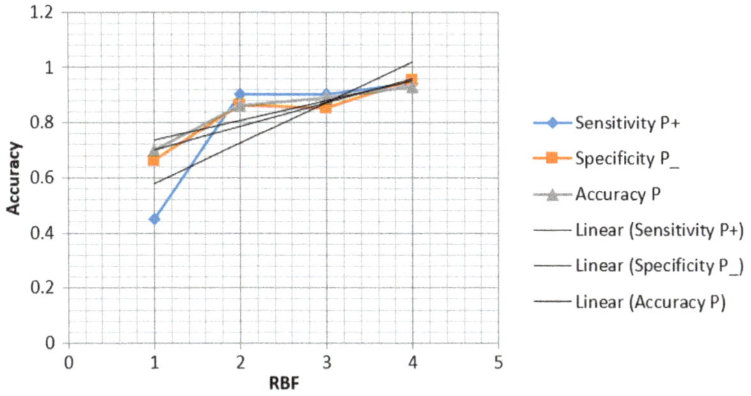

Fig. 8 RBF accuracy tessellated fundus detection

The results predicted focus on an average accuracy of 98% with the linear SVM and 97% with RBF Kernel SVM, highlighting the effectiveness and reliability of our proposed method. As an initial study on automatic computerized tessellated fundus detection that we are aware of, the research could serve as a preprocess for myopic maculopathy automated categorization and grading represented in Figs. 7 and 8. The graphical result analysis of accuracy is shown in Fig. 9, which represents the detection results along with histogram of oriented gradients and local binary patterns.

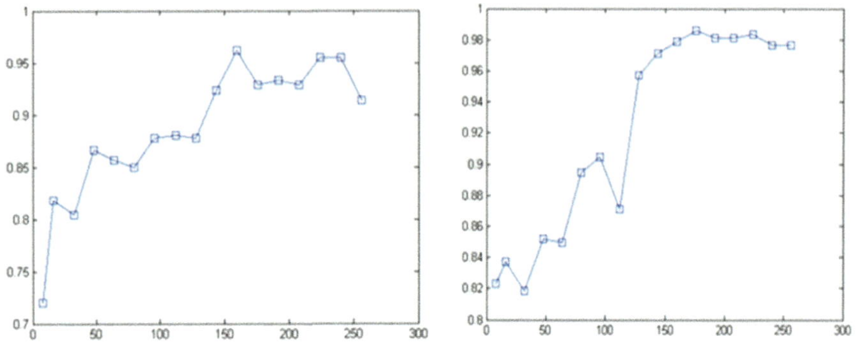

Fig. 9 Result analysis of accuracy of local binary patterns (LBP)

5 Conclusion

Our study introduces an automated technique for detecting tessellated fundus in fundus images by leveraging color and texture information. We employed various color fundus image representation methods, including LBPs, Histograms of Oriented Gradients, and color moments. The identification of tessellated fundus utilized an SVM classifier trained on extracted features, comparing linear and RBF kernels. Evaluation on 836 fundus images revealed the linear SVM's 98% accuracy with 0.99 sensitivity and 0.98 specificity. RBF kernels achieved 97% accuracy with 0.99 sensitivity and specificity. These findings showcase the effectiveness of capturing tessellated fundus essence through color and texture properties, offering promise for automating its detection and potentially enhancing diagnostic procedures for myopic maculopathy and associated ocular conditions. We can also predict some of the tessellated fundus images using deep learning methodologies and classification algorithms.

References

1. Adatia, F.A., Luong, M., Munro, M., Tufail, A.: The other CNVM: a review of myopic choroidal neovascularization treatment in the age of anti-vascular endothelial growth factor agents. Surv. Ophthalmol. **60**(3), 204–215 (2015)
2. Ohno-Matsui, K., Kawasaki, R., Jonas, J.B., Cheung, C.M.G., Saw, S.M., Verhoeven, V.J., Klaver, C.C., Moriyama, M., Shinohara, K., Kawasaki, Y., et al.: International photographic classification and grading system for myopic maculopathy. Am. J. Ophthalmol. **159**, 877–883 (2015)
3. Tan, N., Liu, J., Wong, D., Lim, J., Zhang, Z., Lu, S., Li, H., Saw, S., Tong, L., Wong, T.: Automatic detection of pathological myopia using variational level set. In: 2009 Annual International Conference of the IEEE Engineering in Medicine and Biology Society, pp. 3609–3612 (2009)

4. Cheng, J., Tao, D., Liu, J., Wong, D.W.K., Tan, N.M., Wong, T.Y., Saw, S.M.: Peripapillary atrophy detection by sparse biologically inspired feature manifold. IEEE Trans. Med. Imaging **31**, 2355–2365 (2012)
5. Dalal, N., Triggs, B.: Histograms of oriented gradients for human detection. In: Computer Vision and Pattern Recognition, vol. 1, pp. 886–893. IEEE (2005)
6. Muja, M., Lowe, D.G.: Fast approximate nearest neighbors with automatic algorithm configuration. In: VISAPP, vol. 1, no. 2 (2009)
7. Hattori, M., Fujiyama, A., Taylor, T.D., Watanabe, H., Yada, T., Park, H.-S., Toyoda, A., Ishii, K., Totoki, Y., Choi, D.-K., et al.: The DNA sequence of human chromosome 21. Nature **405**(6784), 311–319 (2000). https://doi.org/10.1038/35012518
8. Antonarakis, S.E., Epstein, C.J.: The challenge of Down syndrome. Trends Mol. Med. **12**(10), 473–479 (2006). https://doi.org/10.1016/j.molmed.2006.08.005
9. Antonarakis, S.E.: Down syndrome and the complexity of genome dosage imbalance. Nat. Rev. Genet. **18**(3), 147–163 (2017). https://doi.org/10.1038/nrg.2016.154
10. Creavin, A.L., Brown, R.D.: Ophthalmic abnormalities in children with Down syndrome. J. Pediatr. Ophthalmol. Strabismus **46**(2), 76–82 (2009)
11. Kim, J.H., Hwang, J.-M., Kim, H.J., Yu, Y.S.: Characteristic ocular findings in Asian children with Down syndrome. Eye (Lond.) **16**(6), 710–714 (2002)
12. Ojala, T., Pietikainen, M., Maenpaa, T.: Multiresolution gray-scale and rotation invariant texture classification with local binary patterns. IEEE Trans. Pattern Anal. Mach. Intell. **24**, 971–987 (2002)
13. Huang, D., Li, R., Qian, Y., Ling, S., Dong, Z., Ke, X., Yan, Q., Tong, H., Wang, Z., Long, T., Liu, H.: Fundus tessellated density assessed by deep learning in primary school children. Transl. Vis. Sci. Technol. **12**(6), 11–11 (2023)
14. Fei-Fei, L., Perona, P.: A Bayesian hierarchical model for learning natural scene categories. In: Computer Vision and Pattern Recognition, vol. 2, pp. 524–531. IEEE (2005)
15. Bishop, C.M.: Neural Networks for Pattern Recognition. Oxford University Press (1995)
16. Cortes, C., Vapnik, V.: Support-vector networks. Mach. Learn. **20**, 273–297 (1995)
17. Chang, C.C., Lin, C.J.: LIBSVM: a library for support vector machines. ACM Trans. Intell. Syst. Technol. **2**, 27:1–27:27 (2011). Software available at http://www.csie.ntu.edu.tw/cjlin/libsvm
18. Jagadeesh, D.: Tessellated fundus appearance and its association with myopic refractive error. BOptom. MPhil. Opt. (2018)
19. Sahu, B., Mohanty, S., Rout, S.: A hybrid approach for breast cancer classification and diagnosis. EAI Endorsed Trans. Scalable Inform. Syst. **6**(20) (2019)
20. Rout, S.K., Sahu, B., Regulwar, G.B., Kavididevi, V.: Deep learning in early prediction of sepsis and diagnosis. In: 2023 International Conference for Advancement in Technology (ICONAT), pp. 1–5. IEEE (2023)

Task Assignment of Cooperating Robot in Multi-Robot Environment

Smita Bakshi, Bandita Sahu, and Sanjay Kumar Kuanar

Abstract Task allocation in multi-robot environment is capable of allocating the tasks to working robots in the multi-robot environment. Selection of a suitable robot is a crucial task for allocation of the task. The problem addressed in this paper focuses on selecting a suitable robot and execution of the task. The proposed research work considers the multi-task multi-robot environment. A working robot is selected for a task to be executed. Multiple tasks are there to be allocated among multiple mobile robots that exist in the environment. Sine cosine algorithm is implemented here to plan a path for task execution and nearest neighbor selection for identifying a robot to be assigned with the task. The proposed research work has been simulated through computer programming language. It has been observed in the simulation that the proposed approach outperforms over existing approaches in terms of waiting time, turn-around time, shortest path, and overall performance.

Keywords Task allocation · Mobile robot · Nearest neighbor · Path planning

1 Introduction

Robotics and automation have been the recent trends in our daily work [1]. Smart home, autonomous car, automated parking system, automatic door, house cleaning, and for other services, robot for automation has an important role in human life. Robots are automatic devices embedded with sensor, camera, and other hardware equipment to perform specific tasks [2]. The task may be divided into several types such as service providing task, information collection task, etc. In this present context, the service provider task is assigned to the robots. Several robots are allowed to move in an environment to form a multi-robot environment [3]. All the robots are

S. Bakshi · S. K. Kuanar
GIET University, Gunupur, India

B. Sahu (✉)
Birla Global University, Bhubaneswar, India
e-mail: bandita.sahu@gmail.com

S. Kumar Udgata et al. (eds.), *Intelligent Systems*, Lecture Notes in Networks and Systems 1149, https://doi.org/10.1007/978-981-97-8160-7_8

autonomous in nature and move freely in the environment. Such robots are equipped with sensor, camera, wheel, LED, etc. to perform several tasks and travels from one position to another position, However, an information provider robot may or may not have the wheel based on the task assigned.

Task allocation and robot selection is a crucial part of path planning [4]. From the set of mobile robots, the service provider robots are selected randomly and assigned a task. The robots are provided with the information of their starting state and goal state. Each selected robot has to plan its path towards the goal. The path may be planned locally or globally. Before transition, the robots compute a path avoiding the obstacle position present in the environment such planning is called static planning or global planning. However, if the robot plans its path by examining each stem while transition is called dynamic planning or local planning [5].

The paper proposes an intelligent technique of robot selection in the multi-robot environment for task execution. The process can be accomplished in several steps such as (i) identifying the autonomous robot in working condition, (ii) selection of the robot nearer to the position where the task execution will be initiated, (iii) path planning of the robot using sine cosine algorithm for its simplicity, (iv) validation of the proposed algorithm through computer simulation, (v) comparison of the proposed approach through several metrics such as time to waiting time, task completion time, total execution time, path length, etc.

The rest part of the paper comprises four sections. The second section represents the literature study related to the proposed work. Section 3 explains the proposed model for robot selection and task allocation. In section 4, the computer simulation and results have been represented. Section 5 describes the conclusion and future direction of the research work.

2 Literature Study

A plethora of algorithms have been developed by the researchers for task assignment and path planning of the mobile robot. Task allocation for robot in multi-robot environments is considered as a non-deterministic polynomial time algorithm that incurs larger time in finding an optimal solution [6]. For achieving a defined goal, multiple robots are assigned in completing a task. The efficient selection affects the types of work to be performed such as rescue, search, transportation, etc. [7] The task allocation problem can be solved using several approaches such as auction-based method [8], integer linear programming approach [9], and graph-based method [10]. Such approaches solve the optimization problem using combinatory theory. But these approaches are unable to provide an optimal solution in large-scale system when there is an exponential growth in the combinations. They are also unable to produce a generic solution to the complex problem. Recent developments in the field of deep learning have been applied for many real-time applications such as task allocation, thermal process, mobile network, and energy system [11–13]. The approach solves complex problem due to its high dimensional data representation. However, its prime bottleneck includes high-dimensional data and labeled data in a larger volume to produce an optimal solution. Reinforcement learning interacts with the environment for making decision and implements the combinatory theories to solve vehicle routing and traveling salesman problem [14]. However, its computational cost becomes high as it requires labelled data which is too expensive in collection, extraction, and processing. As well, it only solves the above-mentioned problems and overlooked the task allocation issue.

Many researchers have suggested the solution to the path planning problem of mobile robot for providing services. Particle swarm optimization (PSO) approach addresses the path planning of mobile robot by selecting the best next position [15]. It is a simple algorithm that uses less number of parameters but suffers from low convergence rate. Sine cosine algorithm (SCA) implements the sine and cos value in selecting the next best position towards the goal [16]. The algorithm exhibits better exploitation through the use of selection parameter. The authors have implemented the semi-supervised algorithm in Q-learning approach for planning a path [17, 18]. The algorithm implements the learning parameter in selecting the next best position. But, for complex problem space with increased number of mobile robots it fails to produce an optimal solution.

The research paper focuses on the robot selection and path planning of the selected robots. The prime objective of the paper is to allocate the task by selecting a working robot. The task allocation problem is addressed in the first part that reduces the waiting time of the task in the multi-robot environment. The second phase addresses the collision-free path computation of the selected robot. A simple SCA algorithm has been implemented here with better exploitation to compute the path planning and the nearest neighbor approach is used to obtain the location of the nearest working robot to assign the task. The proposed model embeds the task allocation and path planning efficiently and produces an optimal solution.

3 Robot Selection for Task Assignment

Task allocation to mobile robot maximizes the utility in providing services. Task allocation can be represented through many perspectives based on the environment such as robot's ability to handle the number of tasks simultaneously and task's requirement for the robots count [19]. The problem can also be formulated as single task allocation, allocation of multiple tasks, single robot task allocation, task allocation for multiple robots, dynamic task allocation, and time-based task allocation [20, 21]. In this paper, multiple task assignments to multiple mobile robots have been considered. Each time a task to be initiated search for a robot that is nearer to the starting position. This has been accomplished through robot localization. The robot positions are acquired by the LED equipped on the robot. Euclidean distance is computed from the predefined start position where the task is to be initiated with the robots moving in the multi-robot environment. The nearest robot is selected and allocated with the task. In this research work, object transportation has been considered as the task to be done. Algorithm 1 describes the robot selection process step-wise.

Algorithm 1: NR – selection (S)
Input: Starting position where the task is to be initiated
Output: Nearest robot position

```
Initialize the starting position as S
Initialize the number of robot available in the environment as n
for i = 1 to n
        P^i = getposition(R^i)
       If (P^i == S)
          Return R^i
     Else
          Compute the Euclidean distance
```

$$D^i = \sqrt{(S_x - R_x^i)^2 + (S_y - R_y^i)^2}$$

```
          Select the minimum distance D^i
          Return R^i
     End if
End for
```

The selected robot R^i executes the SCA algorithm to plan its path towards the goal. Both the starting and goal locations are defined in this context. At each step, the robot selects the next best position based on the selection parameter. The algorithm takes the initial position of the robot as ty input and produces a collision-free path as the output. The selection parameter finds a new next position based on the value computed cosine and cosine value. Each time the position is updated by considering the presence of obstacles in that position. If the computed position is occupied with an obstacle the next position is recomputed. The parameters used in the algorithms represented unique definition as shown in Table 1. Algorithm 2 describes the path planning of the selected robot for collision-free transition to the goal location.

Table 1 Parameters used and their representation

Parameter	Representation	Parameter	Representation
S	Task initial position	$a1$	Direction control parameter
R^i	ith Robot	$a2$	Distance control parameter
D^i	Euclidean distance of ith robot	$a3$	Optimality control parameter
S_x, S_y	Position vector of S	$a4$	Selection parameter
R_x^i, R_y^i	Position vector of ith robot	P_i^{best}	Best position of ith robot
T	Maximum iteration	Obs^x, Obs^y	Position vector of the obstacle

Algorithm 2: $\mathbf{R-Path(R^i, S)}$
Input: Initial robot position, goal position
Output – Collision free path

Initialize robot position and starting state
Initialize max iteration = T
For t = 1 to T assumed T = 30
  Compute the next position as

$$R_x^i(t+1)=\begin{cases} R_x^i(t) + a1 * \sin(a2) * \left(a3 * P_i^{best} - R_x^i\right), & a4 < 0.5 \\ R_x^i(t) + a1 * \cos(a2) * \left(a3 * P_i^{best} - R_x^i\right), & a4 \geq 0.5 \end{cases}$$

  if $((R_x^i(t+1), R_y^i(t+1))! = (Obs^x, Obs^y)$
    Update $P_i^{best} = (R_x^i, R_y^i)$
  Else
    Compute a new position
  End if
  Update t
End for
Return position vector

The algorithm works in an initialized population and dimension. At each dimension, the robots are allowed to compute the next best position with maximum number of iterations as mentioned in the algorithm. The parameters used in the algorithms are denoted in Table 1.

4 Result and Simulation

The proposed algorithm has been implemented in computer using programming in C language. Some other existing approaches as discussed in literature such as PSO and Q-learning have also been exercised to evaluate the result in terms of several metrics such as execution time, path length, turnaround time, and waiting time. The algorithm has also been compared with self without implementing the robot selection methods and with robot selection module. The result obtained through computer simulation in terms of the above-mentioned matrices has been shown in this section. The waiting time in selecting a random robot is more due to the randomness and the worst-case time occurs when a robot is selected at a distant from the initial state. The waiting time for 10 such execution has been represented in Table 2. Similarly, the turnaround time that is the combination of waiting time and execution time has been denoted in Fig. 1. It has been observed that the proposed approach incurs less time as it resolves the randomness from Fig. 2.

Figure 3 describes the path length obtained by executing the three algorithms as mentioned in the literature. It has been noted that SCA exhibits minimum path length as it implements both the exploration and exploitation. The position update function value determines the exploration and exploitation for sin function value in the range [−1, 1] and cos function value in the range [−1, 1] respectively.

Table 2 Execution time with and without randomness in seconds

Task execution number	With randomness	Without randomness
1	78	67
2	82	76
3	65	56
4	67	62
5	76	72
6	66	59
7	67	60
8	75	70
9	73	65
10	68	62

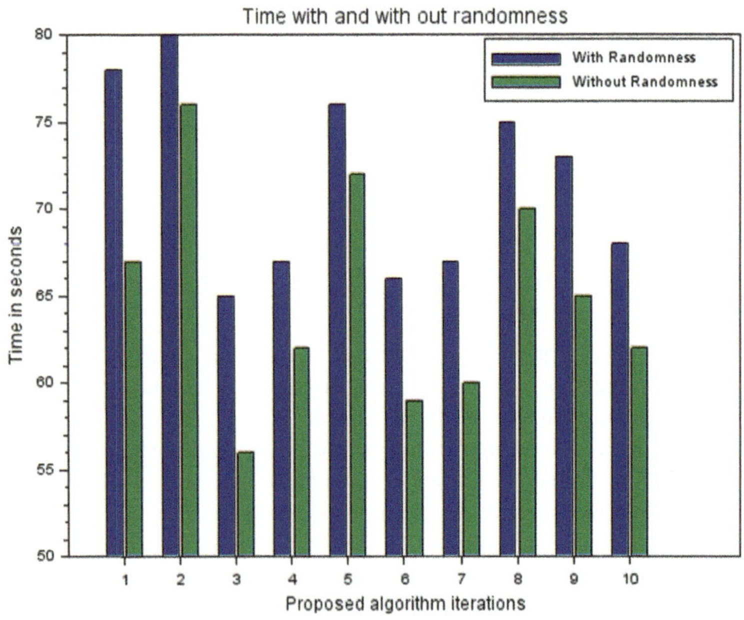

Fig. 1 Execution time with and without randomness

Number of turns and steps incurred by the competing approaches are denoted in Fig. 4. Re-computation time for the existing approaches if the new position is occupied with an obstacle is denoted in Fig. 5. Similarly, the performance has been evaluated based on path length, turns, and time incurred by the existing approaches and represented in Fig. 6.

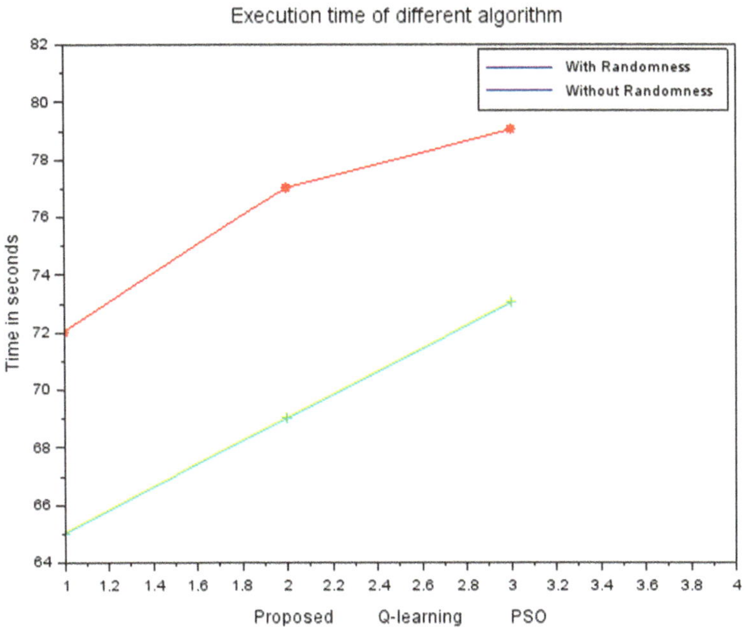

Fig. 2 Execution time of different algorithms

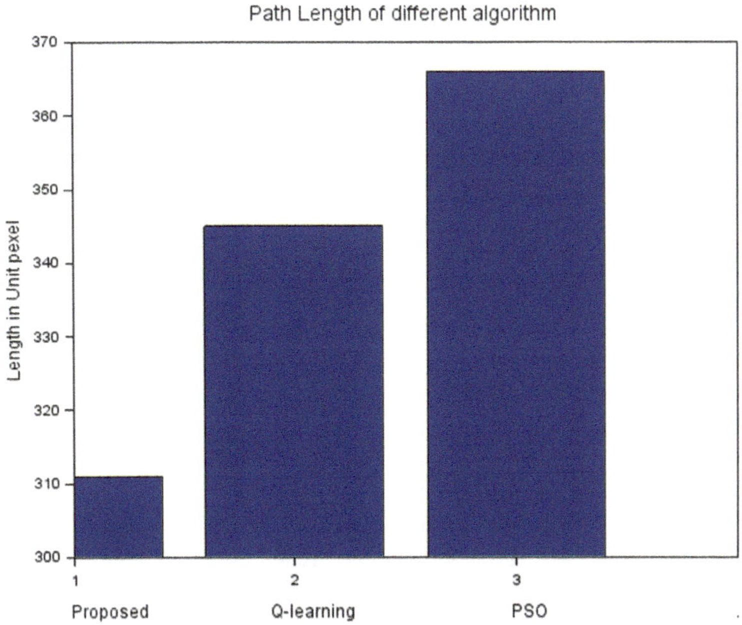

Fig. 3 Path length comparison

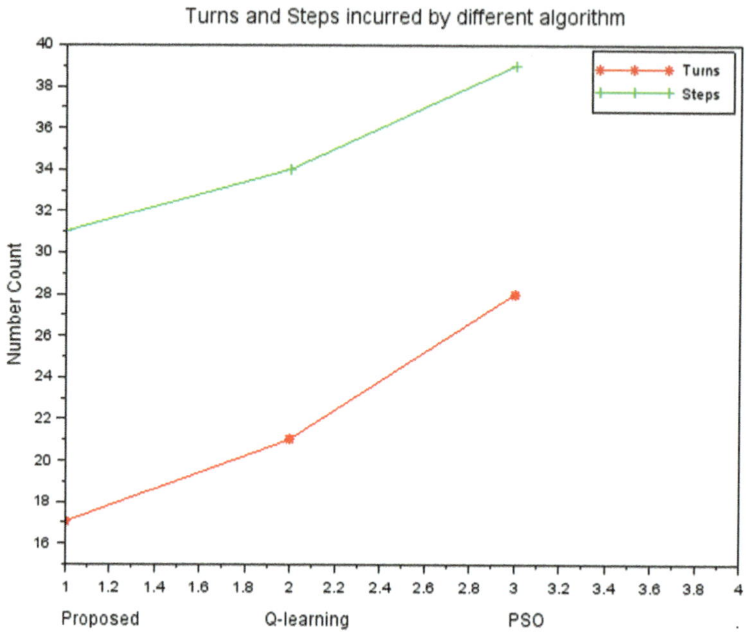

Fig. 4 Turns and steps incurred

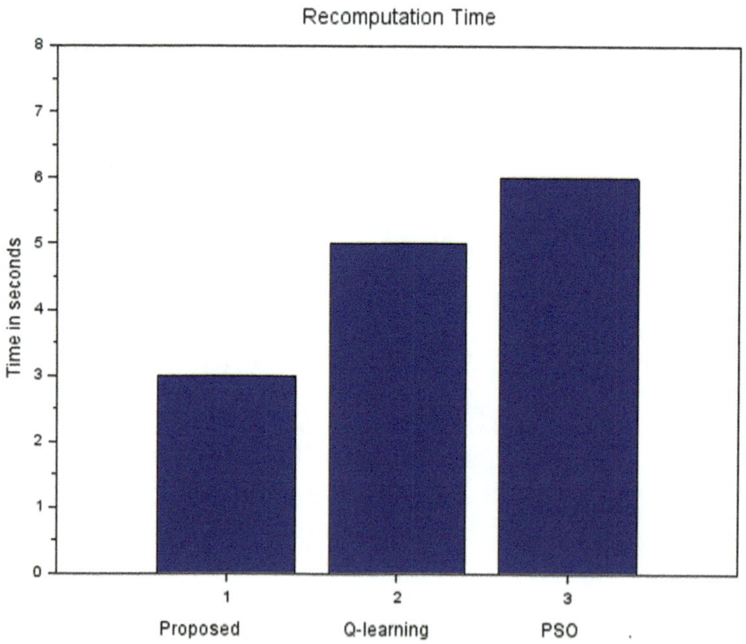

Fig. 5 Recomputation time of different algorithms

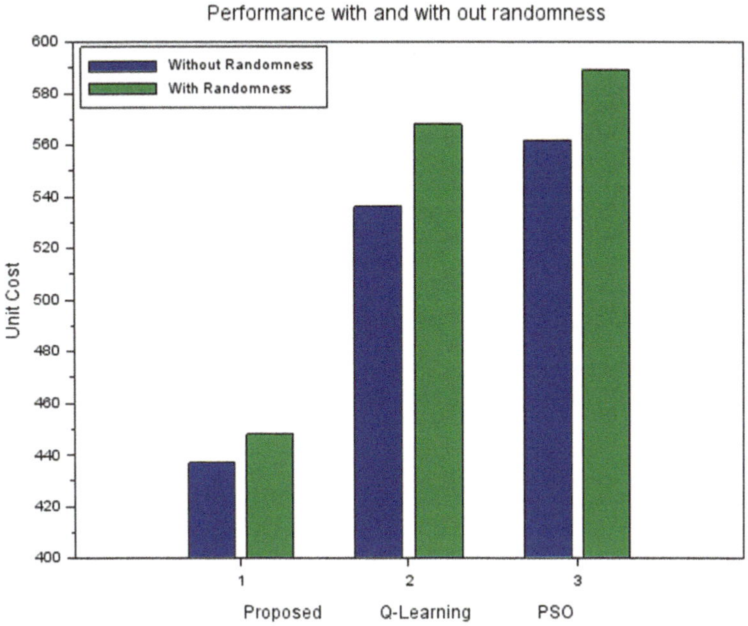

Fig. 6 Performance comparison

5 Conclusion

Allocation of task is a crucial part in task management and task execution. The proposed research work addresses the task allocation by selecting a suitable robot. The research work resolves the task allocation problem in a multi-task multi-robot environment. The proposed research work considers the multi-task multi-robot environment. A working robot is selected for a task to be executed using the nearest neighbor approach. Sine cosine algorithm has been implemented for path planning as it has balanced exploration and exploitation capability. The result of the work has been simulated through computer programming language and compared with other existing approaches. The future direction of the proposed work concentrate on task allocation of multiple mobile robot in dynamic environment.

References

1. Latombe, Jean-Claude.: Robot motion planning. Springer Sci. & Bus. Media, **124**, (2012)
2. Olcay, Ertug, Fabian Schuhmann, Boris Lohmann.: Collective navigation of a multi-robot system in an unknown environment. Robot. Auton. Syst. **132**, pp. 103604 (2020)
3. Zhou, Lifeng, Pratap Tokekar.: Multi-robot coordination and planning in uncertain and adversarial environments. Curr. Robot. Rep., pp. 1–11 (2021)

4. Ntakolia, C., Moustakidis, S., Siouras, A.: Autonomous path planning with obstacle avoidance for smart assistive systems. Expert Syst. Appl. **213**, 119049 (2023)
5. Sahu, Bandita, Pradipta Kumar Das, Manas Ranjan Kabat. Multi-robot co-operation for stick carrying application using hybridization of meta-heuristic algorithm. Math. Comput. Simul. **195**, pp. 197–226 (2022)
6. Qin, Hongwei, et al. "Review of autonomous path planning algorithms for mobile robots. Drones 7.3: **211** (2023)
7. Antonyshyn, Luke, et al.: Multiple mobile robot task and motion planning: A survey. ACM Comput. Surv. **55.10**, pp. 1–35. (2023)
8. Bai, Xiaoshan, et al.: Group-based distributed auction algorithms for multi-robot task assignment. IEEE Trans. Autom. Sci. Eng. **20.2**, pp. 1292–1303 (2022)
9. Nugraha, Raden Cahyadi, et al.: A mixed-integer linear programming formulation for assembly line balancing problem with human-robot shared tasks. Journal of Physics: Conference Series. IOP Publishing, **1858**(1), (2021)
10. Seenu, N., et al.: Review on state-of-the-art dynamic task allocation strategies for multiple-robot systems. Ind. Robot.: Int. J. Robot. Res. Appl. **47**(6), 929–942 (2020)
11. Alitappeh, Reza Javanmard, Kossar Jeddisaravi.: Multi-robot exploration in task allocation problem. Appl. Intell. **52.2**, pp. 2189–2211 (2022)
12. Notomista, Gennaro, et al.: An optimal task allocation strategy for heterogeneous multi-robot systems. 2019 18th European Control Conference (ECC). IEEE, (2019)
13. Rahman, Akhlaqur, et al.: Energy-efficient optimal task offloading in cloud networked multi-robot systems. Comput. Netw. **160**, pp. 11–32 (2019)
14. Tiniç, Gizem Ozbaygin, et al.: Exact solution approaches for the minimum total cost traveling salesman problem with multiple drones. Transp. Res. Part B: Methodol. **168**, pp. 81–123 (2023)
15. Kennedy, James, Russell Eberhart.: Particle swarm optimization (PSO). Proc. IEEE international conference on neural networks, Perth, Australia. **4**(1), (1995)
16. Mirjalili, S.: SCA: a sine cosine algorithm for solving optimization problems. Knowl.-Based Syst. **96**, 120–133 (2016)
17. Bandita, Sahu, et al.: Prevention of Covid-19 affected patient using multi robot cooperation and Q-learning approach: a solution. Qual. Quant. **56.2**, pp. 793–821 (2022)
18. Sahu, Bandita, Pradipta Kumar Das, Manas ranjan Kabat.: (2022) Multi-robot cooperation and path planning for stick transporting using improved Q-learning and democratic robotics PSO. J. Comput. Sci. **60**: 101637 (2022)
19. Lerman, Kristina, et al.: Analysis of dynamic task allocation in multi-robot systems. Int. J. Robot. Res. **25.3**, pp. 225–241 (2006)
20. Quinton, F., Grand, C., Lesire, C.: Market approaches to the Multi-Robot task allocation problem: a survey. J. Intell. Rob. Syst. **107**(2), 29 (2023)
21. An, Xing, et al.: Multi-Robot systems and cooperative object transport: communications, platforms, and challenges. IEEE Open J. Comput. Soc. (2023)

Enhancing Security and Efficiency in Signcryption Schemes Using Elliptic Curve Cryptography

Sudarshan Basu, Sourabh Debnath, and Ramesh Kumar Mohapatra

Abstract One of the main goals of computer and information security research is the delivery and storage of secure and verified messages. The current accepted approach to achieve this goal is "encryption followed by (digital) signature." This paper addresses the cost of storing and delivering secure and verified messages: Can messages of different lengths be delivered or stored in a secure and authenticated manner for less money than what "signature followed by encryption" requires? Since the development of public key cryptography, it appears that this question has never been answered in the literature. Next, we provide a favorable response to the query. Specifically, we identify a new cryptography primitive called "signcryption" that satisfies the needs of public key encryption and digital signatures in a logical single step, at a cost much less than that of "signature followed by encryption." In addition to message confidentiality, authentication, integrity, unforgeability, and non-repudiation, it provides forward secrecy for message confidentiality and public verification. Under the suggested plan, in the event of a dispute, the court might independently confirm the signer's signature without needing the sender's private key. Our scheme's cheap computing and communication costs make it more suitable for usage in mobile communication environments.

Keywords Cryptography · Signcryption · Elliptic curve

S. Basu (✉) · S. Debnath · R. K. Mohapatra
National Institute of Technology Rourkela, Odisha 769008, India
e-mail: sudarshanbasu97@gmail.com

S. Debnath
e-mail: sourabh_debnath@nitrkl.ac.in

R. K. Mohapatra
e-mail: mohapatrark@nitrkl.ac.in

S. Kumar Udgata et al. (eds.), *Intelligent Systems*, Lecture Notes in Networks and Systems 1149, https://doi.org/10.1007/978-981-97-8160-7_9

1 Introduction

Cryptographic techniques are essential in contemporary communication systems to guarantee data integrity, confidentiality, and authenticity. These security services are provided by core primitives like public key encryption and digital signatures. Nevertheless, using these primitives on a message in that order—signing first, then encrypting—is computationally costly and results in a large amount of transmission overhead.

Zheng (1997) proposed the idea of signcryption, a cryptographic system that satisfies both encryption and signing in a single logical step, in order to address these inefficiencies. Signcryption offers confidentiality, unforgeability, integrity, and non-repudiation at a reduced computational cost and transmission overhead by combining these features. For protecting communication in contexts with little computing power, such mobile networks, wireless sensor networks, and the Internet of Things, signcryption is therefore quite appealing. Integer factorization, discrete logarithms, elliptic curves, and more recently post-quantum cryptography have all been used as the foundation for constructing efficient and safe signcryption algorithms since the field's birth. Over the past 20 years, numerous signcryption systems have been presented to meet the needs of various application settings.

People's ability to communicate securely and authentically has been completely transformed by the discovery of public key cryptography almost two decades ago. Any message (text, binary files, or documents) encrypted with the public key can only be unlocked by using the corresponding private key and the same algorithm. The matching public key is the only way to decrypt a message that has been encrypted using the private key. Since the keys are meant to be public, you can stop worrying about sending them over the Internet. However, one drawback of asymmetric encryption is its slower speed in comparison to symmetric encryption. The message's content must be encrypted and decrypted with far more processing power. Before the message is delivered to the recipient, the sender must sign it as verification of the message's authenticity, or the proof of originator.

The most recent signcryption techniques are thoroughly reviewed in this study. We give a summary of signcryption systems' key characteristics and security concept. Surveyed are the main categories of signcryption systems, such as identity-based, certificateless, attribute-based, and certificate-based. We evaluate several systems according to their computational costs, security proofs, and cryptographic underpinnings. There are also several highlighted open difficulties and potential research directions in the development of signcryption methods.

2 Literature Work

This section discusses about a cryptographic primitive called signcryption, which combines encryption and digital signature features into a single operation. With the help of this hybrid concept, secure communication in a variety of applications should be made easier and more effective. Over time, researchers have explored the theoretical underpinnings, security characteristics, real-world applications, and practical implementations of signcryption, leading to an evolution of the literature in this area.

Using signed encryption, Ren-Junn Hwang [1] has presented as a more effective method than traditional encryption and signature schemes for accomplishing security and authentication for message delivery. It goes over the accepted procedure for delivering messages in a secure and verified manner utilizing encryption and digital signatures. The cost of this is the same as the total of the encryption and signature costs. The question of whether secure and authenticated message delivery can be accomplished at a lower cost than signature + encryption is raised in this paper. In a single logical step and at a reduced cost, it presents a new cryptographic primitive called "signcryption" that fulfills both the encryption and signature functions. Comparing signcryption to standard signature + encryption schemes, analysis reveals that it has 50% lower computation cost and 85–96% lower communication overhead, with savings growing for higher security levels. Fast, secure, and authenticated key transport in a very small message is made possible by signcryption. It can also be used for things like health cards and digital currency. Signcryption security arguments about confidentiality, non-repudiation, and unforgeability are given. Verifiability's limitations are also covered.

In this paper, Bashir et al. [2] present signcryption, a technique that combines encryption and signature functionality at a lower overhead than if they were applied sequentially. Along with security proofs and an analysis of the savings over conventional methods, concrete schemes are provided. The ElGamal signature scheme is modified to create two concrete signcryption schemes, known as SCS1 and SCS2. Unlike earlier signcryption proposals, these offer extra features like forward secrecy and public verifiability. According to security analysis, the suggested schemes offer non-repudiation through zero knowledge proofs and unforgeability based on the discrete log problem's hardness. Reduction to the confidentiality of a related encryption scheme serves as proof of confidentiality.

In this paper, Devarajan M. and Sasikaladevi [3] proposed signcryption scheme simultaneously meets security goals while being efficient for implementation on resource-constrained mobile devices due to use of ECC. They have put forth a plan for unified signcryption that combines signature (ECDSA) and ECC encryption (ECIES) techniques. The scheme offers attributes including forward secrecy, public verifiability, unforgeability, integrity, confidentiality, and nonrepudiation. Initialization, signcryption, unsigncryption, and judge verification are its four stages for settling disputes. The session key is established between the sender and the recipient using ephemeral-static ECDH key agreement. Both parties' identifiers are also used in the key derivation process.

The efficient and secure certificateless signcryption scheme proposed by Mohsen Toorani and Yu [4] in this paper achieves the objectives of confidentiality and authentication. In order to maximize efficiency, the scheme shares randomness by combining signature schemes and certificateless encryption. It guarantees forward secrecy, security against insider threats, confidentiality, unforgeability, integrity, and non-repudiation. The scheme uses hash functions modeled as random oracles and is based on the bilinear groups and gap bilinear Diffie-Hellman (GBDH) assumption. Algorithms for setup, user key generation, private key setting, partial private key extraction, signcryption, and unsigncryption are all defined by the scheme.

In this work, Shehzad Ashraf [5] show how the signcryption scheme can be completely broken through a series of attacks, demonstrating that it falls short of the anticipated security objectives. The deficient key establishment mechanism and the absence of security validations are the main causes of the vulnerabilities. The paper examines the security of the elliptic curve-based signcryption scheme developed by Hwang et al. It points out a number of weaknesses that an adversary could exploit to compromise confidentiality, unforgeability, non-repudiation, and other security features. The only security provided by the scheme is the random number r's secrecy. The sender's long-term private key can be easily recovered in the event that r leaks. This totally undermines security objectives. The impersonation attacks are made possible by the scheme's improper validation of public keys and certificates.

In order to achieve security objectives, A. K Patro et al. lightweight HECC-based signcryption scheme [6] considerably lowers computational expenses and communication overhead when compared to ECC schemes. Mobile devices and other environments with limited processing power and communication capabilities are better suited for it. In comparison to current ECC-based schemes, the suggested scheme requires fewer costly HECC scalar multiplication operations, and a computational cost analysis is provided that takes into account the time complexity of various arithmetic operations. Using smaller key sizes also results in less communication overhead.

In addition to achieving fundamental security objectives, Singh [7] has introduced a novel, all-encompassing elliptic curve-based signcryption scheme in this paper. It also achieves forward secrecy, public verifiability, and encrypted message authentication. For efficiency, this paper combines signature algorithms with shared randomness with elliptic curve encryption. Using a random integer and the recipient's public key, the sender signs the message. transmits the recipient the ciphertext and signature values (R, s). Using their own private key and the sender's public key, the recipient decrypts and confirms signature by examining a connection. Without having to decrypt the message, a judge or other third party can directly verify the signature using the sender's public key.

3 Methodology

The simulation setup for the experiment consists of five different scenarios consisting of defining the equation of the curve and the generator points of the curve at first. Then using the generator points, Alice and Bob both derive their private and public key pairs. Using these pairs Alice signcrypt the message with AES encryption and get a hash value of the message and send it to Bob. Bob then decrypts the message by using his private key public key pair. Three algorithms are usually included in a signcryption scheme: key generation, signcryption, and unsigncryption. For both signcryption and unsigncryption, the public keys needed are generated by the key generation algorithm. Additionally, it creates Bob and Alice, the key pair. Bob will receive a signcrypted message (c, r, s) generated by the signcryption scheme. In the unsigncryption stage, Bob, the verifier, decrypts the communication and verifies its legitimacy.

A signcryption scheme should have the following properties:

1. Correctness: The signcrypted message's plain text can be recovered using one of the existing unsigncryption schemes.
2. Confidentiality: Without the sender's and recipient's private keys, it should be impossible for an eavesdropper to decipher the signcrypted message and obtain any information.
3. Integrity: The message's content can only be changed by the intended or authorized user.
4. Unforgeability: Two signcrypted messages that contain the same plain text should not exist. If not, a skilled attacker could produce a legitimate signed text that the unsigncryption algorithm could accept.
5. Non-repudiation: Neither Alice nor Bob may retract their statements after sending or receiving a message. Alice should not dispute that she sent the message.

3.1 Why Elliptic Curve Cryptography?

Based on the provided table, we can infer that the RSA and Diffie-Hellman required key sizes grow far more quickly than the elliptic curve cryptosystems required key sizes if the symmetric key size increases. Therefore, compared to either RSA or Diffie-Hellman public key systems, elliptic curve systems provide greater security with each bit increase in key size (Table 1).

Table 1 Key sizes for equivalent security labels (in bits)

Symmetric key encryption size	ECC key size	RSA key size
80	163	1024
128	283	3072
192	409	7680
256	571	15,360

3.2 Experimental Setup

1. System Architecture: When employing elliptic curve cryptography for signcryption, an experimental setup usually consists of a sender and a recipient corresponding via a network. ECC is used by both parties for encryption, digital signatures, and key generation.
2. Key Generation: Creating keys using elliptic curve cryptography is the first step. A set of public and private keys is generated by both the sender and the recipient. The security requirements determine the selection of the elliptic curve parameters, including the prime modulus and base point.
3. Parameter Selection: For security, elliptic curve parameters must be chosen carefully. It is necessary to select parameters that can withstand possible attacks, like those that take advantage of holes in the curve structure. Selecting a safe elliptic curve for the experimental configuration is crucial.
4. Digital Signature Generation: Using the message's hash and their private key, the sender creates a digital signature. For this reason, the elliptic curve digital signature algorithm (ECDSA) is frequently used in ECC-based signcryption.
5. Encryption Process: Next, the sender uses the recipient's public key to encrypt the message. A shared secret can be created between the sender and the recipient via elliptic curve Diffie-Hellman (ECDH) key exchange, and this secret can then be used for symmetric encryption.
6. Implementation Platform: Depending on the intended use, the experiment can be put into practice on a microcontroller or a computer, for example. To guarantee precise and effective implementation, the choice of programming language and libraries is essential (Fig. 1).

3.3 Initialization Phase

Both Alice and Bob will initialize the following parameters: Select a large prime number [1].

$$q > 2^{160}$$

and Select a and b that satisfies the equation [1]

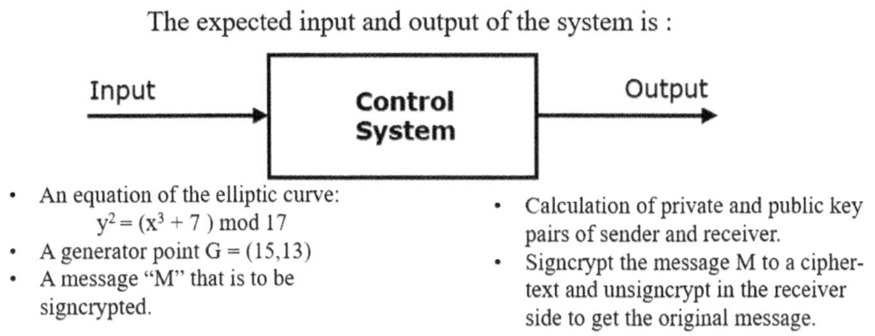

Fig. 1 The expected input given to the system and output generated by the system

$$4a^3 + 27b^2 mod q \neq 0$$

After initialization, they will take a base point G in the elliptic curve and a one-way hash function H.

Alice: Select integer d_a as private key and compute Public key $U_a = d_a * G$ [1] Bob: Select integer d_b as private key and compute Public key $U_b = d_b * G$ [1]

3.4 Signcryption Phase

This is the second phase of the signcryption process[1]. In this phase, Alice computes R = r*G =(r_1, r_2)

K = r*U_b =(k, l) Ciphertext C = $E_k(M)$

One-way hash function[1] to generate h = H(M || r_1) s = d_A− h.r mod n

Alice sends the signcrypted text (C, R, s) to Bob[1].

3.5 Unsigncryption Phase

This is the last phase of the signcryption process where Bob decrypts the encrypted message[1].

Bob computes K = $d_B * R$ =(k,l)[1] Bob generates Plaintext M = D_K(C) [1]
Bob uses one way hash function to compute h = H(M || r_1)[1]
Bob verifies whether s*G + h * R = U_A and returns True or False [1]

4 Results

We employed the one-way hash function, SHA-1, for the message digest in our thesis work. A fixed length output of 160 bits will be produced by this compression function. It can handle messages up to $2^{64}–1$ in size, and blocks up to 512 bits in size. A total of eighty numbers of rounds with a 32-bit word size have been used. Depending on the curve's parameters, the program will produce the points on the elliptic curve. It accepts a string as input, generates (c, T, s), signs the message, and sends it to the intended recipient. The recipient on the other end is responsible for decryption of the message. After verifying the sender's legitimacy, the recipient accepts the string.

In Fig. 2, we have defined an elliptic curve with which we will be encrypting the message. Using that curve, we are getting the "generator points" by which we can calculate the public and private keys of sender and the receiver. In Fig. 3, we get the corresponding private and public keys by using the generator points of the elliptic curve. After generating the public and private keys, the sender sends the message to the receiver using signcryption scheme mentioned above and the receiver get back the original message using unsigncryption.

A comparison of two schemes, signcryption and unsigncryption, in a range of scenarios is shown in Fig. 7. A bar chart is used to display the analysis; the y-axis shows the average computational time in milliseconds, and the x-axis shows the various scenarios. The graph allows for the following observations to be made (Figs. 4, 5, 6):

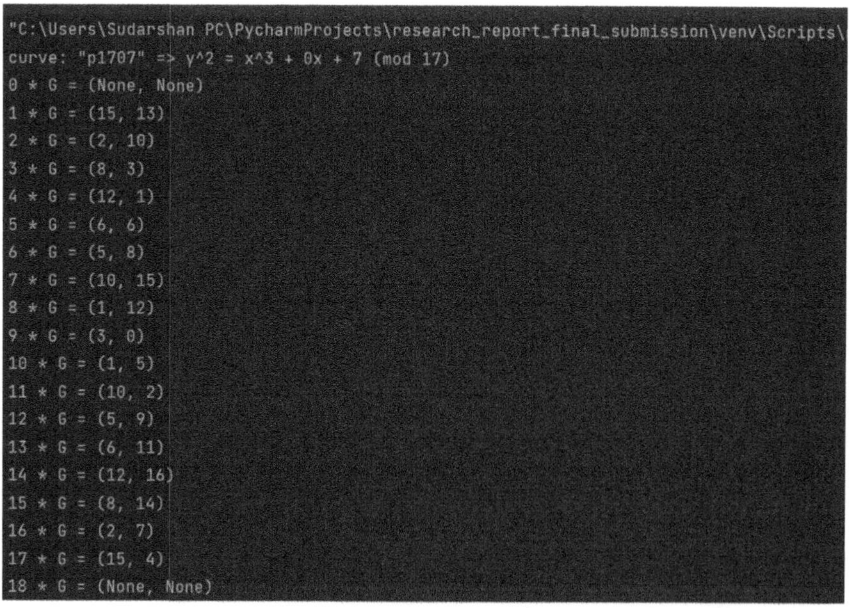

Fig. 2 We define the equation of the curve and get the Generator points of the curve

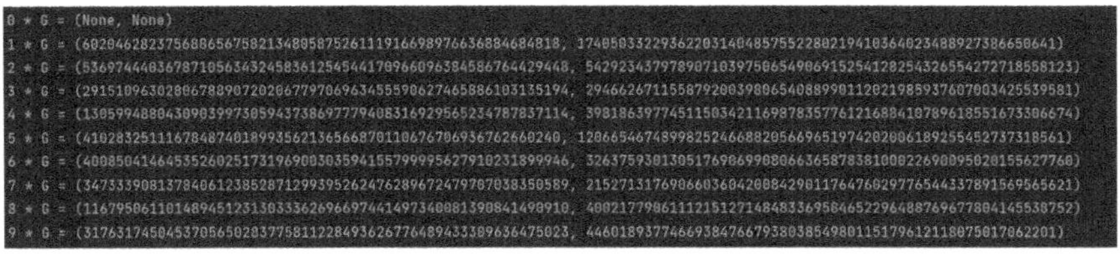

Fig. 3 We get the corresponding private and public keys of the given Generator points

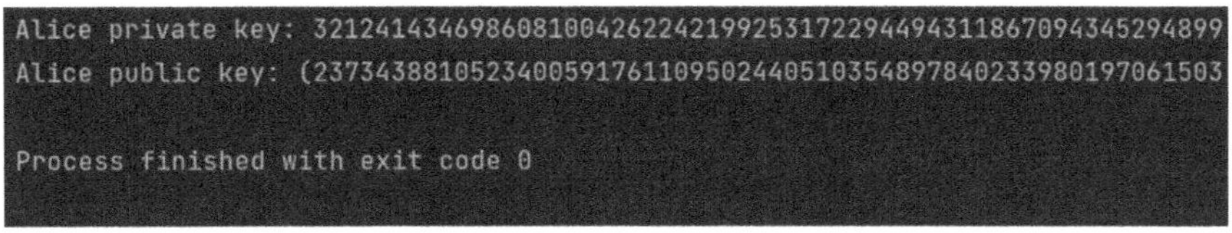

Fig. 4 Alice selects a private key-public key pair

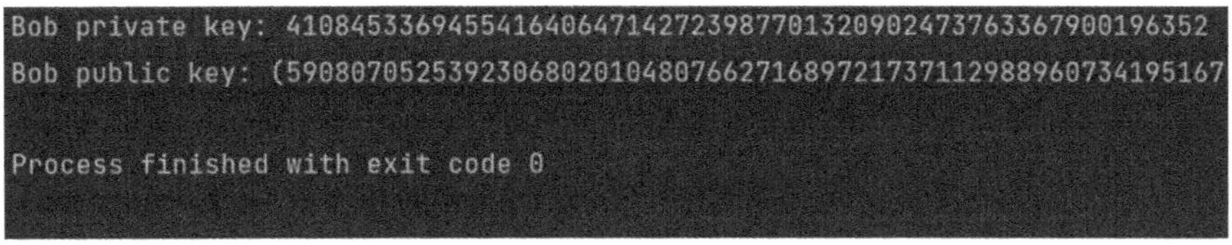

Fig. 5 Bob selects a private key-public key pair

Fig. 6 We signcrypt the original message and get back the decrypted message

1. Zheng (1997): The signcryption scheme is more computationally intensive in this scenario because it takes longer to compute than unsigncryption.
2. Hwang (2000): In this case, signcryption has a computational time that is much greater than unsigncryption, highlighting the signcryption scheme's higher computational cost.
3. Baek (2019): In this case, the computational difference between signcryption and unsigncryption increases, with signcryption requiring more resources.
4. Ke et al. (2020): This scenario continues the trend seen in the earlier scenarios, with signcryption's computational time being longer than unsigncryption's.
5. According to Md. H. et al. (2022), signcryption takes longer to compute than unsigncryption, but this difference is actually somewhat closer than in some of the earlier cases.

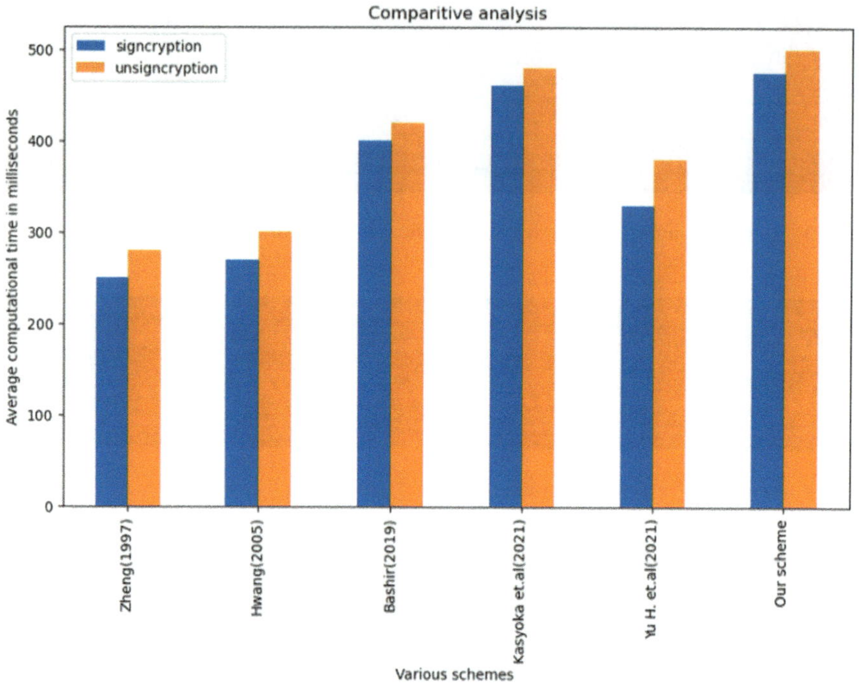

Fig. 7 Performance analysis of our scheme

6. Our proposed scheme: Although signcryption takes longer to compute in this scenario than unsigncryption does, the difference between the two schemes is still quite small when compared to some of the other scenarios.

Overall, the analysis indicates that the signcryption scheme consistently requires more computational time than the unsigncryption scheme for all the scenarios examined. The amount of the difference, however, varies depending on the scenario; some schemes show a greater difference in computational times between signcryption and unsigncryption than others.

5 Conclusion and Future Enhancement

To sum up, the investigation of signcryption through the use of elliptic curve cryptography (ECC) has demonstrated important progress and promise for the field of secure communication. Signcryption's combination of encryption and signature operations into a single computational process simplifies the communication protocol and lowers computational overhead. The overall performance of signcryption schemes is further improved by the inherent efficiency and security benefits of elliptic curve cryptography.

The study has explored both the theoretical underpinnings and real-world applications of ECC-based signcryption. When compared to conventional methods, elliptic curves' cryptographic strength allows for robust security with less computational resources needed. ECC-based signcryption is especially attractive for resource-constrained environments, like mobile and Internet of Things (IoT) devices, because of the smaller key sizes and faster computations. The most recent ECC signcryption schemes have achieved security notions like ciphertext anonymity under adaptive chosen ciphertext attacks, whereas the earlier schemes had limitations regarding message recovery or security assumptions. Modern schemes also show efficiency benefits by having short communication overheads and quick processing times. All things considered, ECC-based signcryption is a promising public key technique to realize secure and reliable communication mechanisms for environments with limited processing and bandwidth. The real-world deployment will be strengthened by ongoing research to analyze security against emerging computing attacks and optimize performance.

References

1. Ren-Junn Hwang, Chih-Hua Lai, Feng-Fu Su.: An efficient signcryption scheme with forward secrecy based on elliptic curve. Appl. Math. Comput., **167**(2):870–881, (2005)
2. Bashir, M.Z.U., Ali, R.: Cryptanalysis and improvement of blind signcryption scheme based on elliptic curve. Electron. Lett. **55**(8), 457–459 (2019)
3. Malathi Devarajan, Sasikaladevi, N.: An hyper elliptic curve based efficient sign-cryption scheme for user authentication. J. Intell. & Fuzzy Syst., **39**(6):8487–8498, (2020)
4. HuifangYu, Runtao Ren.: Certificateless elliptic curve aggregate signcryption scheme. IEEE Syst. J., **16**(2):2347–2354, (2021)
5. Manoj Kumar and Pratik Gupta.: An efficient and authentication signcryption scheme based on elliptic curves. MATEMATIKA: Malays. J. Ind. Appl. Math., pages 1–11, (2019)
6. Shehzad Ashraf Ch, Nizam Uddin, Muhammad Sher, Anwar Ghani, Husnain Naqvi, Azeem Irshad.: An efficient signcryption scheme with forward secrecy and public verifiability based on hyper elliptic curve cryptography. Multimed. Tools Appl., **74**:1711–1723, (2015)
7. Anuj Kumar Singh, BDK Patro.: Elliptic curve signcryption based security protocol for rfid. KSII Trans. Internet & Inf. Syst., **14**(1), (2020)

Brain Tumor Classification Using Vision Transformer Model (ViT)

V. Rajeshwari, P. Ezhilarasi, D. Beaulah Princiba, and S. Rajesh Kannan

Abstract Brain tumors are a serious health issue that thousands of people worldwide are affected. For brain tumor to be treated successfully and patients have an enhance quality of life. Thus, early and precise tumor diagnosis is essential. Several imaging methods are employed for identification of brain tumors. The most popular of these methods is MRI scan. In order to get around the drawback of conventional methods, computer -aided analysis of brain picture has emerged as a viable method for precise and accurate brain tumor classification in recent years. In the current study, we constructed an improved vision transformer model that correctly detects and classifies brain tumors. This model was able to classify the brain tumors, when it was trained using MRI dataset, which included 5712 pictures of the tumors. The accuracy of Vision Transformer (ViT) model was 98.13%. To diagnose rarer and more complicated tumors with greater accuracy, further research may be done.

Keywords Brain tumor · Classification · ViT · Diagnosis · Accuracy · MRI · Dataset

1 Introduction

Brain tumor is an uncontrolled growth cell inside the brain and surrounding tissues. The brain cells themselves, which are referred as brain tumor or tissues cause (secondary brain tumor), will be the source of this benign or malignant development. According to the CBTRUS data report, over 7 lakh people are now suffering from brain. Number of new cases recorded and diagnosed yearly is about 84,000. All ages are prone to brain tumors [1], however, children and older people are the most common sufferers of these diagnoses [2]. The detection of brain tumors is a significant challenge to medicine as early and accurate classification is necessary for efficient treatment planning. The complexity of existing techniques for classifying

V. Rajeshwari (✉) · P. Ezhilarasi · D. B. Princiba · S. R. Kannan
Department of Electronics and Communication Engineering, St. Joseph's College of Engineering, Chennai, India
e-mail: ragavi2701vk@gmail.com

S. Kumar Udgata et al. (eds.), *Intelligent Systems*, Lecture Notes in Networks and Systems 1149, https://doi.org/10.1007/978-981-97-8160-7_10

brain tumors, which frequently depend on human interpretation of medical pictures, highlights the urgent need for innovative methodologies. Deep learning models have completely changed the field of medical image analysis in recent years, and vision transformers have become extremely useful tools for challenging picture categorization problems. The three most symptoms of brain tumors are headaches, seizures, and cognition. The common symptom, headaches, is often characterized as either dull or acute. People suffering from brain tumors also encounter cognitive abnormalities, including difficulties with memory, focus, and thinking. Additional signs of brain tumors include trouble speaking or understanding language, limb weakness, and impaired eyesight or hearing [3]. For classification and identification brain tumors, doctors consider a number of features. These features include tumor size and tumor location in brain [4]. These are two type of brain tumor include glioblastomas and gliomas are malignant brain tumors. Brain has an important organ in it, the pituitary gland is the site of origin of another type of cancer. Function of various body glands is controlled by the pituitary gland. Physicians are capable of accurately diagnosing and treating brain tumors [5].

Various imaging methods are used to identify tumors like magnetic resonance imaging, or MRI, [6] were commonly used methods. Computed tomography, or CT scanning, is another method. This method looks for brain tumors using X-rays. Positron emission tomography, or PET, is another method that can be used to identify brain tumors. A radioactive material is injected into the blood to facilitate this imaging procedure. A biopsy is the most successful surgical approach for the discovery of brain tumors. This type of brain tumor is detected by examining a small portion of the tumor under a microscope. Brain tumor identification is successful with these imaging modalities. Nevertheless, there are a number of drawbacks to these traditional approaches. These imaging methods are costly and require a lot of time. For patients who need frequent follow-up scans, it might be challenging.

The location, tumor size, and tissues can all affect the accurate of these methods. The model's confusion matrix may have errors in its representation of actual results. Treatment delays and inaccurate diagnoses may result from this. To circumvent the aforementioned issues, a number of machine and deep learning models are also utilized in the diagnosis of brain tumors. MRI scan and picture data from patients are utilized as input data to train the model utilizing algorithms. Based on the given picture, the algorithm then forecasts the presence or absence of brain tumors. Brain tumors may be effectively identified by supervised learning algorithms (SVM) using segmentation and classification models. Random forest (RF) is a learning method that can do both predictive analysis (regression) and categorical analysis (classification) for tumor identification. A deep learning technique called CNN, or conventional neural network, is also used to detect brain tumors. CNN carries out an image categorization operation. This program identifies brain tumors by taking characteristics out of brain scans and classifying them according to those traits.

Originally developed for natural language processing applications, vision transformers have shown impressive results in picture classification, especially on large-scale datasets. Self-attention processes enable them to record complicated spatial connections within pictures, which makes them very useful for the complex patterns seen in medical imaging. Vision transformers provide a unique paradigm for feature extraction and representation learning by partitioning an image into smaller patches and enabling the model to take relationships among all patches into account at the same time. The field of image categorization has undergone a paradigm change with the introduction of the vision transformer (ViT). ViT, in contrast to its convolutional forebears, relies on a graceful dance of self-attention rather than the strict grid-based processing. Consider every pixel in an MRI scan as a storyteller that whispers information about a tumor's prognosis or health. ViTs weave together these murmurs through a dynamic web of interactions rather than in a preset sequence. Every story has resonance with other stories, highlighting common themes and drowning out opposing voices. Intricate connections between far-off pixels take center stage in this symphony of attention, exposing patterns hidden from the stiff CNN gaze.

2 Related Work

In [7], using the pretrained CNN model—also referred to the VGG-16 model—as a basis, it was adjusted to recognize various kinds of brain tumors. The CE-MRI dataset, which consists of MRI pictures of brain tumors divided into four different groups, was the dataset utilized for this investigation. A total of 3064 MRI scan and picture samples from 233 patients make up the dataset. The study's results showed that the suggested approach was sufficient, classifying brain tumor pictures into the four unique categories with an overall accuracy of 94.82%. By using data from a single group to train a (CNN) algorithm, they tested the algorithm using data from two more groups to investigate if differences in data collecting and processing between colleges would affect its effectiveness. The authors [8] employed the TCIA dataset.

The researchers used a series of MRI scans of patients whose tumors were identified using the BAT approach to train a CNN model that have three convolutional layers and two fully connected layers. The dataset has MRI imaging data from patients with gliomas, which was used by the researchers. With a 90% sensitivity, 87% specificity, and 92% accuracy, the model functioned effectively [10].

BraTS is often used to identify, classify, and segregate issues associated with brain tumors. Because it grows more slowly than other brain tumors, low-grade glioma or LGG has a better decision. The dataset known as Deletion Dataset includes MRI scans and pictures of people with LGG. For the categorization and division of LGG brain tumors, it is extensively utilized. Glioblastoma MRI scans and pictures (T1 and T2 weighted and FLAIR scans) with certain alterations are available in other databases [11]. Another dataset including MRI pictures and glioma genetic information is TCGA-GBM.

3 Proposed Method

In this study, we take advantage of data from MRIs and other medical exams as well as brain imaging. The brain tumor figures are processed before and tagged to detect whether or not they display a brain tumor. The dataset, which consists of several hundred images, is large and representative of the patient population diagnosed with brain tumors. Photo labels in the collection generally indicate the presence and location of tumors. Radiologists and other medical professionals can manually annotate documents, or computer algorithms can automate the process. The MRI scan data that were utilized in this investigation are publicly available. The proposed brain tumor classification method relies heavily on the diversity and representativeness of the datasets used for training and evaluation.

MICCAI Brain Tumor Image Segmentation Challenge (BraTS): Incorporating diverse datasets from different institutions, BraTS offers a platform for benchmarking and testing the generalization capabilities of the proposed model. Four types of tumors were included in this dataset, which was collected from different Chinese hospitals between 2005 and 2020: gliomas, meningiomas, pituitary tumors, and no tumor at all. Meningiomas form from the surrounding meninges tissues, whereas gliomas, a frequent kind of brain tumor, originates from glial tissues surrounded by neurons. The pituitary gland is located in the back of the nose, grows abnormally and can cause pituitary tumors. The dataset's picture dimensions are shown as 512 $\times$ 512 pixels.

Google AI created the Vision Transformer (ViT) model, a cutting-edge neural network architecture, for the purpose of classifying images. ViT uses self-attention processes to directly process the spatial characteristics of pictures, unlike typical convolutional neural networks (CNNs), which rely on spatial convolutions to extract information from images. This makes it possible for ViT models to detect long-range connections between pixels, which is crucial for correctly categorizing intricate pictures like MRI scans of brain tumors. The proposed system architecture of a vision transformer model for brain tumor classification consists of input, encoder, attention, decoder, feed-forward network, classification, loss, optimizer, data augmentation, model evaluation, and interpretability enhancement components.

The model receives a brain tumor MRI scan as an image tensor and passes it through encoder blocks that extract features using attention layers and feed-forward networks. The features are then passed through a classification layer to assign the MRI scan to a specific brain tumor type. The data augmentation method is developed to artificially increasing the size and training dataset diversity, while interpretability enhancement techniques such as gradient-based visualization, SHAP analysis, and feature attribution maps are used to understand the model's decision-making process. By combining these components, the vision transformer model demonstrates the potential of accuracy, robust, and interpretable brain tumor classification from MRI scans, contributing to improved diagnosis and treatment decisions in neuroradiology. [12] The architecture of ViT model for brain tumor classification is shown in Fig. 1.

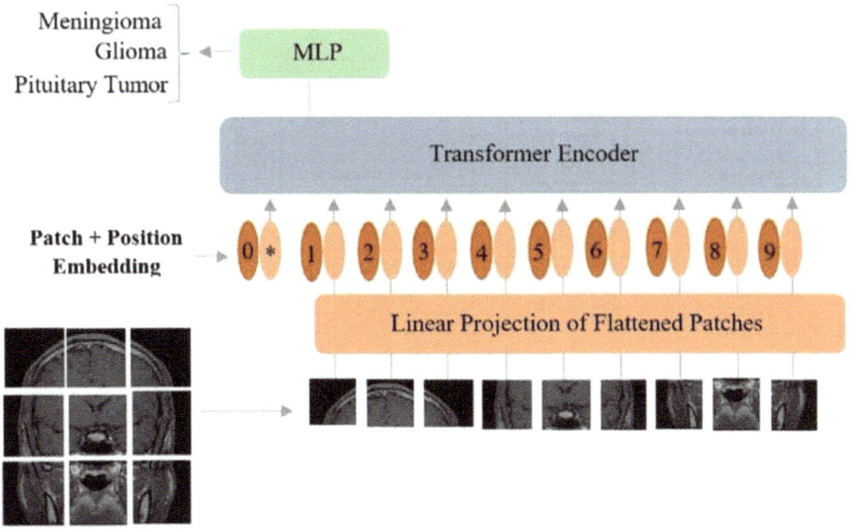

Fig. 1 Architecture of the proposed method

One of the key advantages of the ViT model lies in its ability to handle both local and global information simultaneously. Traditional CNNs often struggle with capturing long-range dependencies in images, particularly in medical scenarios where subtle details may play a crucial role in accurate diagnosis. The self-attention algorithm in ViT enables whole image while making predictions, facilitating the recognition of nuanced patterns indicative of different brain tumor types. Training the ViT model for brain tumor classification involves leveraging large datasets with diverse MRI images. The model learns to associate patterns and features present in these images with their corresponding tumor types through a process of supervised learning.

The brain tumor MRI scans and pictures are divided into image patches by the backend algorithm of the Vision Transformer architecture for brain tumor classification. This stage is required because the input picture can be too big for the model to handle in its entirety. The supplied picture is split into a regular patch of nonoverlapping patches in order to accomplish the splitting process. After that, every single patch is flattened linearly into a single sequence vector [13]. After that, model receives the flattened patches for additional processing is shown in Fig. 2.

The input MRI image dataset is represented by Matrix Y, and the expression for dividing the input pictures into a number of patches is as follows:

The Y sized input MRI picture is H × W × L.

Size of patch H × W × L

Step S

$$Nrow = floor((W - H)/u) + 1 \tag{1}$$

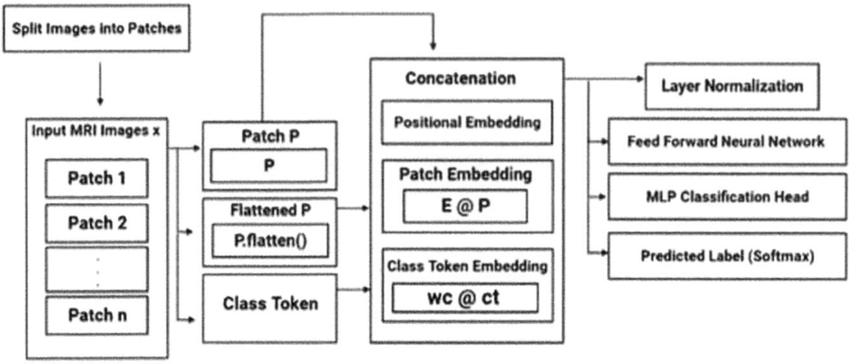

Fig. 2 Patching of images in each block of transformer encoder layer

$$Ncol = floor((O - H)/u) + 1 \tag{2}$$

The MRI brain tumor patches must be flatted into a sequence vector once they have been separated into patches. The step is important to convert the incoming patches to a format that is accepted by the self-attention module for further processing. Mathematically speaking, a patch P-sized H × W × L (where L channels in image) that is flattened down using a matrix can be briefed as follows:

$$patch = H.flatten() \tag{3}$$

This approach converts the H × W × L matrix into a 1D vector of length H × W × L. After every patch has been flattened, the remaining collection of flatted patches is sent into the ViT model's self-attention mechanism for additional processing. In the Vision Transformer (ViT) design for brain tumor diagnostics, the next step is to create 1D linear embeddings from the flatted patches. The token class projection and embedding projection patch are the two linear transformations that the flattened patches go through to finish this operation. This operation can be stated mathematically as follows:

$$patch = E * patch\_flat \tag{4}$$

The class token projection is the second linear transformation that converts a learnable class token vector linearly into a feature vector with a lower dimension. The ViT model uses the class token vector is a learnable parameter, to display the full picture in a fixed length. The following equation is displayed:

$$class\_token\_embedding = Wc * class\_token \tag{5}$$

where class token is a D-dimensional vector, and "Wc" is a matrix of size (D, D) that is a learnable parameter.

After concatenation, the patch image and class token embeddings are passed to the transformer encoder for processing. The Vision Transformer (ViT) architecture uses the flattened patches and class token to create lower-dimensional linear embeddings that are then enhanced with positional embeddings to locate brain tumors. In order to give the ViT model spatial location data for every patch inside the original picture, this step is required. By same matrix as W-sized (D, L), the positional embeddings are added to the concatenated embeddings. L is the maximum length of the concatenated embeddings series, and D is the dimensionality of the patch embeddings.

$$embed = [patch; classtoken; positional] \tag{6}$$

$$normalized = Layer(embed) \tag{7}$$

The sequence picture that results from adding positional embeddings to the flattened patches is then used for input for multilayer in encoder block. Each single layer of the transformer encoder is made up of the FFN module and the MSA module. The model returns the sum of the weighted values sequence using dot product attention. It then processes the output through a residual connection, passes it through an operation (layer normalization), and finally passes it to the FFN block. The FFN have two linear layers and activation mechanism inside the transformer encoder. Before the output from the FFN block is processed by the next layer, it is first routed via a sequence of LN stages that are repeated. The last layer produces a series of feature vectors, each of which represents a different input picture patch. After that, the earlier feature vectors that represent the entire image are assembled to create a unified feature [13]. The diagram of Transformer Encoder is shown in Fig. 3.

The input vector sequence is {p1, p2,..., pn}. The output is calculated by the module in this manner: First, query, key, and value vectors are created from the input vectors:

$$Qu = Va * HA, Ke = V * HB, Va = V * HC \tag{8}$$

where H_A, H_B, and H_C are trainable weight matrices and V is the input sequence. The following equation is used to determine the dot-product attention scores:

$$Attention(Qu, Ke, Va) = softmax(Qu * Ke^T / sqrt(dim)) * Va \tag{9}$$

where dim is the dimensionality of the key vector. The weighted values [14] using the attention are the model's output. Principles:

$$Output = scores\_attention(Q, K, V) * Z \tag{10}$$

Z is a learnable weight matrix in this case.

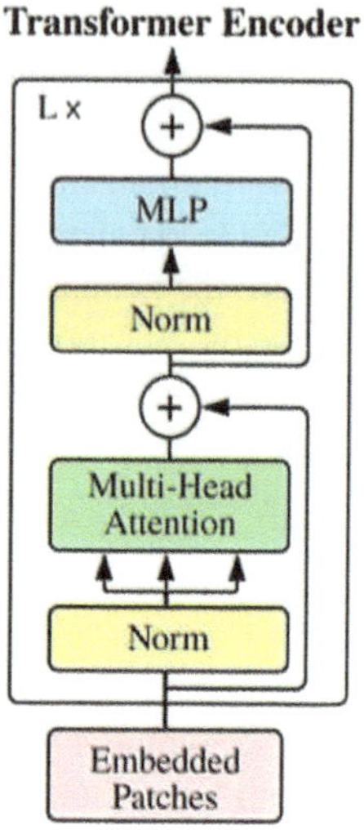

Fig. 3 Transformer encoder block diagram

4 Result and Discussion

The end goal of the project was to provide an effective method for classifying and identifying brain tumors utilizing Vision Transformers (ViT). The aim of the study was to tackle the pressing necessity of precisely and consistently identifying the different forms of brain tumors in order to facilitate diagnosis and treatment planning. The method sought to obtain accuracy in diagnosing several brain tumor classifications, such as pituitary tumor, no tumor, meningioma tumor, and glioma tumor, by utilizing the Vision Transformers. The study aims to improve brain tumor identification in the field of medical field by assessing the performance of the suggested ViT model via thorough analysis and comparison with current methods.

The success of the ViT model in classifying brain tumor pictures was assessed through the use of many measures namely recall, precision, accuracy, and F1 score. The model's remarkable validation accuracy of 98.12% showed that it could correctly identify brain tumors. In order to illustrate the model's learning process, training vs. validation accuracy and loss graphs were created, as seen in Fig. 4. The training accuracy graph showed a steady increase in accuracy over the duration of the training epochs, culminating in a remarkable 0.9813 value. Throughout the training phase, the validation accuracy fluctuated from its high starting point. Nevertheless, it finally reached a stable value of 0.97. The loss graphs for training and validation showed comparable patterns, with a decline in loss.

The confusion matrix provides the correct and incorrect categorized cases, therefore assessing the performance. Out of the 857 photos in the validation set, 841 images were properly recognized by the model whereas 16 images were wrongly predicted, as seen in Fig. 5.

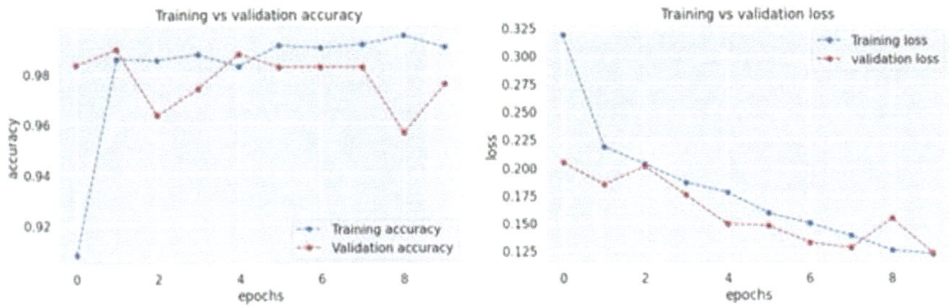

Fig. 4 Training versus validation accuracy and loss graph

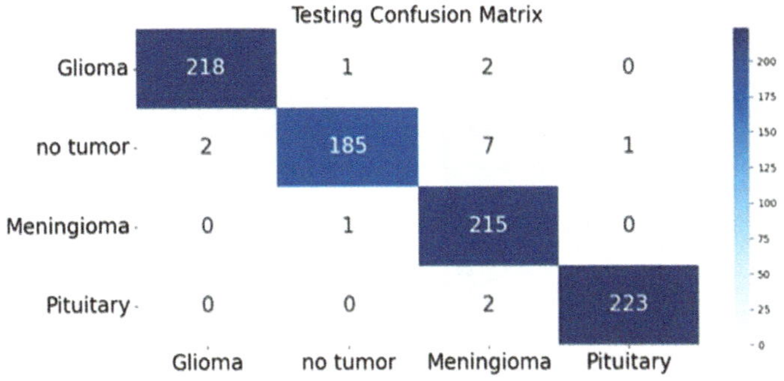

Fig. 5 Confusion matrix of ViT

Table 1 Brain tumor classification comparison

Model and author	Classifier	Precision %	Recall %	F1-Score %	Accuracy %
Fined-tuned VGG19 (Khan et al., 2021)	Softmax	91.05	89.09	90.01	94.06
EfficientNetB3 (Gopika and Samitha, 2022)	SVM	94.68	94.15	94.56	93.1
Proposed method	ViT	92.16	96.0	97.26	98.12

Table 1. Provides a numerical evaluation of brain tumor classification between an existing and the proposed system. The performance of Existing and proposed system is provided using recall, accuracy, F1-score, and precision. These parameter metrics are examined. The comparative analysis helps out in evaluating how well the proposed method improves the accuracy of brain tumor classification.

The classification result of the proposed model is demonstrating right and wrong classifications of brain tumor classes is predicted by giving test image to classify the class type is shown in Figs. 6 and 7.

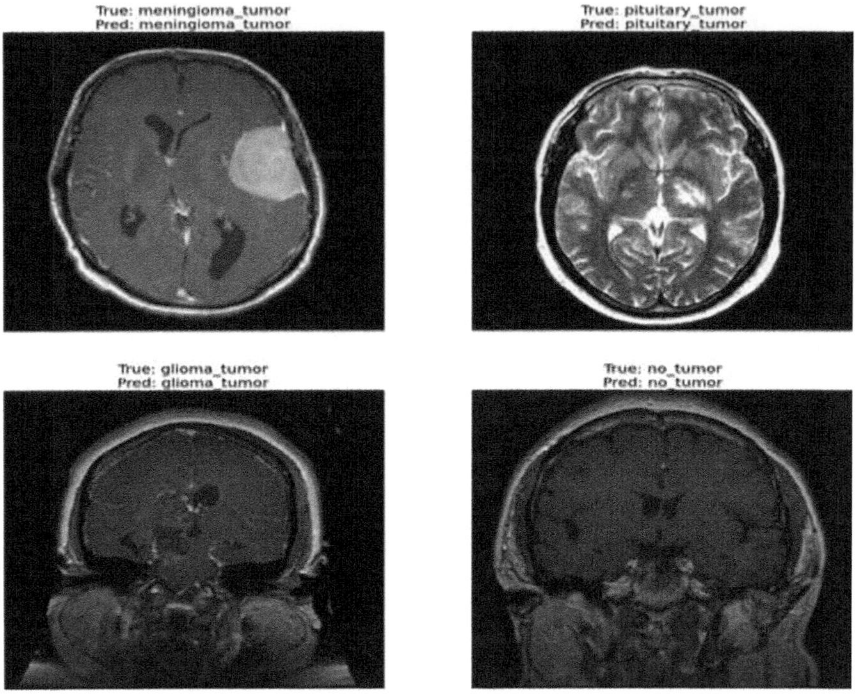

Fig. 6 Right prediction of ViT model

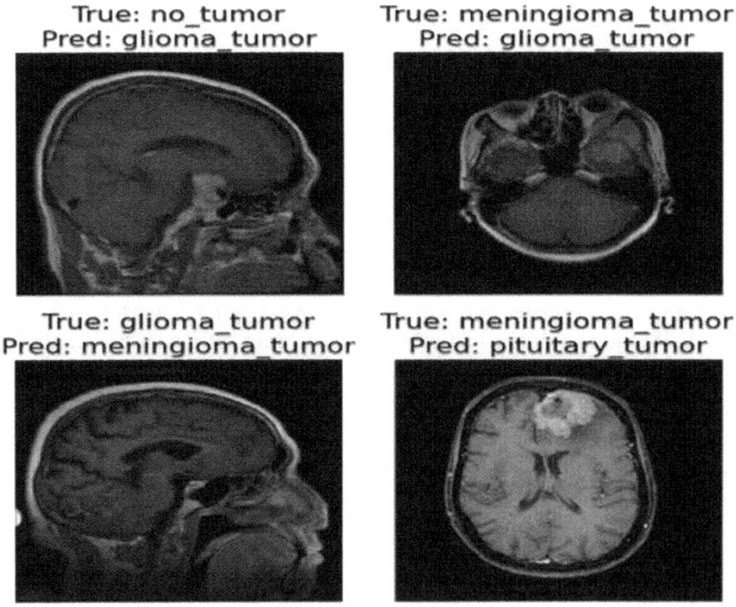

Fig. 7 Wrong prediction of ViT model

5 Conclusion

In conclusion, the outcomes of this study show how successful the suggested method is in detecting brain tumors from MRI data. This method makes use of the Vision Transformer (ViT). Four different kinds of brain tumors exist. They are glioma tumor, meningioma tumor, pituitary tumor, and no tumor that were accurately recognized by the model, which on the brain tumor dataset achieved an accuracy of 98.12%.

The accuracy metric was employed to confirm the output of the ViT model. It determines the correct percentage of recognized images with the total images. On the validation set, our model exhibits an incredible 98.12% accuracy. To graphically portray the results, two graphs were made. The train vs. validity accuracy graph shows the ViT's performance throughout testing and training, along with the percentage of accuracy reached in each phase. The word "epochs" means to the total number of times the model has run over the training dataset. The accuracy of the model is represented by the proportion of correctly predicted tumor class on the y axis of this graph.

References

1. Pichaivel, M., Anbumani, G., Theivendren, P., Gopal, M.: An overview of Brain Tumor. In Brain Tumours; IntechOpen, p. 1, (2022)
2. Ostrom, Q., Patil, N., Cioffi, G., Waite, K., Kruchko, C., Barnholtz-Sloan, J.: CBTRUS statistical report: primary brain and other central nervous system Tumors Diagnosed in the United States in 2013–2017. Neuro Oncol, **22**, pp. iv1–iv96, (2020)
3. Park, J., Park, Y.: Brain Tumor rehabilitation: symptoms, complications, and treatment strategy. Brain Neurorhabilit, **15**, e25, (2022)
4. Bosman, F.: Integrative molecular tumor classification: a pathologist's view. In encyclopedia of Cancer. Academic Press: Oxford, pp. 279–285, (2019)
5. La Rosa, S., Uccella, S.: Pituitary tumors: pathology and genetics. in encyclopedia of Cancer. Academic Press: Oxford, p. 241–256, (2019)
6. Zhang, L., Liu, Y., Huang, H., Xie, H., Zhang, B., Xia, W., Guo, B.: Multifunctional nanotheranostics for near infrared optical imaging-guided treatment of brain tumors. Adv. Drug Deliv, p. 19, (2022)
7. Swati, Z., Zhao, Q., Kabir, M., Ali, F., Ali, Z., Ahmed, S., Lu, J.: Brain tumor classification for MR images using transfer learning and fine-tuning. Comput. Med. Imaging Graph. **75**, 34–46 (2019)
8. AlBadawy, E., Saha, A., Mazurowski, M.: Deep learning for segmentation of brain tumors: Impact of cross-institutional training and testing. Med. Phys. **45**, 1150–1158 (2018)
9. Havaei, M., Davy, A., Warde-Farley, D., Biard, A., Courville, A., Bengio, Y., Pal, C., Jodoin, P.-M., Larochelle, H.: Brain tumor segmentation with deep neural networks. Med. Image Anal. **35**, 18–31 (2017)
10. Thaha, M., Kumar, K., Murugan, B., Dhanasekeran, S., Vijayakarthick, P., Selvi, A.: Brain tumor segmentation using convolutional neural networks in MRI images. J. Med. Syst. **43**, 294 (2019)
11. Polly, F., Shil, S., Hossain, M., Ayman, A., Jang, Y.: Detection and classification of HGG and LGG brain tumor using machine learning. In: Proceedings of the 2018 International Conference on Information Networking (ICOIN), Chiang Mai, Thailand, (2019)

12. Dosovitskiy, A., Beyer, L., Kolesnikov, A., Weissenborn, D., Zhai, X., Unterthiner, T., Dehghani, M., Minderer, M., Heigold, G., Gelly, S., et. Al.: An image is worth 16 × 16 words: Transformers for image recognition at scale. arXiv, (2020)
13. Tummala, S., Kadry, S., Bukhari, S., Rauf, H.: Brain tumor classification using a combination of variational autoencoders and generative adversarial networks. Biomedicines **10**, 203 (2022)
14. Tummala, S., Kadry, S., Bukhari, S., Rauf, H.: Classification of Brain Tumor from magnetic resonance imaging using vision transformers ensembling. Curr. Oncol. **29**, 7498–7511 (2022)
15. Asiri, A., Aamir, M., Shaf, A., Ali, T., Zeeshan, M., Irfan, M., Alshamrani, K., Alshamrani, H., Alqahtani, F., Alshehri.: Block-Wise neural network for brain tumor identification in magnetic resonance images. Comput. Mater. Contin, **73**, pp. 5735–5753, (2022)

Machine Learning and Sensor-Based Real-Time Fire Detection Using YOLOv8

Ipsita Ankita Hota and Ranjan Kumar Dash

Abstract Fire causes numerous disasters and is a giant waste of resources and daily goods for humans. Therefore, automatic fire detection is needed to save mortal life. Here, proposed model adopt object detection using YOLOv8 with the help of a machine-learning approach has been designed which uses a temperature sensor and Arduino for sensing a certain range of temperature. The sensor senses the temperature of the surroundings if it crosses the desired range, an alarm is sounded and an image is captured using a camera, which is then sent to the email address of the user and also sent an SMS call notification. The performance of the proposed YOLOv8 has been evaluated by considering different metrics. The design of the approach is discussed elaborately followed by the flowchart and its working block diagram. A real-time experiment has been conducted to show the efficiency of the proposed approach to detect the fire.

Keywords Object detection · Temperature sensor · Arduino · YOLOv8 · Alarm

1 Introduction

Traditional fire safety systems typically consist of manual devices and procedures. These may include fire extinguishers, fire alarms, smoke detectors, and sprinkler systems. These systems rely on human intervention to detect and respond to a fire incident. For example, when a smoke detector detects smoke, it triggers an alarm, alerting occupants to the potential danger. It is then up to individuals to take appropriate action, such as evacuating the building or using fire extinguishers. On the other hand, automated fire safety systems utilize advanced technologies and automation to detect, control, and suppress fires.

These systems often incorporate intelligent sensors, networked devices, and centralized control panels. They can automatically detect fires, activate alarms, and

I. A. Hota (✉) · R. K. Dash
Odisha University of Technology and Research, Bhubaneswar Odisha, Bhubaneswar, India
e-mail: ipsitaankita2000@gmail.com

S. Kumar Udgata et al. (eds.), *Intelligent Systems*, Lecture Notes in Networks and Systems 1149, https://doi.org/10.1007/978-981-97-8160-7_11

initiate suppression mechanisms without human intervention. For instance, an automated system may use heat or flame detectors to identify a fire and automatically activate sprinklers or fire suppression agents. Automated fire safety systems offer several advantages over traditional systems. They can provide faster response times, as they can detect fires and initiate appropriate actions immediately. They can also minimize human error and ensure a more consistent and reliable response. Additionally, automated systems can be integrated with other building management systems, such as HVAC or access control, for enhanced safety and efficiency. This study embarks on the development of a state-of-the-art Fire Detection System, employing advanced sensor technologies, to provide a more efficient and reliable means of identifying and alerting to potential fire hazards. Here is a novel approach leveraging the YOLOv8 architecture, a state-of-the-art object detection framework for accurate and rapid-fire detection. Surveillance cameras are commonly used nowadays, vision-based cameras are deliberated in existing surveillance cameras and help to inspect fire as well as smoke.

The escalating demand for effective fire detection systems necessitates innovative solutions that leverage both computer vision and sensor technologies. The proposed system introduces a fusion of YOLOv8, renowned for its robust object detection capabilities, with specialized temperature sensors for comprehensive fire detection. Fire detection lends a helping hand to not only humans but also every creature. It looks replica-like a safeguard worldwide. The primary goal is to develop and evaluate Machine learning and sensor-based fire detection systems that harness the strengths of YOLOv8 and sensor technology to enhance accuracy, speed, and reliability in identifying potential fire accidents.

2 Literature Survey

Nowadays technology strengthens the quality of research and assists better development in different research sectors. The authors [1] proposed a model that used Raspberry Pi, which distinguishes the flames and HSV color combination used for flame color texture study, it detects some other object in color combination which causes a false alarm notification.

A cost-effective fire detection model is proposed but some future work will be done in the field of robust privacy and data mining prevention by the authors [2]. The authors [3] have established a system consisting of tightly secure remote flame recognition. The authors [4] proposed a decent methodology that is concomitant of computer vision and HSV color space property.

Using the SFF, automatic fire extinguisher system was proposed by authors [5]. The work [6] used YOLOv4 for automatic fire detection for blind and visually impaired people. Computer vision and smoke sensors have been used [7] for detecting fire.

The work carried out in [8] used BLE RSSI and accessible pedestrian signals for fire detection for visually impaired people. The authors [9] proposed a Finger Sight: A Vibrotactile Wearable Ring.

A fire detection mechanism for indoor guided navigation has been proposed in [10] for blind people. An adaptive intelligent alarm system has been designed [11] for wireless sensor networks. The work [12] proposed an approach for an outdoor navigation system.

3 Proposed YOLOv8-Based Fire Detection Model

The proposed model uses YOLOv8 for real-time detection of fire from the captured images.

3.1 YOLOv8

YOLOv8 is an object detection algorithm that represents a significant advancement in the field of computer vision. YOLOv8 combines efficiency and accuracy in detecting objects within images or video frames. Its distinguishing feature lies in its ability to process images in real-time, making it well-suited for applications such as security systems, and various other tasks. It processes the entire image in a single forward pass. This enables it to achieve remarkable speed and efficiency in real-time applications, without compromising on accuracy.

3.2 Data Set Used

Data have been collected from Kaggle, which is in Yolo format. Yolo format is a specific data format that helps to implement an object detection model. https://www.kaggle.com/datasets/ankan1998/fire-detection-in-yolo-format. The data set is taken from this. In Yolo format, each image contains a text file that consists of class labels and bounding box coordinates.

3.3 Training and Validation

The YOLOv8 architecture is adapted to fire detection tasks, undergoing rigorous training with the datasets. In the proposed model, 80% of training and 20% of testing data are taken for better validation.

Fig. 1 Temperature sensor pins

It partitions the given input images into grid dimensions when the center of an object drops within a specific grid cell that cell is tasked with the responsibility of detecting that object. Within each grid cell, Yolo makes predictions for bounding boxes and assigns confidence scores to assess the validity of these predictions. After successfully training the model, it provides "best.pt" file which supports fire detection.

4 Proposed Sensor-Based Real-Time Detection of Fire

4.1 Sensor

In the proposed system, a DHT11 temperature and humidity sensor is used. It stands as a testament to simplicity and efficiency in temperature and humidity measurement. This compact device is highly accurate for reading temperature. It delivers highly accurate data, which makes it an indispensable tool. It contains three pins, Ground, Data, and Vcc, as shown in Fig. 1.

These pins are connected with Arduino to provide accurate temperature information. An LED light is present in the temperature sensor as shown in Fig. 2. Here Led light is turned on when it is effectively connected to the system.

4.2 Arduino

Arduino is an open-source hardware and software ecosystem centered around a microcontroller, which is a small programmable chip that acts as the main part of the system. These boards are equipped with input and output pins as shown in Fig. 3.

Fig. 2 Temperature sensor with LED

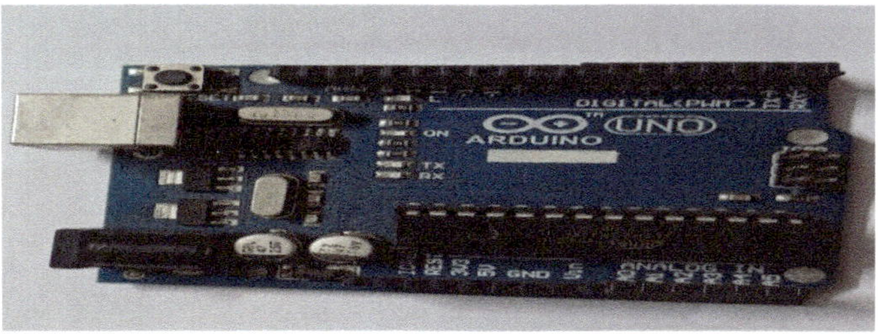

Fig. 3 Arduino UNO board

Arduino is a user-friendly Integrated Development Environment (IDE). In the proposed system, Arduino is an essential tool for detecting temperature through the sensor.

4.3 Jumper Wires and Data Cable

In the proposed system, jumper wires are essential for connecting the temperature sensor with Arduino. Jumper wires play a pivotal role in simplifying the ease of circuit design and eliminating crucial soldering. Data cable is also known as data cord or data wire. It transmits digital information between electronic devices. In the proposed system, USB A-B cable is used.

4.4 Experimental Setup for Fire Detection

In Fig. 4, the temperature sensor is connected with jumper wires with Arduino. Due to a successful connection LED light is turned on. Arduino is connected to the system (computer, laptop) by a USB A-B cable. This architecture helps to distinguish the current temperature effectively.

In the depicted Fig. 5, the schematic illustrates the basic configuration of the proposed system, featuring the temperature sensor DHT11 denoted as "Tmp." The sensor's data point pin establishes connections with an Arduino, while a 5V power supply and ground connection are linked to the corresponding ports on the Arduino board. This setup enables the Arduino to interface with the DHT11 sensor for temperature-sensing functionalities.

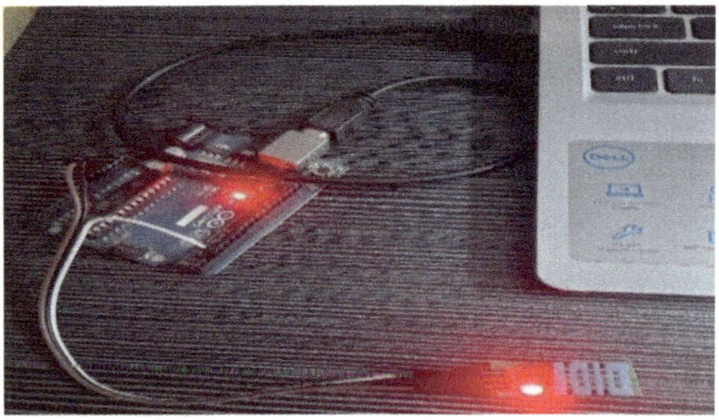

Fig. 4 Arduino and temperature sensor connection

Fig. 5 Circuit diagram

4.5 Flowchart

Figure 6 shows how the proposed system works efficiently. In the first step, the system asks for an email address, and then the connected camera is opened and ready to detect the fire, if the fire is not detected then the camera continues its detection process Otherwise if the fire is detected then check the temperature, if it is unendurable and high (temp > 47) then the camera captures the fire-causing images and send them to the email address with SMS, call notification and rings the alarm, if the temperature is not high then only alarm rings.

4.6 Block Diagram

In Fig. 7, block diagram displays the flow of data, initially, data are collected in Yolo format then it undergoes training by using YOLOv8 and preparing an efficient model which helps to detect fire, and a best.pt file is generated, which is used in Python for implementation. In the next fire prediction stage, camera will open and check whether the fire is present or not if fire is detected then go to the next phase temperature sensor connection where Arduino coordinates the temperature sensor for detecting the temperature of the surrounding. In the final stage, system checks the temperature above 47 degrees then a captured image is sent to the email address. An alert SMS and call notification will go to the user.

5 Results and Discussion

5.1 Performance of the Proposed Fire Detection System

The following metrics are used to evaluate the performance of the proposed YOLOv8 model:

1. Box_loss—It is a loss that measures the overlapping level of predicted bounding boxes and the ground truth objects.
2. Obj_loss—It stands for objectness loss. It is a measure of the capacity of the model to predict the location and class of an object.
3. Dfl_loss—It stands for distribution focal loss, which is a focal loss that helps improve model performance.
4. Cls_loss—It stands for classification loss. It is the sum squared error during the prediction of the classes.
5. Precision—It is a measure indicating the number of true positives versus the number of positive predictions.
6. Recall—It is the true positive rate of a classifier.

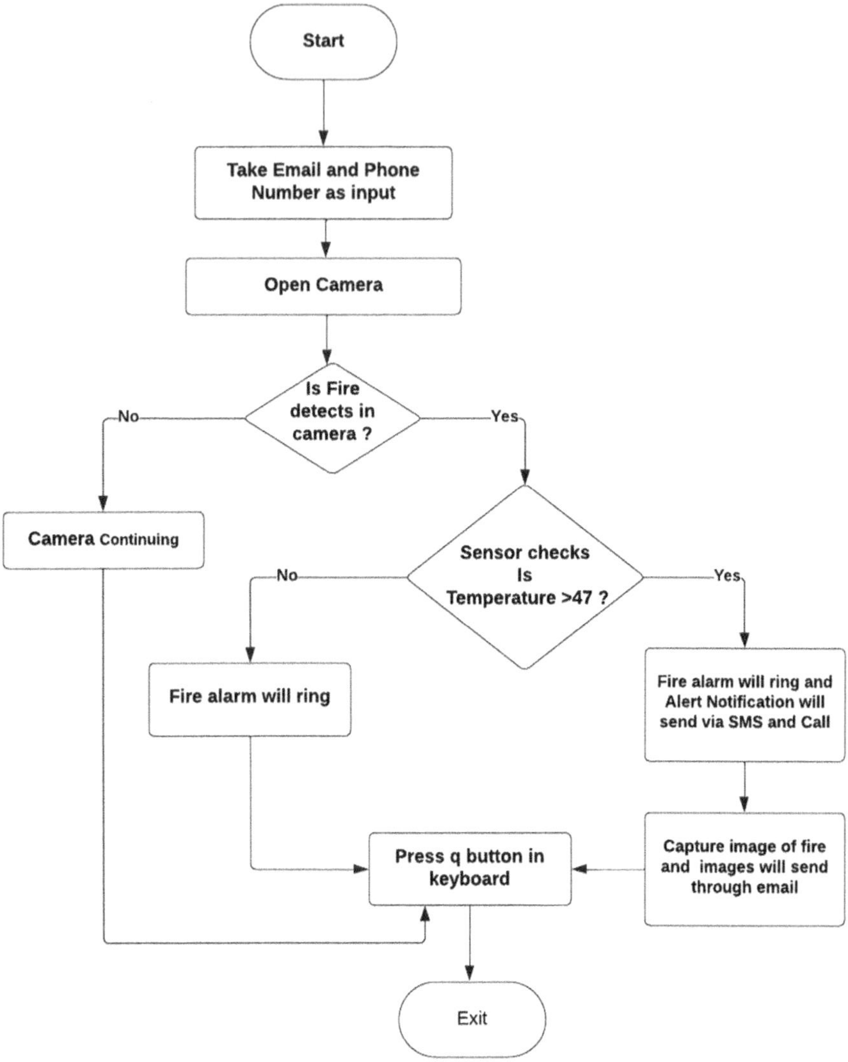

Fig. 6 Flowchart of the proposed system

7. Mean average precision (mAP)—It is a metric used to measure the performance of a model for an object detection task (Fig. 8).

The below metrics are responsible for evaluating the model in different confidence levels and distinguishing the optimized performance of the model.

Fig. 9 represents how confidence threshold affects the recall rate. It shows the relationship between recall and confidence score. Increasing confidence lowers the recall rate.

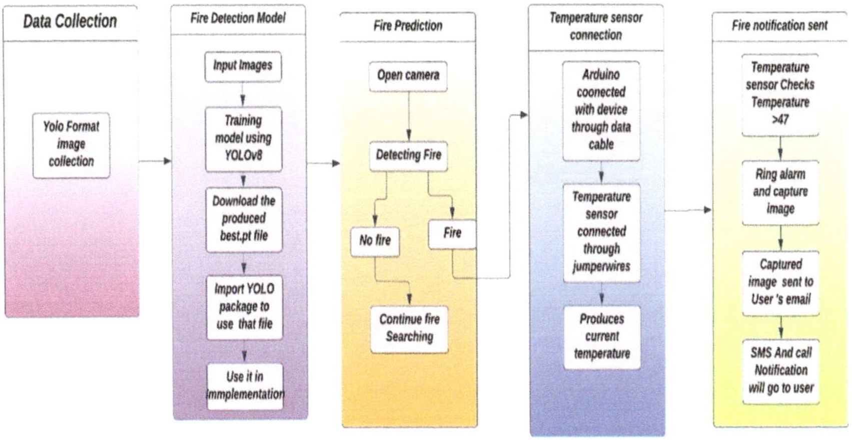

Fig. 7 Block diagram of the proposed system

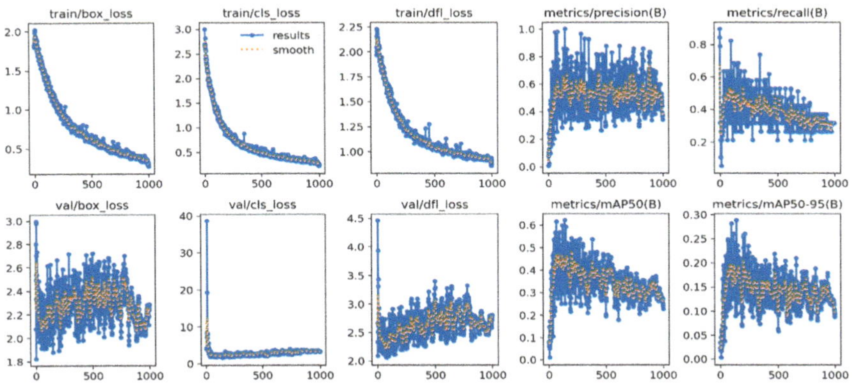

Fig. 8 Performance of the proposed YOLOv8 model

Fig. 10 illustrates the relationship between precision and confidence scores. Precision represents the accuracy of positive predictions among all predicted positives at varying confidence thresholds. Precision-confidence curve adjusting the confidence threshold directly influences the precision. Increasing threshold confidence leads to higher precision. It helps to analyze the optimal confidence threshold that balances precision.

The F1 score is a harmonic mean of precision and recall. Fig. 11 shows how changes in confidence threshold affect the trade-off between precision and recall. High F1 scores indicate better overall performance in object detection.

Fig. 12 curve showcases the trade-offs between precision and recall at varied thresholds.

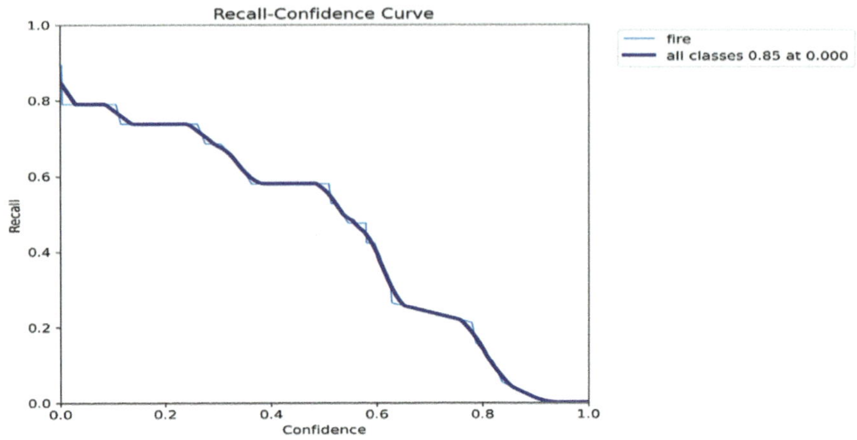

Fig. 9 Recall-confidence curve

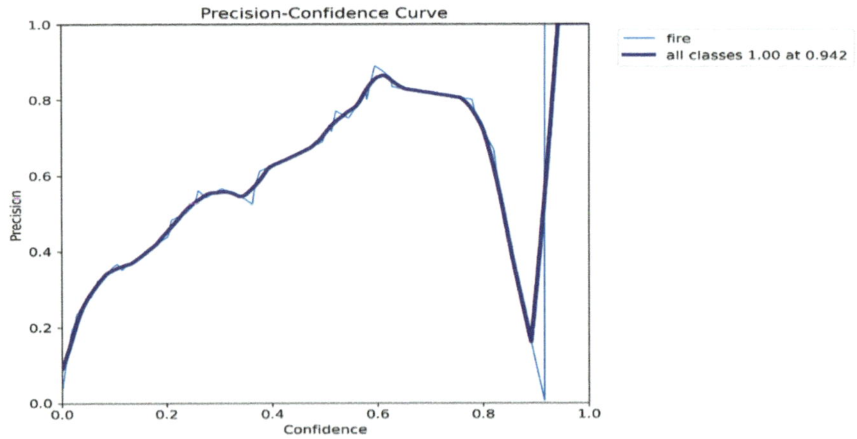

Fig. 10 Precision-confidence curve

In Fig. 13, matrix identifies the false alarm rate that is 0.26, which means the number of incorrectly predicted positive cases.

5.2 Performance Comparison of YOLOv8 Against Some State of Art Models

For comparison, the average precision is calculated by using the following.

formula [6]:

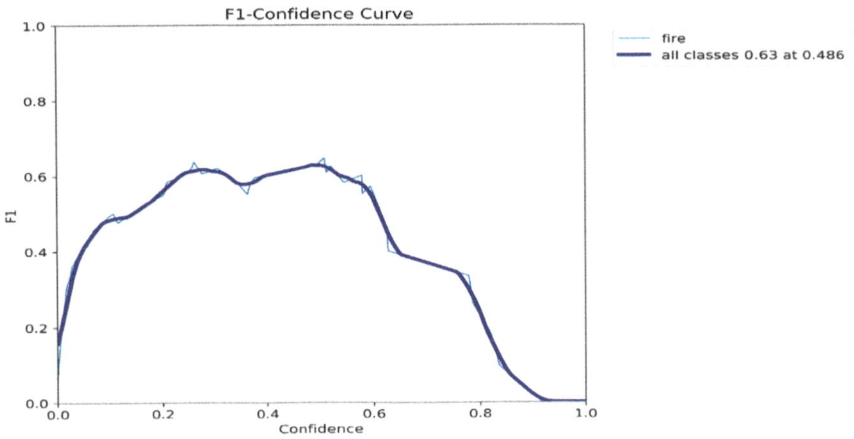

Fig. 11 F1-confidence curve

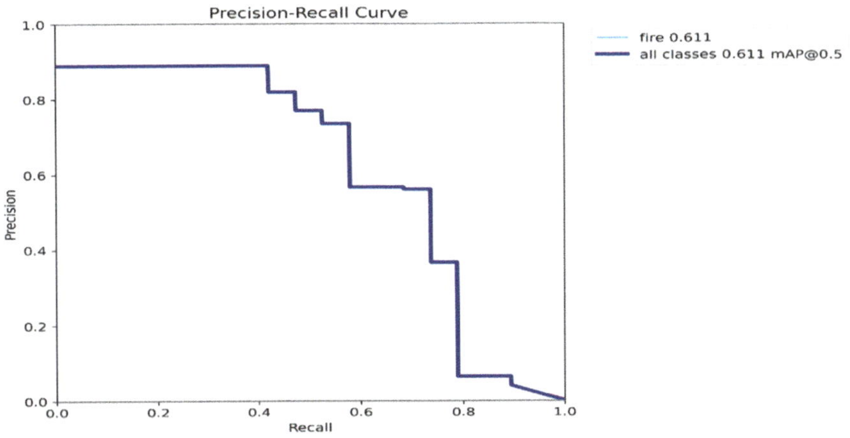

Fig. 12 Precision-Recall curve

$$AP = \frac{1}{n} \sum_{i=1}^{n} P_i \quad (1)$$

The proposed work is compared against some existing models like YOLOv3 tiny, YOLOv4 tiny, YOLOv3, and YOLOv4 [6] in terms of average precision (AP) during the training and validation of the models (Table 1).

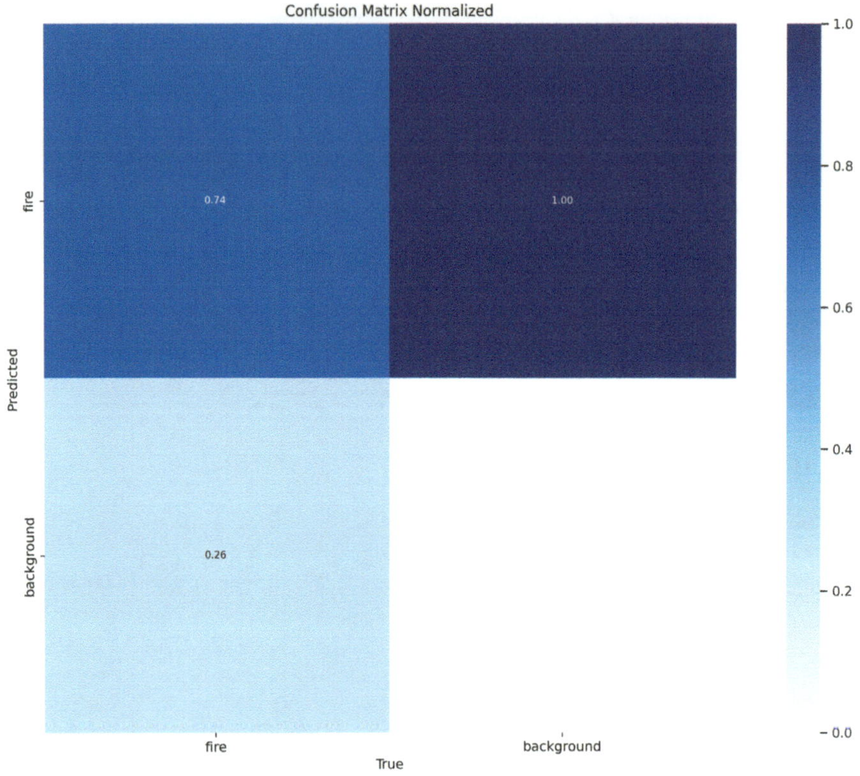

Fig. 13 Normalized confusion matrix

Table 1 Comparison against the existing models

Models	AP (%)	Training time
YOLOv3	39.6	110h
YOLOv3-tiny	21.7	15h
YOLOv4	47.2	102h
YOLOv4-tiny	26.4	12h
Improved YOLOv4	48.6	97h
Proposed YOLOv8	51.4	1.649h

5.3 Real-Time Detection of Fire

The proposed approach is used to detect fire in real time. The sensor senses the temperature and triggers the webcam which is fitted with the computer. The proposed YOLO model detects the fire and subsequently, an email is sent to the appropriate recipient. The GUI used in this work is designed by using the Qt Designer tool and implemented in the proposed system with the help of a package named PyQt5, which enhances communication with the user. Twilio is used in Python for enhancing the alerting performance by SMS and call Features.

In Fig. 14, fire is detected through the camera; the temperature sensor measures the temperature of the room if it is unendurable more than 47 degrees then the system captures the images and shows the fire-detecting image name as shown in Fig. 15.

The captured image sent into the email address is shown in Fig. 16 and alert SMS notification is placed as shown in Fig. 17 and a call alert notification is also placed.

Fig. 14 Fire-detected images

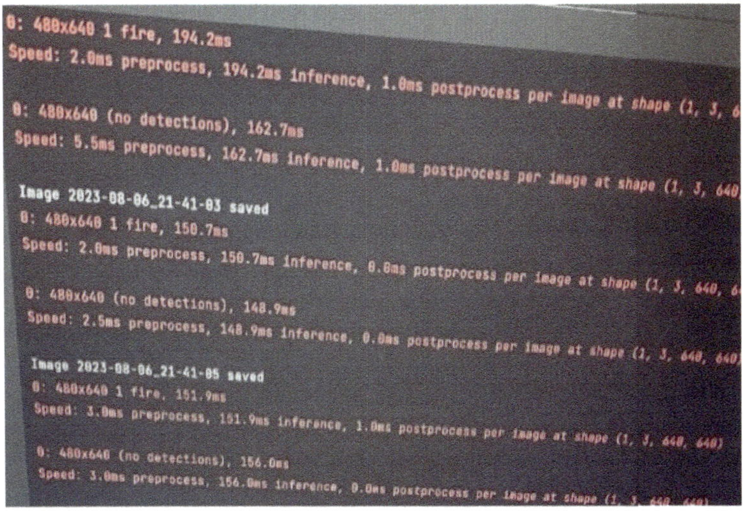

Fig. 15 Image saved in device

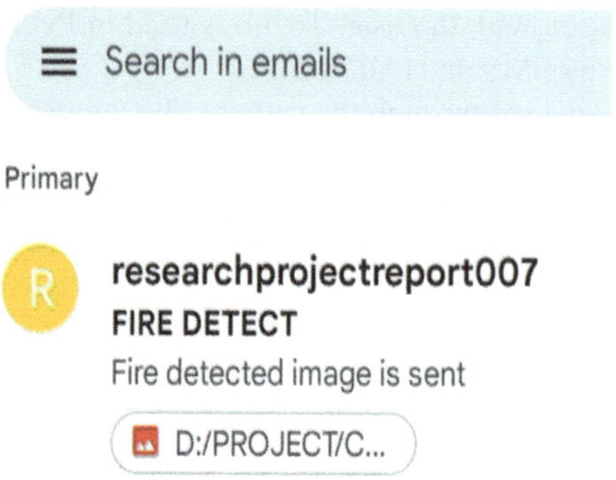

Fig. 16 Captured image sent via email

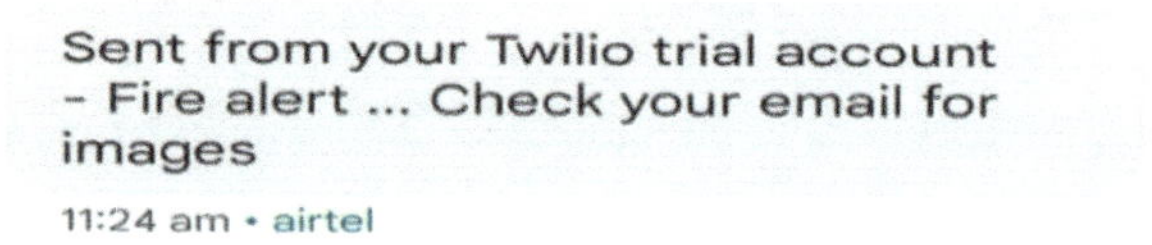

Fig. 17 SMS notification sent

6 Conclusion and Future Work

In this era, technology spread its brilliance all around. Lots of effort has been taken to make the system more flexible. User interaction is simplified through the design of intuitive graphical user interfaces (GUIs), which promote ease of use and seamless engagement with the system. This technological advancement holds promising potential for assisting individuals dealing with fire accidents in their daily lives, offering a cost-effective solution that significantly reduces false alarms. In the contemporary era, the pervasive influence of technology has illuminated various facets of our lives. Significant strides have been made to enhance system adaptability, with a keen focus on user-friendly graphical user interfaces (GUIs) designed for effortless interaction. This innovation not only proves cost-effective but also holds the potential to assist individuals encountering fire accidents in their daily lives. Notably, it contributes to a reduction in false alarms, optimizing the system's reliability. The proposed system contains an implementation of capturing images of fire and sent via email which will show the actual situation of a fire accident. Provision of user-friendly GUI (Graphical User Interface) will interact with users easily.

Looking toward the future, there are plans to further improve the system's security and accuracy, ensuring timely awareness and response. Additionally, there are considerations to integrate automated fire extinguisher activation within the system for emergency purposes. When a fire is detected, the system autonomously initiates the fire extinguisher, swiftly containing the flames and mitigating potential damage. Simultaneously, an emergency call would be placed to a designated emergency number, promptly alerting the required responders or authorities for immediate action. These envisioned developments signify the commitment to continually enhance the capabilities of the fire detection system, aiming for heightened security and accuracy. By incorporating features such as automated fire extinguishers, and emergency call activation, the system becomes an even more comprehensive and proactive solution, offering advanced response measures in critical situations. Through these future improvements, the system will provide enhanced protection, reducing response times and minimizing potential harm caused by fire incidents. The seamless integration of technology and proactive measures in fire safety demonstrates a commitment to safeguarding lives and property, fostering a safer and more secure environment.

Due to the limited images available in the dataset, the average precision can be improved by taking real-time image datasets in the future. It is a quite complex methodology because capturing the real-time images is not sufficient for training the YOLOv8 model that will be properly converted into Yolo format.

References

1. Khan, M.N.A., Tanveer, T., Khurshid, K., Zaki, H., Zaidi, S.S.I.: Fire detection system using raspberry pi. In 2019 International Conference on Information Science and Communication Technology (ICISCT). IEEE. pp. 1–6 (2019)
2. Khan, R.H., Bhuiyan, Z.A., Rahman, S.S., Khondaker, S.: A smart and cost-effective fire detection system for developing country: an IoT based approach. Int. J. Inf. Eng. Electron. Bus. **10**(3), 16 (2019)
3. Khalaf, O.I., Abdulsahib, G.M., Zghair, N.A.K.: IOT fire detection system using sensor with Arduino. AUS **26**, 74–78 (2019)
4. Qi, X., Ebert, J.: A computer vision-based method for fire detection in color videos. Int. J. Imaging **2**(S09), 22–34 (2009)
5. Mobin, M.I., M. A. R., Islam, M., & Hasan, M.: An intelligent fire detection and mitigation System Safe from Fire (SFF). Int. J. Comput. Appl. **133**, 86–92 (2016)
6. Mukhiddinov, M., Abdusalomov, A.B., Cho, J.: Automatic fire detection and notification system based on improved YOLOv4 for the blind and visually impaired. Sensors **22**(9), 3307 (2022)
7. Chowdhury, N., Mushfiq, D.R., Chowdhury, A.E.: Computer vision and smoke sensor-based fire detection system. In 2019 1st International Conference on Advances in Science, Engineering and Robotics Technology (ICASERT). IEEE. pp. 1–5 2019
8. Shin, K., McConville, R., Metatla, O., Chang, M., Han, C., Lee, J., Roudaut, A.: Outdoor localization using BLE RSSI and accessible pedestrian signals for the visually impaired at intersections. Sensors **22**(1), 371 (2022)
9. Satpute, S.A., Canady, J.R., Klatzky, R.L., Stetten, G.D.: FingerSight: A vibrotactile wearable ring for assistance with locating and reaching objects in peripersonal space. IEEE Trans. Haptics **13**(2), 325–333 (2019)
10. Plikynas, D., Indriulionis, A., Laukaitis, A., Sakalauskas, L.: Indoor-guided navigation for people who are blind: Crowdsourcing for route mapping and assistance. Appl. Sci. **12**(1), 523 (2022)

11. Salman, A.D., Khalaf, O.I., Abdulsahib, G.M.: An adaptive intelligent alarm system for wireless sensor network. Indones. J. Electr. Eng. Comput. Sci. **15**(1), 142–147 (2019)
12. Martínez-Cruz, S., Morales-Hernández, L.A., Pérez-Soto, G.I., Benitez-Rangel, J.P., Camarillo-Gómez, K.A.: An outdoor navigation assistance system for visually impaired people in public transportation. IEEE Access **9**, 130767–130777 (2021)

Leafy Tor-Cube: A New Scalable Hybrid Interconnection Network for High Performance Computers

Rashmita Padhi and Nibedita Adhikari

Abstract This paper presents a new scalable interconnection topology called the Leafy Tor Cube (LTC) for high end computing system. As compared to the other interconnection networks LTC is found to be more attractive in terms of topological attributes like degree connectivity, edges, average node distance and message traffic density etc. It helps in improving the node packing density for high performance computing. Our proposed network is highly scalable and also robust in nature. The different types of performance metrics show that the LTC is a better candidate for distributed processing and high performance computing than its predecessors. As the number of nodes of the proposed network is sufficiently more so this network is a preferable one for high performance computing (HPC) with larger volume of communication. Also it is shown that LTC network is more scalable than other important topologies. With increase in node count, the message density for LTC is less compared to some other popular topologies like hyper cube(HC), crossed cube(CC), leafy cube(LC). Less message traffic density also proves that the LTC is an efficient network when traffic congestion is considered. The degree of LTC is more because of child nodes which can be a preferable backup in high performance computing scenario. This property shows that LTC network is more tolerant to fault and reliable than LC. LTC has decreased diameter as compared to LC. The routing and broadcasting algorithms for LTC are proposed with lesser time complexity.

Keywords Fault tolerant · Packing Density · Message traffic density · Terminal Reliability · Routing and broadcasting · Scalability

R. Padhi (✉)
Integrated M.C.A, Utkal University, Bhubaneswar, India
e-mail: rimi.padhi@gmail.com

N. Adhikari
Post Graduate Department of Computer Science and Applications, Utkal University, Bhubaneswar, India

S. Kumar Udgata et al. (eds.), *Intelligent Systems*, Lecture Notes in Networks and Systems 1149, https://doi.org/10.1007/978-981-97-8160-7_12

1 Introduction

The interconnection network, which establishes the means of data interchange among numerous isolated processing units, is an essential platform of high performance computing systems. The interconnection network topology of these systems plays a key role in their overall performance. Thus, in order to create a highly dependable high performance computing system, it is necessary to ensure that the interconnection network being utilized is highly reliable, meaning that it should continue to function well for a said amount of time even when few links or processing units fail. Additionally, these systems should utilize interconnection networks economically [8–13]. The cost of the connectivity network can be calculated by multiplying node degree with network diameter. In other words, it may be claimed that designing such systems requires the deployment of an interconnection network that is both extremely reliable and cost-effective. Hypercube and its variants like Torus, Mesh etc. are some examples of interconnection networks [1, 2, 14–19]. Due to its characteristics, including its tiny diameter, increased degree connectivity, symmetry, expandable for higher dimension, partition ability, fault tolerance, and dependability, the hypercube (HC) [25], has been very popular. The Hypercube has undergone numerous modifications in the past, either to increase its dependability or lower its cost. The Researchers have suggested a variety of hierarchical connectivity networks using the hypercube, including extended hypercubes, hierarchical crossing cubes, and hierarchical cube networks [4–7, 25]. The inherent scale constraint of the hypercube and its derivatives, however, is the fundamental issue. Mesh and torus are the two popular topologies that offer a great level of scalability. In the fields of massive calculations, grid computing, distributive computing, etc., these networks are discovered to be commonly used. The architecture of a massively parallel computer system cannot use a mesh network due to its huge link complexity. It is additionally challenging to deploy on actual parallel machines due to the network's high cost and difficulties in setup and maintenance. Torus is more dependable and has higher connection than mesh [14–19]. With topological criteria including dimension, lower diameter with equal packing density, average node distance, and message traffic density, the Tor cube is seen as a better network. It is still difficult to create a network that is highly scalable, affordable, highly dependable, fault resistant, and has good connectivity with less complex technology. Hence, there stems a need to suggest a more superior and scalable topology. In an order to meet such demand the current paper attempts to suggest an improved topology called Leafy Tor-Cube (LTC) which is based on the properties of the popular interconnection topology Tor Cube [23] and Leafy cube [24]. Most of the significant Tor Cube and hypercube features are inherited by the suggested topology LTC.

However, some of the key considerations that must be made in the design of any interconnection architecture are terminal reliability, error free, and various performance metrics including cost factor, effectiveness of cost and time-cost-effectiveness. This demand prompts the current study to suggest the Leafy Tor-Cube, a new hybrid scalable and fault-tolerant network (LTC). This paper is primed as follows: The

Segment II presents the background of the current work. and the construction of the proposed network. In segment III the topological parameters are offered by using graph theoretical symbolizations. The reliability, routing and broadcasting analysis also presented in this section . In Segment IV a detailed comparison of various parameters with that of the predecessor networks is done. Then Segment V concludes the paper with a scope for further research.

2 Proposed Interconnection Topology: Leafy TOR-Cube (LTC)

The presented topology is a hybrid network which uses a condition by taking the union of two standardized topologies for building a new hybrid interconnection network. The two standard base networks are Tor-Cube(TC) and Leafy Cube(LC). Tor-Cube [23] is the parent network which uses an HC as its predecessor. Few extra edges from a torus topology are added to the HC topology to form TC. To construct a three dimensional TC, initially we have to consider a 3D hypercube (HC) which is presented in the Figure 1, and then additional four edges are added to the base HC similar to the torus interconnection topology. The extra edges are linked to the adjacent opposite processing elements of TC which is presented in the Figure 3 (Fig. 2).

Fig. 1 Hypercube

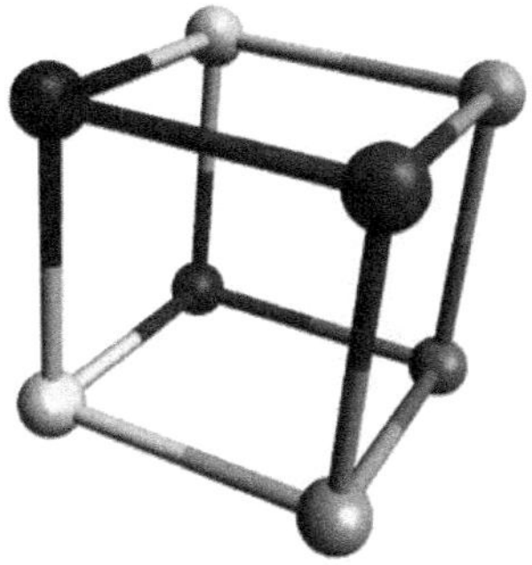

Fig. 2 Torus

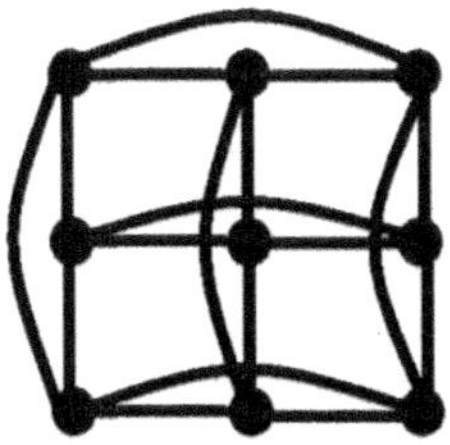

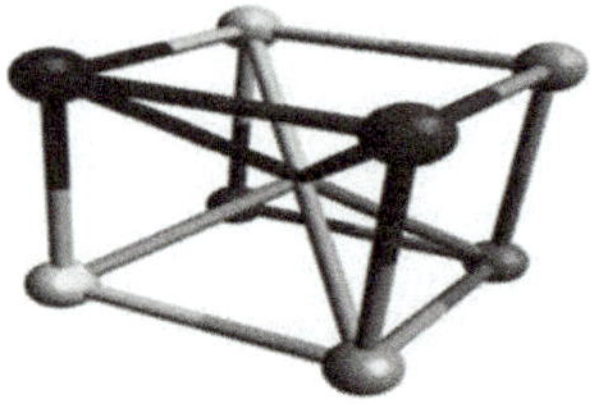

Fig. 3 Tor-Cube

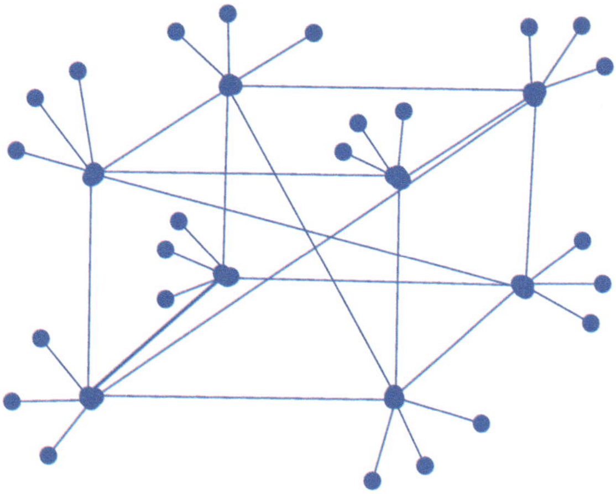

Fig. 4 Leafy Tor cube of dimension 3, LTC(3)

In the proposed network topology, n number of child nodes are added to each vertex of the Tor-Cube network, to derive a new network. Thus it is the hybrid of Tor-Cube similar to the recently proposed network Leafy Cube [24]. So, it is designated as Leafy Tor-Cube(LTC).

Leafy Tor-Cube Interconnection Network Architecture

In this section, for construction of LTC, n dimensional Tor-Cube interconnection network is considered as the base. In this structure additional n numbers of nodes are linked to each processing element. The derived structure which is designated as LTC is shown in the Figure 4.

3 Topological Parameters of Leafy Torcube Interconnection Network

The LTC can be regarded as an undirected graph $\varphi(\nu, e)$, where, $\boldsymbol{\nu}$ represents the set of nodes and e represents the set of communicating links which connects the processing elements. The computational capabilities of the LTC are described through various attributes in this part.

A. **Node Degree**

The degree of a topology is nothing but the degree of the processing elements which has a extreme number of edges associated to it. In other words in a graph the degree of nodes is defined as the total number of connection it has to other nodes or these are the immediate nodes at one hop distance.

Proposition 1:
The degree of node of as Leafy Tor-Cube of dimension n that is LTC(n) is $d_g = 2n+1$

Proof:

The node degree of Tor-Cube is $n + 1$ Incase Leafy Tor-Cube there are additional n number of child nodes are attached to each vertex of the basic Tor-Cube. So in LTC (n) each vertex will have $n + 1$ neighbors and n leaf nodes. Hence, the degree of the proposed LTC(n) is $n + n + 1$

$$d_g = 2n + 1 \tag{1}$$

B. **Node**

The node of an interconnection network is fundamental unit of the graph.

Proposition 2:
The total collection of nodes of Leafy Tor-Cube is $V = 2^n(n + 1)$

Proof:

In the basic structure that is Tor-Cube (n) network there are 2^n number of nodes. In Leafy Tor-Cube, n numbers of child nodes are attached to each vertex of the base Tor-Cube structure. So, in total there are $n * 2^n$ child nodes. Hence the total number of nodes in LTC(n) is derived as follows:

$$\begin{aligned} V &= 2^n + n \times 2^n \\ &= 2^n(n + 1) \end{aligned} \tag{2}$$

C. **Links**

The communicating link can be defined as edge connecting two processing elements within a topology.

Proposition 3:
The sum of communicating links in Leafy Tor-Cube is

$$E = \begin{cases} n2^n(n+1)\ \mathit{if}\ n = 3 \quad \mathit{and} \\ n * \left(2 \times L_{n-1} + 2^{n-1}\ \mathit{if}\ n > 3 \right. \end{cases}$$

Proof:

In the Tor-Cube (n) network, the total number of edges or links can be defined as follows:

$$L_n = \begin{cases} 2 \times L_{n-1} + 2^{n-1} & for\, n > 3 \\ 2^{n+1} & for\, n = 3 \end{cases} \quad and$$

where L_n is the number of links in the network

n is the dimension of the network

In LTC there are 'n' no of extra edges added to connect the child nodes to the existing TC nodes.

Hence total number of links can be computed as follows:

Case I :

$$if\ n = 3 \quad \begin{aligned} Ln &= n \times 2^{n+1} + n \times 2^n \\ &= n \times 2^{n(n+1)} \end{aligned} \tag{3}$$

Case II :

$$if\ n > 3 \quad Ln = n\left(2L_{n-1} + 2^{n-1}\right) \tag{4}$$

D. **Diameter**

The diameter of any undirected graph is the shortest distance between any two farthest processing elements. It is treated as an important performance metric to evaluate the efficiency of any network. For message broadcasting to every other processing elements in the topology, diameter should be retained less.

Proposition 4:
The diameter of LTC(n) is $\mathfrak{D} = n + 1$.

Proof:

As we know that the diameter of the base structure that is Tor-cube is $n - 1$. In the Leafy Tor-Cube i.e LTC (n) each node has n number of leaf nodes at one hop distance. So any two farthest nodes in LTC will take 2 plus the diameter of the Tor-Cube because the farthest nodes are possibly the leaf nodes. So the distance between the two farthest nodes is $n - 1 + 2$ which is equal with $n + 1$.

E. **Cost**

Cost of a network is another important performance metric to evaluate the efficiency of any network. The cost of a network can be computed by multiplying node degree with the diameter of the specified network. Cost should be remained minimum as it is directly proportional to the hardware complexity of any network that means hardware complexity increases with increase in cost.

Mathematically, Cost = Degree × Diameter

Proposition 5:
The cost of Leafy Tor-Cube is $\xi = 2n^2 + 3n + 1$

Proof:

The cost of LTC can be calculated as below

$$\text{Cost} = \text{Degree} \times \text{Diameter}$$

$$= (2n + 1) \times (n + 1) = 2n^2 + 3n + 1$$

$$\xi = 2n^2 + 3n + 1 \tag{5}$$

F. **Bisection Width**

It is the minimum number of edges, whose elimination from the topology results in two distinct and equivalent subnetworks.

Proposition 5:
The Bisection width of Leafy Tor-Cube is 2^n *if* $n = 3$ *and* 2^{n-1} *if* $n > 3$

Proof:

When $n = 3$, minimum 2^n number of edges are to be detached to split the network into two discrete sub-parts. However, when $n > 3$ the least number of edges that are essential to be detached are the links joining the TC with dimension $n - 1$. Thus, the bisection width(B_W) can be formulated as

$$B_w = 2^n \textit{ if } n = 3 \textit{ and } 2^{n-1} \textit{ if } n > 3$$

$$B_W = 2^n \textit{ if } n = 3 \textit{ and } 2^{n-1} \textit{ if } n > 3 \tag{6}$$

G. **Average Distance**

The average distance of a network is another important performance metric to evaluate the efficiency of any network with respect to communication. Average node distance can be computed by sum of distances to every other node from a specified source node of a given network. Average distance should be remained minimum as it is directly proportional to the reliability of a network that means reliability increases with increase in average distance. The average distance of LTC is statistically denoted as $\overline{d}$ which can be calculated as follows:

$$\overline{d} = \sum_{k=1}^{n} L_k = L_1 + L_{2+}L_3 + \cdots \ldots\ldots\ldots\ldots + L_n$$

where L_k *represnts nnumber of nodes situated at k distance*

Preposition 6:
The average node distance of Leafy Tor Cube is defined as follows:

$$\overline{d} = \left(\frac{n}{2^n}\right) \sum\nolimits_{k=1}^{n} L_k = L_1 + L_{2+}L_3 + \cdots \ldots\ldots\ldots\ldots + L_n$$

where L_k *represnts number of nodes situated at k distance*

Proof:

In Leafy Tor Cube, *n* number of child nodes are added to each vertex of the Tor-Cube network as child node. So we have to add n to each hop distance of Tor-Cube for calculating average node distance. As the average node distance of TC is

$$\overline{d} = \left(\frac{1}{2^n}\right) \sum\nolimits_{k=1}^{n} L_k = L_1 + L_{2+}L_3 + \cdots \ldots\ldots\ldots\ldots + L_n$$

where L_k *represnts number of nodes situated at k distance*

Hence the average node distance of LTC is

$$\overline{d} = \left(\frac{n}{2^n}\right) \sum\nolimits_{k=1}^{n} L_k = L_1 + L_{2+}L_3 + \cdots \ldots\ldots\ldots\ldots + L_n$$

where L_k *represnts number of nodes situated at k distance* (7)

H. **Message Traffic Density**

The message traffic density is another important factor to measure the actual performance of the network. For an efficient network message traffic density should be kept minimum. The message traffic density can be represented by P/E, where P = Average Distance × Total number of nodes and E = Total number of links

Preposition 7:
The message traffic density of Leafy Tor Cube is defined as follows:

$$\mathcal{G} = \frac{\left\{\left(\frac{n}{2^n}\right)\sum_{k=1}^{n} L_1 + L_{2+}L_3 + \cdots \ldots\ldots\ldots\ldots + L_n \times 2^n(n+1)\right\}}{2 \times L_{n-1} + 2^{n-1}}$$

where L_k represnts number of nodes situated at k distance

Proof:

The message traffic density of LTC can be represented by $\left(\frac{\overline{d}*N}{E}\right)$

where $\overline{d}$ = Average Distance

N = Total number of nodes

E = Total number of links

The average distance of LTC is stated in equation (6)

The total number of nodes is stated in equation (1)

The sum of communicating links of Leafy Tor-Cube is presented in Equ.2.

Hence by using Equ. 1, 2 and 6 the message traffic density can be computed for LTC as below

$$\mathcal{G} = \frac{\left\{\left(\frac{n}{2^n}\right)\sum_{k=1}^{n} L_1 + L_{2+}L_3 + \cdots \ldots\ldots\ldots\ldots + L_n \times 2^n(n+1)\right\}}{2 \times L_{n-1} + 2^{n-1}}$$

where L_k represnts number of nodes situated at k dist ance (8)

Performance Analysis Inleafy Tor-Cube

(a) **Cost Effectiveness Factor**

Any parallel algorithm design must consider both time- and cost-effectiveness in order to be successful. A parallel algorithm's cost effectiveness takes into account both processors cost and communication links cost. It includes multiprocessor cost and how processor used by parallel algorithms.

Preposition 7:
The cost effectiveness of Leafy Tor Cube(n) is shown as below

$$\frac{\mathbf{1}}{\left\{\mathbf{1} + \boldsymbol{\rho} \times \frac{\left(\boldsymbol{n}(\mathbf{2} \times \boldsymbol{L}_{\boldsymbol{n}-\mathbf{1}}) + \mathbf{2}^{\boldsymbol{n}-\mathbf{1}}\right)}{\mathbf{2}^{\boldsymbol{n}}(\boldsymbol{n}+\mathbf{1})}\right\}}$$

Proof:

The cost effectiveness of Leafy Tor Cube(n) is expressed as below

$$\boldsymbol{CEF(p)} = \frac{\mathbf{1}}{\mathbf{1} + \boldsymbol{\rho} \times \boldsymbol{g(p)}}$$

where $\boldsymbol{\rho} = \frac{\textit{link cost}}{\textit{processor cost}}$

$$\boldsymbol{p} = \textbf{\textit{Sum of processors}}$$

$$\boldsymbol{g(p)} = \frac{\textbf{\textit{Sum of Links}}}{\textbf{\textit{Sum of processors}}}$$

In general the number of links is a function of the number of nodesthatis $\boldsymbol{E} = \boldsymbol{f(p)}$
The total number of processors is expressed in equation (1)
Total number of processors $= 2^n(n+1)$
And the total number of links in Leafy Tor Cube(n) is given by as follows :

$$Q = n2^n(n+1)\ \textit{if}\ n = 3\ \textit{and}\ n(2 \times L_{n-1}) + 2^{n-1}\ \textit{if}\ n > 3$$

As $g(p) = \frac{Q}{P}$
Where Q is the total number of links in Leafy Tor Cube (n)
P is the total number of processors
So $g(p) = \frac{((2 \times L_{n-1}) + 2^{n-1})}{2^n(n+1)}$
As we know that $\boldsymbol{CEF(p)} = \frac{1}{1+\rho g(p)}$
The cost effectiveness of Leafy Tor Cube(n)is defined as follows:

$$\begin{aligned}\boldsymbol{CEF(p)} &= \frac{\mathbf{1}}{\mathbf{1} + \boldsymbol{\rho g(p)}} \\ &= \frac{\mathbf{1}}{\left(\mathbf{1} + \boldsymbol{\rho} \times \frac{(\boldsymbol{n}(\mathbf{2} \times \boldsymbol{L_{n-1}}) + \mathbf{2}^{\boldsymbol{n}-\mathbf{1}})}{\mathbf{2}^{\boldsymbol{n}}(\boldsymbol{n}+\mathbf{1})}\right)} \\ &= \frac{\mathbf{1}}{\left\{\mathbf{1} + \boldsymbol{\rho} \times \frac{(\boldsymbol{n}(\mathbf{2} \times \boldsymbol{L_{n-1}}) + \mathbf{2}^{\boldsymbol{n}-\mathbf{1}})}{\mathbf{2}^{\boldsymbol{n}}(\boldsymbol{n}+\mathbf{1})}\right\}}\end{aligned} \tag{9}$$

(b) Factor of Time Cost Effectiveness

The time it takes to solve an issue is a parameter for this measurement. The TCEF believes that a delayed solution is in no way advantageous. For Leafy Tor Cube or LTC(n) the TCEF is expressed as follows

$$TCEF(P) = \frac{(1+\sigma)}{1 + \rho g(p) + \left(\frac{\sigma}{p}\right)}$$

Preposition 8:
The time cost effectiveness factor in LTC(n) is expressed as follows

Proof:

The *Time Cost Effectiveness Factor* in LTC(n) is derived as follows

$$TCEF(P) = \frac{(1+\sigma)}{1+\rho g(p)+\left(\frac{\sigma}{p}\right)}$$

where $\rho = \frac{\textit{link cost}}{\textit{processor cost}}$

$$p = \textit{Sum of processors}$$

$$g(p) = \frac{\textit{Sum of Links}}{\textit{Sum of processors}}$$

$$\sigma = \frac{\textit{penalty cost}}{\textit{processor cost}}$$

In general the number of links is a function of the number of nodes that is $\boldsymbol{E} = \boldsymbol{f}(\boldsymbol{p})$
The total number of processors is given by as follows:
Total number of processors = Number of Processing Elements
$=2^n(n+1)$
And the total number of links in Leafy Tor Cube (n) is given by as follows:

$$Q = n2^n(n+1)\ \textit{if}\ n = 3\ \textit{and}\ n(2\times L_{n-1}) + 2^{n-1}\ \textit{if}\ n > 3$$

As $g(p) = \frac{Q}{P}$
Where Q is the total number of links in Leafy Tor Cube (n)
P is the total number of processors
So $g(p) = \frac{(n(2\times Ln-1)+2^{n-1})}{2^n(n+1)}$
As we know that

$$\begin{aligned} TCEF(P) &= \frac{(1+\sigma)}{1+\rho g(p)+\left(\frac{\sigma}{p}\right)} \\ &= \frac{(1+\sigma)}{1+\rho\times\frac{(n(2\times L_{n-1})+2^{n-1})}{2^n(n+1)}+\left(\frac{\sigma}{2^n(n+1)}\right)} \\ &= \frac{(1+\sigma)}{1+\rho\times\frac{(n(2\times L_{n-1})+2^{n-1})}{2^n(n+1)}+\left(\frac{\sigma}{2^n(n+1)}\right)} \end{aligned} \quad (10)$$

4 Results and Discussions

This section focuses on topological and performance attributes of LTC and a detailed study is made by comparing LTC with popular topology like hyper cube (HC), crossed cube (CC), leafy cube(LC) and tor cube(TC). Various topological characteristics are compared in Table 1. The factors that are mostly considered for comparison are nodes, edges, degree, diameter, average distance, message traffic density, bisection width, and cost. From Table 1, it can be shown that LTC has higher degree than HC, CC, LC and TC. Higher in degree is better in connectivity among the nodes. In case of diameter, the LTC offers the shortest diameter than the equal sized network Leafy Cube. There is always a need for a larger size network with a low diameter and low cost, and the three desirable and significant qualities of any interconnection network are degree, diameter, and cost. According to Table 1, the proposed topology Leafy TOR-CUBE has a larger cost than other networks that are recommended, but LTC demonstrates scalability, which is a requirement for high speed computing in the modern day. As the node degree increases, the proposed network has more node disjoint paths than the parent networks. The proposed network LTC is also shown to have superior average distance and fault tolerance than its preceding networks, HC, CC, and LC.

The Figure 5 shows a relationship between the degree and dimension of different networks. It demonstrates that the proposed interconnection network Leafy Tor-Cube has a higher degree than both its base topology, the Tor Cube, and various other significant network topologies considered here, such as HC, CC, and LC. This network is preferred for high performance computing since it has a higher degree of LTC with more degrees of freedom to move in case of more traffic.

The Figure 6 shows a relationship between communication edges and the dimensions of various types of networks. It demonstrates that the planned interconnection network Leafy Tor Cube's communication links extend beyond its core topologies of TC, HC, CC, and LC. As the degree of LTC is more so this network is a preferable one for high performance computing. This property shows that LTC network is more tolerant to fault and reliable one than some commonly used topologies like HC, CC and LC.

The relationship between number of nodes and network dimension of popular networks are designed and presented in Figure 7. It confirms that the number of nodes of LTC is higher than its predecessor topologies like HC, CC and TC where the latter three have same number of nodes. It packs equal number of nodes with LC. This property also proved that LTC is having enhanced architecture and suitable candidate for parallel and distributed computing.

In similar manner, Figure 8 shows a relationship between diameter and network dimension that takes diverse network topologies into account. The figure demonstrates that the LTC has the smallest value of diameter than its parent structure and also the LC. The LTC having same number of nodes bears lower diameter and hence travelling in LTC will be more beneficial.

Table 1 Comparison of topological attributes of various networks

Parameter	HC	CC	LC	TC	LTC
Node	2^n	2^n	$2^n + n2^n$	2^n	$2^n(n+1)$
Edges	$n2^{n-1}$	2^{n-1}	$n2^{n-1} + n2^n$	$2L_{(n-1)} + 2^{(n-1)}$	$\begin{cases} n2^n(n+1)\ if\ n=3 \quad and \\ n*(2 \times L_{n-1} + 2^{n-1}\ if\ n>3 \end{cases}$
Degree	n	n	$2n$	$n+1$	$2n+1$
Diameter	n	$\left[\frac{n+1}{2}\right]$	$n+2$	$n-1$	$n+1$
Average Distance	$\frac{n}{2}$	$\frac{n}{2}$	$\frac{n}{2}$	$\left(\frac{1}{2^n}\right)\sum_{k=1}^{n-1} L_k$	$\left(\frac{1}{2^n}\right)\sum_{k=1}^{n} L_k = L_1 + L_{2+}L_3 + \cdots\ldots\ldots\ldots\ldots + L_n$ *where* L_k *represnts number of nodes situated at k distance*
Message Traffic Density	1	Almost 1	$\left(\frac{n+1}{3}\right)$	$\frac{\left(\frac{1}{2^n}\right)\sum_{k=1}^{n-1} L_k 2^n(n+1)}{2} \times Ln-1+2^{n-1}$	$\frac{\{\left(\frac{n}{2^n}\right)\sum_{k=1}^{n} L_1+L_{2+}L_3+\cdots\ldots\ldots\ldots+L_n \times 2^n(n+1)\}}{2\times Ln-1+2^{n-1}}$
Bisection Width	$\frac{2^n}{2}$	$\frac{2^n}{2}$	$\frac{2^n}{2}$	$\frac{2^n}{2}$	$2^n\ if\ n=3\ and\ 2^{n-1}\ if\ n>3$
Cost	n^2	$n\left[\frac{n+1}{2}\right]$	$2n(n+2)$	n^2-1	$2n^2+3n+1$

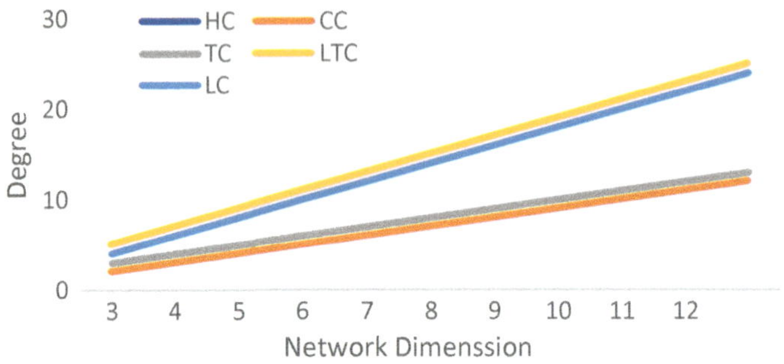

Fig. 5 Comparison of degree wrt network dimension

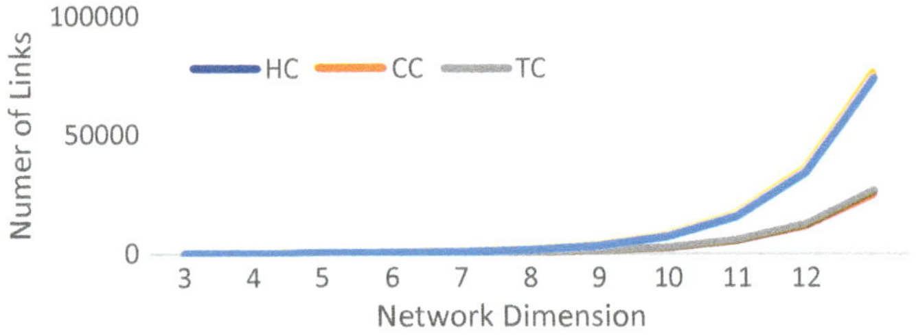

Fig. 6 Comparison of links wrt network dimension

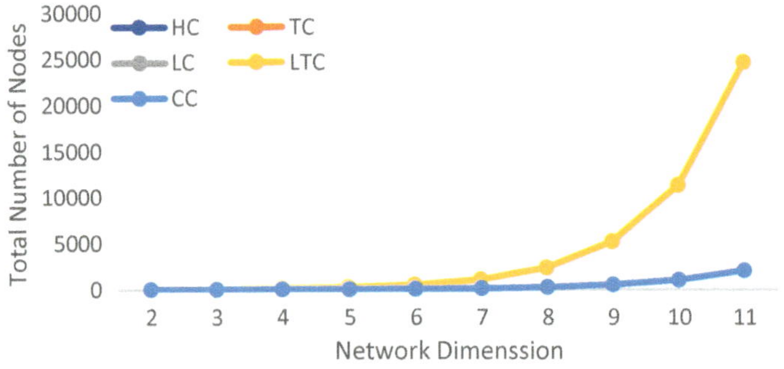

Fig. 7 Comparison of total number of nodes wrt network dimension

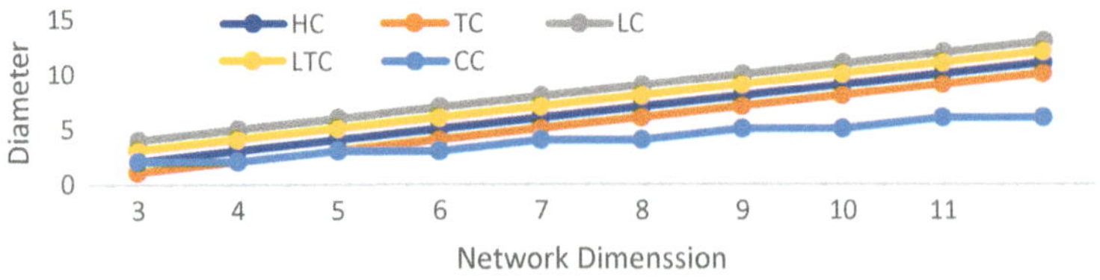

Fig. 8 Comparison of diameter wrt. network dimension

Fig. 9 Comparison of cost wrt network dimension

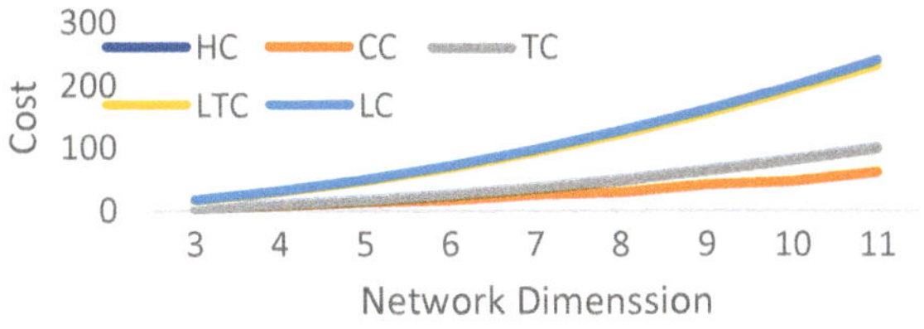

A graph is also designed between cost of the network with respect to network dimension of various network topologies and offered in the Figure 9. The figure displays that the LTC has a reduced cost than LC but higher than that of HC, CC and TC. But from the recent study of ACM Survey in the year 2015 we can conclude that cost should not affect the scalability of a network. This concludes that for a large scalable network which is the demand of today's VLSI implementation, cost does not matter at all. For high scale computing, the hardware cost can be negligible.

Next in the Figure. 10, bisection width of the proposed network is compared against dimension which is same as TC. In the Figure 11, average distance of LTC is compared against dimension taken in x-axis. The Figure 12 shows the comparison of message traffic density for all the networks. With increase in node count, the message density of LTC is less compared to HC, CC and LC. Less message traffic density also proves that the LTC is an efficient network.

The cost effectiveness factor and time cost effectiveness factor with respect to network dimension are described in the Figures 13 and 14 respectively. Similar to the Hypercube, it is a monotonically decreasing function of p. As a result, the network's cost effectiveness and time cost effectiveness decrease as its size increases. Hence the proposed LTC network definitely a cost-effective as well as a time cost-effective topology.

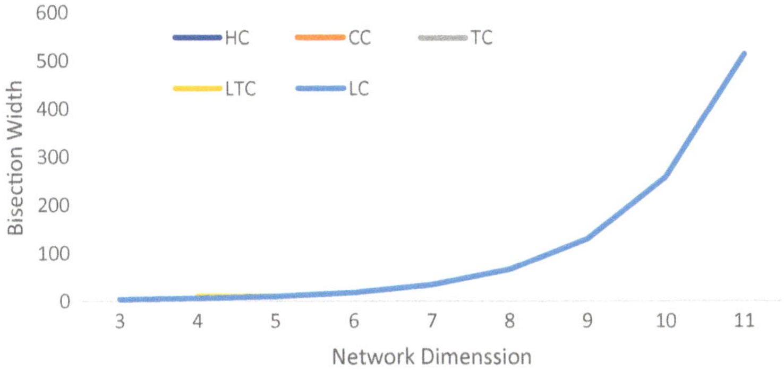

Fig. 10 Comparison of bisection width wrt network dimension

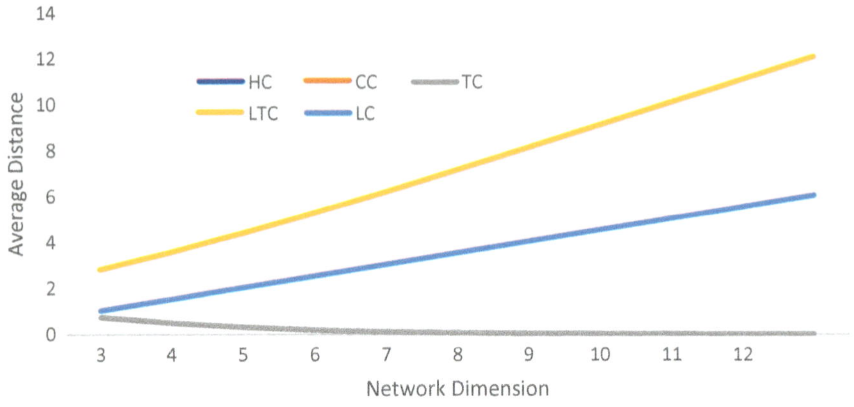

Fig. 11 Comparison of average distance wrt network dimension

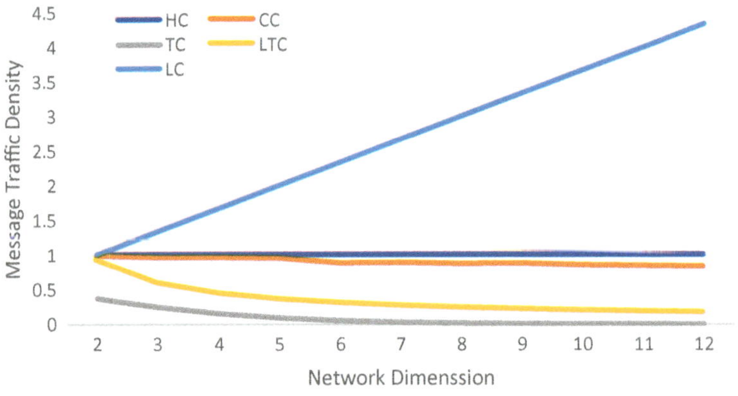

Fig. 12 Comparison of message traffic density wrt network dimension

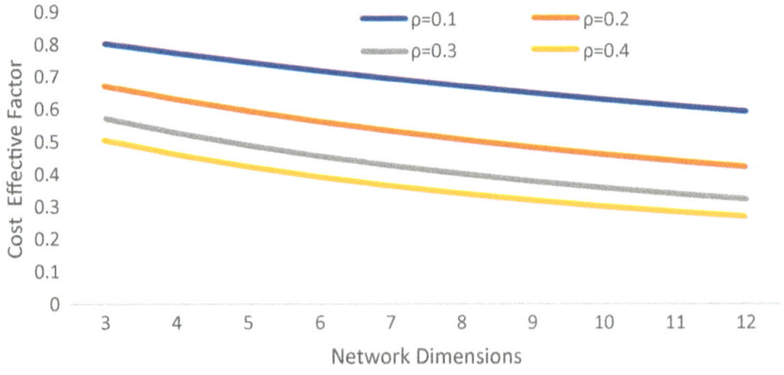

Fig. 13 Comparison of CEF wrt network dimension

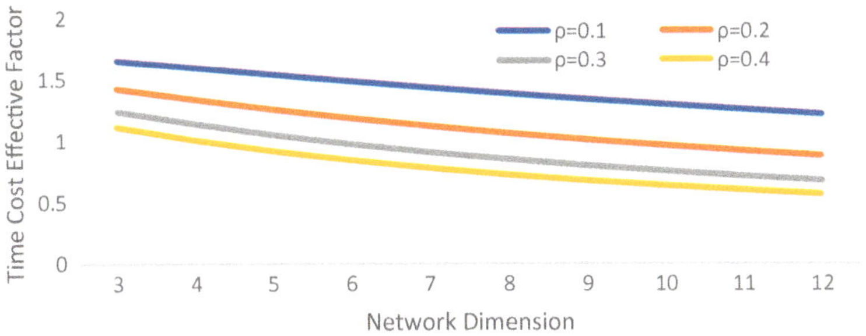

Fig. 14 Comparison of TCEF wrt network dimension

5 Conclusion and Future Scope

In the current paper a new hybrid type interconnection network for massive computing system called Leafy tor cube is introduced. This new network is recursive in nature while retaining the basic structure. So construction of higher dimensional network is very easy. With low link complexity it can pack a large number of nodes nearly double the size. Thus Leafy tor cube is found to be an enhanced architecture for large scale computing systems. With higher packing density, LTC network can be greatly preferable for implementing high performance and large scale computing systems. It is discovered to have greater and more reliable efficiency regardless of the amount of the input compared to HC, CC and LC. It is clear that for an LTC network, speedup is inversely correlated with packing density. LTC can be implemented with parallel computing system with extremely great scalability. The efficiency changes as the number of PEs and input size grow. With in crease in node count, the message density for LTC is less compared to HC,CC and LC. Less message traffic density also proves that the LTC is an efficient network. This property shows that LTC network is more tolerant to fault and reliable one than other popular cube based network topologies. In future fault tolerant aspect of LTC are to be highlighted. The various reliability metric of the LTC like network reliability and k-terminal reliability can be estimated and compared contrary to that of other topology of interest in future. The research presented in this study could be expanded upon to suggest a brand-new hierarchical interconnection network that makes use of LTC as its foundation. More research should be done to determine whether LTC interconnection network is appropriate to use as the parallel architecture for a massively parallel database management system. The technology of big data computing can be implemented for LTC.

References

1. Adiga, N.R., Almási, G., Almasi, G.S., Aridor, Y., Barik, R., Beece, D., Bellofatto, R., Bhanot, G., Bickford, R., Blumrich, M. et al.: An overview of the Bluegene/l Supercomputer. In: Supercomputing, ACM/IEEE 2002 Conference, pp. 60–60 (2002)
2. Chang, C.-P., Sung, T.-Y., Hsu, L.-H.: Edge congestion and topological properties of crossed cubes. IEEE Trans. Parallel Distrib. Syst., 11(1), pp. 64–80, (2000)
3. Cvetanovic.: Performance snalysis of the Alpha 21364- Based HP GS1280 multiprocessor. In: Proceedings of 30th Annual International Symposium on Computer Architecture, pp. 218–228, (2003)
4. Dash, R.K., Tripathy, C.R.: A New Fault-tolerant interconnection topology for parallel systems. IE(I) Journal-CP, 89, pp. 8–13, (2008)
5. Rahman M.M.H., Horiguchi, S.: HTN: a new hierarchical interconnection network for massively parallel computers. In: IEICE Trans. Inf. Syst., E86-D, pp. 1479-1486, (2003)
6. Hsu, H.-C., Tsai, C.-H., Tsai, C.-H., Wu, G.-J., Liu, S.-H., Lai, P.-L.: Hierarchical crossed cube: a new interconnection topology. In: Proceedings of 25th Workshop on Combinatorial Mathematics and Computation Theory, Taiwan, pp. 70–79, (2008)
7. IBM, "IBM System Blue Gene Solution," http://www-03.ibm.com / servers / deepcomputing / bluegene.html, (2008)
8. Imani, N., Sarbazi-Azad, H.: Properties of a hierarchical network based on the star graph. Inf. Sci. **180**(14), 2802–2813 (2010)
9. Rajabzadeh, P., Sarbazi-Azad, H., Zarandi, H.R., Khodaie, E., Hashemi-Najafabadi, H., Ould-Khaouae, M.: Performance modeling of n-dimensional mesh networks. Perform. Eval. **67**(12), 1304–1323 (2010)
10. Razavi, S., Sarbazi-Azad, H.: The triangular pyramid: routing and topological properties. Inf. Sci. **180**(11), 2328–2339 (2010)
11. Sabbaghi-Nadooshan, R., Modarressi, M., Sarbazi Azad, H.: A Novel High-Performance and Low-Power Mesh-Based NoC. In: IEEE International Symposium on Parallel and Distributed Processing IPDPS, pp. 1–7, (2008)
12. Szlachcic, E.: Fault tolerant topological design for computer networks. In: Proceedings of the International Conference on Dependability of Computer Systems (DEPCOS-RELCOMEX), pp. 150-159, (2006)
13. Bhardwaj, M.: C2Torus new interconnection network topology based on 2D Torus, american journal of networks and communications. Special Issue: Ad Hoc Networks **4**(3–1), 1–4 (2015)
14. Yang, Y., Funahashi, A., Jouraku, A., Nishi, H., Amano, H., Sueyoshi, T.: Recursive diagonal torus: an interconnection network for massively parallel computers. IEEE Trans. Parallel Distrib. Syst. **12**(7), 701–715 (2001)
15. Liu, Y.-H., Zhu, M.-F., Xiao, L.-M., Wang, J.: Asymmetrical topology and Entropy-based Heterogeneous Link for Many-core massive data communication. Cluster Computers. **16**(4), 679–691 (2013)
16. Andujar-Munoz, F.J., Villar-Ortiz, J.A., Sanchez, J.L., Alfaro, F.J., Duato, J.: N-Dimensional twin torus topology. IEEE Trans. Comput. **64**(10), 2847–2861 (2015)
17. Yadav, S., Krishnai, C. R.: CCTorus : a new torus topology for interconnection networks. In Proc.of International Conference on Advanced Computational Technologies and Creative Media, pp. 8–14, (2014)

Sports Prediction for Cricket Match Using Grid Search and Extreme Gradient Boosting Classifier

Soumya Ranjan Mahanta and Mrutyunjaya Panda

Abstract Sports prediction has experienced a sharp increase in popularity over the last two decades with the implementation of machine learning (ML) methods. The objective of this paper is to examine the research trends in Sports prediction using machine learning approaches and then to identify effective ones amongst those for assessing the accurate sports predictions especially cricket with key characteristics of such successful applications. In this paper, five machine learning algorithms such as: Support Vector Machine (SVM), Logistic Regression (LR), Multinomial Naïve Bayes (MNB), Random Forest (RF) and Extreme Gradient Boosting (XG Boost) are evaluated for effective sports prediction of Cricket after data pre-processing is done through grid search, mean encoding based feature engineering. The experiments uses team sports data such as cricket which is collected from GitHub and the results shows whether the difficulty level of predicting outcomes varies across sports considering team data only or in combination with player. The experimental research obtained are interesting which may help the sports enthusiasts and analysts in making informed decisions while predicting outcomes of team sports.

Keywords Cricket match · Sports prediction · Grid search · Mean encoding · XG boost · Performance measures · T-Test

1 Introduction

Physical education is a scientific field concerned with the application of scientific ideas and processes to improve athletic performance. Since the beginning, there has been a link between sports results and various aspects of data. Certain factors, such as the type of sport, have a direct impact, while several other factors affect the environment and the goals of the player. Some teams choose not to use any predictive tactics, while others have long relied on the expertise and instinct of their players. However, groups looking for more reliable predictions tend to take a

S. R. Mahanta (✉) · M. Panda
Utkal University, Bhubaneswar, Odisha, India
e-mail: dipusoumyaranjan019@gmail.com

S. Kumar Udgata et al. (eds.), *Intelligent Systems*, Lecture Notes in Networks and Systems 1149, https://doi.org/10.1007/978-981-97-8160-7_13

more conservative approach. To solve this, use of statistics in the decision-making process might be a popular choice but still it is in an initial stage [1]. It is also observed that predicting the outcomes of matches is gaining popularity among sports enthusiasts worldwide, particularly in football, cricket, and basketball [2] but with obvious challenges particularly in team sports. Sports prediction is considered as a challenging task for the analyst and software developers to develop methodologies to deal with the large sports related data including: statistical data and information about players, teams, matches, and seasons [3]. At the same time, threats arises out of possible teams emotions and system malfunctions are to be dealt with proper care. Analysing past team sports statistical data and prediction of game outcomes with program schedule, commentators, and pre- and post-match analyses falls under the realm of multi-class classification task. To achieve good sports outcome prediction, several sports television channels are being established such as: Microsoft's Bing accurately predicting World Cup results in 2014 but the real worry is that the details know how of these correct prediction by Microsoft's Bing is never being disclosed to the public [4]. It may also be noted that sports analytics become vital for gamblers where the venture company offering for betting tries to influence the client by offering better chance of making profits and to help the client to make a better gambling decision.

Motivation

The identified research gap lies in the limited utilization of statistical analysis and machine learning techniques for effective and efficient plan evaluation in sports prediction in terms of formulations of game winning strategies, tactics, and game plans in sports. This needs further investigation to bridge the gap between predictive analytics and plan evaluation by enabling data-driven evaluations and adjustments to enhance performance and increase the chances of success in sports competitions. Additionally, studying the impact of contextual factors, such as crowd emotions, social media trends, and external conditions, on team performance and match outcomes may also contribute to a more comprehensive understanding of the complex dynamics influencing sports outcomes and the development of more accurate prediction models.

Hence, looking into these, this paper explores the technical details with team sports data for better analysis and corrects predictions of the game outcome. The rest of the paper is organised into several sections. Section 2 provides a literature review summarizing previous studies on the team sports prediction. Section 3 describes the datasets used followed by proposed methodologies in Sect. 4. Section 5 illustrates the experimental design and evaluation of the model to illustrate the research findings. Finally, Conclusions are presented in Sect. 6 with future scope of research, emphasizing the importance of ongoing exploration and advancements in integrating machine learning techniques for accurate sports prediction.

2 Related Work

Iyer and Sharda [5] propose neural networks to predict the performance of cricket players with recommendation using these assessments to select players for the 2007 World Cup squad. Lamsal et al. [6] uses a multivariate regression- based approach to evaluate players in the Indian Premier League (IPL), where players are assigned with points based on past performance and the team is given weightage based on seven crucial factors that were identified as impacting match outcomes. Jhanwar and Paudi [7] adopted a distinctive strategy for anticipating the result of cricket games by assessing the performances of individual players and assessing the relative strengths of the teams involved. To assess a player's potential, they created algorithms that simulated the abilities of bowlers and batters, taking into account both recent and lifetime performances. Bhattacharjee and Pahinkar [8] focused on evaluating the effectiveness of bowlers in the IPL by utilizing a combined bowling rate and identified several additional factors that contribute to a bowler's performance using a multivariate regression model.

Munir et al. [9] uses decision tree methods to forecast the outcomes of 20 cricket matches, using data from international T20 matches and IPL competitions till 2015 with match prediction outcomes of 78% accuracy for the first team and 75% accuracy for the second team. Nimmagadda et al. [10] presented a study on real-time prediction of T20 cricket match outcomes using several statistical and machine learning approaches to anticipate the winning team in a cricket match. The model considers various factors, such as runs scored per over in innings, scores at regular intervals during the innings, and the final scores.

Our Contributions

The main contributions of this research are as follows. At First, we propose an efficient feature engineering method such as: mean encoding to have quality features from the team sports cricket dataset. Secondly, accurate and practical method for cricket match winning predictions are performed based on coaches' decisions in team selection and individual player qualities using the information from the model, which has not yet been emphasized in previous studies. The third point is to provide a new direction by considering past features of cricket playing and then embedded into the proposed machine learning model to obtain higher prediction probabilities. Fourth point is that five machine learning approaches such as: Multinomial naïve Bayes, Logistic Regression, Support Vector Machine, Random Forest and Extreme Gradient Boosting for cricket winning prediction. Next, Grid search CV is used for fine tune the hyper- parameters of the machine learning model automatically for better prediction. At last, statistical significance test using T-test carried out to understand the best machine learning model to predict the cricket match winning team.

3 Datasets

The sports prediction dataset is collected from GitHub and ESPN Star Sports, a platform containing datasets obtained from the cricket community. Initially, data sets for sports prediction were obtained from multiple online sources. Cricket match and player statistics were collected from various websites [11, 12]. Additionally, a free database was collected from sports analytics hackathon done through stats. espncricinfo.com with an objective to analyse the ODI matches between India and Australia and predict the results of a past played series to understand who won the series, who will be the man of the series, who will hit maximum number of sixes etc.

The collected data was extensive and hence after re-organization, they are segregated into various CSV files based on the information required for each experiment. These CSV files were then processed using Jupyter for data processing. Data pre-processing is done through cleaning the data by eliminating null values, duplication, and noise. Mean encoding is used to pre-process the sports data to convert it a quality data suitable for application. While mean encoding envisages of improving the classification task through decreasing cardinality but there is always a risk of trap in over fitting because of possible data leakage with feature bias while performing encoding. This may be eliminated by using some regularization method such as grid search. A sample dataset with different labels of individual player used for this research is provided in Table 1. From Table 1, it can be observed that for player Virat Kohli, rank is mentioned as number 1 with rating points 887 and the career best rating points as 911 with rating difference for himself is 911- 887=24 but with others is zero being placed in first rank. Accordingly all other player's details are presented in Table 1, for understanding.

Table 1 Different labels of a dataset

Sl no	Rank	Rating	Name	Career best	Self_Rating	Rating_Diff
1	1	887	V Kohli	911	24	0
2	2	854	RG Sharma	883	29	33
3	10	744	S Dhawan	813	69	143
4	17	688	MS Dhoni	836	148	199
5	21	659	AJ Finch	744	85	228
6	31	606	GJ Maxwell	735	129	281
7	35	590	KM Jadhav	591	1	297
8	42	574	AT Rayudu	574	0	313
9	43	572	SE Marsh	581	9	315
10	44	564	MP Stonis	616	52	323
11	79	472	HH Pandya	497	25	415

4 Proposed Methodology

In this section, several classification techniques, including Random Forest, Multinomial Naive Bayes, Support Vector Machine, Logistic Regression, and XG Boost, are discussed with feature engineering methods such as: Mean encoding and Grid Search algorithm.

4.1 Support Vector Machines

The Support Vector Machine is a well-known and widely used machine learning method that can effectively overcome various challenges and obstacles in regression problems [13, 14] and is a popular classification model due to its high accuracy, ability to work in high dimensional space and having low memory requirements. Despite the several advantages of Support Vector Machines (SVM), they do have some limitations. One drawback is the need for regularization to prevent over fitting of kernel functions when the number of features exceeds the number of samples. Additionally, SVMs typically cannot provide probability estimates, requiring computationally intensive n-fold cross- validation.

4.2 Multinomial Naive Bayes (MNB)

The Multinomial Naive Bayes algorithm [15, 16], a modified version of the Naive Bayes algorithm is commonly utilized for classification purposes, especially when dealing with discrete features such as word frequencies in the analysis of textual data. Though the algorithm usually operates with integer feature counts, it can also take advantage of tf-idf vectors to indicate fractional counts. The Bayes theorem is a mathematical approach that allows for the calculation of the posterior probability of an event occurring, given the prior probability and new evidence. The MNB is easy to implement, efficient, accurate and scalable too which makes it a popular choice for classification problems. However, it does have limitations. One such limitation is the assumption that attributes belong to independent vectors. Additionally, the algorithm sets the probability value to zero for unseen categories, but this can be addressed using a smoothing method.

4.3 Logistic Regression (LR)

Logistic regression [17] is known for its straightforwardness and simplicity in the field of machine learning, making it easy to use. Moreover, it requires minimal processing power and is highly resistant to overfitting. However, in real- world scenarios, the assumption of a linear relationship between the independent and dependent variables may not hold true, which is one of the limitations of logistic regression. Furthermore, it is best suited for discrete functions and may not perform well when there are more features than observations. Logistic regression is well-suited for low-latency applications that have a smaller data dimension.

4.4 XG Boost

Extreme Gradient Boosting (XG Boost) [18] is a distributed, scalable machine learning approach that uses gradient- boosting decision trees for supervised learning problems with multiple features. Boosting is used to gradually add trees to the ensemble and optimize them to address the prediction errors of previous models. The XG Boost uses gradient descent technique to fit the model and to minimize the loss function. Time complexity for learning each tree is O(n log n) due to sorting being the most time-consuming step in the tree learning algorithm.

4.5 Random Forest (RF)

Random Forest classifier [19] combines two similar algorithms to create a multi-choice tree and a forest of trees. Compared to other algorithms, Random Forest often provides more accurate results for classification tasks. The algorithm generates multiple decision trees by creating a forest from input data, and produces a professional model for precise prediction and comparison. The Random Forest algorithm's computational complexity can vary depending on its different phases: testing, training, and space usage. To elaborate on the computational complexity of the algorithm, during the testing phase, the time required for prediction can be expressed as O(m*k'), where m represents the number of features and k' is the number of trees in the model. However, during the training phase, the time complexity is O(k'nlog(n)*m), where n denotes the number of training samples. The space complexity can be defined as O(k'*depth of the tree) where k' is the number of trees and the depth of the tree refers to the maximum number of nodes from the root to a leaf in the decision tree.

4.6 Mean Encoding

The process of converting the categorical attributes to its numeric counterpart is called as Encoding. Since encoding removes redundancies from the data, the size of the data gets smaller hence needs a lesser memory which ultimately makes the faster input speed when data is being saved for exploratory analysis. In mean encoding, apart from encoding task, the probability of the target attribute conditioned on each value of the attribute is calculated and new encoded feature is produced.

4.7 Feature Selection

Feature selection plays a significant role in improving the prediction model's accuracy. By selecting the optimal combination of features that can accurately forecast match results, the model's ability to adapt is enhanced. This helps mitigate the potential negative impact of extending the number of features on the classifier's overall accuracy.

4.8 Grid Search

A grid search [20] is an efficient method used to fine tune the hyper-parameter configurations form a grid of given parameters we are interested in while building the classifiers. In machine learning model development, there are two ways by which a model's parameters are being defined. In one case, the parameter is learned through the learning model itself while in the other case; parameters are referred as hyper parameter that is being passed to the machine learning model. In case of Random Forest, the hyper-parameters are the random number of decision trees in the forest and the number of features by each tree while splitting a node is chosen manually. Similarly, for SVM, the hyper-parameters are type of kernels used, in K-NN algorithm, it is number of K and for XGBoost, the most important hyperparameter settings are the depth of the tree, learning rate, subsample training instances, minimum number of observations and minimum loss reduction etc. This random selection of hyper-parameters sometimes presents difficulty in choosing the best machine learning approaches as changing the hyper-parameters may perform worse or better than the other. Hence, an automatic selection of such hyper-parameters are suggested which is achieved in this paper by using Grid Search method. However, the Grid Search algorithm may result slow processing for its large automatic test combinations and sometimes is expensive due to cross validation used by Grid Search.

5 Experimental Set-Up

In this, different machine learning methods for predicting cricket match results [21] are investigated and compared. The objective here is to develop a predictive model that takes into account the influence of a home run and the outcome of a coin toss in order to forecast game results. When deciding whether to bat first or field first, the winning team takes into account factors such as weather, playing surface, and outfield. One of the models examines the effects of the home field, while the other analyzes the consequences of the coin toss. The experimental setup for our proposed work on sports prediction especially for Cricket is presented in Figs. 1 and 2.

From Fig. 1, one can observe that after input data is taken, data preprocessing is used to remove all incomplete elements to ensure that the input dataset doesn't contain

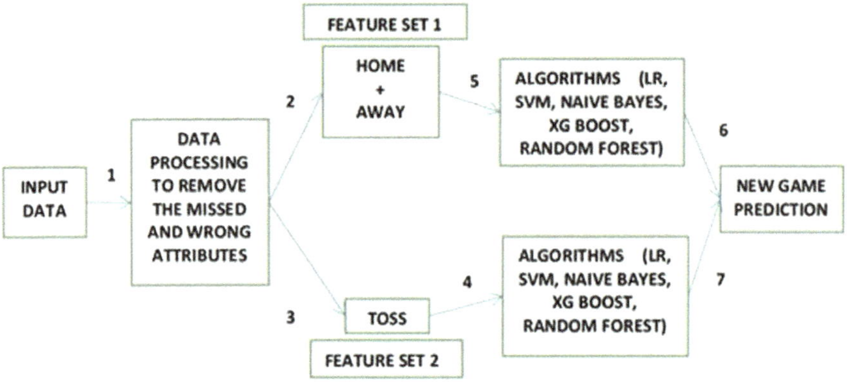

Fig. 1 Cricket analysis model

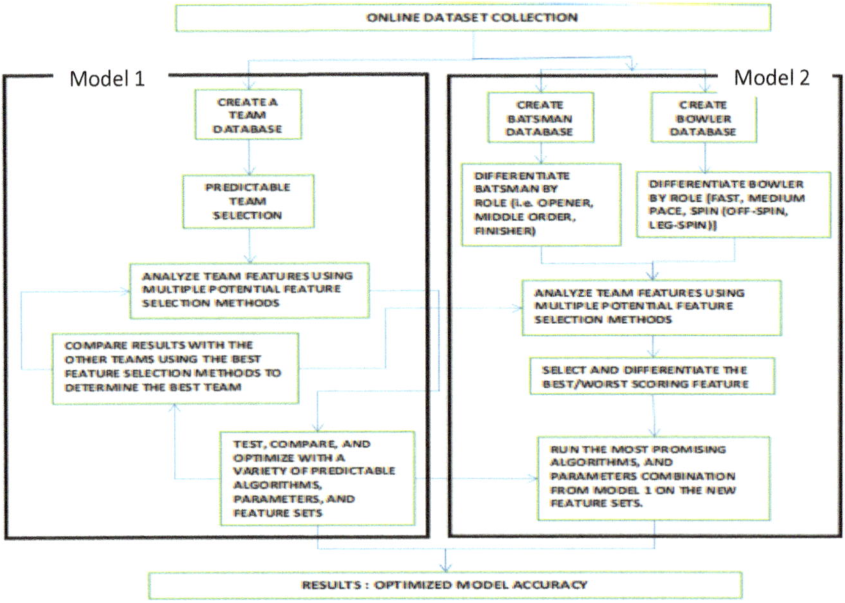

Fig. 2 Different levels of optimized model

any missing values. Further, irrelevant data relating to match number, date, and location were eliminated. Two distinct sets of features were extracted from the input datasets – (1) a dataset with feature set 1 representing the cricket ground both home & away from home and (2) the other part of the dataset with feature set 2 representing the decision made during the coin toss. Multiple learning techniques were then implemented on these sets to construct models capable of accurately forecasting game outcomes. To create classifiers, we utilized Grid Search cross-validation method. The potential for using five different machine learning models in this research rests on accurate forecasting of future cricket matches based on the above two criteria. By leveraging the insights gained from the constructed classifiers, the model aims to provide valuable predictions and insights into cricket matches, assisting stakeholders in making informed decisions and strategizing effectively.

Figure 2 shows the two different optimized models for sports prediction using machine learning algorithms. While Model 1 discusses about the prediction model using Team data only whereas the model 2 represents the procedures followed by considering both Team and individual players together.

Model 1: Team Data Only

A team's skill level or "form" can be represented by statistical characteristics. These basic performance characteristics are categorized at three different levels, where Level 1 represents base feature characteristics, while combinations of two base features from Level 1 characteristics forms net characteristics at Level 2, and finally, combinations of two Level 2 characteristics forms difference features at Level 3. The hierarchy of these characteristics is illustrated in Fig. 3.

As a result of this categorization, for every game in the given cricket dataset, a total number of 31 attributes are computed based on the preceding games of each team. Further, these attributes are categorized into three levels: 18 attributes in level 1, level 2 with 8 attributes, and 5 attributes for level 3, as demonstrated in Fig. 3.

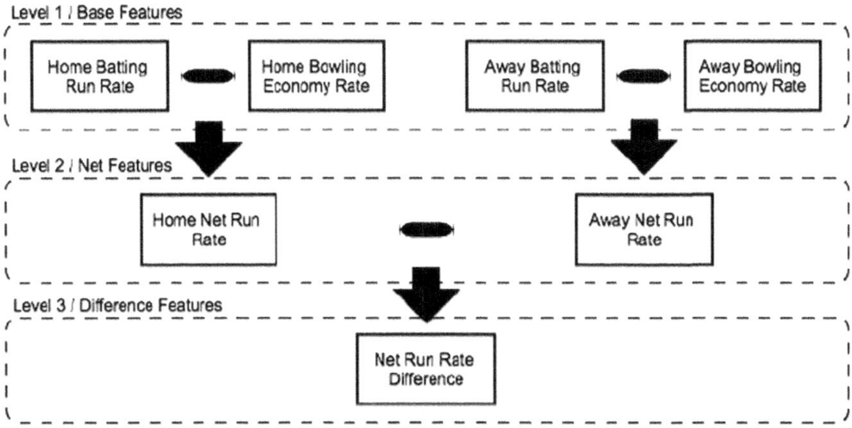

Fig. 3 Feature details in cricket dataset

In Level 1, out of 18 attributes, two attributes are assigned each for winning percentage, batting run rates, bowling run rates, bowling strike rate, economy rate, and batting wicket rate, respectively; four attributes for averages and two attributes for Batting Index and Bowling Index. To generate these features, nine distinct data sources were utilized as collected from Github [22], resulting in the computation of eighteen key attributes for both the playing at home and away from home teams. While many of these attributes are commonly used in similar analyses, the inclusion of the batting wicket rate attribute provided a more comprehensive coverage of the data. It is important to note that these level 1 attributes can be further enhanced by extracting more complex features. Additionally, net features were introduced at Level 2, which involved computing the difference between a team's batting and bowling performance statistics. For example, the net run rate attribute is obtained by subtracting the bowling economy rate from the batting run rate. Each team has four level 2 features, and the set of eight attributes does not include win percentages. Distinctive Characteristics, at Level 3, represent the differences between the net statistics and win percentages of the two teams. The variance between the net run rates of the home and away teams is referred to as the "net run rate difference," and these five level 3 features capture such distinctions. The details of these team features with their descriptions are summarized in Table 2.

Model 2: Player & Team Data

In model 2, a large number of features were generated, exceeding 500 in total. The features obtained from individual batting and bowling statistics of each player are categorized into a hierarchical structure consisting of three levels, namely Level 1, Level 2, and Level 3. After applying successful strategies from the initial modeling phase, the feature space was reduced, and the best algorithms from Model 1 were re-trained using the new dataset, in an effort to ascertain if any improvements could be achieved.

To generate higher-level features, Level 1 player features were developed and employed. The approach used to calculate these features from the raw data is

Table 2 Team features

Team features	Description
Win Percentage	The percentage of games won
Games won Batting Average	Mean runs scored per wicket fallen
Batting Run Rate	Mean runs scored per six legal deliveries
Batting Index	Product of Batting Run Rate and Batting Average
Batting Average Batting Wicket Rate	Mean wickets lost per balls faced
Bowling Average	Mean runs given away per wicket taken
Bowling Economy-Rate	Mean runs given away per over bowled
Bowling Strike-Rate	Mean deliveries bowled per wicket taken
Bowling Index	Bowling Average multiplied by Bowling Economy-Rate

Table 3 Level 1 batting features

Batting Features	Description
Batting Average	Runs scored per dismissal
Batting Strike-Rate	Runs scored per ball faced

Table 4 Level 1 bowling features

Bowling features	Description
Bowling Average	Runs conceded per wicket
Bowling Economy-Rate	Runs conceded per over
Bowling Strike-Rate	Balls bowled per wicket

detailed in Tables 3 and 4, which display the Level 1 batting and bowling attributes, respectively.

Next, by combining the batting features such as: batting average and batting strike-rate from Table 2, net batting features i.e. batting performance index and alternative batting index are created for Level 2 which is shown in Table 5. Similarly, net bowling features are created for Level 2 by combining the several features mentioned in level 1 from Table 4, which are shown in Table 6.

Based on this, a total of 228 level 2 features were generated by calculating logical combinations of players based on their individual statistics. These combinations included different groups of batsmen and bowlers, such as the top three batsmen and the top five bowlers. In order to further increase the number of level 1 features, a variety of logical combinations of players were calculated for both batting and bowling. In the batting analysis, a total of 28 different combinations of batsmen were considered, including individual players, as well as groups of 2, 3, 4, 5, 6, 7, 8, 9, 10, and all 11 players. Additionally, there were specific groupings such as numbers 5 to 8, 4 to 6, 7 to 9, 3 to 4, 5 to 6, 7 to 8, and 9 to 10 that were also included in the analysis. In the bowling analysis, 11 different combinations were analyzed, including 6 individual bowlers and groups consisting of the 2, 3, 4, 5, and all 6 bowlers. This

Table 5 Level 2 net Batting Features created using combinations of Level 1

Batting Features	Descriptio
Average + Strike-Rate	Batting performance index
Average x Strike-Rate	Alternative batting index

Table 6 Level 2 net bowling features created using combinations of Level 1

Batting Features	Description
Average + Economy-Rate + Strike- Rate	Bowling performance index
Average x Economy-Rate xStrike- Rate	Alternative to batting index
Average + Economy-Rate	Alternative 2 to bowling index
Average x Economy-Rate	Alternative 3 to bowling index

led to a total of 89 level 1 features for each team, combined with the 28 level 1 batting features, which resulted in a total of 178 features.

To create level 2 player features, the mean and standard deviation of each player's level 1 feature were computed for both the home and away teams. The level 2 features were then generated by subtracting the corresponding away team feature from the home team feature using the difference method or dividing the home team feature by the away team feature using the ratio method. An example of this is the creation of a level 2 feature through the computation of the variance between the mean values of Batsman 1 in the home team and the away team. This process resulted in the creation of 28 level 2 features for the away team's batsmen.

Finally, Combination Metrics are created by combining two or more level 1 element to create a new performance metric for a specific player or group of players. For instance, Home Batsman 1's Strike Rate with Home Batsman 1's Average are combined to create a new metric for prediction analysis. Same way, level 3 player features were derived by analyzing the level 2 combination features between the home teams and away teams. An example of such level 3 feature creation from combination of several level 2 features by combining the performance of the team and top three bowlers at home ground with the performance of the team and top three bowlers at foreign ground (i.e. away from home) using one metric 'Average x Economy Rate'. This way, the difference between the two teams resulted in an additional 100 level 3 features, making a total of 506 features.

6 Experiments and Evaluations

A predictive model for sports prediction used for our experiments is presented in Figs. 1 and 2. All the experiments are conducted in a HP Pavilion Intel Core i5 PC with 1 TB HDD, 4 GB RAM in Windows environment using Python. The following steps are taken to experimental analysis.

Step-1: Best Team Selection

With List of the squad (15 members or players) and out of the 15 members, 11 players is selected based on their recent performances.

Step-2: Predictable Team Selection to

Analyze team features using multiple potential following features selection methods considering batting Average and Rankings for Batsman taken together.

- Batting positions for 1st, 2nd, and 3rd are kept for the players who have a batting average of 35 or more.
- 4th and 5th positions of Batsman are selected whose batting averages are between 28 to 35 or more.

- 6th and 7th positions are kept for All-rounders (Those who are capable of both batting and bowling together) and the players are selected based on their batting average of 28 to 35 or more and a minimum average of 20 in bowling.
- Finally selection of Bowlers are selected for rest of the team position i.e. 8th, 9th,10th, & 11th, with less than 20 average strike rate.
- Based on these, the probable best team out are being selected.

Step-3: Mean Encoding and Feature Engineering

In this step, the dataset is grouped by [Team, Opposition, and Year] aggregate with Mean win-difference, Mean Average, Mean run per over (RPO), Total Home to Home (H2H) Matches, Total H2H Won, Total H2H Lost, Total H2H Tied.

The following definitions are used for calculation of ratios for HT (Home Team) and AT (Away Team). A sample of this is shown in Table 7, for the sports data (Cricket) collected from Github.

- Won_Ratio_H2H = Ratio between Total H2H Won and Total H2H Matches
- Lost_Ratio_H2H = Ratio between Total H2H Lost and Total H2H Matches
- Tied_Ratio_H2H = Ratio between Total H2H Tied and TotalH2H Matches
- Won_Ratio_HT_Year = Ratio between Total HT_Year Won and Total HT_Year Matches
- Lost_Ratio_HT_Year = Ratio between Total HT_Year Lost and Total HT_Year Matches
- Tied_Ratio_HT_Year = Ratio between Total HT_Year Tied and Total HT_Year Matches
- Won_Ratio_AT_Year = Ratio between Total AT_Year Won and Total AT_Year Matches
- Lost_Ratio_AT_Year = Ratio between Total AT_Year Lost and Total AT_Year Matches
- Tied_Ratio_AT_Year = Ratio between Total AT_Year Tied and Total AT_Year Matches

Step-4: To Predict the Winner of the Series Two Countries [Case Study: India vs Australia]

- Predict the series output proportion as the venues, as well as squad are declared from earlier,
- Data Understanding
    1. Form current ranking in the dataset, current best team is being figured out
    2. Data processing, where two teams which will play between them are taken into consideration.
    3. Feature engineering and mean encoding are applied to the dataset to obtain best features for classification and prediction of the cricket match based on win, lose or draw.

Table 7 Feature details in mean encoding

	0	1	2	3	4
Team	Australia	Australia	Australia	India	Australia
Opposition	Afghanistan	Bangladesh	England	England	Misc
Year	2011–2012	2011–2012	2011–2012	2011–2012	2011–2012
mean_win_diff	1.0	3.0	0.5	1.0	1.5
mean_ave	20.6	40.77	44.445	33.395	20.5675
mean_RPO	4.69	4.89	5.15	5.69	5.0025
Total_H2H_ Mat	1	3	11	9	22
Total_H2H_Won	1	3	6	5	14
Total_H2H_Lost	0	0	5	3	8
Total_H2H_Tied	0	0	0	1	0
Total_HT_year_Mat	53	53	53	52	53
Total_HT_year_Won	33	33	33	31	33
Total_HT_year_Lost	19	19	19	18	19
Total_HT_year_Tied	1	1	1	3	1
Total_AT_year_Mat	1	3	20	20	42
Total_AT_year_Won	1	3	11	11	26
Total_AT_year_Lost	0	0	8	8	14
Total_AT_year_Tied	0	0	1	1	2
Won_Ratio_H2H	1.0	1.0	0.545455	0.555556	0.636364
Lost_Ratio_H2H	0.0	0.0	0.454545	0.333333	0.363636
Tied_Ratio_H2H	0.0	0.0	0.0	0.111111	0.0
Won_Ratio_Ht_Year	0.622642	0.622642	0.622642	0.596154	0.622642
Lost_Ratio_Ht_Year	0.358491	0.358491	0.358491	0.346154	
Tied_Ratio_Ht_Year	0.018868	0.018868	0.018868	0.057692	0.018868
Won_Ratio_At_Year	1.0	1.0	0.55	0.55	0.619048
Lost_Ratio_At_Year	0.0	0.0	0.4	0.4	0.333333
Tied_Ratio_At_Year	0.0	0.0	0.05	0.05	0.047619

Step-5: Data Visualization

In this step, a graphical representation is presented in Fig. 4 to predict the winner of the cricket match series between two teams (Here, between India and Australia) for better understanding of the datasets being used for Win/Loss analysis.

Data visualization in Bar Chart showing the number of wins, losses, and ties for each team in different venue and conditions is presented in Fig. 5. In this, each bar represents a specific team and venue condition combination, and the height of the bar indicates the count of wins, losses, and ties. By analyzing the bar plot, one can gain insights into the strengths and weaknesses of different teams in different venue

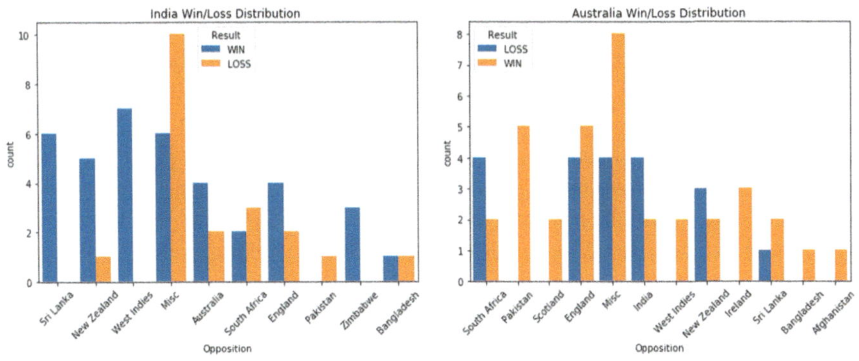

Fig. 4 Comparison of Win/Loss distribution: India Versus Australia

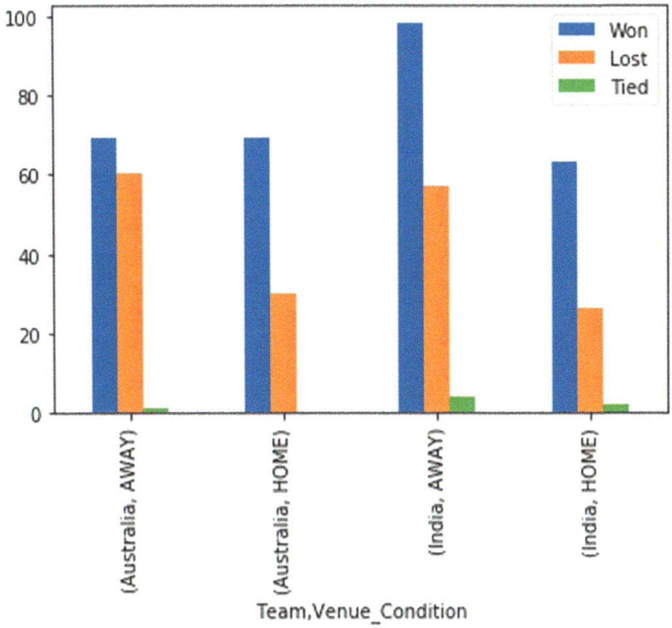

Fig. 5 Analyzing team performance in different venue conditions

conditions, which seems to be a valuable information in assessing team performance, identifying patterns, and making strategic decisions based on the observed trends.

Step-6: Model Modelling and Prediction

The objective of this data modeling approach is to predict the winning team in cricket matches using machine learning techniques that consider multiple independent variables. These variables encompass runs scored, wickets lost, number of deliveries, and dot balls. Notably, the momentum of the game, evaluated through each quarter's 5-over performance, is considered a crucial factor in the model. Consequently, these variables are integrated into the proposed model, avoiding sole reliance on the total score. In instances of missing values for run scores or fall of wickets, imputation is carried out using a constant strategy with a fill value of 0. Despite the potential for a large number of variables resulting from appropriate scaling, feature reduction is executed through principal component analysis (PCA). Hyper-parameter optimization is conducted using Grid Search. Subsequently, model prediction is performed using various classifiers, namely (i) Logistic Regression, (ii) Support Vector Machines (SVM), (iii) Multinomial Naive Bayes, (iv) XG Boost, and (v) Random Forest classifiers. Each classifier undergoes a Grid Search Cross-Validation to determine the best hyper-parameters. The optimal model parameters for each classifier are stored separately. The selection of the best classifier is based on performance measures such as accuracy, mean absolute error (MAE), and mean squared error (MSE), taking runtime considerations into account. The baseline predictor assumes a 50% chance of winning, regardless of the runs scored. The obtained results are presented in Tables 8, 9, and 10.

Based on the dataset with previous records of win/loss between India-Australia series in bilateral cricket matches, the following predictions are obtained using XGBoost classifier. India has a 76.68% probability of winning the 5-match home cricket series against Australia, with a predicted outcome of 4–1. The chance of losing the series is 12.06%, and a draw is anticipated at 11.26%. Conversely, there is a 56.33% probability of India losing the home-away series against Australia in Australia, with a 42.8% chance of winning the series with a predicted outcome of 4–1.

Table 10 illustrates the comparative performance of various algorithms. Notably, the Random Forest algorithm exhibited superior performance, achieving an impressive accuracy score of 93% on the testing dataset. Following closely is the XG Boost

Table 8 Sports prediction of India as a Home Team

Venue_ Condition	Rating_ diff	Mat	mean_ Win_diff	mean_ Ave	Mean_ RPO	Won_ Ratio_ H2H	Lost_ Ratio_ H2H	Tied_ Ratio_ H2H
1	−22	5	−2.0	32.485	5.56	0.25	0.75	0.0

Won_Ratio_ HT_Year	Lost_Ratio_ HT_Year	Tied_Ratio_ HT_Year	Won_Ratio_ AT_Year	Lost_Ratio_ AT_Year	Tied_Ratio_ AT_ Year
0.275862	0.724138	0.0	0.25	0.75	0.0

Probability of Home Team India Winning 0 match in the series is 7.46.
Probability of Home Team India Winning 1 match in the series is 7.47.
Probability of Home Team India Winning 2 match in the series is 8.72.
Probability of Home Team India Winning 3 match in the series is 11.91.
Probability of Home Team India Winning 4 match in the series is 56.33.
Probability of Home Team India Winning 5 match in the series is 8.1

Table 9 Sports prediction of India as a Home-Away Team

Venue_Condition	Rating_diff	Mat	mean_Win_diff	mean_Ave	Mean_RPO	Won_Ratio_H2H	Lost_Ratio_H2H	Tied_Ratio_H2H
0	22	5	2.0	32.485	5.56	0.75	0.25	0.0

Won_Ratio_HT_Year	Lost_Ratio_HT_Year	Tied_Ratio_HT_Year	Won_Ratio_AT_Year	Lost_Ratio_AT_Year	Tied_Ratio_AT_Year
0.732143	0.232143	0.035714	0.75	0.25	0.0

Probability of Home Team Australia Winning 0 match in the series is 14.78.
Probability of Home Team Australia Winning 1 match in the series is 42.8
Probability of Home Team Australia Winning 2 match in the series is 11.41.
Probability of Home Team Australia Winning 3 match in the series is 10.8
Probability of Home Team Australia Winning 4 match in the series is 10.16.
Probability of Home Team Australia Winning 5 match in the series is 10.06.

Table 10 Error performance of classifiers for Cricket match prediction

Name of the algorithms	Algorithms accuracy	(MAE)	(MSE)
Multinomial naive bayes	0.733	0.333333	0.333333
Support vector machines	0.82	0.458333	0.458333
Logistic regression	0.82	0.25	0.25
XG boost	0.923	0.291666	0.375
Random forest	0.934	0.291666	0.291666

classifier with an accuracy of 92.3%, creating a competitive landscape among the top-performing algorithms. SVM and logistic regression share the next position in terms of accuracy.

Moreover, logistic regression demonstrated noteworthy superiority in error measures, showcasing the lowest Mean Absolute Error (MAE) and Mean Squared Error (MSE) at 25%, outperforming the other techniques. Despite differences in accuracy and error measures, all models demonstrated accuracy well above the baseline predictor, which stood at 50%.

To further assess the significance of the observed differences, a two-tailed T-test was conducted between the Random Forest and XG Boost classifier at a 0.05 significance level. The results indicated that both models are equally significant at a p-value of less than 0.05 in accurately predicting cricket match outcomes. This statistical analysis reinforces the efficiency and reliability of both Random Forest and XG Boost classifier in the context of cricket match prediction.

7 Conclusion

Cricket is known for its unpredictability, making even slight accuracy improvements crucial for enhancing the predictive model. The inherent unpredictability of cricket underscores the significance of marginal accuracy enhancements in refining the prediction model. This is particularly crucial in the context of the Cricket match, where strategic decisions heavily depend on accurate forecasts. To address this, machine learning classifiers are employed to advance the prediction model. The analysis scrutinizes various features deemed essential for team management in making effective decisions regarding team selection. The primary goal is to maximize the team's probability of winning.Utilizing five classifiers, the study incorporates preprocessing techniques such as Principal Component Analysis (PCA) with grid search Cross-Validation (CV), taking into account both team and individual player performances. The performance of the model is meticulously presented in terms of accuracy, mean square error, and mean absolute error. Findings reveal that Random Forest and XG Boost classifiers demonstrate equal prowess in accurately predicting match-winning scenarios. However, it is acknowledged that there is room for improvement, and the model's performance can be further enhanced by incorporating additional relevant variables and exploring alternative machine learning algorithms. In conclusion, continual refinement and exploration of the model are deemed crucial for improving its accuracy and predictive capabilities. The dynamic and ever-changing nature of cricket necessitates a proactive approach in adapting the model to evolving scenarios, thereby enhancing its effectiveness. This sets the stage for the future scope of work, paving the way for ongoing advancements in cricket match prediction models.

References

1. Kapadia, K., Abdel-Jaber, H., Thabtah, F., Hadi, W.: Sport analytics for cricket game results using machine learning: An experimental study. Appl. Comput. Inform., (2020)
2. Passi, K., Pandey, N.: Increased prediction accuracy in the game of cricket using machine learning. arXiv preprint (2018). arXiv:1804.04226
3. Vistro, D.M,, Rasheed, F., David, L.G.: The cricket winner prediction with the application of machine learning and data analytics. Int. J. Sci. Technol. Res., 8(9) (2019)
4. Horvat, T., Job, J.: The use of machine learning in sport outcome prediction: A review. Wiley Interdisciplinary Reviews: Data Mining and Knowledge Discovery **10**(5), e1380 (2020)
5. Iyer, S.R., Sharda, R.: Prediction of athlete's performance using neural networks: An application in cricket team selection. Expert Syst. Appl. **36**(3), 5510–5522 (2009)
6. Lamsal, R., Choudhary, A.: Predicting outcome of Indian Premier League (IPL) matches using classification based machine learning algorithm, (1809). https://arxiv.org/abs
7. Keshtkar, M., Yamaghani, M.: Sports result prediction based on machine learning and computational intelligence approaches: A survey. J Adv Comput Eng Technol **5**(1), 27–36 (2019)
8. Jayalath, K.P.: A machine learning approach to analyze ODI cricket predictors. J Sport Anal **4**(1), 73–84 (2018)

9. Munir, F., Hasan, M., Ahmed, S.: Predicting a T20 cricket match result while the match is in progress (Doctoral dissertation, Brac. University) (2015)
10. Nimmagadda, A., Kalyan, N.V., Venkatesh, M., Teja, N.N.S., Raju, C.G.: Cricket score and winning prediction using data mining. Int J Adv Res Dev **3**(3), 299–302 (2018)
11. Bunker, R.P., Thabtah, F.: A machine learning framework for sport result prediction. Appl Comput Inform **15**(1), 27–33 (2019)
12. Kampakis, S., Thomas, W.: Using machine learning to predict the outcome of english county twenty over cricket matches. arXiv preprint. (2015). arXiv:1511.05837
13. Vayatis, N., Azencott, R.: How to estimate the Vapnik-Chervonenkis dimension of support vector machines through simulations? support vector machines: theory and applications," eds.: Paliouras, G., Karkaletsis, V., ACAI, 99 (1999)
14. Evgeniou, T., Pontil, M.: Support vector machines: Theory and applications. In Adv. Course. Artif. Intell. Springer, Berlin, Heidelberg. pp. 249–257 (1999)
15. Abbas, M., Memon, K.A., Jamali, A.A., Memon, S., Ahmed, A.: Multinomial Naive Bayes classification model for sentiment analysis. IJCSNS Int J Comput Sci Netw Secur **19**(3), 62 (2019)
16. Wickramasinghe, I.: Naive Bayes approach to predict the winner of an ODI cricket game. J Sport Anal **6**(2), 75–84 (2020)
17. Peng, C.Y.J., Lee, K.L., Ingersoll, G.M.: An introduction to logistic regression analysis and reporting. J. Educ. Res. **96**(1), 3–14 (2002)
18. Hu, T., Song, T.: Research on XGboost academic forecasting and analysis modelling. In Journal of Physics: Conference Series. IOP Publishing 1324(1), p 012091 (2019)
19. Patil, N.: Dalgade D Cricket prediction using random forest regression
20. Belete, D.M., Huchaiah, M.D.: Grid search in hyperparameter optimization of machine learning models for prediction of HIV/AIDS test results. Int. J. Comput. Appl. Taylor and Francis. **44**(9), 875–886 (2022)
21. Pathak, N., Wadhwa, H.: Applications of modern classification techniques to predict the outcome of ODI cricket. Procedia Computer Science **87**, 55–60 (2016)
22. Github Sports prediction dataset. https://github.com/jsaikmr/Sports_Analytics_Hackathon (Accessed on 05.06.2023)

Nanometer-Range Optimization in a 4x4 Array of 7T SRAM Cell

Deepak Garg and Devendra Kumar Sharma

Abstract Modern electronic appliances, including logic circuits and memory based devices, are being pushed to their limits by aggressive 2D scaling. However, it is inevitable that leakage will increase as generation of technology forward one to next. As a result, the demand for modelling leakage using various leakage reduction approaches has grown significantly, and the identification and simulation of various leakage contributing components has become a key aspect for estimating various leakage characteristics. In this research, researcher explores and then designs several circuits for self-controllable voltage levels (SVLs) in a 7 T SRAM cell organized in a 4 $\times$ 4 array. This method is a power-conserving model in which the supply voltage is at its highest during the active mode and at its lowest during the standby mode. The circuit has been simulated in cadence virtuoso using 45 nm technology at supply voltages of 1 V, 0.7 V, 0.5 V, and 0.3 V, with the results validated.

Keywords Array · SRAM cell · SVL · Leakage current · Low power · VLSI · Logic circuit

1 Introduction

The continuous increase in scaling of devices in order to accomplish modern high performance integrated circuits has lead to increase in chip density involving hundred million transistors per chip. This has escorted to design complexities with significant increase in leakage currents and power. Along with the leakage current data stability during read and write is also one of the major factors which should be kept in consideration while designing memory design. In order to reduce the number of parameters affected by leakage, the usage of SVL in 7 T SRAM has been suggested. This has prompted the exploration of leakage reduction approaches in the design of low-power circuits. Many common practices exist for creating such circuits. The

D. Garg (✉) · D. K. Sharma
Department of Electronics and Communication Engineering, Delhi-NCR Campus, SRM Institute of Science and Technology, Ghaziabad, India
e-mail: deepakgarg1985@gmail.com

S. Kumar Udgata et al. (eds.), *Intelligent Systems*, Lecture Notes in Networks and Systems 1149, https://doi.org/10.1007/978-981-97-8160-7_14

power supply may be lowered to a level near or below the threshold voltage [1, 2]. However, its functionality is compromised when the nominal voltage is lowered, and the supply voltage suffers as a result, with the cell's stability, noise margin, on/off current ratio, and sensitivity to PVT (process voltage temperature) changes all suffering [3].

The need for electronics that are smaller, more portable, and more easily customized is growing. The primary considerations in creating an array are its power consumption, size, and high speed. The constant shrinking of device sizes has created issues for MOSFETs [2]. Techniques for minimizing leakage in totally exhausted SOI devices at high speeds are discussed. In order to store a large amount of data that can yet be quickly retrieved, memory arrays are used. Flash memory was created by Fujio Masuoka in the late 1970s. The fast circuitry and memory he developed were his contribution. At IBM, DRAM was conceived by Robert Dennard in 1966, and by the mid-1970s, it was almost ubiquitous. However, because to the disadvantages discussed above, SRAM entered the scene, and it is now often used in digital ICs [4]. These days, data storage is a primary consideration in the development of most electronic parts. Memory is required for the storage of information. Its significance may be inferred from its extensive use in semiconductor memories, computers, and high-performance microprocessors where cache memories play a crucial role. RAM (random access memory) allows for arbitrary reading and writing to storage places and is therefore distinguished from ROM (read only memory). Both static random access memory (SRAM) and dynamic random access memory (DRAM) are common forms of RAM. DRAM has a number of limitations, including the fact that it requires a capacitor to store each individual bit, which means that the bits are lost whenever the capacitor is charged or discharged [5, 6]. All the characteristics of Static Random Access Memory (SRAM) are implied by its name. The term "static" indicates that there is no floating node situation in the circuit, and that all components are permanently linked to either VDD or VSS. The term "random" implies that information is always available, regardless of when or where it was last accessed. That which is accessed is sought, and that which is stored is bits. One bit is stored in each cell.

SVL technique is used to reduce leakage power during standby mode. In one of the technique of SVL it supplies maximum supply voltage during and in another technique it raises the ground potential. Analysis of different leakage parameters in 7 T SRAM has been conceded out with different leakage reduction techniques [7] and the results have been compared and same has been used in 4 $\times$ 4 arrays. Power consumption during read and write operation at different VDD of 1 V, 0.7 V, 0.5 V and 0.3 V.

2 4 x 4 Arrays with 7 T SRAM Basic Unit Cell and SVL Technique

Array is shown as 2N x M in Fig. 1, where N is the number of address bits or inputs to a decoder or the depth of the array and M is the number of data bits, which is also the width of the array. 4 × 4 array can be written as 22 × 4, which shows that there will be two address bits (N = 2) and four data bits (the width of bits is 4bit) [8, 9].

All the necessary parts for creating an array are shown in Fig. 1. It demonstrates the essential parts of an array's workings, such as the decoder, pre charge circuit, write driver circuit, and sensing amplifier. There is a special role for each component in the array. A decoder's output consists of rows of lines, which are called word lines, and columns of lines, which are called bit lines. Each SRAM cell is put in a distinct cell, which is formed at the intersection of two rows or columns of lines. In this article, the 7 T SRAM is utilized with various SVL (LSVL, USVL, and LSVL + USVL) methodologies and the results of various parameters are compared with those obtained from a straightforward 7 T SRAM-based array.

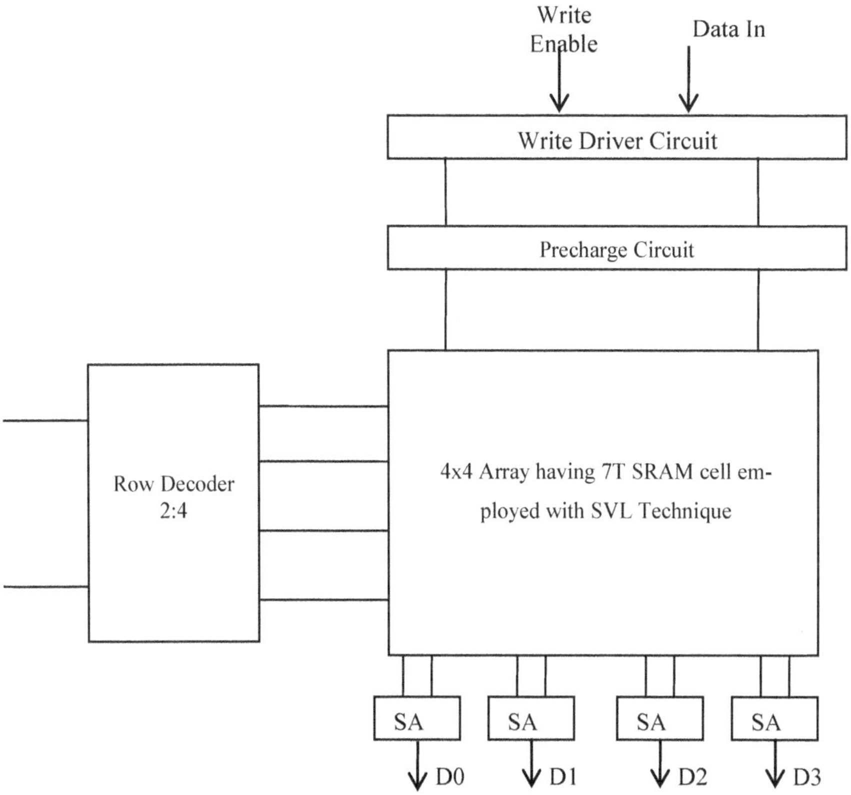

Fig. 1 4 × 4 Array with different SVL technique employed 7 T SRAM cells

2.1 Decoder

A decoder is a tool used to choose a certain select line or word line. Since only two address bits are needed for a 4 × 4 array, a 2:4 decoder is employed in its implementation. Bits 00, 01, 10, and 11 make up the address space. The decoder selects a row whose corresponding output becomes high based on the applied bit [10].

2.2 Precharge and Write Driver Circuit

When the array is not being used for a read or write operation, the bit lines are charged using the pre charge circuit. This circuit makes use of three PMOS transistors. When the WE (Write Enable) is on, the write driver circuit is employed to record the data. When the circuit is not operating, the pre-charge circuit charges the bit lines to high levels; the write driver circuit then releases one of these bit lines [11].

2.3 Sense Amplifier

Research has shown that there are many different ways to build a sense amplifier. They are one of the means via which read operation speed is specified. The sense amplifier's primary function is to enhance the weak signal by measuring the voltage difference between the two bit lines. Turning on the sense enable (SE) is equivalent to turning on the sense amplifier. When performing a read operation, one of the bit lines may get discharged, necessitating the usage of a sensing amplifier [12, 13].

2.4 SVL Technique Employed 7 T SRAM Cell

Switches may be positioned at the top of the cell, the bottom of the cell, or both the top and the bottom using a method called self-controllable switching (SVL). Figures 2, 3 and 4 shows the circuit diagram of 7 T SRAM with USVL, VSVL and Combined (USVL + VSVL) technique respectively. The supply voltage is lowered by the USVL method, while the ground node voltage is increased via the LSVL method. By combining the USVL and LSVL approaches in a normal 7 T SRAM cell, the supply voltage is lowered while the ground node voltage is increased, resulting in much lower leakage current, leakage voltage, and leakage power. Using this method, the 4 × 4 array's power usage during read and writes an operation is reduced.

In Active mode, switches are enabled or disabled in response to the input supplied to the clock (clk), which acts as a kind of control signal. When CLK is 1, PU1 is

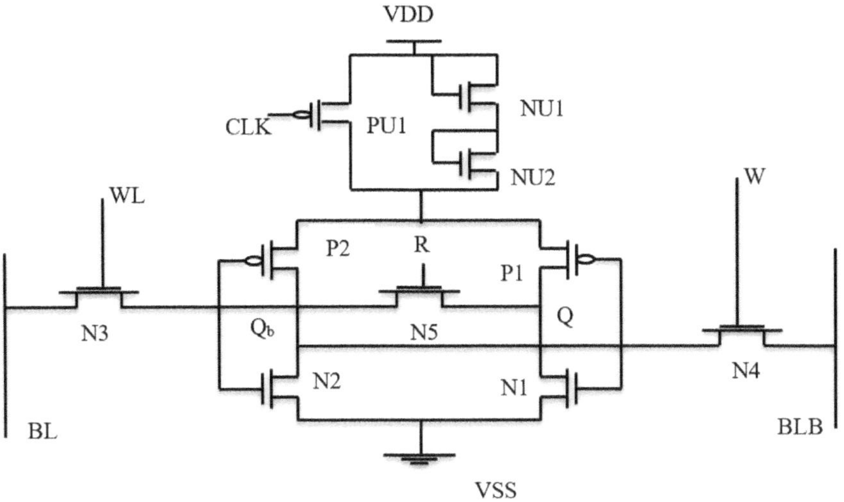

Fig. 2 7 T SRAM cell with USVL technique

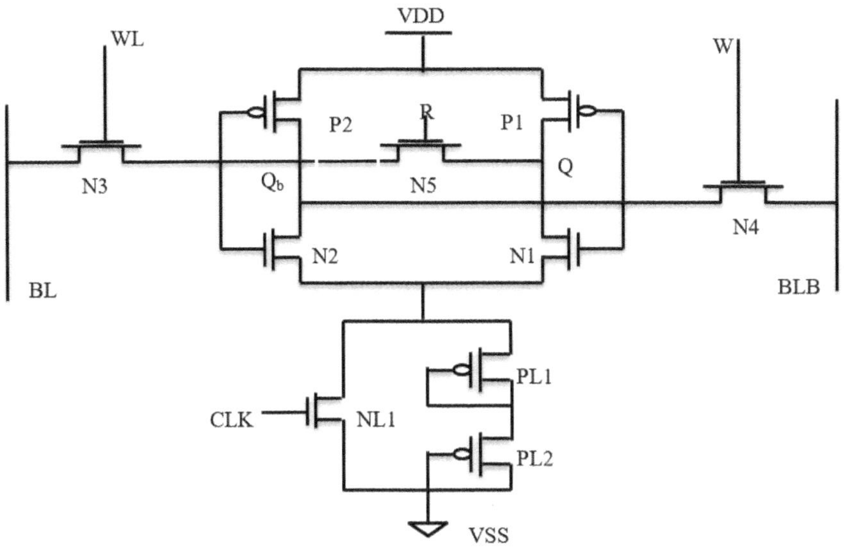

Fig. 3 7 T SRAM cell with LSVL technique

disabled and NU1,2 are switched on weakly, supplying VDD directly via the ON switches, and when CLK is 0, PL1,2 are turned ON weakly and NL1 is disabled, supplying VSS to the standby SRAM.

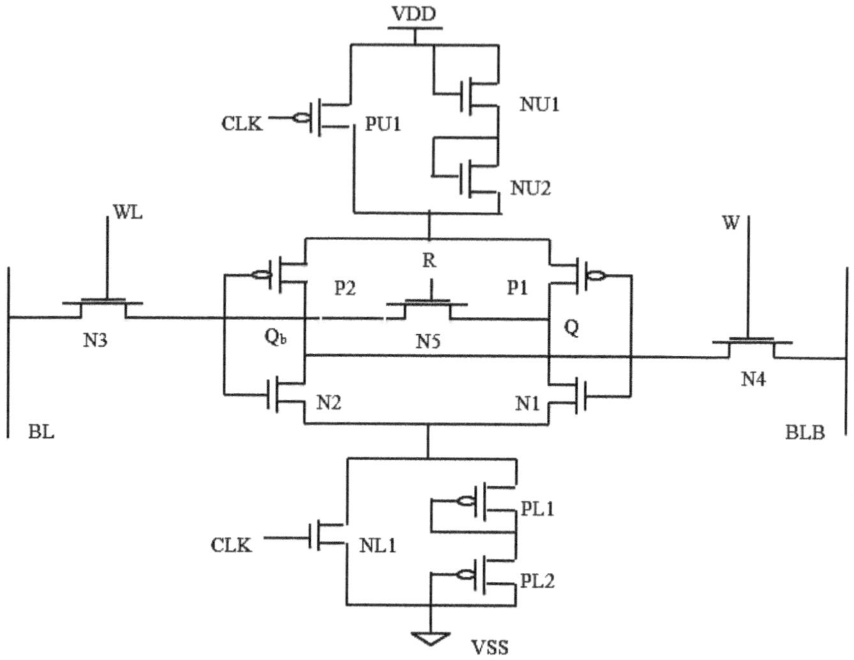

Fig. 4 7 T SRAM cell with USVL and LSVL technique

3 Simulation and Results

At 45 nm technology, the cadence virtuoso tool has simulated a 4 × 4 array in which the SVL approach used a 7 T SRAM cell as the fundamental unit cell. In this paper, the leakage parameters have been calculated for 0.3 V, 0.5 V, 0.7 V, and 1 V supply voltages for USVL, LSVL, and USVL + LSVL, respectively. Here, power consumption of a 4 × 4 array is estimated at several supply voltages and compared the findings as provided by Tables 1, 2 and 3.

All the leakage parameters and power consumed by 4 × 4 arrays at different supply voltages with USVL, LSVL and Mixed SVL techniques during READ operation has been represented in Figs. 5, 6 and 7 respectively.

During a READ operation, the graphs show the current, voltage, and power leakage characteristics of a 7 T SRAM cell using USVL, LSVL, and USVL + LSVL as one of the basic unit cells in a 4 × 4 array.

During the WRITE operations, all the leakage parameters and power consumed by 4 × 4 arrays at different supply voltages with USVL, LSVL and Mixed SVL techniques has been represented in Figs. 8, 9 and 10 respectively.

Power consumption of 4 × 4 arrays with different SVL techniques during READ and WRITE operation is shown in Fig. 11.

Table 1 Leakage parameters during read operation in multiple SVL technique

	Parameters		
	Leakage current	Leakage voltage	Leakage power
0.3V USVL	29.98 Pa	2.106 mV	1.936 nW
0.3V LSVL	630.84 fA	14.30 mV	1.728 nW
0.3V USVL + LSVL	527.63 fA	10.09 mV	1.342 nW
0.5V USVL	2.252 pA	21.63 mV	6.129 nW
0.5V LSVL	613.4 fA	5.727 mV	7.783 nW
0.5V USVL + LSVL	512.3 fA	4.127 mV	5.23 nW
0.7V USVL	4.55 pA	12.33 mV	9.757 nW
0.7V LSVL	513.32 fA	2.009 mV	12.11 nW
0.7V USVL + LSVL	612.93 fA	2.004 mV	10.63 nW
1V USVL	9.738 pA	9.753 mV	34.69 nW
1V LSVL	681.41 pA	2.911 mV	49.06 nW
1V USVL + LSVL	621.34 pA	1.93 mV	45.32 nW

Table 2 Leakage parameters during write operation in multiple SVL technique

	Parameters		
	Leakage current	Leakage voltage	Leakage power
0.3V USVL	34.45 pA	5.245 mV	2.001 nW
0.3V LSVL	731.87 fA	15.08 mV	1.752 nW
0.3V USVL + LSVL	729.73 fA	12.12 mV	2.01 nW
0.5V USVL	63.12 pA	57.03 mV	23.73 nW
0.5V LSVL	649.82 fA	30.18 mV	5.982 nW
0.5V USVL + LSVL	625.21 fA	20.15 mV	7.32 nW
0.7V USVL	67.60 pA	57.03 mV	23.73 nW
0.7V LSVL	670.81 fA	55.10 mV	42.24 nW
0.7V USVL + LSVL	631.52 fA	23.12 mV	21.23 n W
1V USVL	86.09 pA	81.68 mV	79.9 nW
1V LSVL	699.04 fA	87.09 mV	1.027 nW
1V USVL + LSVL	611.12 fA	21.01 mV	39.42 nW

Table 3 Power consumption in 4X4 array during Read and write operation at multiple power supply voltage

	Power supply voltage (VDD)			
	0.3V	0.5V	0.7V	1V
READ USVL	46.33 pW	29.08 pW	39.04 pW	176.52 pW
READ LSVL	36.04 pW	23.12 pW	42.11 pW	169.18 pW
READ USVL + LSVL	40.12 pW	21.23 pW	30.12 pW	153.23 pW
WRITE USVL	46.02 pW	29.72 pW	48.45 pW	236.8 pW
WRITE LSVL	1.752 pW	29.71 pW	46.45 pW	230.23 pW
WRITE USVL + LSVL	44.24 pW	25.31 pW	32.35 pW	201.53 pW

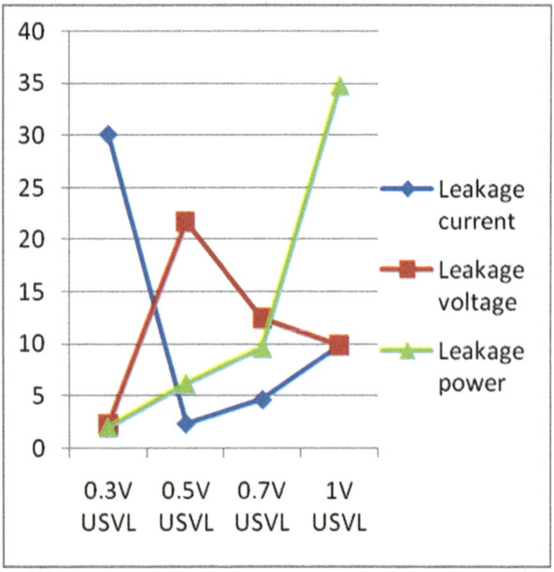

Fig. 5 READ operation (USVL) technique

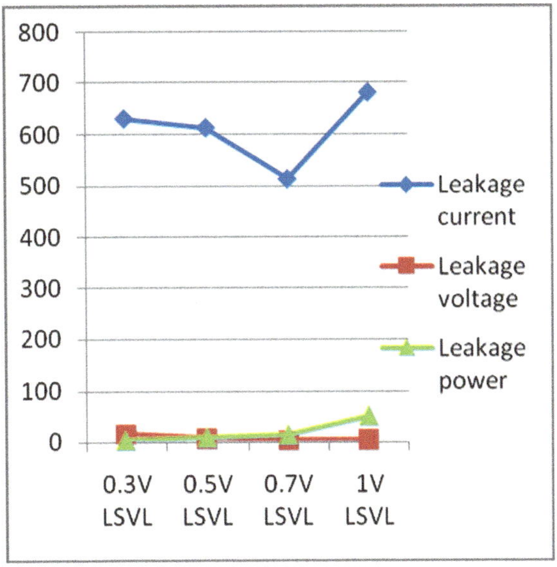

Fig. 6 READ operation (LSVL) technique

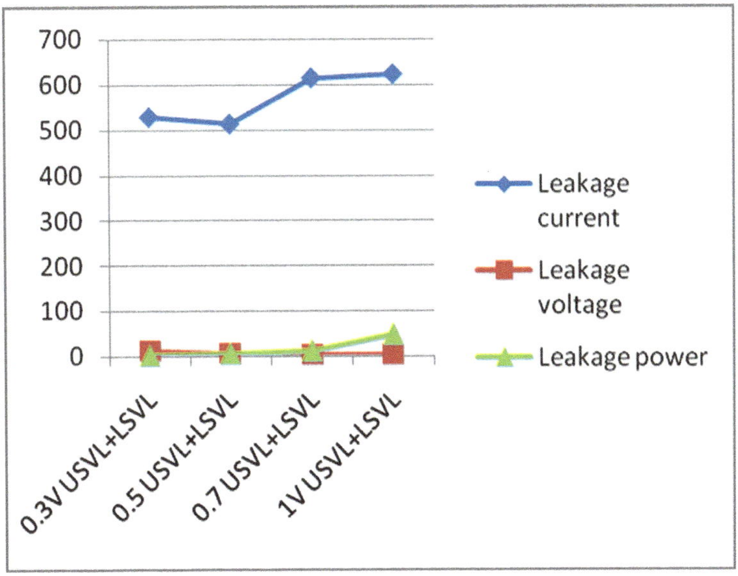

Fig. 7 READ operations (USVL + LSVL) technique of 7 T SRAM cell

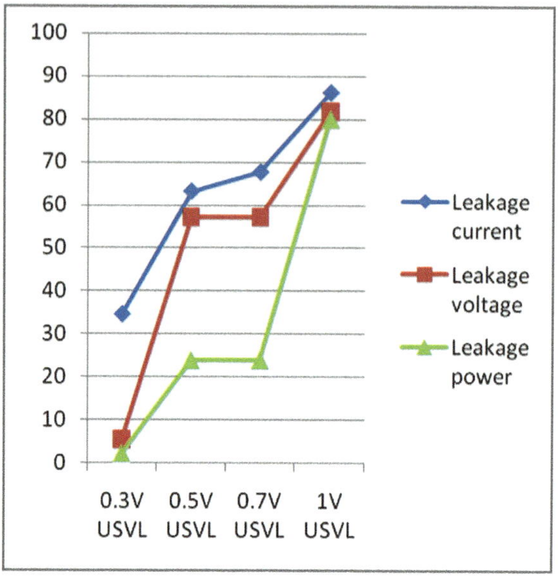

Fig. 8 WRITE operation USVL technique

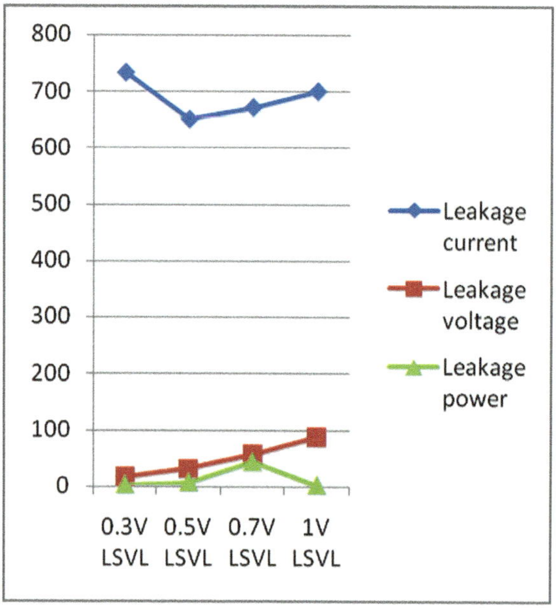

Fig. 9 WRITE operation LSVL technique

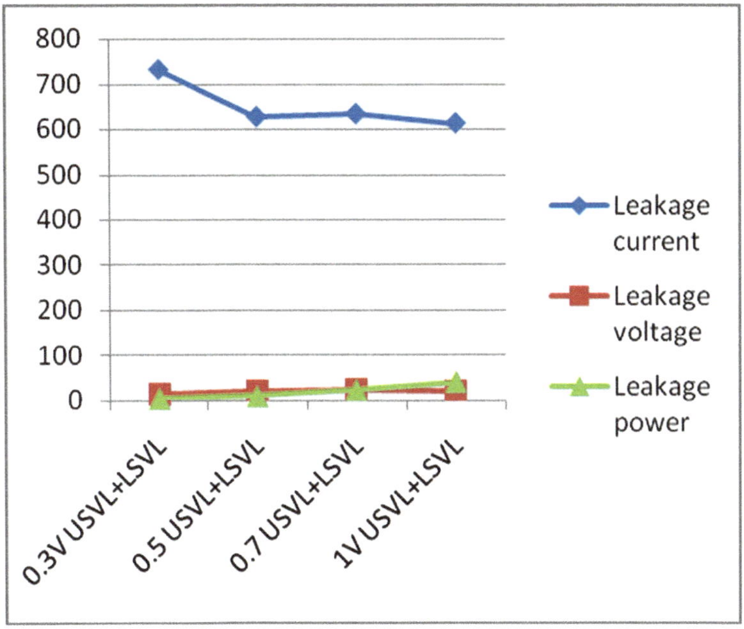

Fig. 10 WRITE operations (USVL + LSVL) technique

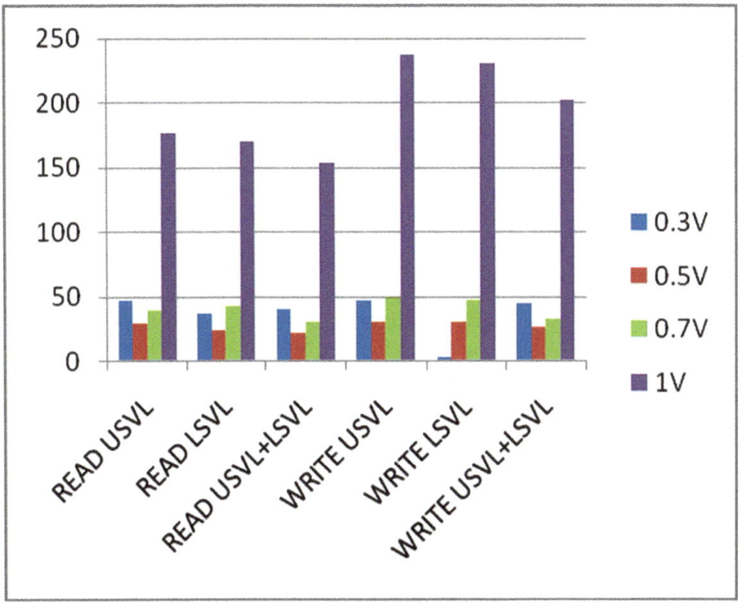

Fig. 11 Power consumption of 4 × 4 arrays with different SVL techniques during READ and WRITE operation

4 Conclusion

Using low Vt and high Vt devices cautiously to avoid compromising on leakage and performance has become a hamster wheel of supply voltage VDD, threshold voltage Vt, power dissipation, and sub-threshold current. Leakage parameters for a 7 T SRAM cell using SVL technique as the basic unit cell in a 4 × 4 array were analyzed using the cadence virtuoso tool, 45 nm technology, and the results showed that combined SVL was superior to USVL and LSVL in terms of reducing leakage. It is also evident from the graphs that the optimal findings are shown at 0.5 V, suggesting that the supply voltage has a significant impact in predicting the leakage values. It is also deduced that the power needed to perform a read operation is much less than the power needed to do a write operation due to the increased complexity of the write driver circuit.

References

1. Gavaskar, K., Priya, S.: Design of efficient low power stable 4-bit memory cell. Int. J. Comput. Appl. **84.1** (2013)
2. Wen, L., Li, Z., Li, Y.: Single-ended, robust 8T SRAM cell for low-voltage operation. Microelectron. J.. J. **44**(8), 718–728 (2013)
3. Calhoun, Benton H., Anantha Chandrakasan.: A 256kb sub-threshold SRAM in 65nm CMOS. In: 2006 IEEE International Solid State Circuits Conference-Digest of Technical Papers. IEEE, (2006)
4. Raj, Nikhil, Rohit Lorenzo.: An effective design technique to reduce leakage power. In: 2012 IEEE Students' Conference on Electrical, Electronics and Computer Science. IEEE, (2012
5. Shyam, A., Sharma, S.: High density and low leakage current based SRAM cell using 45 nm technology. Int. J. Electron. **100**(4), 536–552 (2013)
6. Mishra, S., et al.: Design and simulation of high level low power 7T SRAM cell using various process & circuit techniques. In: 2012 IEEE International Conference on Signal Processing, Computing and Control. IEEE, (2012)
7. Garg, D., Sharma D.K.: Towards the evaluation from low power VLSI to quantum circuits. Quantum-Dot Cell. Autom. Circuits Nanocomputing Appl.. CRC Press. 1–24 (2023)
8. Shyam A., Shastri M., Sharma S.: Multi Vt 7T SRAM cell for high speed application at 45 nm technology. In: International Conference on Nanoscience, Engineering and Technology (ICONSET 2011). IEEE, (2011)
9. Alorda, B., et al.: 8T vs. 6T SRAM cell radiation robustness: A comparative analysis. Microelectron. Reliab.. Reliab. **51**(2), 350–359 (2011)
10. Garg, A., Tony Tae-Hyoung Kim.: SRAM array structures for energy efficiency enhancement. In: IEEE Transactions on Circuits and Systems II: Express Briefs 60.6: 351–355 (2013)
11. Rabaey, Jan M., Anantha P. Chandrakasan, Borivoje Nikolic.: Digital integrated circuits. Vol. 2. Englewood Cliffs: Prentice hall, (2002)
12. Baker, R. Jacob.: CMOS: circuit design, layout, and simulation. John Wiley & Sons, (2019)
13. Evans, Robert Joseph, Paul D. Franzon.: Energy consumption modeling and optimization for SRAM's. In: IEEE Journal of Solid-State Circuits 30.5 571–579 (1995)

Wild Horse Optimization Algorithm-Based Energy-Efficient Task Scheduling for Fog Computing Environment

J. Sengathir, M. Deva Priya, and A. Christy Jeba Malar

Abstract In the recent past, huge amount of data is generated during processing in a Cloud Computing (CC) environment, resulting in a systematic increase in load. Internet of Things (IoT)-based applications in general necessitate the combination of both CC and Fog Computing (FC) to overcome the limitations of response time and latency. In specific, FC facilitates scheduling of tasks over different fog nodes in the network. However, as the resources of fog nodes are limited, achieving optimal task scheduling is identified to be more challenging. In this paper, Wild Horse Optimization Algorithm-based Energy-Efficient Task Scheduling (WHOAEETS) algorithm is proposed for achieving better response time and reduced latency during task execution in an FC environment. WHOAEETS algorithm schedules tasks on the fog server based on fitness evaluation parameters that include execution time, energy and makespan. Wild Horse Optimizer (WHO) is adopted for scheduling tasks to ideal fog servers by using search agents that explore and exploit factors of scheduling in search space. Simulation results of the proposed WHOAEETS algorithm confirm 8.32% minimized energy consumption, 9.61% reduced execution time and 11.42% minimized makespan in contrast to baseline approaches taken for evaluation.

Keywords Energy efficiency · Task scheduling · Fog servers · Wild Horse Optimization Algorithm (WHOA) · Fog computing

---

J. Sengathir (✉)
Department of Information Technology, CVR College of Engineering, Vastunagar, Mangalpally, Hyderabad 501 510, Telangana, India
e-mail: j.sengathir@gmail.com

M. Deva Priya
Department of Computer Science and Engineering, Karpagam College of Engineering, Coimbatore, Tamilnadu, India
e-mail: meetdevapriya@gmail.com

A. Christy Jeba Malar
Department of Information Technology, Sri Krishna College of Technology, Coimbatore, Tamilnadu, India
e-mail: a.christyjebamalar@skct.edu.in

S. Kumar Udgata et al. (eds.), *Intelligent Systems*, Lecture Notes in Networks and Systems 1149, https://doi.org/10.1007/978-981-97-8160-7_15

1 Introduction

Though clouds are capable of handling huge quantities of data, they face problems owing to inadequate resources [1, 2, 3]. Real-time applications demand reduced latency and Cloud Computing (CC) does not support this. To handle as well as process huge volumes of data, CC with Fog Computing (FC) seems to be an emerging solution [4].

Task offloading is inappropriate for delay-sensitive applications as cloud servers are remotely positioned. To deal with this issue, FC also called "Cloud at the Edge" is a balancing solution [5]. Though FC brings computing as well as radio resources closer to mobile devices, the nodes are not capable of sufficiently meeting user demands owing to inadequate computing resources.

Nevertheless, incompetent scheduling of tasks in FC can lead to more delays in contrast to CC [6]. It offers improved QoS to users through CC which is overwhelmed by increased data flow. Transmitting data to cloud and getting them back involves increased latency and demands more amount of bandwidth. FC enhances QoS of delay-sensitive applications by permitting applications to take benefit of reduced latency in contrast to CC. It involves reduced latency and response time for IoT-based applications [7]. It enables users to process data that is close to the end user. It facilitates data collection from devices and pre-processes them before forwarding them for processing in CC environment. This makes data processing quicker. Fog nodes positioned closer to the user focus on processing data by satisfying user demands [8].

FC facilitates scheduling of application tasks on fog nodes in the network. As these nodes have restricted amount of resources when compared to CC, Task Scheduling (TS) among fog nodes is essential. TS is a significant factor that contributes to energy consumption in fog servers. It is necessary to effectively perform LB on fog nodes so as to offer improved Quality of Service (QoS) to IoT applications based on response time and efficient use of fog nodes [9]. TS focuses on task completion in a particular amount of time by using limited amount of resources. Tasks in diverse applications should be efficiently distributed over fog nodes to enhance QoS, specifically response time of tasks.

Task completion within allocated time is challenging in FC due to increased quantity of data to be processed [10]. Hence, scheduling tasks as well as resources is crucial. Further, scheduling tasks on Virtual Machines (VMs) in fog nodes is an NP-hard problem as it involves exponential complexity with increase in quantity of fog as well as sensor nodes. Hence, traditional greedy methods cannot be applied [11].

In this paper, Wild Horse Optimization Algorithm-based Energy-Efficient Task Scheduling (WHOAEETS) is proposed for achieving better response time and low latency during task execution in FC environment. This WHOAEETS algorithm schedules tasks on fog server depending on fitness evaluation parameters that include execution time, energy and makespan. Wild Horse Optimizer (WHO) is adopted for scheduling tasks to ideal fog servers using search agents that explore and exploit factors of scheduling in the search space. The simulation experiments of the proposed WHOAEETS algorithm and benchmarked approaches are evaluated in terms of fitness assessment parameters for varying time and energy trade-off.

2 Related Work

Energy overhead at FC servers is the main constraint for applications as transmitting huge volumes of data involves increased amount of network bandwidth and reduces responsive time. Abdel-Basset et al. [12] have proposed an energy-based model using Marine Predators Algorithm (MPA) for dealing with TS in FC (TSFC) which focuses on improving QoS. Two versions of MPA are proposed. Modified MPA (MMPA) modifies MPA to enhance exploitation capability by employing last updated locations rather than best recent one. MMPA is enhanced by using ranking-based re-initialization along with mutation towards the best. It randomly initializes half of the population after some predefined amount of iterations to avoid getting stuck into local optima. The final half of population is mutated toward best solution determined so far. MPA is continuous, while TSFC is found to be discrete. The proposed scheme uses normalization as well as scaling for converting MPA into discrete one. It includes metaheuristic algorithms and Genetic Algorithm (GA) and performance is analyzed based on makespan, amount of energy consumed, flow time and quantity of CO_2 emission. Enhanced MMPA offers better results in contrast to other schemes.

Hussein and Mousa [13] have used two nature-inspired metaheuristic schedulers for performing TS in fog environment. Particle Swarm Optimization (PSO)-based scheduler deals with effective balancing of IoT tasks on fog nodes and involves reduced communication cost as well as response time. The main focus is to fulfill QoS demands of delay-sensitive applications by offloading tasks involving sensor data onto fog nodes by considering communication costs as well as present load on nodes. The propounded ACO-based scheduler involves reduced response time of IoT applications in contrast to Round Robin (RR) and proposed PSO-based algorithms. The proposed scheduler balances load effectively among fog nodes.

Hosseinioun et al. [14] have propounded an energy-based scheme using Dynamic Voltage and Frequency Scaling (DVFS) method which focuses on reducing the amount of energy consumed. Further, it applies hybrid Invasive Weed Optimization and Culture Algorithm (IWO-CA), an energy-based evolutionary TS scheme for constructing acceptable task sequences. The proposed scheme offers enhanced performance in contrast to present algorithms based on the amount of energy consumed as well as level of resource utilization. To reduce delays in responding to user demands, it is essential to establish a trade-off amid execution of local request on end devices as well as fog environment. Keshavarznejad et al. [15] have dealt with task offloading as an optimization problem by involving multiple objectives with an emphasis on dropping total power as well as delay involved in executing tasks. It formulates multivariate optimization problem with several objectives based on the quantity of energy consumed, delay involved in local execution as well as fog and cloud offloading. Meta-heuristic approaches including Non-Dominated Sorting Genetic Algorithm (NSGA-II) and Bees algorithm are employed to perform delay-based optimization involving reduced delay and energy in heterogeneous fog environment. It is seen that the proposed scheme offers an improved trade-off based on offloading probability and power necessary for transmission of data.

To deal with issues related to resource utilization, latency and response time, ideal and efficient schemes are essential for scheduling. Potu et al. [16] have designed Extended PSO (EPSO) algorithm with additional gradient scheme for optimizing TS in cloud–fog environment. The primary goal is to enhance the efficiency of resources and minimize the time taken for task completion. Performance is analyzed in terms of makespan as well as total cost. Dev et al. [17] have designed a hybrid meta-heuristic method using Grey Wolf Optimization (GWO) algorithm that is combined with PSO (HPSO_GWO) to schedule tasks to VMs and obtain optimized QoS. AL-Amodi et al. [18] have proposed an energy-based TS algorithm that is based on Harris Hawks Optimization (HHO) scheme to enhance QoS with Service Level Agreement (SLA). The proposed scheme outperforms other existing schemes based on the amount of energy consumed and other QoS factors.

Efficient TS reduces computation complexity as well as latency while improving QoS. Yadav et al. [19] have propounded a hybrid scheme combining heuristic and meta-heuristic approaches for scheduling tasks. Fireworks Algorithm (FWA), a meta-heuristic scheme is combined with Heterogeneous Earliest Finish Time (HEFT), a heuristic approach. This bi-objective optimization method reduces makespan as well as cost. Performance is analyzed for distinct scientific workflows. The proposed BH-FWA scheme offers better results in terms of cost, makespan and throughput. Vispute and Vashisht [20] have proposed a scheduling scheme called Energy-Efficient TS-based on PSO (EETSPSO) in FC. It offers better performance in terms of energy, makespan and execution time. Iftikhar et al. [21] have used Gated Graph Convolution Network (GGCN) called HunterPlus that examines the influence of Gated Recurrent Unit (GRU) of GGCN in extending to Bidirectional GRU. It uses Convolutional Neural Networks (CNNs) for optimizing cloud–fog TS. CNN-based scheduler outperforms GGCN-based models in terms of energy consumed per task and rate of job completion. It assesses graph inputs in forward as well as backward directions. CNN model takes host as well as task metrics in the form of pre-processed 2D array and maps them to the energy metric.

3 Proposed Wild Horse Optimization Algorithm-Based Energy-Efficient Task Scheduling (WHOAEETS) for FC Environment

WHOAEETS proposed for achieving TS in the FC environment concentrates on the process of assigning IoT tasks to suitable fog nodes from the set of available candidate fog nodes for achieving efficient use of energy and reduction of time in FC environment.

In this context, IoT tasks in the implementation environment are represented using a set , $T_{IoT} = \{T_1, T_2, \ldots\ldots., T_i, \ldots..T_n\}$, wherein each task "$T_i$" is explicitly presented using a set of attributes. The set of attributes $T_i = \{D_{Task}, T_{Type}, DS_{Input}\}$ represent Deadline of task, Type of delay sensitive tasks and input Data Size respectively. In specific, "T_{Type}" and "DS_{Input}" indicate whether IoT tasks are computational-intensive or delay-sensitive, and the density of computing essential for task execution respectively. Moreover, "D_{Task}" represents the deadline or time incurred for executing a specific IoT task in the FC environment. On the other hand, the fog nodes are denoted as a set $F_{Nodes} = \{f_1, f_2, \ldots\ldots., f_j, \ldots..f_m\}$, in which each node is represented using a set of attributes $f_j = \{FN_{BC}, FN_{CC}, FN_{SC}\}$ that includes the total Battery Capacity, Computing Capacity which is measured in terms of CPU cycles per bit and Storage Capability assessed in terms of bits of fog nodes respectively. Depending on the above definitions, the TS problem in FC concentrates on the process of assigning "n" number of IoT tasks to "m" number of fog nodes such that optimization of QoS factors is achieved. In other words, the process of assigning an IoT task to a fog node can be represented as a function $F_j : TS_i \rightarrow f_j$ based on the quantity of energy consumed and total time. In this case, the total time incurred (TT_{Tasks}) for mapping the incoming IoT tasks "TS_i" to the fog node "f_j" is mathematically represented using Eq. (1).

$$TT_{Tasks}(F_j) = T_{up}(F_j) + T_{down}(F_j) + T_{Execute}(F_j) \tag{1}$$

where "$T_{up}(F_j)$" indicates the time incurred for transmitting and offloading end user input tasks onto fog nodes, and "$T_{down}(F_j)$" represents the time incurred for transmitting the result of executed task from fog nodes to end user.

Moreover, "$T_{Execute}(F_j)$" denotes the time associated with task execution in fog nodes in the implementation environment. However, "$T_{down}(F_j)$" introduces a minor impact over task execution in FC environment since the output data size is comparatively smaller than the input data size. Thus, the value of "$TT_{Tasks}(F_j)$" can be computed using Eq. (2).

$$TT_{Tasks}(F_j) = T_{up}(F_j) + T_{Execute}(F_j) \tag{2}$$

In particular, the value of "$T_{up}(F_j)$" is generally computed depending on the amount of data transmitted from IoT devices to selected fog nodes. In Eq. (2), the value of "$T_{up}(F_j)$" is determined as the ratio between data size of tasks (T_{DS}) and channel transmission capacity (CT_{Cap}) established between IoT devices and fog nodes as presented in Eq. (3)

$$T_{up}(F_j) = \frac{T_{DS}}{CT_{Cap}} \quad (3)$$

Further, "CT_{Cap}" is computed based on the formula of Shannon capacity as mentioned in Eq. (4).

$$CT_{Cap} = B_{SC} \times \log_2\left(1 + \frac{TP_{IoT} \times CG_{IoT-FN}}{G_{NP_C} \times B_{SC}}\right) \quad (4)$$

where "TP_{IoT}" and "CG_{IoT-FN}" represent the transmission power of IoT nodes and channel gain determined between IoT and fog nodes.

"G_{NP_C}" denotes the channel Gauss noise power. Furthermore, "$T_{Execute}(F_j)$" is computed based on the ratio between amount of essential IoT task computation (TC_{IoT}) to computational capability of fog node (FN_{CC}) as specified in Eq. (5).

$$T_{Execute}(F_j) = \frac{TC_{IoT}}{FN_{CC}} \quad (5)$$

The total energy involved in task execution is determined based on "$E_{up}(F_j)$" which indicates the amount of energy utilized for transmitting and offloading end user input tasks on fog nodes, and "$E_{Execute}(F_j)$" represents the amount of energy incurred for task execution on fog nodes in the implementation environment as mentioned in Eq. (6).

$$TE_{Tasks}(F_j) = E_{up}(F_j) + E_{Execute}(F_j) \quad (6)$$

In this context, the value of "$E_{up}(F_j)$" and "$E_{Execute}(F_j)$" are determined based on Eqs. (7) and (8) by including "$FN_{PC_{Watts}}$" which represents power consumption in fog nodes (measured in Watts)

$$E_{up}(F_j) = T_{up}(F_j) \times FN_{PC_{Watts}} \quad (7)$$

$$E_{Execute}(F_j) = T_{Execute}(F_j) \times FN_{PC_{Watts}} \quad (8)$$

3.1 Task Scheduling (TS) Problem Formulation

The core objective of TS focuses on the process of minimizing energy and time consumption by optimally allocating IoT tasks to fog nodes. The cluster manager associated with fog nodes helps in determining the solution represented as a Boolean matrix which is either "1" or "0" based on whether the fog node is selected for a task or not. This problem is formulated as a Linear Integer Problem (LIP) which helps in allocating IoT tasks to suitable or ideal fog nodes as specified in Eq. (9).

$$\underset{\mathrm{TS_{Assign}}}{\mathrm{Argmin}}\,\mathrm{U}(\mathrm{T_{Assign}}) \tag{9}$$

where

$$U\left(T_{Assign}\right) = \sum_{i=1}^{n}\sum_{j=1}^{m} F_j \times QoS_{Score_{i,j}} \tag{10}$$

Subject to

$$\sum_{j=1}^{m} F_j = 1, 1 \leq i \leq n \tag{11}$$

$$\sum_{j=1}^{m} F_j \times T_{Ds} \leq FN_{BC}, 1 \leq j \leq m \tag{12}$$

$$\sum_{j=1}^{m} F_j \times I_{DS} \leq FN_{CC}, 1 \leq j \leq m \tag{13}$$

$$\sum_{j=1}^{m} F_j \times T_{Execute}\left(F_j\right) \leq FN_{SC}, 1 \leq j \leq m \tag{14}$$

$$TT_{Tasks}\left(F_j\right) \leq D_{Task}, 1 \leq i \leq n \text{ and } 1 \leq j \leq m \tag{15}$$

$$F_j \in [0.1], 1 \leq i \leq n \text{ and } 1 \leq j \leq m \tag{16}$$

At this juncture, "$\mathrm{U}(\mathrm{T_{Assign}})$" indicates the comprehensive utility rate achieved during the implementation of a formulated TS solution through the determination of matrix "$\mathrm{T_{Assign}}$" computed based on Eq. (10). In Eq. (10), "$\mathrm{QoS_{Score_{i,j}}}$" represents the score of QoS associated with "$\mathrm{F_j}$". Constraints presented in Eqs. (11) and (12) emphasize that a task can never be assigned to more than one fog node, and the necessitated intensity of computing associated with the IoT tasks set is always allocated to the node which could never be greater than the computation capacity possessed by the node. In addition, the constraints presented in Eqs. (13) and (14) guarantee that the data size necessitated for executing the collection of IoT tasks assigned to a specific fog node can never exceed a nodes' storage capacity. The energy incurred by a node for executing IoT tasks is less than the battery capacity possessed by the nodes. Moreover, Eqs. (15) and (16) represent that the total time involved in IoT task execution in a fog node cannot exceed the deadline associated with an IoT task, and that binary decision variables are associated with the process of mapping tasks and fog nodes.

Moreover, the score of QoS associated with "F_j" is computed based on the weighted sum of time and energy factors as specified in Eq. (17)

$$\mathrm{QoS}_{\mathrm{Score}_{i,j}} = \mathrm{W_{TT}} * \mathrm{TT_{Tasks}}\left(\mathrm{F_j}\right) + \mathrm{W_{TE}} * \mathrm{TE_{Tasks}}\left(\mathrm{F_j}\right) \tag{17}$$

where "W_{TT}" and "W_{TE}" lie between "0" and "1" which specify the weight factors related to time and energy factors.

3.2 *Background of WHOA for TS in FC*

In this section, the background of WHOA algorithm used for TS in FC is presented. The adopted WHOA algorithm is proposed based on WH's behavior including leadership, domination, grazing and mating process.

3.2.1 Initial Population Formation, Horse Group and Leaders' Determination

In this step, based on number of groups (G_n) and number of individuals (I_n), the number of mares and foals (non-leaders) is given by "$I_n - G_n$" with number of leaders represented as "G_n". The stallions' proportion in the entire population (T_{Pop}) is determined based on $\frac{G_n}{I_n}$.

3.2.2 Behavior of Grazing

In this phase of grazing, search agents are used for determining the position of stallion in the center of the grazing area such that they can move in the search space as shown in Eq. (18).

$$WH_{i,j} = 2A_F.\cos(2\pi RA_F) \times \left(WH_{i,j} - ST_{i,j}\right) + WP_{i,j} \tag{18}$$

where "$WH_{i,j}$" and "$ST_{i,j}$" represent the position of group member in a specific group "i" and stallion in "j^{th}" group.

Moreover, the value ranges between "−2" and "2". Further, the Adaptive Factor (A_F) is computed based on Eq. (19).

$$A_F = R_{nd_1}(ID_X) + \overrightarrow{R_{nd_3}}(\sim ID_X), ID_X = (P == 0), P = \overrightarrow{R_{nd_2}} < LDP \quad (19)$$

where "R_{nd_1}" and "$\overrightarrow{R_{nd_3}}$" are random vectors whose values lie between "0" and "1".

Moreover, "$\overrightarrow{R_{nd_2}}$" indicates a random number which lies between "0" and "1". Vector "P" comprising of "0" and "1" denotes problem dimension. Linear Decreasing Parameter (LDP) is computed based on Eq. (20).

$$LDP = 1 - \frac{Iter_{curr}}{Iter_{max}} \quad (20)$$

3.2.3 Behavior of Horse Mating

Horse mating behavior is simulated depending on the mating behavior visualized between WHs by separating foals from their original groups as specified in Eqs. (21) and (22).

$$WH_{i,j} = Crossover\left(WH_{i,k}, WH_{k,j}\right), i \neq j \neq k, p = h = end \quad (21)$$

$$Crossover = Mean \quad (22)$$

where "$WH_{i,j}$" represents a wild horse (p) present in a group (h) which is determined based on the positions of horse (q) in a group (g).

In this traditional algorithm, the crossover probability is assigned a constant value of 0.13.

3.2.4 Leadership Behavior in Groups

This phase is mainly responsible for identifying leaders (stallions) in the groups as specified in Eq. (23)

$$\overleftarrow{ST_{I,J}} = \begin{cases} 2A_F \cos(2\pi RA_F) \times \left(WH_{i,j} - ST_{i,j}\right) + WP_{i,j} \ if \ R_{nd} > 0.5 \\ 2A_F \cos(2\pi RA_F) \times \left(WH_{i,j} - ST_{i,j}\right) + WP_{i,j} \ if \ R_{nd} \le 0.5 \end{cases} \quad (23)$$

where "$\overleftarrow{ST_{I,J}}$" and "$ST_{i,j}$" represent candidate and current leader (IoT tasks) positions determined in the group with "$WP_{i,j}$" as Waterhole Position (fog server).

3.2.5 Selection and Exchange of Leaders

This phase deals with random selection and exchange of leaders by evaluating fitness values. This selection and exchange in leader positions between other individuals is determined based on Eq. (24).

$$ST_{i,j} = \begin{cases} WH_{i,j} \text{ if } f(WH_{i,j}) < f(ST_{i,j}) \\ ST_{i,j} \text{ if } f(WH_{i,j}) \geq f(ST_{i,j}) \end{cases} \quad (24)$$

where "$f(WH_{i,j})$" and "$f(ST_{i,j})$" represent the fitness value determined for WHs (search agents) and stallions used for assigning the incoming IoT tasks to specific fog servers.

4 Results and Discussion

The efficiency of the proposed WHOAEETS approach is validated over the benchmarked EETSPSO, HPSO-GWO and EPSO approaches using the evaluation metrics of time, energy, and time-energy balance for varying number of tasks and iterations considered for implementation. The implementations of the proposed WHOAEETS and baseline approaches are conducted using IFogSim simulator. The FC environment considered for implementation comprises of 20 fog nodes which possess different capabilities in terms of computing energy, distance from manager within each cluster, bandwidth and computing capacity. The number of tasks considered for evaluation varies from 50 to 400, and the number of iterations ranges between 20 and 200. The population size considered for implementation of all the algorithms is 50 and the maximum number of iterations is 200. The results of the proposed WHOAEETS approach and the benchmarked approaches are determined through the mean computed over the results of 30 independent runs.

In specific, Figs. 1 and 2 present the time and energy-based log (fitness value) achieved by the proposed WHOAEETS approach over the benchmarked EETSPSO, HPSO-GWO, and EPSO approaches for varying number of tasks in IoT applications. The proposed WHOAEETS approach based on the number of IoT applications adopt WHOA for balancing the trade-off between the rate of exploration and exploitation with superior solution diversity. The results of WHOAEETS approach for varying number of IoT tasks confirm minimized energy-based log (fitness value) up to a considerable level of 11.86%, 13.44% and 15.42% in contrast to benchmarked EETSPSO, HPSO-GWO and EPSO approaches. The results of WHOAEETS approach for different number of IoT tasks confirm minimized time-based log (fitness value) up to a considerable level of 10.94%, 13.28% and 15.16% when compared to benchmarked EETSPSO, HPSO-GWO, and EPSO approaches.

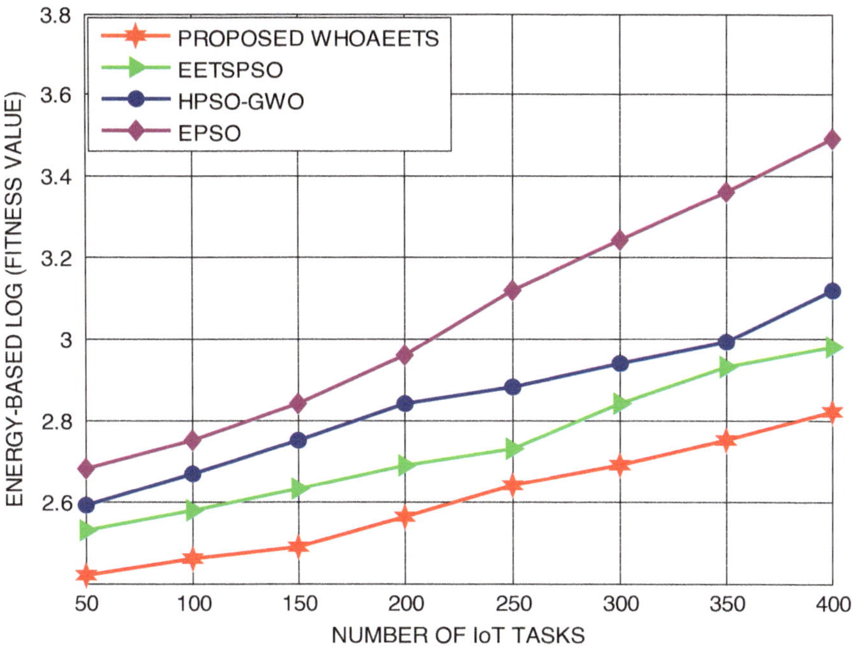

Fig. 1 Energy-based log (fitness value) for varying number of IoT tasks

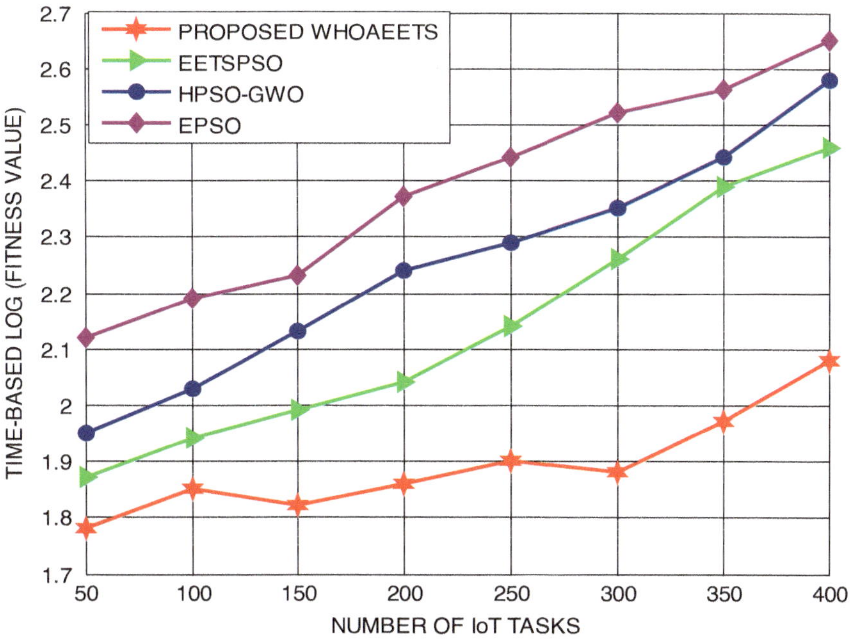

Fig. 2 Time-based log (fitness value) for varying number of IoT tasks

In addition, Figs. 3 and 4 present the time-energy balance-based log (fitness value) achieved by the proposed WHOAEETS approach validated over the benchmarked EETSPSO, HPSO-GWO, and EPSO approaches for varying number of tasks and iterations involved in the implementation. With increase in the number of tasks, the time-energy balance-based log (fitness value) achieved by the proposed WHOAEETS approach is comparatively better than the baseline approaches since it prevents the solution from entering the local point of optimality with delayed convergence. The proposed WHOAEETS approach facilitates better exploration and exploitation of search space such that required IoT tasks are assigned to ideal fog nodes in the network. The results of WHOAEETS approach for varying number of IoT tasks confirm minimized time-energy balance-based log (fitness) up to a considerable level of 12.38%, 15.92% and 17.46% when compared to benchmarked EETSPSO, HPSO-GWO, and EPSO approaches. The results of WHOAEETS approach for varying number of IoT tasks confirm minimized time-energy balance-based log (fitness value) up to a considerable level of 12.38%, 15.92%, and 17.46% in contrast to benchmarked EETSPSO, HPSO-GWO and EPSO approaches.

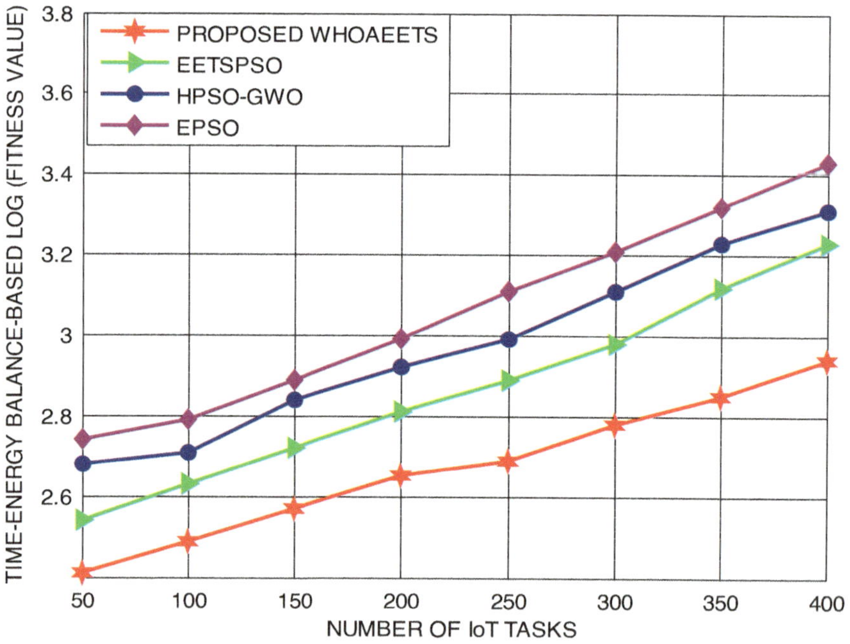

Fig. 3 Time–energy balance-based log (fitness value) for varying number of IoT tasks

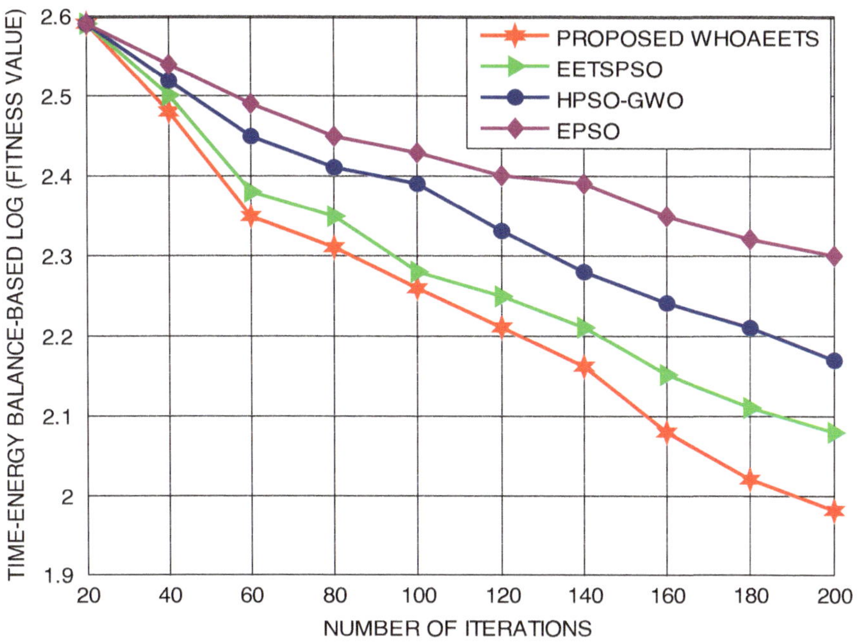

Fig. 4 Time-energy balance-based log (fitness value) for varying number of iterations

5 Conclusion

The proposed WHOAEETS algorithm confirm better response time and reduced latency during task execution in the FC environment. It schedules tasks optimally in the fog server by determining the fitness function which adopts factors of execution time, energy and makespan. It uses search agents that mimics the behavior of WHO and explores the scheduling factors in the system model such that assignment of IoT tasks to ideal fog servers is achieved. The simulation experiments of the proposed WHOAEETS algorithm conducted using IFogSim for different number of IoT tasks confirm a minimized mean time-energy balance-based log (fitness value) up to a considerable level of 14.98% when compared to benchmarked EETSPSO, HPSO-GWO, and EPSO approaches. The results of WHOAEETS approach for different number of iterations also confirm minimized mean time-energy balance-based log (fitness value) up to a considerable level of 15.62% in contrast to benchmarked EETSPSO, HPSO-GWO, and EPSO approaches. As a part of future plan, it is decided to formulate and implement an improved Jellyfish algorithm, and compare it with the proposed WHOAEETS approach with respect to response time and degree of imbalance for varying number of iterations.

References

1. Alizadeh, M.R., Khajehvand, V., Rahmani, A.M., Akbari, E.: Task scheduling approaches in fog computing: a systematic review. Int. J. Commun. Syst.Commun. Syst. **33**(16), e4583 (2020)
2. Sengathir J., Deva Priya M.: Hybrid grey wolf and improved particle swarm optimization with adaptive inertial weight-based multi-dimensional learning strategy for load balancing in cloud environments. Sustain. Comput.: Inform. Syst. ISSN: 2210-5387, **38**, 100875 (2023)
3. Sengathir J., Deva Priya M.: Improved artificial bee colony using monarchy butterfly optimization algorithm for load balancing (IABC-MBOA-LB) in cloud environment. J. Netw. Syst. Manag. Springer, ISSN: 1573-7705, **29**, 39 (2021)
4. Kaur, N., Kumar, A., Kumar, R.: A systematic review on task scheduling in fog computing: taxonomy, tools, challenges, and future directions. Concurr. Comput.: Pract. Exp. **33**(21), e6432 (2021)
5. Yang, X., Rahmani, N.: Task scheduling mechanisms in fog computing: review, trends, and perspectives. Kybernetes **50**(1), 22–38 (2020)
6. Guevara, J.C., da Fonseca, N.L.: Task scheduling in cloud-fog computing systems. Peer-To-Peer Netw. Appl. **14**(2), 962–977 (2021)
7. Abd Elaziz, M., Abualigah, L., Attiya, I.: Advanced optimization technique for scheduling IoT tasks in cloud-fog computing environments. Futur. Gener. Comput. Syst. Gener. Comput. Syst. **124**:142–154 (2021)
8. Janakiraman, S., Deva Priya, M.: Dynamic butterfly optimization algorithm-based task scheduling for minimizing energy consumption in distributed green data centers. Self-Powered Cyber Phys. Syst. 91–113 (2023)
9. Vijayalakshmi, R., Vasudevan, V., Kadry, S., Lakshmana Kumar, R.: Optimization of makespan and resource utilization in the fog computing environment through task scheduling algorithm. Int. J. Wavelets Multiresolut. Inf. Process.Multiresolut. Inf. Process. **18**(01), 1941025 (2020)
10. Singh, S.P., Nayyar, A., Kaur, H., Singla, A.: Dynamic task scheduling using balanced VM allocation policy for fog computing platforms. Scalable Comput.: Pract. Exp. **20**(2), 433–456 (2019)
11. Nithiavathy, R., Janakiraman, S., Deva Priya, M.: Adaptive guided differential evolution-based slime mould algorithm-based efficient multi-objective task scheduling for cloud computing environments. Trans. Emerg. Telecommun. Technol. **35**(1), e4902 (2024)
12. Abdel-Basset, M., Mohamed, R., Elhoseny, M., Bashir, A.K., Jolfaei, A., Kumar, N.: Energy-aware marine predators algorithm for task scheduling in IoT-based fog computing applications. IEEE Trans. Industr. Inf.Industr. Inf. **17**(7), 5068–5076 (2020)
13. Hussein, M.K., Mousa, M.H.: Efficient task offloading for IoT-based applications in fog computing using ant colony optimization. IEEE Access **8**, 37191–37201 (2020)
14. Hosseinioun, P., Kheirabadi, M., Tabbakh, S.R.K., Ghaemi, R.: A new energy-aware tasks scheduling approach in fog computing using hybrid meta-heuristic algorithm. J. Parallel Distrib. Comput. **143**, 88–96 (2020)
15. Keshavarznejad, M., Rezvani, M.H., Adabi, S.: Delay-aware optimization of energy consumption for task offloading in fog environments using metaheuristic algorithms. Clust. Comput. 1–29 (2021)
16. Potu, N., Jatoth, C., Parvataneni, P.: Optimizing resource scheduling based on extended particle swarm optimization in fog computing environments. Concurr. Comput.: Pract. Exp. **33**(23), e6163 (2021)
17. Dev, K.C., Dash, B.B., De, U.C., Pattnayak, P., Satapthy, R., Patra, S.: Task scheduling in fog assisted cloud environment using hybrid metaheuristic algorithm. In: International Conference on Expert Clouds and Applications, pp. 223–236. Springer Nature Singapore, Singapore (2022)

18. AL-Amodi, S., Patra, S.S., Bhattacharya, S., Mohanty, J.R., Kumar, V., Barik, R.K.: Meta-heuristic algorithm for energy-efficient task scheduling in fog computing. In: Recent Trends in Electronics and Communication: Select Proceedings of VCAS 2020, pp. 915–925. Springer Singapore (2022)
19. Yadav, A.M., Tripathi, K.N., Sharma, S.C.: A bi-objective task scheduling approach in fog computing using hybrid fireworks algorithm. J. Supercomput. Supercomput. **78**(3), 4236–4260 (2022)
20. Vispute, S.D., Vashisht, P.: Energy-efficient task scheduling in fog computing based on particle swarm optimization. SN Comput. Sci. **4**(4), 391 (2023)
21. Iftikhar, S., Ahmad, M.M.M., Tuli, S., Chowdhury, D., Xu, M., Gill, S.S., Uhlig, S.: HunterPlus: AI based energy-efficient task scheduling for cloud–fog computing environments. Internet Things. **21**, 100667 (2023)

YOLOv5 Approach for Pothole Identification: A Case Study

Sunny Kumar, Amlan Dash, Siksha Swaroopa Kar, and Amit Kar

Abstract Potholes are irregular-shaped depressions on road surfaces and represent a pervasive threat to road safety worldwide. Potholes pose significant risks to road users, contributing to accidents, vehicle damage, and fatalities. Statistical data from the Ministry of Road Transport & Highways (MoRTH), India, reveal a staggering toll of incidents, with thousands of fatalities and injuries resulting from pothole-related accidents annually. Moreover, the negative impacts of potholes extend beyond safety concerns, affecting traffic flow, vehicle efficiency, and economic productivity. They are known to cause traffic jams, increase fuel consumption, and degrade vehicle components, leading to decreased fuel economy and heightened maintenance costs. Road transport is like the heart and arteries of a nation when it comes to moving people and things around. Road upkeep is a difficult task in nations like India. The rising number of potholes is causing an annual increase in the number of accidents. Considering these challenges, our research leverages a computer vision system centered around the YOLOv5 algorithm's capabilities to detect potholes efficiently and accurately. By employing deep learning techniques, we aim to develop a robust pothole detection system capable of identifying these hazards in real time. Such a system holds promise for enhancing road safety, reducing accident rates, and mitigating the economic burdens associated with pothole-related incidents. Through this case study, we demonstrate the potential of YOLOv5 as a tool for proactive pothole detection and pave the way for future advancements in road maintenance and safety protocols.

Keywords Pothole detection · YOLOv5 · Machine learning · Computer vision

S. Kumar · S. S. Kar (✉)
CSIR-CRRI, Mathura Road, New Delhi 110025, India
e-mail: siksha.crri@nic.in

A. Dash · A. Kar
Sterlite Technology Ltd. Gurugram, Haryana 122102, India

S. Kumar Udgata et al. (eds.), *Intelligent Systems*, Lecture Notes in Networks and Systems 1149, https://doi.org/10.1007/978-981-97-8160-7_16

1 Introduction

A pothole is a bowl-shaped depression on the top of a pavement surface, constituting one of the most dangerous distresses. The primary causes of pothole formation are the loss of bituminous-wearing coat adhesion due to water seepage into the pavement, creation of voids in the surface, or repetitive movement of heavy loads across the surface [1].

Potholes are a significant cause of accidents, traffic congestion, and negative impacts on vehicle functionality and components. They contribute to increased vehicular repairs and decreased fuel economy, posing a heightened danger for high-speed and nighttime commuters. Referring to statistics released by the Indian Government, between 2020 and 2021, a total of 7189 incidents involving potholes have occurred, resulting in 2952 fatalities and 6167 injuries [2].

Regular inspection and maintenance of roads are essential to overcome the challenges posed by potholes. Traditionally, manual inspection methods are deployed to survey the status of roads, which are often time-consuming, labor-intensive, and subjective to the views of the surveyor [3].

With the evolution of technology, innovative and tech-laden interventions have been introduced to address this challenge. Various methods are being tested worldwide, including those based on vehicle vibration principles [4], computer vision-based systems [5], and even 3D imaging techniques [6] for pothole detection.

With increase in in capacity of computers, significant work in deep-learning techniques and algorithms has created the belief that such techniques can profoundly impact the resolution of real-world problems. The development of computer vision-based methods, capable of automating the pothole detection process, emerges as a trustworthy solution to address the challenges faced by traditional techniques [7]. These techniques leverage deep learning-powered artificial neural networks (ANNs), which have demonstrated high potential in improving the efficiency and accuracy of computer vision tasks.

In this paper, we aim to showcase a system that utilizes the renowned YOLOv5 (You Only Look Once) algorithm to build a computer-vision-powered system. This system facilitates the development of an efficient and accurate model capable of detecting potholes on pavement surfaces.

1.1 The YOLO Algorithm

The You Only Look Once (YOLO) Convolutional Neural Network (CNN) algorithm represents a pivotal advancement in the field of object detection under the domain of computer vision. Developed by Joseph Redmon et al., YOLO has attracted attention and acclaim for its ability to perform real-time object detection at remarkable speeds and high accuracy.

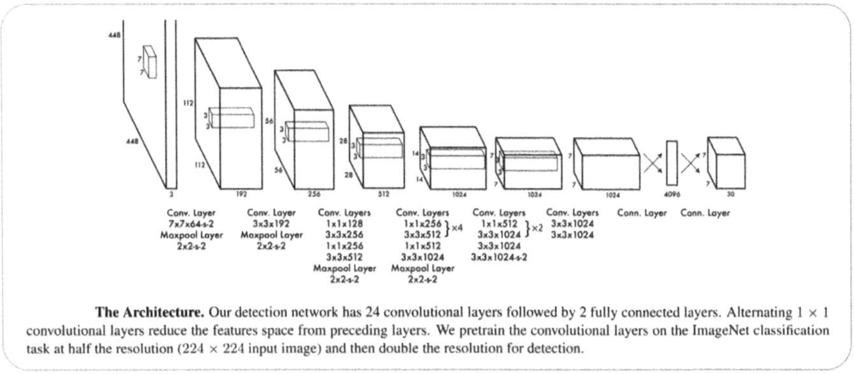

Fig. 1 Representation of the CNN model backbone for you only look once algorithm

Traditional object detection methods typically involve the use of sliding window approaches or region-based convolutional neural networks, which are computation heavy and inefficient for real-time applications. YOLO, however, revolutionizes this paradigm by adopting a fundamentally different approach. Instead of scanning an image multiple times at varying locations and scales, YOLO formulates object detection as a single regression problem applied to the entire image.

At its core, YOLO divides an input image into a grid and predicts bounding boxes and class probabilities directly with reference from this grid. By performing these tasks simultaneously, YOLO achieves impressive speed while maintaining high accuracy. Furthermore, YOLO utilizes one single neural network architecture to make predictions, streamlining the inference process and enabling real-time performance on standard hardware.

YOLO can conduct comprehensive image analysis, encompassing classification and precise localization in a unified, single-step process. It divides the input image into SxS grids and utilizes an individual grid cell to estimate N number of objects of interest, incorporating the object's center (x, y), dimensions (w, h), confidence scores, and class probabilities [8] (Plastiras et al., 2018) (Fig. 1).

1.2 YOLOv5

YOLOv5–in the 5th iteration of the updates to the original algorithm, brings in significant updates from its predecessors, introducing several key improvements and innovations compared to the original YOLO algorithm.

1. Architecture Optimization: YOLOv5 features a streamlined architecture compared to earlier versions, resulting in improved speed and efficiency without sacrificing accuracy. The network architecture has is redesigned to reduce computational complexity while maintaining high performance.

2. Backbone Network: YOLOv5 utilizes a more powerful backbone network, based on the EfficientNet architecture, enabling better feature extraction and representation. This choice of backbone network enhances the model's ability to capture intricate spatial and semantic information from input images.
3. Enhanced Training Techniques: YOLOv5 incorporates advanced training techniques such as mosaic data augmentation, and self-adversarial training. These techniques help improve the model's generalization capabilities and robustness to variations in input data, resulting in better performance on unseen images.
4. Scale-Aware Training: YOLOv5 introduces scale-aware training, which involves training the model on images of varying sizes during the same training batch. This approach enables the model to effectively handle objects of different scales and sizes, leading to more accurate predictions across a wide range of scenarios.

1.3 YOLO for Pothole Detection

For the pothole detection use case, YOLOv5 efficiently identifies and locates these road hazards within images and video feeds, showcasing near real-time object detection capabilities. Operating on a single neural network, YOLOv5 swiftly detects potholes with bounding boxes, providing class probabilities in near real-time. Its effectiveness in rapid and accurate identification makes it indispensable for road maintenance and autonomous driving systems, ensuring prompt repairs and enhancing road safety.

Multiple studies have attempted to display the capabilities of the YOLO family of algorithms by attempting to develop systems dedicated to resolving the challenge of pothole detection.

In the paper titled "Optimizing Pothole Detection in Pavements: A Comparative Analysis of Deep Learning Models," by [9], a comparative analysis of multiple computer vision algorithms, including the YOLO family of models, is presented, specifically on the pothole detection use case. The model utilizing the YOLOv4 algorithm produced the best results for pothole detection in the dataset, with a mAP of 83.2%. Nonetheless, using the Darknet framework as its backbone, makes training a custom model using YOLOv4 and its deployment considerably more difficult. This becomes a barrier to the scalability of the solution and the implementation of the model in production. However, with a little tweaking, YOLOv5 can produce superior outcomes.

Another study aims at enhancing vehicle safety and lowering traffic accidents, focusing on pothole guidance, [10]. By recording the number of potholes on roadways, the suggested approach is designed to keep an eye on the state of the roads. This will facilitate the government's planning process for road upkeep and maintenance activities. It illustrates the use of YOLOv3 and YOLOv5 deep learning algorithms, highlighting the increase in accuracy of object recognition and classification with updated model frameworks. The suggested technique trains the network using the YOLOv5 algorithm, yielding a 97.22% pothole detection accuracy. The YOLOv5 model performs better than the YOLOv3 model for pothole detection in Indian road settings, according to qualitative and quantitative studies.

There have also been attempts to integrate similar systems with autonomous-driving suites that aim to enhance the decision-making abilities of new-age automobiles. One such work, titled "Pothole detection for autonomous vehicles using deep learning: a robust and efficient solution" by [11], introduces a novel, efficient pothole detection solution for autonomous vehicles using YOLO family of algorithms.

The difficulty in pothole detection arises from their varied shapes and sizes, posing a unique challenge, unlike other objects [12]. The YOLOX model offers enhanced pothole detection capabilities even when running on systems that are cheap, readily available, and have relatively low computational power. That paper attempts to perform a comparison between YOLOv3, YOLOv4, YOLOv5, and YOLOX variants. The YOLOX-nano model exhibits superior precision, surpassing even YOLOv5m by 24.06% in Average Precision, while maintaining a smaller model size, albeit with slightly slower inference speed.

2 Proposed Method

For this case study, we have attempted to showcase the entire procedure involved in the creation of a system that can utilize a YOLOv5-based deep learning model for near real-time detection of potholes. The entire process is highlighted in Fig. 2. We have collected video feeds—our raw data for the creation of the dataset, with the help of the National Survey Vehicle (NSV) in the video form. This video was converted into images by use of the OpenCV framework and Python code. All the images sampled from this video feed were resized to 640*640 resolution.

The practice of identifying potholes in the raw data is referred to as annotation and labeling. The process involves marking and labeling of the region of interest in each sample image. We made use of an open-source tool labeling and annotation tool (from MakeSense.Ai) to facilitate the export of labeled data as per YOLOv5 framework input requirement. We were able to perform this activity on 408 images

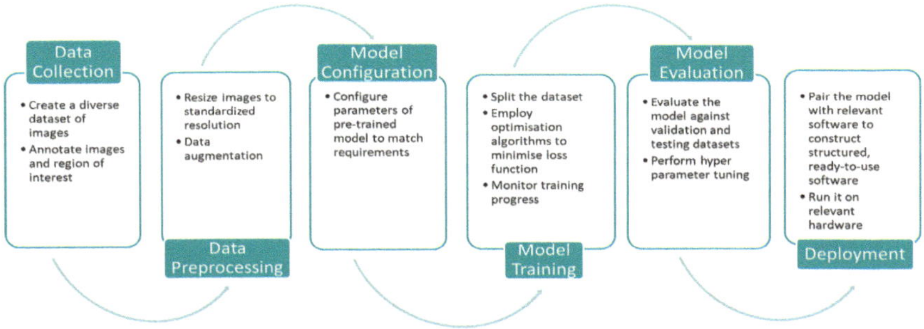

Fig. 2 Methodology of the proposed method

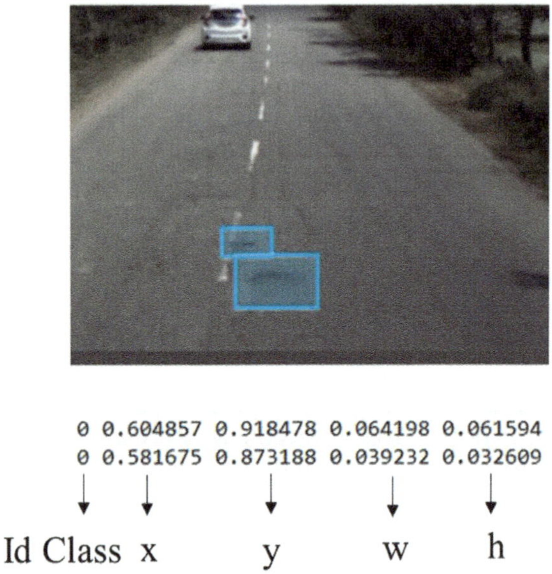

Fig. 3 Annotation and labeling

that were sampled and selected from the raw video feeds. The framework input requires parameters such as bounding box center (x and y), bounding box width and height (w and h), and class_id to understand the marked regions in an image. Generally, for each image in the dataset, we have a corresponding text file that contains these parameters. Pothole information is contained in a file in the ***.txt** format for every image in ***.jpg** format. The bounding box information (coordinates of highlighted section–the pixels in the image) are in normalized format within a 0–1 scale, and the class_id is an integer value representing the label–class of interest.

Figure 3 displays the examples of the labeling and annotation process.

The transfer learning technique was adopted to train the model. Transfer learning is a machine learning practice that leverages knowledge gained from training one model on a specific task and applies it to a different but related task. In the context of computer vision and object detection, transfer learning plays a crucial role in speeding up model training, improving accuracy, and reducing the need for large, annotated datasets.

A base model of YOLOv5 was chosen to start with, and such models are typically trained on large-scale datasets like the COCO (Common Objects in Context) dataset, which contains a diverse range of object categories (80 labels in COCO's case). These pre-trained models have learned to recognize a wide variety of objects across different contexts, making them valuable starting points for transfer learning.

One of the key advantages of using transfer learning is the ability to fine-tune the pre-trained model on a specific dataset with a smaller number of labels or a different distribution of classes. This is particularly beneficial when working with limited annotated data or when the task involves a subset of the classes present in the original dataset.

Transfer learning allows to retrain the YOLOv5 pre-trained model using a dataset containing annotated images of a single category, in this case, only "potholes", thus customizing the model to focus solely on this specific object category.

The advantages of using transfer learning with YOLOv5 for this scenario enabled:

1. **Reduced Training Time**: Starting with a pre-trained model, already capable of identifying general features and having object detection capabilities, helped reduce the training time for the customized model as compared to training from scratch.
2. **Improved Accuracy**: Transfer learning enables the new model to leverage knowledge from the pre-trained weights, leading to improved accuracy and robustness.
3. **Optimized Resource Utilization**: Since transfer learning allows for fine-tuning rather than retraining the entire model, it optimizes resource utilization such as computational power and memory, making it more feasible for deployment in resource-constrained environments.

Post the data annotation and labeling activity, the dataset was divided in two parts– namely, the training dataset (335) and the validation dataset (73). These two datasets were then used to train and validate the new computer vision model. The pothole training process was completed in 10 epochs in 0.851 h.

For evaluation purposes, the model is tested against the validation dataset consisting of 73 annotated and labeled data. The purpose of this test is to obtain the performance value of the model by comparing the detection result with the ground truth.

IoU (intersection over union), precision, recall, and mAP (mean average precision) parameters were used as evaluation metrics for analyzing the model performance.

1. **Precision**: A measure of how many correctly predicted positive instances are relevant among all instances predicted to be positive. In object detection, it denotes the accuracy of the model in detecting objects. It is calculated as the ratio of true positive detections to the sum of true positive and false positive detections.
2. **Recall**: Recall, also known as sensitivity or true positive rate, measures the ability of the model to correctly identify all relevant instances. It also indicates the model's completeness in detecting objects. Calculated as the ratio of true positive detections to the sum of true positive and false negative detections.
3. **Intersection over Union (IoU)**: IoU is a metric used to evaluate the accuracy of object localization by measuring the overlap between the predicted bounding box and the ground truth bounding box. It is calculated as the area of intersection divided by the area of union between the predicted and ground truth bounding boxes.
4. **Mean Average Precision (mAP)**: mAP is a widely used metric in object detection that combines precision and recall across multiple object categories. It calculates the average precision for each class and then averages these values to obtain the mAP score. Higher mAP values indicate better overall performance of the object detection model across different classes.

IoU, Precision and Recall are measured using Eqs. 1, 2, and 3, respectively.

$$IoU = |A \cap B| \div |A \cup B| \tag{1}$$

$$Precision = TP \div (TP + FP) \tag{2}$$

$$Recall = TP \div (TP + FN) \tag{3}$$

where, A = Area of the ground truth bounding box

B = Area of the detection bounding box.

TP = True positive.

FP = False positive.

FN = False negative.

IoU is a term used to describe the extent of two boxes. The greater the region of overlap, the greater the IoU. The overlap area of the ground truth bounding box (A) and the detection bounding box (B). When an object is identified as a pothole, it is correctly detected known as true positive (TP). When anything is falsely identified as a pothole even when it is not one, it is called a False Positive (FP). On the other hand, a (FN) is an object that should be a pothole but is identified as one. On the other hand, Under Curve Area (AUC), also known as mean Average precision or mAP, is a single statistic that indicates a mix of values between accuracy and recall [13]. 0.5 is the IoU that is utilized as the mAP parameter.

3 Results and Discussions

The trained model, YOLOv5 is used to train and validate the pothole images. Loss functions are continuously decreasing up to the 10 epoch. Precision and Recall values are plotted at every confidence level. After 10 epochs in the training cycle, the precision, recall, mAP@50, and mAP@50–95 values were calculated to be 42.3%, 33.2%, 29.8%, and 9.51%, respectively. Figure 6 shows the model evaluation parameter-related graphs.

The YOLOv5-based pothole detection system exhibited consistent enhancement across the training cycle, demonstrating decreased value of loss functions and improved precision, recall, and mAP values. The robust performance across confidence levels and notable mAP@50–95 underscores the efficacy, showcasing a pivotal step in leveraging computer vision for road safety and maintenance (Figs. 4 and 5).

Precision versus confidence score: As the confidence threshold increases (moving from left to right on the x-axis), the number of detected objects decreases because the model becomes more selective in its predictions. Only objects with higher confidence scores are retained.

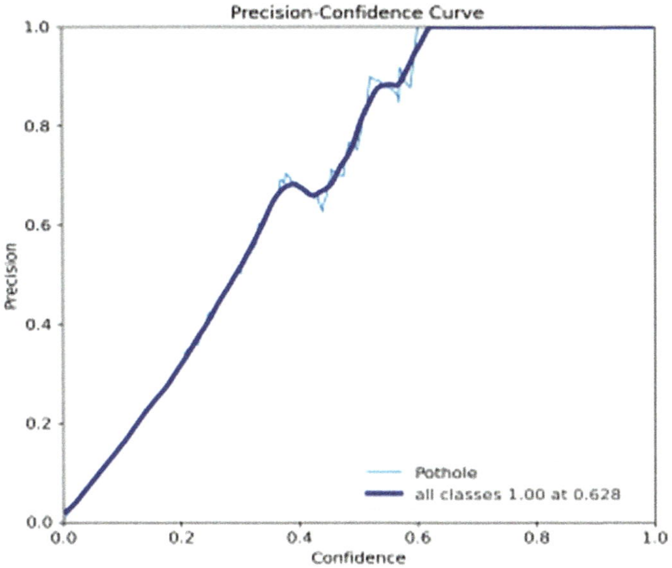

Fig. 4 Variation of precision with confidence value

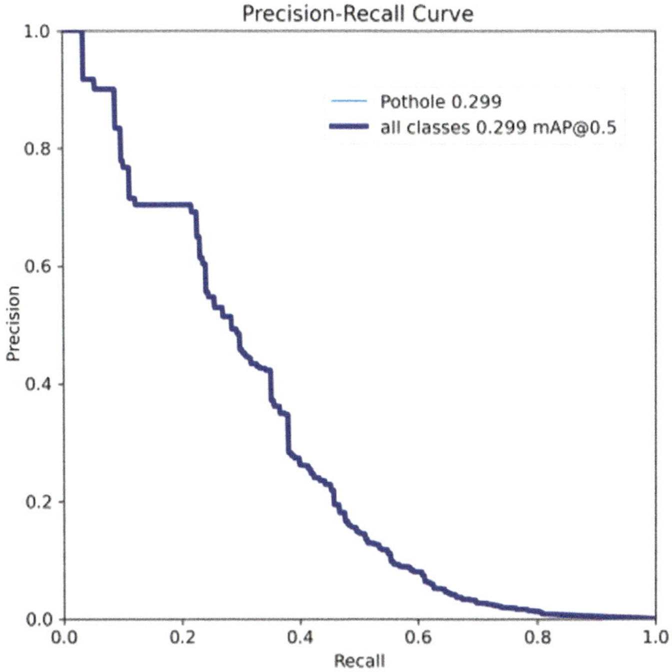

Fig. 5 Variation of precision with recall value

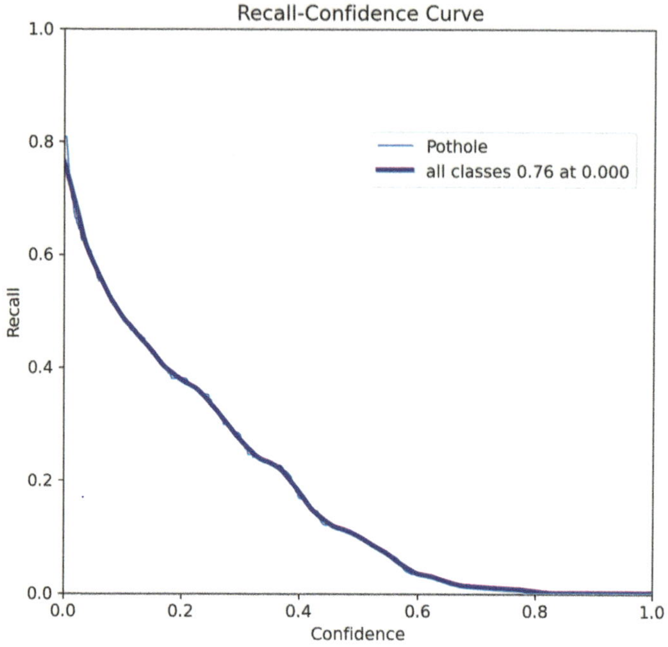

Fig. 6 Variation with recall with confidence level

At higher confidence thresholds, precision tends to increase because the model is more certain about its predictions. It is more likely to make correct detections when it is highly confident about the objects it identifies. However, increasing the confidence threshold too much may lead to a significant reduction in recall (the proportion of true positives that are correctly identified), as the model may miss some objects with lower confidence scores.

Optimal Operating Point: We required that our application detect even the smallest of areas where there is a possibility of potholes. Hence, we allowed more false positives rather than filtering them out completely.

Precision versus recall: An increase in the recall (lowered threshold for considering an object to be detected) will certainly identify more true positives thus increasing the recall but can also increase the number of false positives reducing precision.

Conversely, increase in precision (by increasing the threshold for considering an object as detected), filters out some true positives, thereby reducing recall. This trade-off is often visualized as a curve in the precision-recall space.

We prioritized recall over precision for we aimed at maximizing the number of detected potholes and hence were okay to have some false positives in our results (Fig. 7).

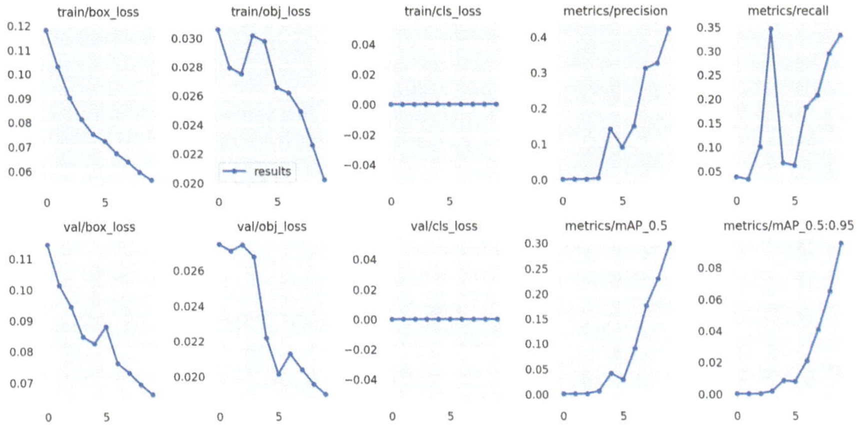

Fig. 7 Variation of different types of losses in training and validation data in 10 epochs

4 Conclusion

The findings from 335 training datasets suggest that the YOLOv5 model detects potholes with a promising 42.3% precision and 33.2% recall. The model's accuracy in identifying potholes within a specific confidence threshold is demonstrated by its mAP@50 of 29.8%, and its consistency over a wider range of confidence levels is indicated by its mAP@50–95 of 9.51%.

Considering the amount of training data used, these metrics point to marginal, yet-promising and effective detection capabilities. Increasing the training data, however, may help the model perform even better by increasing recall and precision rates, resulting in a more reliable and accurate pothole-detecting system.

To conclude, the advancements in deep learning techniques, such as attention mechanisms and transformer architectures, can be further integrated into the YOLOv5 framework to enhance pothole detection accuracy and efficiency. Additionally, exploring the use of multispectral or LiDAR data alongside visual imagery could provide richer contextual information for more robust pothole detection in various environmental conditions.

Furthermore, incorporating real-time monitoring capabilities and integrating the pothole detection system with smart city infrastructure could lead to proactive maintenance strategies, improving road safety and reducing repair costs. Collaborations with transportation authorities and industry stakeholders could facilitate the deployment of automated pothole detection systems on a larger scale, contributing to more sustainable and efficient road maintenance practices.

This case study on pothole detection using YOLOv5 demonstrates the effectiveness of deep learning in addressing critical infrastructure challenges. It also tries to highlight the potential for further advancements and real-world applications in road maintenance and urban management, emphasizing the importance of innovative technologies in creating safer and more resilient transportation networks.

References

1. IRC 82.: Code of practice for maintenance (2015)
2. Heo, D.H., Choi, J.Y., Kim, S.B., Tak, T.O., Zhang, S.P.: Image-based pothole detection using multi-scale feature network and risk assessment. Electron. (Switzerland). **12**(4) (2023). https://doi.org/10.3390/electronics12040826
3. Slavkovic, N., Bjelica, M.: Risk prediction algorithm based on image texture extraction using mobile vehicle road scanning system as support for autonomous driving. J. Electron. Imaging. **28**(03) (2019). https://doi.org/10.1117/1.jei.28.3.033034
4. De Zoysa, K., Keppitiyagama, C., Seneviratne, G.P., Shihan, W.W.A.T.: A public transport system-based sensor network for road surface condition monitoring. In: Proceedings of the 2007 Workshop on Networked Systems for Developing Regions, NSDR'07 (2007). https://doi.org/10.1145/1326571.1326585
5. Bharat, R., Ikotun, A.M., Ezugwu, A.E., Abualigah, L., Shehab, M., Zitar, R.A.: A real-time automatic pothole detection system using convolution neural networks. Appl. Comput. Eng. **6**(1), 750–757 (2023). https://doi.org/10.54254/2755-2721/6/20230948
6. Sathay, R., Saleena, B., Nikhil Anirudh, N.: An analytical study of various pothole detection and prevention techniques. In: IEEE International Students' Conference on Electrical, Electronics and Computer Science (SCEECS) (2023). https://doi.org/10.1109/SCEECS57921.2023.10063088
7. Ma, N., Fan, J., Wang, W., Wu, J., Jiang, Y., Xie, L., Fan, R.: Computer vision for road imaging and pothole detection: a state-of-the-art review of systems and algorithms. Transp. Saf. Environ. **4**(4), 1–16 (2022). https://doi.org/10.1093/tse/tdac026
8. Ma, N., Fan, J., et al.: Computer vision for road imaging and pothole detection: a state-of-the-art review of systems and algorithms. TransportatioTransportation Saf. Environ. Saf. Environ. **4**(4), 2022 (2022). https://doi.org/10.1093/tse/td
9. Tamagusko, T., Ferreira, A.: Optimizing pothole detection in pavements: a comparative analysis of deep learning models (2023) https://doi.org/10.3390/engproc2023036011
10. Chougule, S., Barhatte, A.: Smart pothole detection system using deep learning algorithms. Int. J. Intell. Transp. Syst. Res. (2023). https://doi.org/10.1007/s13177-023-00363-3
11. Khan, M., Raza, M.A., Abbas, G., Othmen, S., Yousef, A., Jumani, T.A.: Pothole detection for autonomous vehicles using deep learning: a robust and efficient solution. Front. Built Environ **9**, 1323792 (2024). https://doi.org/10.3389/fbuil.2023.1323792
12. Mohan, P., Sriharipriya.: Enhanced pothole detection system using YOLOX algorithm B (2022). https://doi.org/10.1007/s43684-022-00037-z
13. Ukhwah, E.N., Yuniarno, E.M., Suprapto, Y.K.: Asphalt pavement pothole detection using deep learning method based on YOLO neural network. In: Proceedings-2019 International Seminar on Intelligent Technology and its Application, ISITIA 2019, pp. 35–40. https://doi.org/10.1109/ISITIA.2019.8937176

Particle Swarm Optimization Based Graph Perturbation to Achieve Localization of Information Spread

Baishnobi Dash and Debasis Mohapatra

Abstract In this information age, the dissemination of information has a pivotal function in creating the adoption of new technologies, marketing of products, etc. However, the localization of information in a particular region is also a problem of the same importance to be addressed. For example, once a rumor is detected, it has to be localized at some parts of the network to obstruct its spreadability. This paper focuses on achieving information localization. The study says that localization can be achieved by manipulating the IPR value achievable through perturbation of the connections. In this paper, a Particle Swarm Optimization (PSO) based perturbation approach is proposed that passes through the steps of evolution to generate a graph with optimal IPR from the original graph. Hence, it guarantees a higher state of information localization in the output graph. The empirical results confirm a better localization by the proposed method than some existing methods. We use average optimal IPR as a parameter of evaluation.

Keywords Inverse participation ratio · Particle swarm optimization · Network adjacency matrix · Localization · Information diffusion

1 Introduction

The research on diffusion modeling to understand the phenomena of spreadability and controllability of influence of one component on the other is of paramount interest in a system ranging from social system to biological system [1–5]. In modeling of information diffusion, often the interest is on controlling the spread of rumors so that the spreadability can be localized at some parts of the network. Information diffusion in a social network [4, 6–9] can be simulated using a variety of models, including Linear Threshold (LT), Independent Cascade (IC), Susceptible-Infected-Susceptible (SIS), Susceptible-Infected-Recovered (SIR), etc. Models like SIS and SIR are derived from epidemic behavior and can be used for simulation of information

B. Dash · D. Mohapatra (✉)
Parala Maharaja Engineering College, Berhampur, Odisha 761003, India
e-mail: debasis.cse@pmec.ac.in

S. Kumar Udgata et al. (eds.), *Intelligent Systems*, Lecture Notes in Networks and Systems 1149, https://doi.org/10.1007/978-981-97-8160-7_17

diffusion. The controllability/localization of information is essential in the case of the spread of rumors. The idea of localization of epidemic is of great use in simulating rumor propagation. In one of the seminal works presented by Goltsev et al. [10], it is shown that the localization of a network can be achieved by regulating the Inverse Participation Ratio (IPR) for an SIS model. As obtaining a network configuration that confirms an optimal IPR is a hard problem, the heuristic-based approach [6–8, 11–15] is affluent to provide some approximate solutions for the same. Hence, several promising heuristic and meta-heuristic-based solutions are developed but still, there is a scope for improvement [15, 16]. This paper provides a novel interpretation of PSO [17–20] for graph perturbation that achieves localization. The overall proposal of this paper can be described by optimality and objective.

Optimality: The localization of the network can be achieved by maximizing the IPR value.

Objective: The objective is to obtain the network configuration that maximizes the IPR value corresponding to the principal eigenvector of the network adjacency matrix.

Major contributions of this paper are:

1. The proposed PSO algorithm for graph perturbation includes graph-level operations.
2. The effectiveness of the proposed work is evaluated with respect to state-of-the-art approaches.

2 Problem Statement

Given a network/graph G, it is transformed to G1 such that the IPR of the principal eigenvector P of the network adjacency matrix is maximized [7, 8, 15]. This can be noted mathematically as:

$$Maximize\ IPR(P) = \sum_{i=1}^{n} P_i^4$$

Subject to

$$|E| = |E1| \tag{1}$$

where P_i is the ith component of the principal eigenvector P. The constraint denotes that the number of edges before and after the perturbations remains the same. Here, one perturbation operation is considered to be two sequential operations, i.e., deletion of an edge followed by insertion of an edge. The numbers of perturbation operations are limited to a small percentage, as the objective is to achieve a high state of localization with less number of perturbations.

3 Methodology

This section describes the proposed Particle Swarm Optimization (PSO) based Perturbation algorithm for obtaining optimal IPR. Here, the approach of converting the actual graph to a graph with the optimal IPR value is elucidated in detail. Figure 1 illustrates the overall process of PSO based perturbation for obtaining the graph with optimal IPR.

The first population is constructed from the initial graph by single perturbation with the help of perturbation and selection. In perturbation approach, m graphs are generated in the initial step. In this phase, a random perturbation approach is used where an edge is selected randomly and removed from the graph. Subsequently, an edge is randomly selected from the complement graph of the original graph and inserted into the graph. These m graphs maintain a balance between the networks with more IPR than the original graph and less IPR than the original graph. This process of evolution continues until the optimal IPR is found at the last iteration. Here, the stopping criterion is considered to be a predefined number of iterations.

PSO algorithm uses the following equations for candidate updation.

$$V_i^{p+1} = V_i^P + c1 * r_1^P\left(C_{pbest,i}^p - C_i^p\right) + c2 * r_2^p\left(C_{gbest,i}^p - C_i^p\right) \quad (2)$$

$$C_i^{p+1} = C_i^p + V_i^{p+1} \quad (3)$$

In the PSO algorithm, the candidate update operation takes place using Eq. (3). Where V_i^p represents the velocity of ith candidate in p^{th} iteration, r_1^p and r_2^p represent

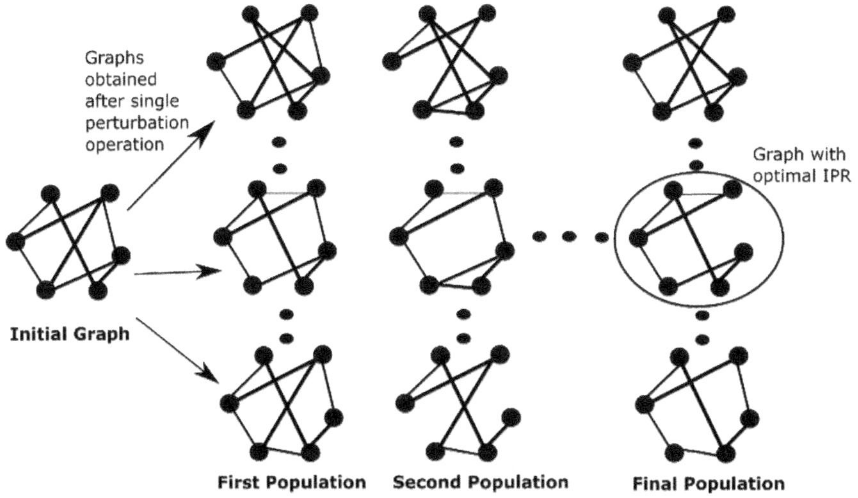

Fig. 1 The overall process of evolution takes place in the proposed approach

the random value ranging from [0, 1] for all candidates. The constants $c1$ and $c2$ represent acceleration constants that can be considered as 1. The candidate with the maximum value as compared to others, is represented as C^{p}_{gbest} (global best value). Similarly, C^{p}_{pbest} represents its own maximum value (personal best value) of the candidate.

For implementing the PSO algorithm in graphs, a part of Eq. (2) is considered in this approach. Thus the velocity is skipped from Eq. (3) and the rest of the equation can be mentioned as Eq. (4). Here, each graph can be considered as a candidate.

$$G^{p}_{i}(updated) = G^{p}_{i} + r^{p}_{1}(G^{p}_{pbest} - G^{p}_{i}) + r^{p}_{2}(G^{p}_{gbest} - G^{p}_{i}) \quad (4)$$

To represent the working of PSO at graph level, a different approach is adopted in this proposed method to get the maximized IPR. G^{p}_{i} is a graph that can be represented as the candidate. Likewise, G^{p}_{gbest} is the candidate with maximum value (global best) and G^{p}_{pbest} is the maximum value (personal best) of the candidate. The steps mentioned below are used to upgrade a candidate in each iteration.

1. $G^{p}_{i}\ (updated) = G^{p}_{i}$
2. Randomly select $e \in E[G^{p}_{i}\ (updated)]$ and remove it
3. $E = E[G^{p}_{gbest}] - E[G^{p}_{i}]$
4. $Q = E[G^{p}_{pbest}] - E[G^{p}_{i}]$
5. Randomly select $x \in (E \cup Q)$ that is present in $complement(G^{p}_{i}\ (updated))$ and include x to $G^{p}_{i}\ (updated)$
6. If IPR of $G^{p}_{i}\ (updated) >$ IPR of G^{p}_{i} then set $G^{p}_{i} = G^{p}_{i}\ (updated)$ otherwise, retain the old value of G^{p}_{i}.

4 Results and Discussion

A system with 12th Gen Intel(R) Core (TM) i7-12700 2.10 GHz processor having 32.0GB of random-access memory and Microsoft Windows 11 Home Single Language 22H2 version OS is used for experimentation. Networkx is used as a predominant tool for graph-level computation and visualization.

The experiment is conducted by considering six different networks. From the six networks four are generated by graph generative models and the rest two are real datasets. Among all the six networks, two are generated by Barabási-Albert (BA) model. The BA model uses a preferential attachment mechanism to generate scale-free networks. One random network dataset is generated by Erdos-Renyi (ER) model. One dataset is created using Powerlaw-Cluster (PC) model that is an extension of BA model. PC model uses preferential attachment with clustering coefficient to build a scale-free network with clustering. The rest two real networks Facebook and Zachary's Karate Club are also used for the evaluation of the proposed algorithm. Table 1 shows the details of the datasets with their basic statistics.

Table 1 Datasets with basic statistics

Dataset	N No. of vertices	M No. of edges	Dataset type	Diameter of the graph	Average clustering coefficient	Average degree of the graph
Zachary's Karate club (KC) [21]	34	78	Real dataset	5	0.572	4.589
Barabasi-Albert 1 (BA-1) [22]	184	895	BA model	4	0.137	7.875
Barabasi-Albert 2 (BA-2) [22]	256	1008	BA model	4	0.120	9.728
Erdos–Renyi (ER) [23]	46	307	ER model	3	0.281	14.696
Power-cluster (PC) [24]	147	705	PC model	4	0.249	9.605
Facebook (FC) [25]	4039	88,234	Real dataset	8	0.605	43.69

The efficacy of the proposed PSO based Perturbation (PSO) is demonstrated against the performance of existing Random Perturbation (RP) and Jaya Perturbation (Jaya) algorithms. These three algorithms are used for optimizing the Inverse Participation Ratio (IPR). An approach of random selection is adopted by all three algorithms that generate distinct outputs for different runs. A pictorial depiction of Optimal IPR obtained in one run is shown in Fig. 2 for KC dataset. Each algorithm is executed 100 times and the average optimal IPRs of 100 runs of the algorithms are compared. The population size for all three algorithms is set to 15 (number of candidates). Table 2 represents the details about the initial and average optimal IPR values obtained by all three algorithms and it is depicted in Fig. 3. The average clustering coefficient of the three algorithms is compared and it is shown in Fig. 4 that all three algorithms report nearly the same average clustering coefficient.

5 Conclusions

In this paper, a PSO based graph perturbation approach is proposed to achieve IPR maximization. The proposed PSO based approach reports a maximum average IPR than the existing RP and Jaya based approaches. The results are presented by implementing the proposed approach on various datasets where some are considered from the generative models and some are real datasets. The comparison is presented by considering the optimal average IPR as an evaluation parameter. In the future, several other heuristic and meta-heuristic approaches can be developed to obtain a better result.

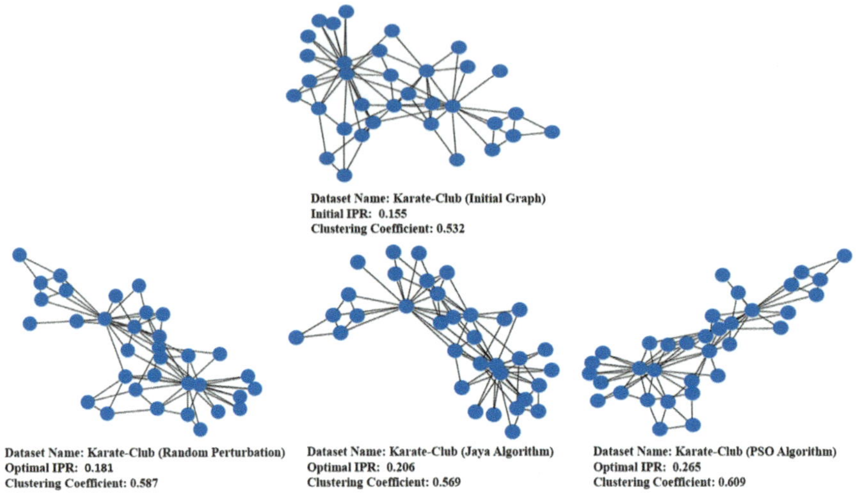

Fig. 2 Optimal IPR achieved by three different approaches in a single run on KC dataset

Table 2 Comparison among RP, Jaya and PSO algorithm

Dataset	Initial IPR	Avg. Optimal IPR (Random perturbation algorithm)	Avg. Optimal IPRs (Jaya algorithm)	Avg. Optimal IPR (PSO algorithm)
KC	0.155	0.165	0.194	0.235
BA-1	0.146	0.200	0.226	0.264
BA-2	0.158	0.168	0.189	0.209
ER	0.049	0.062	0.075	0.096
PC	0.082	0.146	0.185	0.232
FB	0.180	0.181	0.194	0.204

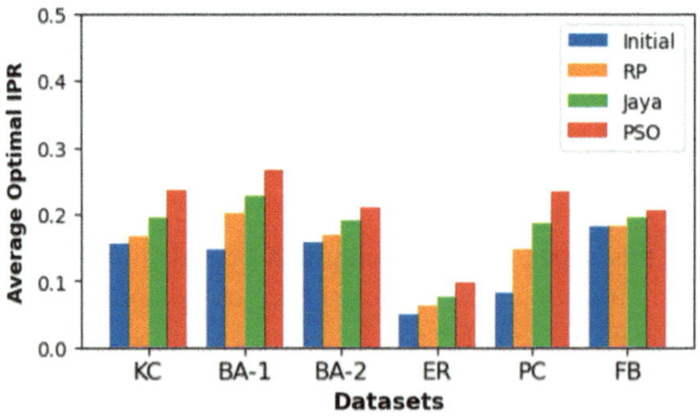

Fig. 3 Optimal average IPR achieved by RP, Jaya and PSO algorithm

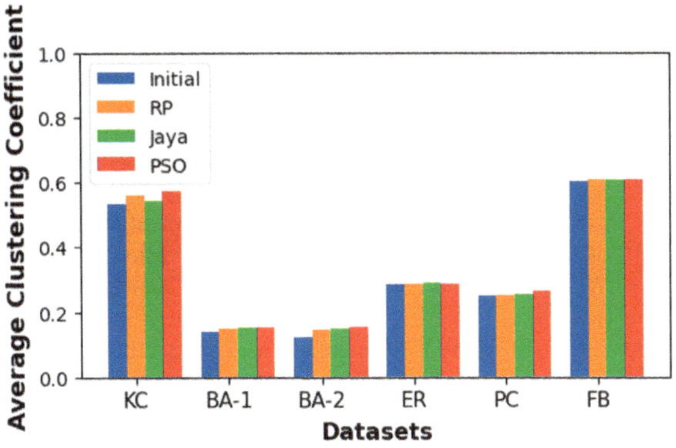

Fig. 4 Comparison of average clustering coefficients

Acknowledgements This work has been funded by OSHEC, Odisha with Grant Number 21SF/CS/2019.

References

1. Ferreira, R.S., da Costa, R.A., Dorogovtsev, S.N., Mendes, J.F.F.: Metastable localization of diseases in complex networks. Phys. Rev. E **94**, 062305 (2016)
2. Kempe, D., Kleinberg, J., Tardos, E.: Maximizing the spread of influence through a social network. In: Proceedings of the Ninth ACM SIGKDD International Conference on Knowledge Discovery and Data Mining, KDD'03, pp. 137–146 (2003)
3. Kitsak, M., Gallos, L.K., Havlin, S., Liljeros, F., Muchnik, L., Stanley, H.E., Makse, H.A.: Identification of influential spreaders in complex networks. Nat. Phys. **6**(11), 888–893 (2010)
4. Newman, M.E.J.: Networks: an Introduction. Oxford University Press, New York (2010)
5. Suweis, S., Grilli, J., Banavar, J.R., Allensian, S., Maritan, A.: Effect of localization on the stability of mutualistic ecological networks. Nat. Commun. **6**, 10179 (2015)
6. Mohapatra, D.: Jaccard guided perturbation for eigenvector localization. New Gener. Comput. **39**, 159–179 (2021)
7. Pradhan, P., Cu, A., Jalan, S.: Principal eigenvector localization and centrality in networks: Revisited. Phys. A: Stat. Mech. Its Appl. **554**, 124169 (2020)
8. Pradhan, P., Yadav, A., Dwivedi, S.K., Jalan, S.: Optimized evolution of networks for principal eigenvector localization. Phys. Rev. E **96**, 022312 (2017)
9. Wang, Y., Chakrabarti, D., Wang, C., Faloutsos, C.: Epidemic spreading in real networks: an eigenvalue view point. In: Proceedings of the 22nd International Symposium on Reliable Distributed Systems, p. 25. IEEE, New York
10. Goltsev, A.V., Dorogovtsev, S.N., Oliveira, J.G., Mendes, J.F.: Localization and spreading of diseases in complex networks. Phys. Rev. Lett. **109**(12), 128702 (2012)
11. Degertekin, S.O., Lamberti, L., Ugur, I.B.: Sizing, layout and topology design optimization of truss structures using the Jaya algorithm. Appl. Soft Comput. **70**, 903–928 (2018)
12. Rao, R.V., Patel, V.: An improved teaching-learning-based optimization algorithm for solving unconstrained optimization problems. Sci. Iran. **20**(3), 710–720 (2013)
13. Rao, R.V., Pawar, R.B.: Constrained design optimization of selected mechanical system components using Rao algorithms. Appl. Soft Comput. **89**, 106141 (2020)
14. Rao, R.V., Savsani, V.J., Balic, J.: Teaching–learning-based optimization algorithm for unconstrained and constrained real-parameter optimization problems. Eng. Optim. **44**(12), 1447–1462 (2012)

15. Mohapatra, D.: Spectral approach to localize information spread in a network using Rao's metaheuristic variants. Soc. Netw. Anal. Min. **13**(1), 32 (2023)
16. Dash, B., Mohapatra, D.: SP: shell-based perturbation approach to localize principal Eigen vector of a network adjacency matrix. In: International Conference on Machine Learning, IoT and Big Data, pp. 369–379. Springer Nature Singapore, Singapore (2023)
17. Xu, H.Q., Gu, S., Fan, Y.C., Li, X.S., Zhao, Y.F., Zhao, J., Wang, J.J.: A strategy learning framework for particle swarm optimization algorithm. Inf. Sci. **619**, 126–152 (2023)
18. Juneja, M., Nagar, S.K.: Particle swarm optimization algorithm and its parameters: a review. In: 2016 International Conference on Control, Computing, Communication and Materials (ICCCCM), pp. 1–5. IEEE (2016)
19. Shi, Y., Eberhart, R.: A modified particle swarm optimizer. In: 1998 IEEE International Conference on Evolutionary Computation Proceedings. IEEE World Congress on Computational Intelligence (Cat. No. 98TH8360), pp. 69–73. IEEE (1998)
20. Shi, Y.: Particle swarm optimization. IEEE Connect. **2**(1), 8−13 (2004)
21. Zachary, W.W.: An information flow model for conflict and fission in small groups. J. Anthropol. Res. **33**, 452–473 (1977)
22. Barabási, A.L., Albert, R.: Emergence of scaling in random networks. Science. **286**(5439), 509–512 (1999)
23. Erdős, P., Rényi, A.: On the strength of connectedness of a random graph. Acta Math. Hungar. **12**(1), 261–267 (1961)
24. Holme, P., Kim, B.J.: Growing scale-free networks with tunable clustering. Phys. Rev. E **65**(2), 026107 (2002)
25. Leskovec, J., Mcauley, J.: Learning to discover social circles in ego networks. Adv. Neural Inf. Process. Syst. **25** (2012)

COVID-19 Severity Detection Using Convolutional Neural Network

Md Abdul Muqueem, Koppula Koushik Kumar, Mohd Sayeed Farhaan, Masna Rashmitha, and Pujari Arjun Goud

Abstract Convolutional Neural Networks (CNNs) show great promise as a quick and accurate way to identify COVID-19-positive individuals. CNNs are a subclass of deep learning algorithms that work especially well on image classification tasks. In order to detect COVID-19, scientists have used CNNs to examine pictures from CT scans or chest X-rays in an effort to find distinctive patterns connected to the virus. A popular approach uses a sizable dataset of chest CT or X-ray pictures that have been classified as positive or negative for COVID-19 to train the CNN. Through this process, the network learns to recognize visual cues that are characteristic of COVID-19, which makes it possible for it to categorize new images as either positive or negative for the virus. The use of CNNs for COVID-19 detection has several benefits, chief among them being their ability to analyze large amounts of data quickly and accurately. CNNs are therefore especially well-suited for use in medical contexts where accurate and timely diagnosis is critical. Furthermore, CNNs are essential for detecting COVID-19 in medical pictures due to their high levels of accuracy in image categorization. In addition, continuous improvements in CNN architectures and techniques keep these neural networks' overall COVID-19 detection efficacy higher. Scholars are presently investigating novel methodologies to enhance the sensitivity and specificity of CNNs, guaranteeing their resilient performance in a range of clinical contexts. This exciting field has the potential to make more advancements and support continuing efforts to mitigate the COVID-19 pandemic's worldwide effects.

Keywords RT-PCR · COVID-19 · CNN · Chest X-ray · VGG-16

M. A. Muqueem (✉) · K. K. Kumar · M. S. Farhaan · M. Rashmitha · P. A. Goud
Department of Electronics and Communication Engineering, Kakatiya Institute of Technology and Science, Warangal, India
e-mail: mq.ece@kitsw.ac.in

S. Kumar Udgata et al. (eds.), *Intelligent Systems*, Lecture Notes in Networks and Systems 1149, https://doi.org/10.1007/978-981-97-8160-7_18

1 Introduction

The pandemic Covid-19, caused by the novel coronavirus, has posed unprecedented challenges worldwide since its emergence in late 2019. Characterized by its high infectiousness and respiratory transmission, the virus presents a spectrum of symptoms ranging from mild to severe, including fever, cough, and respiratory distress. Effective control of the pandemic necessitates the rapid and accurate identification of infected individuals. However, traditional detection methods like Polymerase Chain Reaction (PCR) tests, while accurate, are time-consuming and require specialized equipment, hindering the timely implementation of preventive measures. Since its start in late 2019, the COVID-19 pandemic, which is caused by a new coronavirus, has caused unprecedented challenges on a global scale [1]. Its unique characteristics, such as high infectiousness and respiratory transmission, lead to a range of symptoms, from moderate to severe, that include fever, coughing, and dyspnea. Effective containment of this epidemic requires prompt and accurate identification of those who are afflicted [2].

Leveraging computer-aided approaches for COVID-19 identification has gained popularity as a response to these problems. Convolutional neural networks (CNNs), a subclass of artificial intelligence (AI) algorithms well-known for their skill in image identification tasks, have become one of these techniques' most promising tools. By analyzing several layers of information, CNNs can recognize and learn complex patterns in visual data, which makes it possible to accurately classify objects and scenes in photographs [3]. This also helps in predicting the features of the image in advance.

Diagnosing COVID-19 just from symptoms isn't enough. We need diagnostic examination like RT-PCR and Chest X-rays. But, worldwide field of healthcare struggles to test enough people due to low supplies and high demand for tests. This means many people don't get tested, making it hard to isolate them properly. Also, in stressful situations, the accuracy of tests can be at risk [4]. Most of these days the price of RT-PCR test is also high so that most people neglect to diagnose them. This may lead to an increase in the spread of the disease.

CNNs can be trained to recognize distinctive symptoms linked to COVID-19 in the particular context of lung X-ray or CT scan pictures. Extensive dataset of or X-ray pictures that have been carefully tagged with the presence or absence of COVID-19 is curated as part of the training procedure. CNN is able to identify the unique features of the infection thanks to this labeling. In order for the network to learn and recognize distinct characteristics connected to the infection, this carefully selected dataset is an essential part of the CNN training process. CNNs gain proficiency in identifying patterns and traits unique to COVID-19 in lung pictures by utilizing the labeled data [5, 6]. This method makes a substantial contribution to the advancement of precise and trustworthy diagnostic instruments for the prompt detection and handling of COVID-19 cases.

CNN can effectively categorize fresh photos as positive or negative for COVID-19 once it has been trained on this dataset. The need for alternate testing procedures emerges in this situation in order to increase the accessibility of testing for a greater number of people. Because of their high sensitivity, Chest X-ray reports have become one of the most important diagnostic and detection techniques among other diagnostic tools. But it's important to acknowledge that, considering the increasing number of patients, determining a patient's COVID status based just on analyzing the Chest X-ray poses a challenging assignment. In this situation, extra assistance in making decisions might be beneficial for medical practitioners [7].

The use of CNNs in COVID-19 detection provides a quick and effective way to identify infected people, enabling early isolation and identification to stop the virus from spreading. Beyond COVID-19, CNNs show great promise in medical imaging, with a variety of uses from detecting irregularities in medical pictures to diagnosing other lung disorders like pneumonia or tuberculosis. Additionally, by combining CNNs with other [8] AI techniques, including natural language processing (NLP), it is possible to automate the study of medical reports and other text-based data, giving doctors a more thorough grasp of their patients' conditions and improving the precision of their diagnoses and treatments.

2 Existing Methods

Existing methods include different Architectures and Basic CNN model. Architecture model includes various architectural elements, such as pooling and convolution layers, which make efficiency optimization difficult. Transfer learning, which makes use of ImageNet's pre-trained models, appears to be a crucial tactic in addressing problems. Using their solid base, the suggested model merges four such models. Maintaining the original structure while making necessary adjustments to account for COVID-19 characteristics is known as fine-tuning. By effectively extracting pertinent information, this transfer learning approach advances precise diagnostic tools for the pandemic fight [9].

- **DenseNet**: The main idea behind Dense Net 121 is to alter traditional CNN (Convolutional Neural Network) layers. Dense Net creates connections between every layer and every other layer in the architecture, in contrast to a typical CNN, which only connects each layer to the layer after it. The name Dense Net comes from this networked structure, or Dense [10]. The power and efficacy of a Dense Net design with L layers are greatly increased because the architecture has a total of $L*(L + 1)/2$ connections.
- **MobileNet**: This unique design makes use of a method called convolution with depth-wise separable architecture to enable the development of a deep and lightweight convolutional neural network (CNN) that is very effective for developing a computer vision application. Depth-wise separable filters provide the basis of its structure [11].

This CNN-based model has 88 layers and 4,253,864 parameters in its architecture. Its small size (around 16 MB) distinguishes it from other existing classification models. One reason for MobileNet's remarkable computing efficiency is its smaller size [11].

- **Xception**: Deeply Separable Convolutions are used in place of traditional Inception modules in the new architecture concept introduced by Francois Chollet, a software developer at Google Inc., in his Xception model. This CNN-based method has a large depth of 126 layers with a total of 22,910,480 parameters
- The model is notable for having a file size of about 88 megabytes, which is competitive when compared to other modern models in the field. Xception stands out in its search for effective and efficient convolutional neural network architectures because of its creative application of depth-wise Separable Convolutions. This model's architecture deviates significantly from traditional methods, which adds to its special performance attributes. The Xception model, put forth by Francois Chollet, a software developer at Google Inc., presents a novel architecture by depth-wise Separable Convolutions in place of typical Inception modules [10].

3 Experiment

A. Datasets

Figures 1, 2, and 3 showcase images illustrating distinct classes including Severe Covid, Normal, and Mild Covid.

As a suitable dataset was not identified for this study, a unique dataset was curated by merging three open-source datasets accessible on GitHub and Kaggle The compiled dataset includes X-ray images categorized into three distinct groups:

1. X-ray images of individuals without health disease.
2. X-ray images of individuals with mild COVID-19 symptoms.
3. X-ray images of individuals with severe COVID-19 symptoms.

We implemented safeguards against bias and anomalies in the training and testing datasets due to the different amounts of datasets from the aforementioned platforms, with the limitation of COVID-19 X-ray pictures. As a result, the dataset was meticulously reduced to guarantee equilibrium. The total number of images in the dataset is 9830 of which 3800 images are Severe, 4500 are Normal and 1530 images are Mild.

After splitting the dataset into 80 and 20% as training and testing datasets, the testing dataset is of the range 1994. Out of which 269 are mild, 1002 are normal, and 723 are severe.

B. Experimental Setup

The use of the Python library Keras for the implementation allowed for the smooth integration of pre-built models and simplified neural network development. In this situation, Keras is revealed to be a really useful tool.

Using Google Colaboratory's ability to create and use Jupyter notebooks in a free environment, the implementation was completed. With Colaboratory, developers

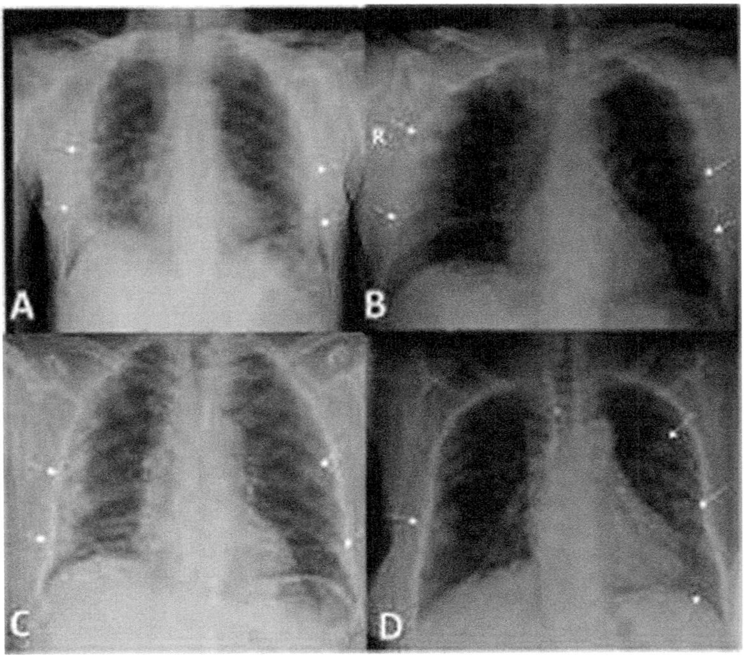

Fig. 1 Sample chest X-ray images of severe covid patients

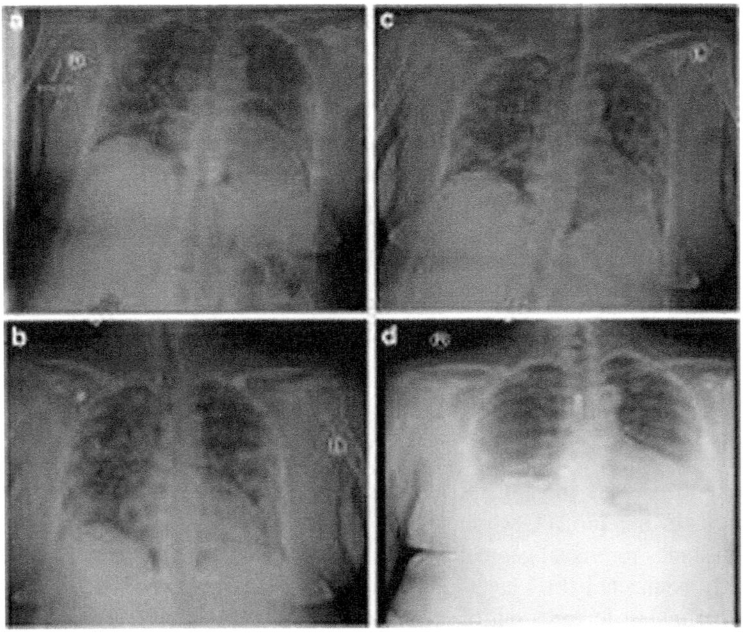

Fig. 2 Sample chest X-Ray images of normal person

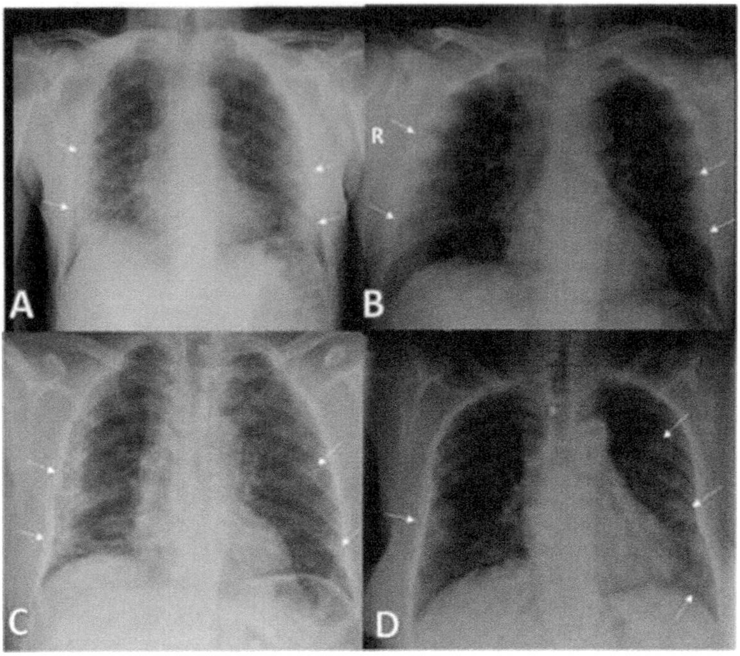

Fig. 3 Sample chest X-Ray image of mild covid patients

have access to a cloud-based writing environment similar to Jupyter notebooks. The service is totally free up to a certain RAM and memory limit. Users have the option to save files remotely or in Google Drive. They can also collaborate and share projects with others. Mini Batch Gradient Descent was selected as the optimization method, and a batch size of 64 and around 20 epochs were taken for training the model.

A dropout layer is added with a 0.25 dropout rate to help prevent overfitting during training and the chosen loss function is Categorical Cross Entropy; it is a commonly used benchmark for this kind of work. The Adam optimizer aids in the optimization process.

4 Proposed Method

In order to classify the Images more accurately, VGG-16 architecture has been proposed. In our project, we have introduced VGG-16 as our proposed method. V66-16 stands for visual geometry group and it consists of 16 layers. Convolutional and fully connected (FC) layers make up its two primary types. The former is in charge of taking features out of the input images and reducing dimensionality by using max pooling layers and filters. In the meantime, the latter helps generate a

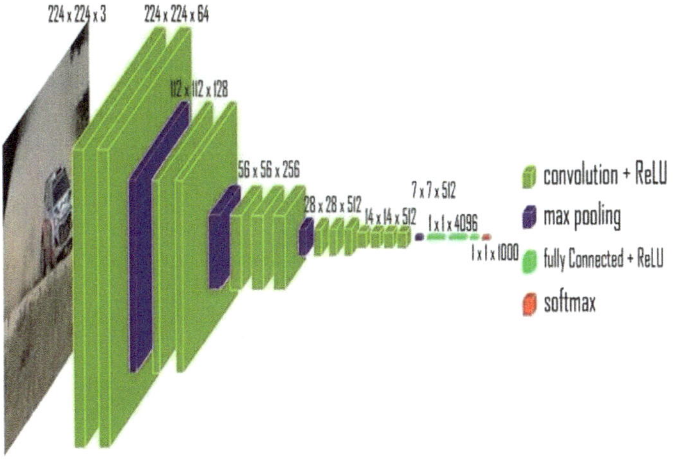

Fig. 4 VGG-16 architecture

probability distribution across classes using a softmax activation function, which is then used to make final predictions based on the extracted characteristics.

The VGG-16, created by the Visual Geometry Group at the University of Oxford, showed success on the ImageNet dataset in 2014. This dataset has 1.2 million images in 1000 categories, and VGG-16 won the ImageNet Large Scale Visual Recognition Challenge. Small 3×3 filters are used in the deep convolutional layers of this architecture to improve feature representation and parameter sharing.

One of the biggest models at the time of its development was VGG-16, which included over 138 million parameters. Even though it performs well in picture classification tasks, its applicability on low-power devices is limited by issues like high computational costs and memory needs. Although VGG-16 has been superseded by more recent models such as ResNet and Inception, it is still a widely used benchmark, especially in deep learning studies including neural style transfer and transfer learning. Because it is open-source, frameworks like TensorFlow and PyTorch can be used to implement it (Fig. 4).

5 Results

The confusion matrix that results from running the suggested VGG-16 models on the testing dataset is shown in Fig. 5. Simultaneously, Tables 1, 2, 3, and 4 provide a thorough analysis of the class-specific recall, precision, and F1-score outcomes derived from the four different models that were used on the testing dataset. These metrics provide a comprehensive assessment of the models' efficacy in the classification task by providing a clear understanding of their performance across different classes.

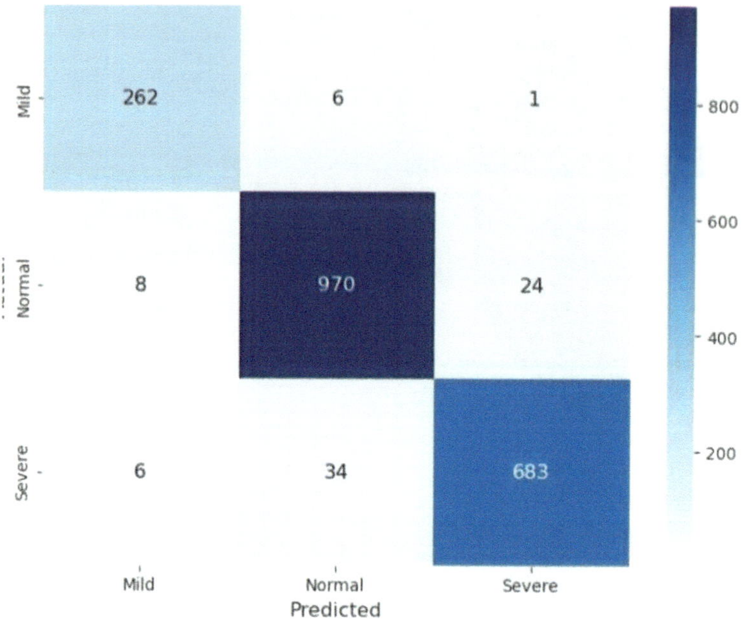

Fig. 5 Confusion matrix analysis for VGG-16

Table 1 Metrices of VGG-16

	Mild	Normal	Severe
Precision	0.994	0.998	0.997
Recall	0.994	0.998	0.997
F1-score	0.994	0.998	0.997

Table 2 Metrices of dense Ne

	Mild	Normal	Severe
Precision	0.991	0.7986	0.991
Recall	0.767	0.991	0.982
F1-score	0.868	0.885	0.987

Table 3 Metrices of MobileNet

	Mild	Normal	Severe
Precision	0.971	0.891	0.999
Recall	0.879	0.991	0.983
F1-score	0.923	0.997	0.991

Table 4 Metrices of Xception

	Mild	Normal	Severe
Precision	0.971	0.817	0.999
Recall	0.879	0.965	0.957
F1-score	0.923	0.885	0.978

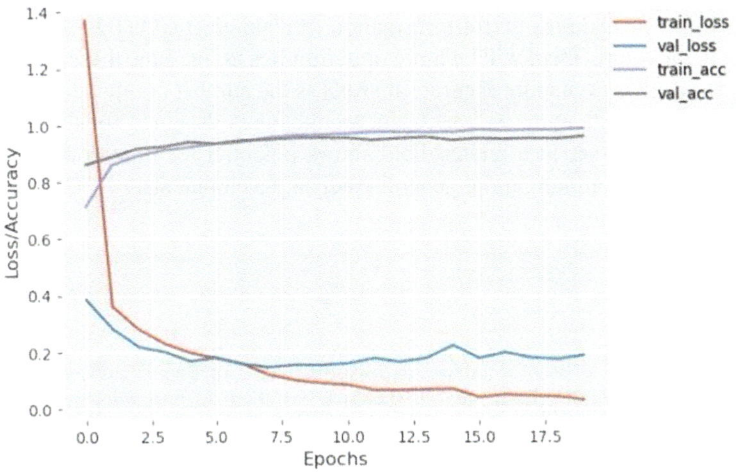

Fig. 6 Training and validation accuracy and loss

From confusion matrix, VGG-16 architecture forecasted the highest count of accurate outputs which is 1915 out of 1994 in which 262 is of mild out of 269,970 is of normal out of 1002, and 683 out of 723 are correctly predicted by the architectural design of VGG-16 (Fig. 6).

6 Conclusion

A proactive approach to testing is required because of the global impact of COVID-19, as it plays a critical role in managing the pandemic. Even while Real Time-PCR is still the gold standard test for COVID-19 detection, it is not considered sufficient to depend just on this approach. In light of the shortcomings of a single testing strategy, our research highlights the significance of diversity. We have carefully examined four different designs in our research, each of which adds to a comprehensive testing plan. By taking a diverse approach, we hope to improve the efficacy of COVID-19 testing and guarantee a thorough understanding of the virus's presence. We hope to contribute to the global effort to mitigate the effects of COVID-19 by investigating different architectures.

Our research has shown that employing the transfer learning technique for weight transfer and using CNN architectures that were first trained on the ImageNet dataset is a reliable method. The data presented above categorically show that, among the four models taken into consideration, VGG-16 is the best option. Notably, our technique was robust since none of the models showed evidence of either underfitting or overfitting.

Our results highlight CNN-based architectures' potential as a substitute testing approach and provide insightful information to counteract COVID-19's long-term effects. In the future, there will be more opportunities to fine-tune these models even further to produce even more accurate answers as the number of input data increases. Furthermore, investigating various CNN architectures designed for the identification of other illnesses in a related field shows potential for improving healthcare diagnostics and promoting more general medical developments.

References

1. Chatterjee, P.: The 2019 novel coronavirus disease (Covid-19) pandemic: a review of the current evidence. Indian J. Med. Res., Suppl. **151**(2–3), 147–159 (2020). https://doi.org/10.4103/ijmr.IJMR_519_20
2. Chatterjee, P., and associates.: The 2019 novel coronavirus disease (Covid-19) pandemic: a review of the current evidence. Suppl., Indian J. Med. Res. **151**(2–3), 147–159 (2020). https://doi.org/10.4103/ijmr.IJMR_519_20
3. Huang, C.: Clinical features of patients infected with 2019 novel coronavirus in Wuhan, China. Lancet. **395**(10223), 497–506 (2020). https://doi.org/10.1016/S0140-6736(20)30183-5
4. Dhaya, R.: Deep net model for detection of Covid-19 using radiographs based on ROC analysis. J. Image Inf. Process. (2020). https://doi.org/10.36548/jiip.2020.3.003M
5. Luz, E., Silva, P.L., Silva, R., Silva, L., Moreira, G., Menot, D.: Towards an effective and efficient deep learning model for COVID-19 patterns detection in X-ray images. [Online]. http://arxiv.org/abs/2004.05717
6. El-Din Hemdan, E., Shouman, M.A., Karar, M.E.: COVIDXNet, a framework of deep learning classifiers to diagnose COVID-19 in X-Ray images (2020). arXiv preprint arXiv:2003.11055
7. Simonyan, K., Zisserman, A.: Very deep convolutional networks for large-scale image recognition (2014). [Online]. http://arxiv.org/abs/1409.1556
8. Elghamrawy, S., Hassanien, A.E.: Diagnosis and predict ion model for COVID-19 patient's response to treatment based on convolutional neural networks and whale optimization algorithm using CT images (2020). PrePrint, https://doi.org/10.1101/2020.04.16.20063990
9. Wang, L., Wong, A.: COVID-Net: a tailored deep convolutional neural network design for detection of COVID-19 cases from chest X-Ray images. [Online]. http://arxiv.org/abs/2003.09871
10. Pereira, R.M., Bertolini, D., Teixeira, L.O., Silla, C.N., Costa, Y.M.G.: COVID-19 identification in chest X-ray images on flat and hierarchical classification scenarios. Comput. Methods Programs Biomed. 105532 (2020). https://doi.org/10.1016/j.cmpb.2020.105532
11. Szegedy, C., Ioffe, S., Vanhoucke, V., Alemi, A.: Inception-v4, Inception-ResNet and the Impact of Residual Connect ions on Learning (2017)
12. Deng, J., Dong, W., Socher, R., Li, L.-J., Li, K., Fei-Fei, Li.: ImageNet: a large-scale hierarchical image database. In: IEEE Conference on Computer Vision and Pattern Recognition, pp 248–255 (2009). https://doi.org/10.1109/cvpr.2009.5206848

13. Szegedy, C., Vanhoucke, V., Ioffe, S., Shlens, J.: Rethinking the incept ion architecture for computer vision. In: The IEEE Conference on Computer Vision and Pattern Recognition (CVPR), pp. 2818–2826 (2016)
14. Farooq, M., Hafeez, A.: COVID-ResNet : a deep learning framework for screening of COVID19from radiographs. arXiv:2003.14395
15. [Online]. https://github.com/lindawangg/COVID-Net (2020)
16. Szegedy, C., Vanhoucke, V., Ioffe, S., Shlens, J.: Rethinking the inception architecture for computer vision. In: The IEEE Conference on Computer Vision and Pattern Recognition (CVPR), pp 2818–2826 (2016)
17. Howard, G., et al.: MobileNets: efficient convolutional neural networks for mobile vision applications. [Online]. http://arxiv.org/abs/1704.04861 (2017)
18. Chollet, F., Xception: deep learning with depthwise separable convolutions. In: IEEE Conference on Computer Vision and Pattern Recognition (CVPR), Honolulu, HI, pp 1800–1807. https://doi.org/10.1109/CVPR.2017.195
19. Browne, P., Chandraratna, D., Angood, C., Tremlett, H., Baker, C., Taylor, B. V., Thompson, A.J.: Atlas of multiple sclerosis 2013: a growing global problem with widespread inequity. Neurol. **83**, 1022—102411 (2014)
20. https://github.com/ieee8023/covid-chestxray-dataset

Alzheimer's Disease Identification and Classification Using the VGG16 Deep Convolutional Neural Network Model

Riyam Patel, Adesh Jothi Prasanna, Borra Sivaiah, Punyaban Patel, Reddyvari Venkateswara Reddy, and Bibhudatta Sahoo

Abstract Different areas of the brain alter as a result of different kinds of brain disorders. Alzheimer's disease is a long-term condition that leads to memory loss due to the degeneration of brain cells. Forgetfulness and confusion are two of the most common cognitive mental health issues in Alzheimer's sufferers. Nowadays, image-based computer-aided diagnosis is being used more and more in the diagnosis of Alzheimer's disease. A VGG16-based model is suggested in this study to improve AD categorization. The original sources of the brain scans were internet databases like the AD Neuroimaging Initiative (ADNI). In order to reduce the dimensionality of the extracted features, the ReliefF technique is used to choose the active features from the retrieved feature values. ReliefF is a very effective method for handling real-time, noisy data. The Manhattan distance metric, which only requires a few features to accurately define the data, is used by the method. This methodology facilitates precise location selection to improve illness identification. The experimental findings demonstrate that the suggested technique outperformed the LSTM, NN, and Deep neural network models in illness prediction.

Keywords Deep Neural Network (DNN) · Alzheimer's Disease (AD) · Long Short-Term Memory (LSTM) · VGG16 · Neural Network (NN)

R. Patel · A. J. Prasanna
Department of Computational Intelligence, College of Engineering and Technology, SRM Institute of Science and Technology, Chennai, India

B. Sivaiah
Department of Computer Science and Engineering, CMR CET, Hyderabad, India

P. Patel (✉) · R. V. Reddy
Department of Computer Science and Engineering-Cyber Security, CMR CET, Hyderabad, India
e-mail: punyaban@gmail.com

R. V. Reddy
e-mail: venkatreddyvari@cmrcet.ac.in

B. Sahoo
Department of CSE, NIT Rourkela, Rourkela, Odisha, India

S. Kumar Udgata et al. (eds.), *Intelligent Systems*, Lecture Notes in Networks and Systems 1149, https://doi.org/10.1007/978-981-97-8160-7_19

1 Introduction

Dementia is a major public health concern in aging societies, with over 48 million global sufferers in 2015 and an estimated \$819 billion in associated costs. The World Health Organization predicts a rise to 79 million by 2030 and 140 million by 2050, with 11 million new cases annually. Dementia is characterized by cognitive impairment and disability, necessitating increased monitoring and support for daily activities. Alzheimer's disease is the basis, with over 80% of cases attributed to AD. Sensor-based wearable devices provide a precise and reliable solution, as data collection for a full day is necessary to determine the exact cause of dementia [1].

Deep learning methods are extensively employed for the analysis of multidimensional data, image recognition, and time series classification, often used to identify Alzheimer's disease (AD) from neuroimaging information. Additionally, these methods have been used to look at possible genetic bio-markers for AD. To diagnose AD, Zhang et al. integrate neuroimaging data with clinical and cognitive evaluation outcomes using a multi-modal deep learning model. Using large-scale gene expression and DNA methylation data, deep learning-based systems have also been able to identify AD, averting common side effects including falls and memory loss. Daily activities and gait analysis are two additional information sources that are employed by both classic and deep learning algorithms to diagnose AD and explore physical activity resistance. Convolutional Neural Networks can identify people and facilitate gait recognition using 3D pictures, even at different viewing angles.

Using a smartphone accelerometer to track patients' daily activities, researchers extend neuroimaging and clinical analysis by examining the connection between mobility patterns and AD stage. The strength of the work is to identify and classify Alzheimer's disease patients. The weakness of the work is not support the clustering of classified Alzheimer's disease patients.

This paper is organized as Sect. 2: reviews on previous works, Sect. 3: explain our VGG16-based proposed method to classify Alzheimer's disease, Sect. 4: performance measures, Sect. 5: indicate the experimental results and analysis and; the conclusions are given in Sect. 6.

2 Literature Review

By providing a greater range of settings and the collecting of massive quantities of data from users' daily activities, technological innovations are increasing healthcare. Sensor-based devices, particularly those related to dementia, offer the advantage of not requiring patient interaction, allowing them to be used at home for basic everyday tasks like treatment reminders or safety monitoring like fall identification. Wearable sensors, such as tri-axial accelerometer-based methods, have shown benefits in identifying falls, a significant cause of health issues and disability among dementia patients.

Tracking behavioral abnormalities is an exciting potential use case for wearable accelerometers. It has been shown that in several studies it is more effective in identifying apathy, which is one of the most prevalent neuropsychiatric symptoms associated with AD. Furthermore, the authors as in [2] used wearable sensors for a seven-day assessment of patients with severe dementia in two assisted living facilities. The total severity of difficult behaviors, as determined by a clinical scale, was positively linked with the accelerometric motion score (AMS), which was computed using the collected data. Accelerometric signals have been utilized by some authors to estimate gait parameters, which allows for the distinction of dementia patients from healthy controls, as well as moderate cognitive impairment [3, 4]. Furthermore, recent research identified substantial variations in seven estimated gait parameters to classify dementia subtypes, including Alzheimer's disease [5], dementia with Lewy bodies, and dementia associated with Parkinson's disease. Accelerometry has also been used to assess circadian rhythm abnormalities, which are commonly observed in AD dementia [6], but are not clearly understood in the beginning stages of the illness.

A circadian rhythm study of 189 cognitively unimpaired people found preclinical Alzheimer's disease. The study found increased intradaily changeability, a measure of disruption of the regular pattern of rest and activity, regardless of age or gender. Sensor-based devices classify Alzheimer's disease patients into three functional stages. El Maachi et al. [7] use Gait data analysis, has been utilized to diagnose Parkinson's disease using convolutional neural networks, but dementia staging has not.

The authors have developed an innovative multi-modal classification technique that utilizes MRI data for the identification of AD. Firstly, a nonlinear graph fusion approach was employed to generate a separate graph for each. After that, an iterative cross-diffusion process was used to create a unified graph, which further strengthens the link between the graphs. Until the unified graph converged, the iterative cross-diffusion process was carried out repeatedly. Finally, the unified graph that was obtained was used to carry out the categorization. The ADNI dataset was used in this study to confirm the effectiveness of the presented method. The deep CNN approach was used by Islam and Zhang [8] to identify and categorize AD. The created method's effectiveness was evaluated using an unbalanced database (OASIS). As was previously indicated, if there is a decrease in the amount of input data acquired, the deep learning models perform poorly. The authors in [9] have been modeled for Alzheimer's disease detection is quite successful. Three modules make up the created model: face, object, and medical recognition. Faster robust features and oriented gradient feature histograms were used in this literature to extract features. Ultimately, a multi-SVM classifier was used to classify diseases. Convolutional neural networks (CNNs) were used to classify AD. The ADNI dataset was used to evaluate CNN's efficacy in classifying AD. For the AD classification, it has been proposed CNN and gradient-based optimization techniques. The effectiveness of the proposed CNN classification was evaluated using the ADNI dataset. The VGG16 model, developed by Huong Hoang Luong et al. [10], utilizes transfer learning and fine-tuning methods to distinguish between brain scans of every person with Alzheimer's disease and those without the disease, achieving high accuracy and F1-scores. Arafa et al. [11] developed two methods for analyzing Alzheimer's disease (AD), using a simple CNN architecture and a pre-trained VGG16 model. The new technique is faster and needs fewer labeled training samples and domain expertise. Gupta et al. [12] used two MRI datasets for early AD identification using a deep Triplet network algorithm, achieving high accuracy, sensitivity, precision, and F1-score in both datasets.

3 Proposed Methodology

Alzheimer's disease (AD) is a kind of illness when the central nervous system's cells malfunction that involves synapse dysfunction, structural abnormalities, and cognitive decline. This study suggests a guided framework for early Alzheimer's identification. All of the following models—NN, LSTM, DNN, and VGG16—are used to classify the abnormalities and normalities associated with Alzheimer's disease. The ReliefF approach is a method used in imagining analysis to improve learning performance and understand important pixels for classification. As a result of its ability to efficiently minimize the dimension of retrieved features, it is suitable for dealing with noisy and real-time data. Manhattan distance, a significant advantage, requires just limited features for exact neighborhood selection, which enables better disease recognition through precise neighbor selection. This is only possible because of the Manhattan distance.

i. **Deep Neural Network (DNN) Classification**

DNN is used to classify Alzheimer's disease and goes through different stages [13] using images used for training and testing. The correctness of classification is primarily influenced by training. The DNN approach provides precise transformation parameters regardless of the original image changes. The input pattern is sent for quick output. The network consists of 3-layers: input, hidden, and the output layer. The input layer uses neurotransmitters to send data to hidden layer, and the output layer sends the data to the output layer.

ii. **LSTM (Long Short-Term Memory)**

A type of neural network called LSTM, may link past knowledge to present. Thus, an LSTM-network [14] is fully connected with activation layers that encode the temporal relationship between characteristics and AD progression. The sequence of repeated neural network modules in LSTM avoids long-term dependence. An LSTM neural network has four interacting layers: forget, input, update, and output. This "forget gate" decides whether to abandon cell formation. Pre-Fully Connected, Cells, and Post-Fully Connected layers are shown in Fig. 1. Pre-completely connected has one fully connected layer and a ReLU function, whereas cells have one LSTM layer and a Dropout Wrapper., and Post-Fully Connected has one and a SoftMax layer. Equation 1 represents the "forget gate," which decides whether to discard cell state information. The sigmoid and tanh layer 'input gate' is calculated using Eqs. 2 and

3. An 'update gate' refreshes the old cell state with the input gate's value represented in Eq. 4. The 'output gate' in Eqs. 5 and 6 selects the layer's output value.

$$f_t = \sigma\left(W_f.\left[h_{t-1}, x_t\right] + b_f\right) \tag{1}$$

where, $W_f = weight\ \ matrix, b_f = bias\ vector,\ and\ 0 < f_t < 1$

where, 0 assigned for forget and 1 assigned for the keep.

$$i_t = \sigma\left(W_i.\left[h_{t-1}, x_t\right] + b_i\right) \tag{2}$$

$$\tilde{C}_t = tanh\left(W_C.\left[h_{t-1}, x_t\right] + b_C\right) \tag{3}$$

where, $W_i\ and\ \ W_C = weight\ matrics,$

$b_i\ and\ b_C = vectors, \tilde{C}_t = output\ \ of\ \ two\ \ equations$

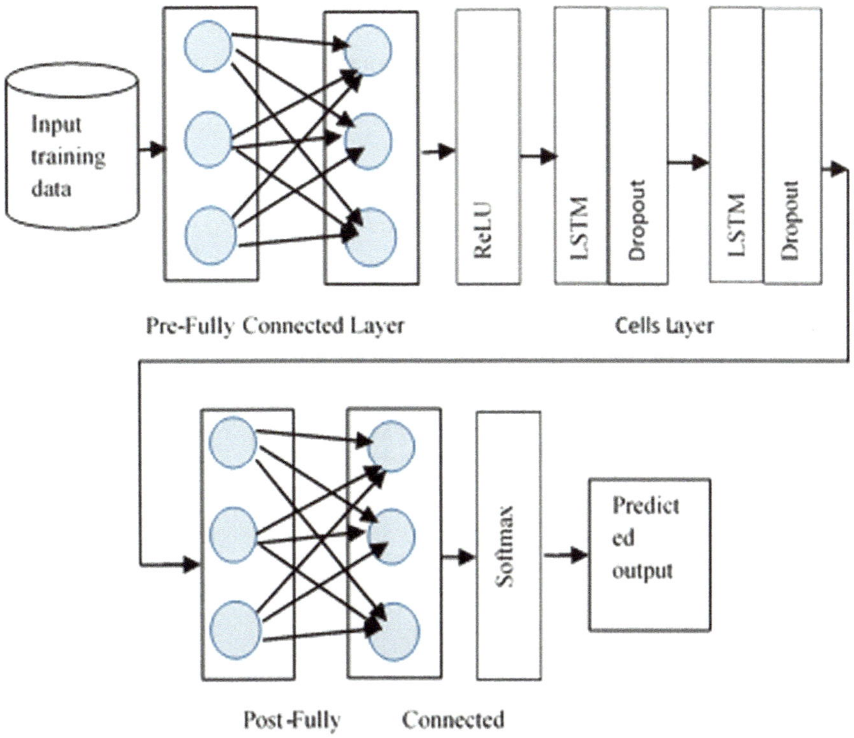

Fig. 1 LSTM model for Alzheimer's disease

$$C_t = f_t * C_{t-1} + i_t * \tilde{C}_t \tag{4}$$

iii. **VGG16 Model**

The CNN, based on VGG16, is a significant deep convolutional neural network with 160 million parameters. Google researchers created In-ceptionv3 in 2015, reducing processing costs with pre-trained architectures like VGG16.

iv. **Neural Network**

Neural networks use convolution layers with activation functions to transform linear data tensors into nonlinear ones. ReLU is commonly used in deep learning. Max Pooling downsizes the picture while retaining crucial elements. Flattening creates a one-dimensional array. Full connection builds connections, activation layers follow, and a classifier classifies the picture as normal or abnormal.

4 Performance Measures

The measures used in the experimental results and analysis are given by Eqs. 5, 6, 7, 8, 9 and 10. Here, TP = True-Positive, TN = True-Negative, FP = s False-Positive, and FN = False-Negative.

$$Accuracy = \frac{TP + TN}{(FN + TP + FP + TN)} \times 100 \tag{5}$$

$$F - score = \frac{2TP}{(2TP + FP + FN)} \times 100 \tag{6}$$

$$Error\ rate = 100 - accuracy \tag{7}$$

$$Miss\ rate = \frac{FN}{(FN + TP)} \times 100 \tag{8}$$

$$\text{Precision(P)} = \frac{\text{TP}}{(\text{TP} + \text{FP})} \tag{9}$$

$$\text{Recall}(R) = \frac{\text{TP}}{(\text{TP} + \text{FN})} \tag{10}$$

5 Experimental Results and Analysis

The images have been chosen from the Alzheimer's Disease Neuroimaging Initiative(ADNI) database [2]. We experimented with VGG16, NN, DNN, and LSTM models in Python. The model VGG16 visualization is shown in Fig. 2. The classification report of the VGG16 model is in Table 1 [1].

The confusion matrix is illustrated in Fig. 3. Four classes of Alzheimer's data exist. Mild, normal, moderate, and very mild patients are classified. The ADNI database is used to evaluate the proposed method in this part. Table 2 compares the suggested framework's accuracy, f-score, precision, and recall. This scenario involves quantitative brain image analysis of 150 scans (50 AD, 50 normal, and 50 MCI) with 80% training and 20% testing. Performance is also investigated using LSTM, neural networks, and DNN classification methods. The model loss and accuracy graph is in Fig. 4. Comparison of NN, LSTM, DNN, and VGG16 models is illustrated in Fig. 5.

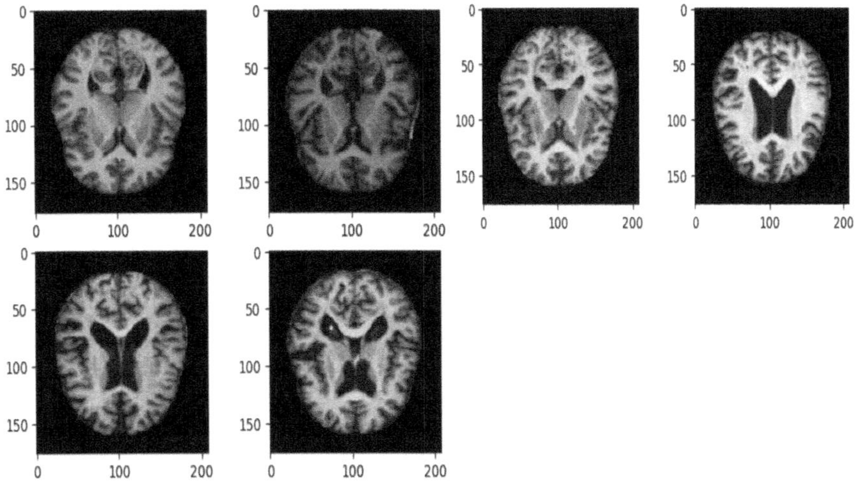

Fig. 2 VGG16 model visualization

Table 1 Performance metrics of the VGG16 model

Support	Precision	Recall	F1-score
176	0.99	0.99	0.99
13	1	1	1
651	0.99	0.99	0.99
416	0.99	0.99	0.99

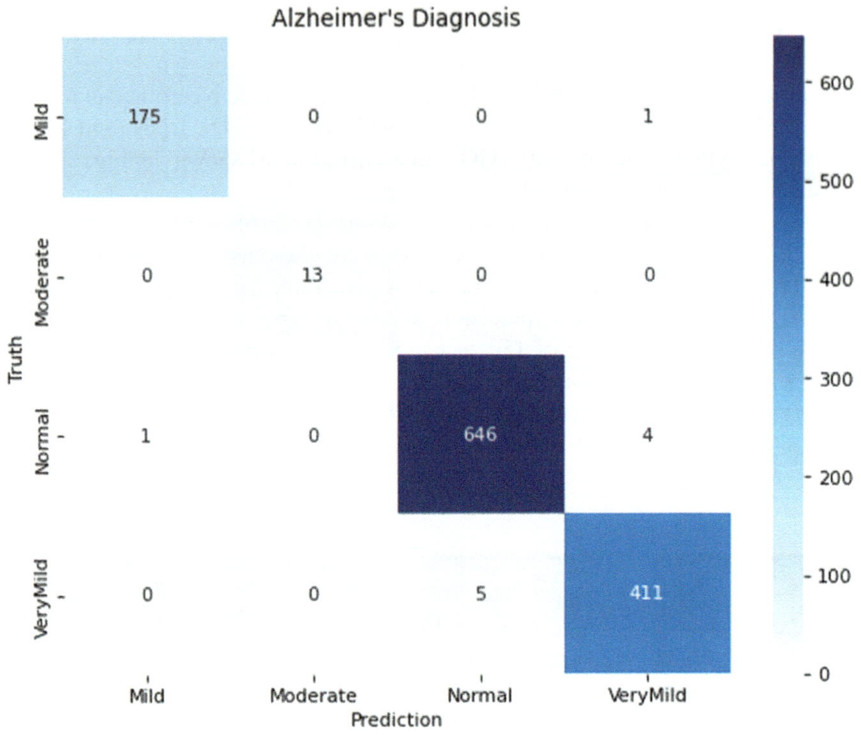

Fig. 3 Confusion matrix

Table 2 Model performance visualization

Model name	F1-score	Accuracy
NN	73.7	69.47
LSTM	95.81	96.54
DNN	98.34	97.02
VGG16	99.12	99

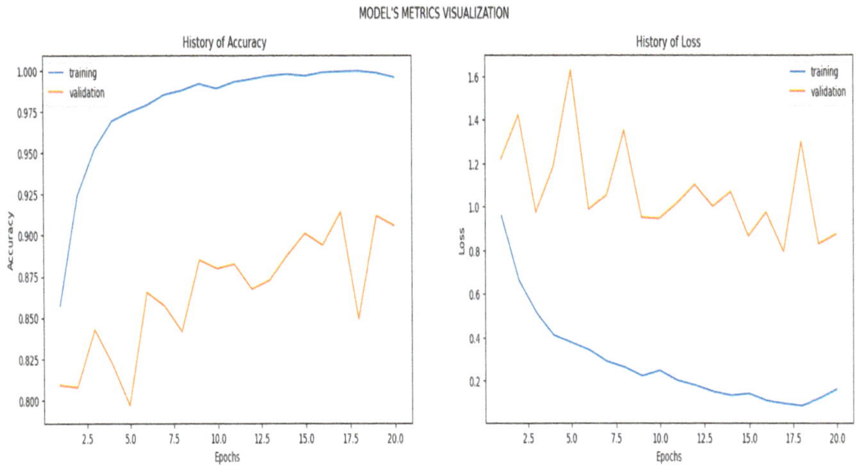

Fig. 4 Model loss and accuracy graph

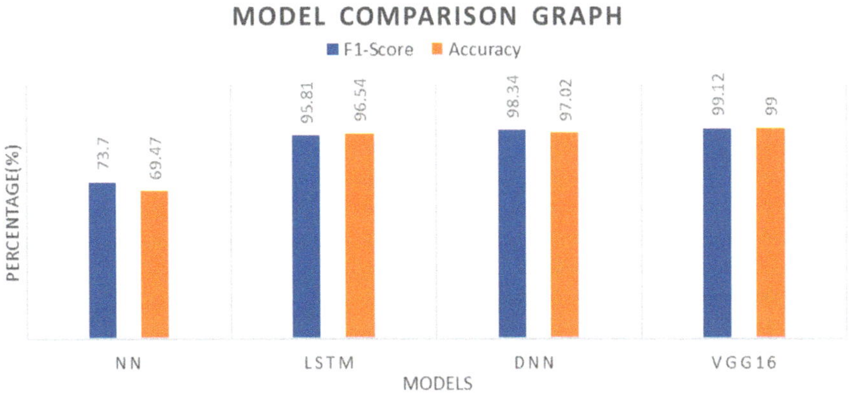

Fig. 5 NN, LSTM, DNN, VGG16 models analysis

6 Conclusion

This research focuses on MRI-based identification and categorization of Alzheimer's disease (AD) using the ADNI dataset. The ReliefF method is used for handling noisy and real-time data, using the Manhattan distance metric for neighborhood selection. Pre-processed images are used, and the irregularity and normalcy of AD are categorized through neural networks like NN, LSTM, DNN, and VGG16. The proposed framework achieved superior accuracy in AD prediction compared to previous efforts. Future work will use feature extraction methods for larger AD datasets. This research is crucial for improving the field of medicine.

References

1. Li, F., Liu, M.: Alzheimer's disease neuroimaging initiative. Alzheimer's disease diagnosis based on multiple cluster dense convolutional networks. Comput. Med. Imaging Graph. **70**, 101–110 (2018)
2. Goerss, D., Hein, A., Bader, S., Halek, M., Kernebeck, S., Kutschke, A., Heine, C., Krueger, F., Kirste, T., Teipel, S.: Automated sensor-based detection of challenging behaviors in advanced stages of dementia in nursing homes. Alzheimer's & Dement.: J. Alz-Heimer'S Assoc. (2020). https://doi.org/10.1016/j.jalz.2019.08.193
3. Hausdorff, J.M., Hillel, I., Shustak, S., Del Din, S., Bekkers, E.M.J., Pelosin, E., Nieuwhof, F., Rochester, L., Mirelman, A.: Everyday stepping quantity and quality among older adult fallers with and without mild cognitive impairment: initial evidence for new motor markers of cognitive deficits? J. Gerontol. Ser. A, Biol. Sci. Med. Sci. **73**(8), 1078–1082 (2018). https://doi.org/10.1093/gerona/glx187
4. Weissova, K., Bartos, A., Sladek, M., Novakova, M., Sumova, A.: Moderate changes in the circadian system of Alzheimer's disease patients detected in their home environment. PloS One **11**(1) (2016). https://doi.org/10.1371/journal.pone.0146200e0146200.
5. Tong, T., Gray, K., Gao, Q., Chen, L., Rueckert, D.: Alzheimer's disease neuroimaging initiative: multi-modal classifcation of Alzheimer's disease using nonlinear graph fusion. Pattern Recognit. **63**, 171–181 (2017)
6. Mc Ardle, R., Del Din, S., Galna, B., Thomas, A., Rochester, L.: Differentiating dementia disease subtypes with gait analysis: feasibility of wearable sensors? Gait & Posture **76**, 372–376 (2019). https://doi.org/10.1016/j.gaitpost.2019.12.028
7. Maachi, I.E., Bilodeau, G.-A., Bouachir, W.: Deep 1d-convnet for accurate Parkinson disease detection and severity prediction from gait. Expert Syst. Appl. **143**, 113075 (2020). https://doi.org/10.1016/j.eswa.2019.113075
8. Islam, J., Zhang, Y.: Brain MRI analysis for Alzheimer's disease diagnosis using an ensemble system of deep convolutional neural networks. Brain Inform. **5**(2), 2 (2018)
9. Sultan, S., Javed, A., Irtaza, A., Dawood, H., Dawood, H., Bashir, A.K.: A hybrid egocentric video summarization method to improve the healthcare for Alzheimer patients. J. Ambient. Intell. Humaniz. Comput. **10**(10), 4197–4206 (2019)
10. Luong, H.H., Vo, P.T., Phan, H.C., Tran, N.L.D., Le, H.Q., Nguyen, H.T.: Fine-tuning VGG16 for Alzheimer's disease diagnosis. In: Conference on Complex, Intelligent, and Software Intensive Systems, pp. 68–79. Springer Nature Switzerland, Cham (2023)
11. Arafa, D.A., Moustafa, H.E.D., Ali, H.A., Ali-Eldin, A.M., Saraya, S.F.: A deep learning framework for early diagnosis of Alzheimer's disease on MRI images. Multimesdia Tools Appl. **83**(2), 3767–3799 (2023)
12. Gupta, A., Gupta, D., Gupta, S.S.: Identification of Alzheimer's disease from MRI image employing a probabilistic deep learning-based approach and the VGG16 (2023)
13. Jo, T., Nho, K., Saykin, A.J.: Deep learning in Alzheimer's disease: diagnostic classifcation and prognostic prediction using neuroimaging data. Front. Aging Neurosci. **11**, 220 (2019)
14. Hochreiter, S., Schmidhuber, J.: Long short-term memory. Neural Comput. **9**(8), 1735–1780 (1997)

Prediction of Cardiac Arrest Using Artificial Intelligence Methods

Subrata Kumar Nayak, Sujogya Mishra, Subhalaxmi Das, Sipali Pradhan, and Geetanjali Pradhan

Abstract The abrupt, potentially fatal stop of heart action is known as cardiac arrest. Predicting cardiac arrest early on is crucial because it gives you time to take the appropriate precautions or act quickly when it starts. Big data and artificial intelligence (AI) technologies are increasingly employed to improve the capacity to anticipate and prepare for patients in danger. The objective of this research is to explore the implementation of AI tools for the prediction of cardiac arrest. To achieve the desired results, this study has used Rough Set Theory (RST), one of the significant tools to handle vague and imprecise data; this study has been divided into three categories: category 1 includes the application of RST reduct algorithm to find the significant attribute, category 2 includes an AI warming technique using several studies. Several studies have also been carried out to forecast future cardiac arrest in category 2, and the third category classifies the healthy person as a person suffering from cardiac arrest. The last section of this study used statistical techniques to validate the claim.

Keywords Cardiac arrest · AI · RST · Classifications · Statistical technique

1 Introduction

Cardiac arrest and Sudden Cardiac Death (SCD) are both the same from the physician's perspective. In both cases, the ability of the heart to pump blood is severely affected. Due to the rapid halt of blood flow, critical organs like the heart and brain need quick attention. Intervention delays can result in permanent problems or even

S. K. Nayak (✉)
GVHSS, Sishu Ananta Mahavidyalaya, Balipatna, Odisha, India
e-mail: subratnayaksap@gmail.com

S. Mishra · S. Das · G. Pradhan
Odisha University of Technology and Research, Bhubaneswar, India
e-mail: sdacse@outr.ac.in

S. Pradhan
Department of Computer Science, RBVRR Women's College, Hyderabad, India

S. Kumar Udgata et al. (eds.), *Intelligent Systems*, Lecture Notes in Networks and Systems 1149, https://doi.org/10.1007/978-981-97-8160-7_20

death. 78% of out-of-hospital cardiac arrest (OHCA) cases end in death before the patient does any research on the hospital, indicating a notably high global rate of mortality following cardiac arrest, as discussed by Berdowski et al. [1]. Another noteworthy statistic is the quantity of cardiac arrest deaths that take place within hospital settings. Over 290,000 in-hospital cardiac arrests occur annually in the United States alone. Survival rates range from zero to 36.2%, with a small fraction of these cases having a satisfactory neurological prognosis. as discussed by Chan et al. [2]. Every day, healthcare is being transformed by artificial intelligence (AI). With the increasing usage of electronic health records, advances in computing power, continuous monitoring systems, and the availability of large data, artificial intelligence (AI) technologies are in a great position to develop and flourish, as discussed by Rong et al. [3]. This crucial clinical decision-making tool makes personalized diagnosis, treatment, prognosis, and future health outcome prediction possible, which helps physicians and other stakeholders make the best decisions for their patients, as discussed by Francisco et al. [4]. Cardiology is one area of medicine where AI technology is advancing quickly, much like any other, as discussed by Jimenez et al. [5]. Cardiovascular medicine research, clinical practice, and public health have all advanced thanks to AI-guided diagnosis and therapy selection, as Johnson et al. Discussed [6]. Further improving tailored treatment, machine learning models have been demonstrated to perform better than classic statistics models in identifying sex differences in cardiovascular illness, as discussed by Sarajlic et al. [7]. It also significantly contributes to improving care for cardiac arrest patients. By early risk factor identification, as discussed by Jayasree and Rao [8], and early detection, as discussed by Kumari et al. [9], better care (such as efficient cardiopulmonary resuscitation) as studied by Didon et al. [10], and prognostication analysis of post-cardiac arrest patients by Cronberg et al. [11], artificial intelligence (AI) technologies are being used to avoid cardiac arrest. Predicting cardiac arrest before it happens is a major focus of cardiac arrest research because it allows clinicians to better prepare their patients and make better decisions. This study faces many challenges in analyzing cardiac-related medical data. These challenges are due to the persistent research gap in the classification process; some researchers used various data sets of cardiac-related issues from a specific hospital, as cardiac-related issues are a global threat.

1.1 Contemporary Research and Challenges

Contemporary research data analysis usually includes normalizing data without mentioning the outcomes. In general, machine learning tools are applied to a local data set, which is not universally true. Analyzing vague and imprecise data using contemporary machine learning tools is always a challenge. Data preparation using a local data set is questionable. The above research gaps can be overcome using RST [12].

2 Background for Data Analysis

The data for this research are collected from various medical sources. The medical sources include various private hospitals in various districts of the state of Odisha. This research designed a centralized system with centralized administration; for every input of raw data, the data set is automatically grouped according to dissimilarity. This technique needs to be more robust towards the outliers. It will perfectly work for the new data sets. The data was collected from districts like Sambalpur, Koraput, Sundergarh, Bolangir, Puri, Cuttack, and Dhenkanal. A total of 1,70,000 data sets are collected only concerning cardiac-related disease (Table 1).

The data are grouped according to their dissimilarity using a model resembling logistic regression.

2.1 Logistic Regression

Logistic regression involves determining the likelihood of a categorical outcome based on an input variable. Logistic regression often yields a binary output result. Logistic regression is applicable to multinomial classification. Logistic regression is a useful analytical method for addressing classification problems. It is used to determine the group to which a new sample most accurately belongs. Logistic regression is a valuable analytical technique for addressing classification issues in several aspects of cyber security, such as threat detection. Logistic regression is mathematically modeled by an exponential function as shown in Fig. 1. Logistic regression transforms input values into output outcomes expressed as fractions or probabilities.

Table 1 Initial data table

SL	Districts	Cardiac asthma	Pneumonia	Cardiac arrest	Lungs infection	Bronchiolitis	Total
1	Sambalpur	1500	1500	2000	500	500	6000
2	Koraput	2000	2000	2000	500	500	7000
3	Sundergarh	15,000	3000	2000	1500	1500	23,000
4	Bolangir	10,000	10,000	500	500	1000	22,000
5	Puri	10,000	15,000	15,000	1500	500	42,000
6	Cuttack	10,000	10,000	5000	5000	10,000	40,000
7	Dhenkanal	5000	5000	5000	5000	10,000	30,000

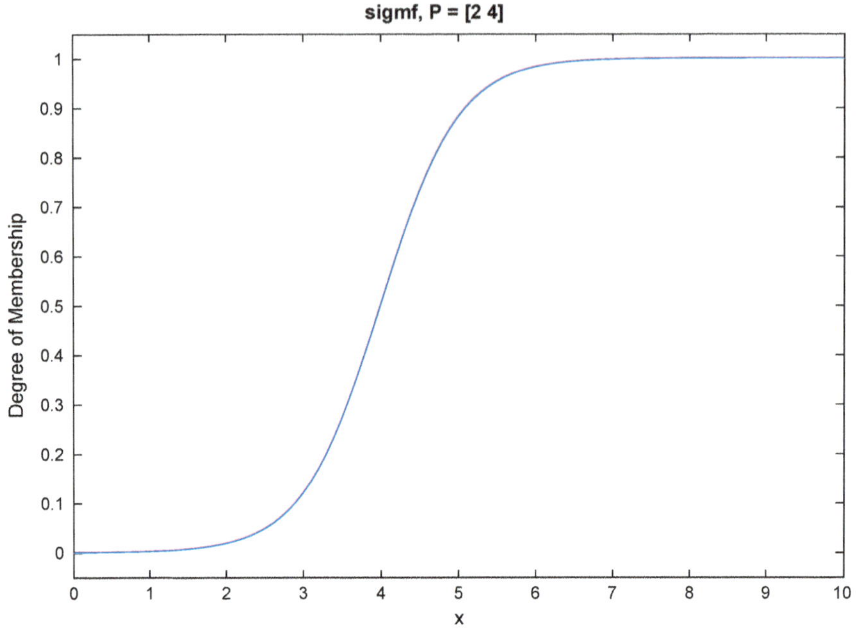

Fig. 1 Sigmoid function used to avoid Boolean results

$$F(z) = \frac{1}{1 + e^{-z}} \tag{1}$$

The sigmoid function is a very useful tool for classification and regression, and the sigmoid function is a useful tool for outliers. This research used a new concept to regroup the data according to dissimilarity.

2.2 *Algorithm for Regrouping Using Logistic Regression.*

Step-1 Initialize two parameters a_0, a_1, then form the equation

$$z = a_0 + a_1 x \tag{2}$$

For binary classification

Step-2 Use of sigmoid function to avoid Boolean results

$$\varphi(z) = \frac{1}{1 + e^{-z}} \tag{3}$$

Table 2 Initial Information

SL	tt	tk	tr	Tl	tm	D
1	aa	aa	aa	aa	aa	1
2	bb	aa	aa	aa	bb	1
3	aa	bb	aa	bb	aa	0
4	bb	bb	bb	bb	bb	0
5	aa	aa	bb	bb	aa	1
6	bb	bb	bb	aa	aa	0

Step-3.

Calculate the z values for new points, then find $\varphi(z)$ and classify the points accordingly.

Step-4.

Find cross entropy, i.e.,

$$J(K) = \frac{1}{N}\left(\sum k_i \log\left(p(k_i) + \sum(1 - k_i)\log(p(1 - k_i))\right)\right) \quad (4)$$

Use the cross entropy to classify the points correctly.

Step-5 Continue steps 1 to 5 till all points are correctly classified.

Using the above algorithm, this study had the following results (Table 2).

Cardiac Asthma–tt, Pneumonia–tk, Cardiac Arrest–tr, Lungs Infection–tl, Bronchiolitis–tm, D–Decision.

The values aa–significant, bb-insignificant, 1–Boolean value agree, i.e., significant, 0–Bolean value Dis agree, i.e., Insignificant.

This study found a set of reducts using the RST reduct algorithm; certain concepts are useful in finding reduct, indiscernibility, and upper and lower approximation.

2.3 *Various Components of RST*

i. Lower approximations L (x) = {X ⊆ U, U is the universal set, collection of all objects, possibly belongs to X}.
ii. Upper Approximations L'(x) = {Collection of all objects which belong to X, where X ⊆ U}
iii. Indiscernibility–: {(x,y)∈ $U \times U$, a(x) = a(y)}
iv. Equivalence classes–: If m,n,p are three objects, then $m^r m, n^r n, p^r p$, $m^r n \rightarrow n^r m$, similarly a combination of n, p, transitive, i.e., $m^r p$, $p^r n \rightarrow m^r n$

Algorithm to find the reduct

Input:

Define the conditional attributes as R, Decision attributes as K.

Output: Feature subset is L.

1. J: = {}, L: = {}
2. repeat
3. J: = L
4. ∀ y ∈ {R\L}
5. if $\gamma_{L\cup\{x\}}$ (D)>γ_J (D)
6. J: = L U {x}
7. L: = J
8. until γ_J(D) = γ_R(D)
9. return L

Using the above algorithm

$$IND(tt) = \{(1, 3, 5), (2, 4, 6)\} \tag{5}$$

$$IND(tk) = \{(1, 2, 5), (3, 4, 6)\} \tag{6}$$

$$IND(tr) = \{(1, 2, 3), (4, 5, 6)\} \tag{7}$$

$$IND(tl) = \{ (1, 2, 6), (3, 4, 5)\} \tag{8}$$

$$IND(tm) = \{(1, 3, 5, 6), (2, 4)\} \tag{9}$$

$$IND(tt, tk) = \{ (1, 5), (4, 6), 2, 3\} \tag{10}$$

$$IND(tt, tr) = \{(1, 3), (4, 6), 2, 3\} \tag{11}$$

$$IND(tt, tl) = \{(3, 5), (2, 6), 1, 4\} \tag{12}$$

$$IND(tt, tm) = \{(1, 3, 5), (2, 4) 6\} \tag{13}$$

$$IND(tk, tr) = \{(1, 2), (4, 6), 3, 5\} \tag{14}$$

$$IND(tk, tl) = \{(1, 2), (3, 4), 5, 6\} \tag{15}$$

$$\mathrm{IND}(\mathrm{tk}, \mathrm{tm}) = \{(1,5), (3,6), 2, 4\} \tag{16}$$

$$\mathrm{IND}(\mathrm{tr}, \mathrm{tl}) = \{(1,2), (4,5), 3, 6\} \tag{17}$$

$$IND(tr, tm) = \{(1,3), (5,6), 2, 4\} \tag{18}$$

$$IND(tl, tm) = \{(1,6), (3,5), 2, 4\} \tag{19}$$

$$IND(tt, tk, tr) = \{(4,6), 1, 2, 3, 5\} \tag{20}$$

$$IND(tt, tr, tl) = \{1, 2, 3, 4, 5, 6\} \tag{21}$$

$$IND(tt, tl, tm) = \{(3,5), 1, 2, 4, 6\} \tag{22}$$

Similarly, all indiscernibility by combining 3 and 4 attributes combination to find the reduct, reduct are those attribute combinations which produce distinct values, i.e., IND (tt, tr, tl) is a reduct, similarly, this research produces several reducts, and taking the intersection of the reduct we have the core tr, i.e., cardiac arrest.

2.4 *Application of Artificial Intelligence for Data Analysis*

Artificial intelligence (AI) is often used to predict the occurrence of cardiac arrest in various patient settings. The use of technology is expected to greatly improve the field of cardiac medicine. Further inquiries are necessary to have a comprehensive understanding of the difficulties linked to the utilization of AI technology in the therapeutic environment. The environment. Conducting research on the most efficient strategies to enhance medical practitioners' understanding, integration, and use of technology is of utmost importance. Cardiac arrest is the cessation of cardiac activity that may be life-threatening. Accurate anticipation of cardiac arrest is essential as it enables proactive intervention and therapy to minimize the advancement of the disease. Artificial intelligence and big data technologies are being used to enhance the ability to predict and prepare for patients who are at danger. This exhaustive investigation, which thoroughly examined over 700 sources, arrived at the conclusive determination of AI.

3 Different Research Gaps

Multiple studies have shown the use of artificial intelligence (AI) in the field of cardiac care, namely in the accurate forecasting of cardiac arrest. Nevertheless, ongoing research must improve its efforts in thoroughly describing the characteristics of AI technology, datasets, and data sources. To discover research gaps in the existing literature and adhere to relevant standards, researchers and healthcare practitioners must summarize new findings. A study was encountered which scrutinized the implementation of artificial intelligence (AI) in cardiac arrest prediction. Nevertheless, the evaluation remained incomplete until 2018, resulting in the omission of a substantial number of papers published in the preceding two years. As a result, it is critical to conduct a comprehensive evaluation that investigates in detail the various AI technologies utilized to identify cardiac arrest in different situations. As described in the literature, this exploratory review investigates the properties and applications of artificial intelligence (AI) systems used to predict cardiac arrest. The physician and medical practitioner will get benefit from the discoveries of these research. This study investigated the use of artificial intelligence (AI) in analyzing data connected to the cardiovascular system. This research encompasses the information provided in Table 3.

For the prediction of cardiac arrest, several researchers used various machine learning techniques like Support Vector Machine (SVM), Artificial Neural Network (ANN), and Convolutional Neural Network (CNN). However, the above methods are useful for a particular pattern of data.

3.1 Data Set Processing

This study uses the following techniques:

a. Pre-processing Phase
b. Design of the model

Phase of pre-processing: labeled data encoding, duplicate value removal,

Model created: Validating the model and training and testing the data set are included in this phase.

The Gradient Boosting Classifier is often used in supervised learning tasks. Ensemble learning is a method that enhances model performance by using the

Table 3 Proposed methods

Norms	Significant norms
1	Only focus on primary implementation
2	The models are being tested concerning various classifiers

strengths of several weak classifiers. The technique involves iteratively incorporating weak classifiers into the model, with each classifier being trained based on the errors made by the previous classifiers. The main objective of the Gradient Boosting Classifier is to optimize a loss function that quantifies the disparity between the actual and predicted values of the target variable.

Nevertheless, the evaluation was ultimately concluded in 2018, disregarding a significant influx of articles in the preceding two years. Therefore, it is vital to do a scoping evaluation that specifically examines various AI technologies employed in different situations to identify cases of cardiac arrest. This scoping review examines the properties and applications of artificial intelligence (AI) systems used for predicting cardiac arrest, as described in the literature. The physician and medical practitioner will get significant benefits from the discoveries made in these studies.

3.2 Description of Methodology

The approach first trains a weak classifier on the training set to make predictions about the target variable. Subsequently, it calculates the deviation of a value from both the predicted and actual values using a single node. The term "base" or "weak learner" refers to the fundamental model in question.

The prediction process involves using the training data and relies entirely on the base learner. The relationships between the characteristics are established by correlating the actual values with the projected values.

- Optimization: The objective of this research is to minimize the residuals in order to develop a new model that outperforms the previous one. The new model is essentially a recently trained decision tree that is used to forecast the residuals.
- Integration: In order to get a substantial outcome, the two aforementioned models are combined, resulting in an ensemble model that incorporates the combined forecasts of each individual model. The ultimate outcomes of this ensemble model are being included for examination.
- Iteration: The aforementioned methods are repeated a preset number of times until the ensemble model achieves the desired level of performance.
- Regularization: This work has included several regularization strategies, such as subsampling, to prevent overfitting. During this phase, a subset of training data samples has been randomly selected to train several models concurrently. Another technique is used to reduce the impact of each model's contribution to the ensemble.
- Optimization: The performance of the Gradient Boosting Classifier may be improved by adjusting the hyperparameters, such as the number of iterations, learning rate, and decision tree depth.
- Evaluation: The final model's ability to apply its knowledge to new data is measured by measuring its performance on a separate dataset.

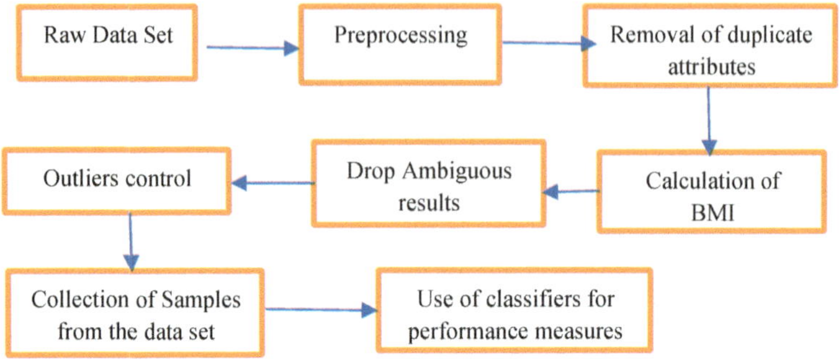

Fig. 2 The flow diagram

3.3 Analysis and Details About the Universal Data Set

The universal data set, i.e., Kaggle repository, is considered in this study. This data set has 13 columns or attributes and a total of 7000 + records. The attributes are age, Id, gender, weight, height, aphi, aplo, cholesterol, gluc, smoke, alco, active, cardio. The 11 features selected for this research are collected from the universal data set; these traits could consist of lifestyle elements, medical measurements, and demographic data. The particular characteristics could change based on the dataset. However, age, gender, blood pressure, cholesterol, smoking status, and body mass index (BMI) are typical examples. This research used target values in Boolean form, i.e., Cardiovascular diseases present or not (Fig. 2).

4 Result Analysis

The results of this study include all pre-processing steps for the vague data set. Analysis of the choice of classifiers with respect to their performances. The simulator used for the study includes Matlab-19 for the performance graph (Bar graph). The accuracy was given in the following table, along with its comparison measures.

A confusion matrix for logistic regression is given

$$\begin{Bmatrix} 33 & 2 \\ 9 & 17 \end{Bmatrix}$$

The accuracy of logistic regression is approximately 82% (Table 4).

The accuracy of Naïve Bayes classifiers is approximately 79%.

The confusion matrix for Naïve Bayes classifiers is given as (Table 5).

Table 4 Analysis using logistic regression

Sl	Precision	Recall	F1-score	Support
0	0.79	0.94	0.86	35
1	0.89	0.65	0.76	26
Accuracy			0.82	61
Macro avg	0.84	0.80	0.81	61
Weighted avg	0.83	0.82	0.81	61

Table 5 Analysis using Naïve Bayes classifiers

Sl	Precision	Recall	F1-score	Support
0	0.79	0.86	0.82	35
1	0.78	0.69	0.73	26
Accuracy			0.79	61
Macro avg	0.79	0.77	0.78	61
Weighted avg	0.79	0.79	0.78	61

$$\begin{Bmatrix} 30 & 5 \\ 8 & 18 \end{Bmatrix}$$

The confusion matrix for Random Forest classifiers is given as

$$\begin{Bmatrix} 31 & 4 \\ 9 & 17 \end{Bmatrix}$$

Accuracy levels of the Random forest model given as 79% (Table 6).
Confusion Matrix for Extreme Gradient Boosting

$$\begin{Bmatrix} 33 & 2 \\ 9 & 17 \end{Bmatrix}$$

Accuracy for Extreme Gradient Boosting is approximately 82% (Table 7).

Table 6 Analysis using random forest model

Sl	Precision	Recall	F1-score	Support
0	0.78	0.86	0.83	35
1	0.81	0.65	0.72	26
Accuracy			0.79	61
Macro avg	0.79	0.77	0.78	61
Weighted avg	0.79	0.79	0.78	61

Table 7 Analysis using extreme gradient boosting

Sl	Precision	Recall	F1-score	Support
0	0.79	0.94	0.86	35
1	0.89	0.65	0.76	26
Accuracy			0.82	61
Macro avg	0.81	0.80	0.81	61
Weighted avg	0.83	0.82	0.81	61

Confusion matrix for K-Nearest Neighbor (K-NN) (Table 8).

$$\begin{Bmatrix} 26 & 6 \\ 10 & 16 \end{Bmatrix}$$

Confusion matrix for Decision Tree (K-NN)

$$\begin{Bmatrix} 27 & 8 \\ 12 & 14 \end{Bmatrix}$$

Accuracy for Decision Tree is around 69% (Table 9).

Table 8 Analysis using K-NN

Sl	Precision	Recall	F1-score	Support
0	0.83	0.83	0.78	35
1	0.73	0.62	0.67	26
Accuracy			0.74	61
Macro avg	0.74	0.72	0.73	61
Weighted avg	0.74	0.74	0.73	61

Table 9 Accuracy comparison of various classifiers

Sl	Model	Accuracy
0	LR	82
1	NB	79
2	RF	77
3	XGB	82
4	K-NN	74
5	DT	68
6	SVM	78

5 Conclusion and Future Work

This study includes RST combined with several machine learning algorithms to find significant cardiac-related diseases and performance measures of cardiac-related diseases, especially cardiac arrest. Initially, this research found cardiac-related diseases by analyzing vague cardiac-related diseases using RST. This research found that cardiac arrest was the most significant disease, and then this research used a universal data set for performance measures. This analysis can be extended into the entertainment, sports, and business sectors.

References

1. Berdowski, J., Berg, R.A., Tijssen, J.G.P., Koster, R.W.: Global incidences of out of hospital cardiac arrest and survival rates: systematic review of 67 prospective studies. Resuscitation **81**(11), 1479–1487 (2010). https://doi.org/10.1016/j.resuscitation.2010.08.006
2. Chan, P.S., Krein, S.L., Tang, F., Iwashyna, T.J., Harrod, M., Kennedy, M., Lehrich, J., Kronick, S., Nallamothu, BK.: American heart association's get with the guidelines–resuscitation investigators. Resuscitation practices associated with survival after in-hospital cardiac arrest: a nationwide survey. JAMA Cardiol. **1**(2), 189–97 (2016). https://doi.org/10.1001/jamacardio.2016.0073. Erratum in: JAMA Cardiol. **3**(9), 898 (2018). PMID: 27437890; PMCID: PMC5745254
3. Rong, G., Mendez, A., Assi, E.B., Zhao, B.: Artificial Intelligence in Healthcare. Elsevier, Engineering (2020)
4. Lopez-Jimenez, F., Attia, Z., Arruda-Olson, M., Carter, R., Chareonthaitawee, P., Jouni, H., Kapa, S., Lerman, A., Luong, C., Medina-Inojosa, J.R., Noseworthy, P.A., Pellikka, P.A., Redfield, M.M., Roger, V.L., Sandhu, G.S., Seneca, C., Friedman, P.A.: Artificial Intelligence in Cardiology: Present and Future, https://doi.org/10.1016/j.mayocp.2020.01.038
5. Lopez-Jimenez, F., Attia, Z., Arruda, A.M., Carter, R., Chareonthaitawee, P., Jouni, H., Kapa, S., Lerman, A., Luong, C., Medina-Inojosa, J.R., Noseworthy, A., Pellikka, P.A., Redfield, M.M., Roger, V.L., Sandhu, G.S., Conor Senecal
6. Johnson, K.W., Torres Soto, J., Glicksberg, B.S., Shameer, K., Miotto, R., Ali, M., Ashley, E., Dudley, J.T.: Artificial intelligence in cardiology. J. Am. Coll. Cardiol. **71**(23), 2668–2679 (2018). https://doi.org/10.1016/j.jacc.2018.03.521. PMID: 29880128
7. Sarajlic, P., Plunde, O., Franco-Cereceda, A., Bäck, M.: Artificial intelligence models reveal sex-specific gene expression in aortic valve calcification. JACC Basic toTransl. Sci. **6**(5), 403 (2021). https://doi.org/10.1016/J.JACBTS.2021.02.005
8. M. Jaya Sree and L. Koteswara Rao, "Survey on-identification of coronary artery disease using deep learning," Mater. Today Proc. (2020). https://doi.org/10.1016/j.matpr.2020.09.526
9. Usha Kumari, C., Sampath Dakshina Murthy, A., Lakshmi Prasanna, B., Pala Prasad Reddy, M., Kumar Panigrahy, A.: An automated detection of heart arrhythmias using machine learning technique: SVM. Mater. Today Proc. (2020). https://doi.org/10.1016/j.matpr.2020.07.088
10. Didon, J.P., Ménétré, S., Jekova, I., Stoyanov, T., Krasteva, V.: Analyze whilst compressing algorithm for detection of ventricular fibrillation during CPR: a comparative performance evaluation for automated external defibrillators. Resuscitation. **160**, 94–102 (2021). https://doi.org/10.1016/j.resuscitation.2021.01.018
11. Cronberg, T., Greer, D.M., Lilja, G., Moulaert, V., Swindell, P., Rossetti, A.O.: Brain injury after cardiac arrest: from prognostication of comatose patients to rehabilitation. Lancet Neurol. **19**(7):611–622 (2020). Lancet Publishing Group. https://doi.org/10.1016/S1474-4422(20)30117-4.
12. Pawlak, Z.: Rough Set Theoretical Aspect. Kluwer Academic Publisher (1991)

Design of Cell-Free Massive MIMO for Beyond 5G Systems with MMSE and RZF Processing

Sunita Dhalbisoi, Arunanshu Mohapatra, and Ashima Rout

Abstract A massive MIMO with Cell Free (CF-mMIMO) is an efficient technology for growing users' requirement and better data rates beyond 5G systems. CF-mMIMO contains distributed Access Points (APs) in large number which serves few numbers of users at same time or else frequency. APs connect with users in 5G cell-free network by using joint initial access and pilot assignment. The main goal is to analyze technology to cooperate between APs. Mainly, spectral efficiency is analyzed with four different levels of cell-free implementation. It compares CF-massive MIMO with small cell and cellular massive MIMO network. In this paper, Regularized Zero Forcing (RZF), Maximum ratio (MR), and Minimum Mean Square Error (MMSE) are used to optimize spectral efficiency. By using this precoding, it reduces fronthaul signaling in compare to cellular network. It is likely better performance than cellular and small cell massive MIMO network. In our paper, we discuss to centralize implementation by using MMSE and MR. It maximizes spectral efficiency with less fronthaul signaling. We also discuss the variation of small-cell signal and cell-free signal with a fixed number of user equipment.

Keywords Massive MIMO · MMSE · RZF · MR · Spectral efficiency · 5G

S. Dhalbisoi · A. Mohapatra
Department of ETC Engineering, VSSUT, Burla, Odisha, India
e-mail: sunita.muskan1@gmail.com

A. Mohapatra
e-mail: arun227@gmail.com

A. Rout (✉)
Department of ETC Engineering, IGIT Sarang, Sarang, Odisha, India
e-mail: ashimarout@igitsarang.ac.in

S. Kumar Udgata et al. (eds.), *Intelligent Systems*, Lecture Notes in Networks and Systems 1149, https://doi.org/10.1007/978-981-97-8160-7_21

1 Introduction

CF-mMIMO is one of the important concepts used in 5G wireless networking environment. Base station with arrays of antennas is used to improve spectral efficiency more than ten times than conventional network and each BS serves thousands of user equipment. It can improve SE by reducing size of cell. In CF massive MIMO, huge amount of access points is distributed over a geographical area [1]. All APs are connected to CPU through fronthaul line. Central processor unit acts as a cloud edge processor.

Different types of signal processing methods are used. Mean square error (MSE), Regularized Zero Forcing (RZF), Normalized Mean Square Error (NMSE), Minimum Mean Square Error (MMSE), Local-MMSE, and Maximum Ratio (MR) are different signal processing methods used to avoid interference at base station or access point.

All APs conjoins with phase in backhaul path at same time or frequency. APs connects in Time-Division Duplex (TDD) concept with no cell boundaries [2]. APs are used in particular region and provides network to specific user equipment's. Small cells are also types of cell-free mMIMO with less performance [3].

In CF-mMIMO, it has central processing unit, but then the data exchange between the CPU and APs is restricted to this payload data, and coefficient of power control changes gradually. Channel State Information (CSI) sharing takes place between the APs or CPU. Altogether channels are formed through uplink pilots and channel estimation takes place at each AP. During downlink, the channel estimation precodes transmit data and also performs detection of data during uplink. All over we emphasize each user's throughput instead of sum of throughput. To this terminal, we employ max–min optimization for power control.

In belief, CF-mMIMO is also an embodiment of general concepts identified as "virtual network MIMO," "system network MIMO," "distributed MIMO," Distributed Systems of Antenna (DAS), and "Cooperative multipoint joint of processing" (CoMP). The main objective behind this is to provide backhaul connection in advanced to accomplish coherent signal processing [4]. Distributed APs antennas

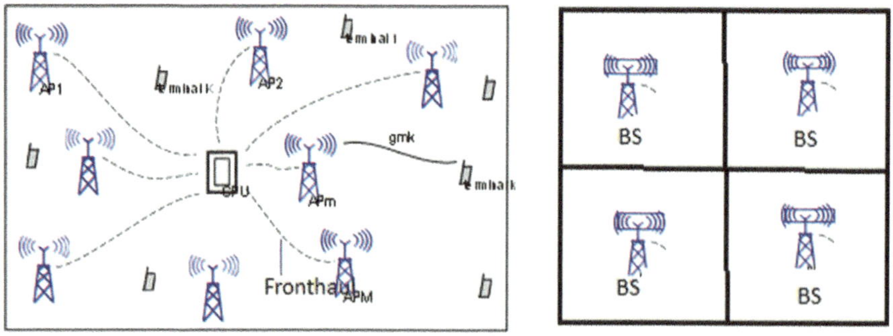

Fig. 1 Difference between cell-free network and cellular network

across geographically area provides consistently good facility for each and every user in wireless network. The unresolved aspect of CF-mMIMO is their operating regime: huge number of single-antenna APs instantaneously serve few numbers of users, using computationally simple signal processing commonly called conjugate beamforming. These are enabling the manipulation of the features like favorable signal propagation and hardening perception of channel are used. This technique can also enhance the Spectral Efficiency (SE) in terms of hundred times as compare to cellular structure networks [5]. It can also enhance the speed by upgrading the hardware components base station as a replacement for installing one base station position. The enhanced SE gain derives from that BSs which are compacted arrays of antennas place with hundred antennas or even more. These are used in different techniques like digital technique for beamforming and also spatially multiplexing. Here, several UEs are used at same frequency–time resource [5]. If we compare the characteristics of conventional MIMO with mMIMO, then every AP or BS have a greater number of antennas than our UE in that cell. Signal processing approaches like MMSE precoding combining in uplink, can be required at each BS for overpowering the problem of interference in both the same and different cells [1, 6] instead of using any of BS support. The CF-mMIMO concept not only helps in placements but also spatially spread-arrays in every cell-area [7]. The setup of DAS and CoMP are same with standing effect and separate cooperation clusters. They are all different in configuration such as cellular networks. Another network configuration is measured over the term of CF-mMIMO. Main work is to install a very big number of antennae distributed over APs and all APs connected to the CPU. This concept is C-RAN, commonly known as a cloud radio access wireless network, as well as edge-cloud processor of data center. CPU connects the system with network mMIMO with no specific cell like other network. They conjointly assist the UEs with coherently combined transmission as well as reception. Unlike old-style network mMIMO, unsettled feature of CF-mMIMO is the operation management with the help of APs than UEs. In systematic viewpoint, a significant originality is that imperfect-CSI and performance analysis was considered. It supported the use of MR processing locally generated at each AP and for achieving higher spectral efficiency, partially or fully centralized the processing at the CPU.

2 Motivation

The main aim of these papers [3–5] was only in comparing CF-mMIMO with small-cell network. Specifically, this AP is arranged in geographical area and dedicated to its own limited set of user equipment. Small cells provide less performance in compared to CF-mMIMO. The motivation behind research in this field is scarcity of bandwidth and other parameters. Research demonstrated a large improvement in intermediate value and probably 95% of SE. Partially, it says the detail that an inefficient service in small-cell network was measured. In detail, it demonstrates a more accurate processing method than MR. It is required in CF-mMIMO to constantly beat small cells. Our paper aims to compare a CF-mMIMO with a predictable cellular-mMIMO and its main goal is to treasure the utmost inexpensive CF execution.

The large alterations make evaluation of none trivial and deliver motivating efforts into analysis and 5G design process and beyond-5G networks. Traditional cellular-mMIMO advantages over channel hardening concept and spatial interference concept of suppression. But cell-boundary area UEs can have poor channel conditions. Conversely, CF-mMIMO advantages from durable macro-diversity and its interference destruction ability extremely depends on how it is operated. Previously, it said that channel hardening may also appears in CF-mMIMO, then it was later revealed that capacity limits that imagines hardening can greatly underestimate practical SE [8]. To accomplish a practically fair comparison, our paper emphasizes on uplink and assumes that transmission of data is preceded by pilot signal-based channel-estimation phase. All UEs are transmitted data with equal amount of powers for different levels and different network topologies.

3 Background

The authors [1] discuss CF-mMIMO in a brief way. They are trying to explain cell-free technique. CF-mMIMO is measured as a suitable technology for sustaining the growing user number, in addition to high data rate requirements in 5G and also beyond-5G wireless networks. The main purpose is to place distributed APs in network which communicate among themselves. This paper [1] provides an all-inclusive study of the technology with different steps of support among all the APs. It supports the network with signal processing in joint coherent process.

The implementations of spectral efficiencies in uplink processing are analyzed in cell-free network. It uses linear processing arbitrarily and fades spatially. It tries to outperform present orthodox cellular-mMIMO networks and small-cell a widespread margin. It basically uses global with local combining method known as MMSE. This paper is contradicted with present literature, which also supports for combing like maximum ratio (MR). It demonstrates that implementation with centralize and optimum MMSE method maximizes spectral efficiency, and also basically decreases fronthaul signaling capacity related to the average distributed approaches. This helps to prefer right way to operate CF-mMIMO. Non-linear method of interpreting is considered, therefore made known to take insignificant developments.

Wireless communication by using joint signal processing technique is a probable network organization from large distributed APs, known as CF-mMIMO. The goal of the paper is to provide all-inclusive comparison with cellular mMIMO. The spectral efficiencies in uplink uses four different cell-free implementations. A centralized processing implementation maximizes performance. And also reduces the requirement of fronthaul signaling.

In paper [2], authors are trying to give the basic idea behind the requirement of cell-free network. A network in CF-mMIMO system incorporates an identically large quantity of distributed APs which instantaneously serve a much lesser number of user equipment at same time or frequency resources based on directly measured channel characteristics. The APs and users have only one antenna each. The APs attain CSI over Time-Division Duplex (TDD) operation and the response of uplink pilot signals at receiver transmitted by the users. The APs perform multiplexing and demultiplexing through conjugate beamforming technique on the downlink and matched filtering method on the uplink. Similar expressions for every individual user throughput in uplink and downlink lead to optimize the algorithm of max–min power control. This algorithm of max–min power control guarantees uniformly good service throughout the area of coverage. A pilot assignment algorithm helps to moderate the things of pilot contamination, but also power control is a very important parameter in that era.

In paper [9], authors are explaining different configurations used in 5G wireless network. Cell-Free mMIMO has a significantly better quality of performance with respect to our present conventional small-cell network scheme. Here, each user equipment is attended by a focused APs. Every user throughput is more likely 95% and helps to provide better immunity to shadow fading for spatial correlation. During uncorrelated shadow fading situations, the CF system provides almost 5 times improvement in throughput in compare to the small-cell method and 10 times improvement in correlated shadow fading.

These are the papers which motivate to work in this field. I have followed the above paper and tried to find problem in this research area. After our research, we found some problems. Our contribution towards this field is finding better spectral efficiency and trying to find better solution by using different precoding methods. It has been tried to solve those issues using this approach.

Previously, cell free was compared with small cell. APs are placed on a particular area. In this paper cell free structure can be implemented in two different steps [2]. The main aim is to design the modest cell-free network in Fig. 1. It compares a cellular network with cell-free configuration. Cellular network has the ability to suppress interference and better channel hardening. But UEs at cell-edge have some problems. CF has good quality of channel handling. But sometimes it is difficult to handle interference. So, our work is to obtain better spectral efficiency with probability distribution function [3].

In the first step, four different types of levels are designed. In the second step, spectral efficiency is enhanced [9]. To obtain optimal solution, spatial multiplexing is used for APs with N number of antennas each. So, we compare cell free with small cells and cellular network. The four different levels are Level 1, Level 2, Level 3, and Level 4. Level 1 is preferred for distributed network. In Level 1, one AP is for one UE and detection is achieved at each AP. It is a type of small cell where it does not exchange with CPU.

Level 2 and Level 3 are similar. Detection is performed by local estimating at CPU. It makes CPU denser which may reduce the amount of information. In this level, AP estimates a channel for UE and combines at CPU [5]. They detect the linearly processed data. Level 4 is suitable for network MIMO. It is preferable for a fully centralized structure where APs receive the pilot signals and data. Pilot signals and data are received at APs. It is collected at CPU through fronthaul link [6].

4 Methodology

In this paper, we consider a cell-free network with N user equipment (UE), K number of access points (APs) each with L antennas, and each UE with one antenna. APs are connected to CPU through fronthaul. Cell-free network makes a channel between UE and AP known as H_{nk} which is constant in coherence block τ_c. It uses the block fading model.

Again, in a block, Rayleigh fading distribution is obtained

$$H_{nk} \to \mathrm{N}_c(0, R_{nk})$$

where R_{nk} indicates the spatial correlation matrix.

τ_c consists of τ_p for pilots and $\tau_c - \tau_p$ for data and based on TDD. H_{nk} performs two different operations, that is, pilot transmission by estimating channel and transmission of uplink data. Pilot channels τ_p are required for estimation of channel and pilot signals are randomly assigned to UEs that is more than one.

When pilots transmitted by UEs, signal received at AP is

$$Z_K = \sum_{i=1}^{N} \sqrt{p_i} H_{iK}\, \Phi_{t_i}^{T} + \mathfrak{N}_k \tag{1}$$

where p_i is the UEs' transmitting power, and

$\mathfrak{N}_k$is the noise at receiver.

Due to sharing of pilot, interference occurs which is known as pilot contamination and causes degradation in system performance.

In the second phase, the received signal Y_K at AP K during uplink transmission of data is given by

$$Y_K = \sum_{i=1}^{N} H_{iK} S_i + n_k \tag{2}$$

where S_i is the UEs' transmitted signal, and

n_k is the independent noise at receiver.

We implemented three precoding methods for CF, small cell, and cellular configuration. These are RZF, MMSE, and MR. Regularized zero forcing commonly called RZF is a method of precoding used in communication systems. It is considered to moderate the effects of interference and improve performance. It is improved method of ZF which eliminates interference for better signal broadcast. Minimum mean square error called MMSE is a signal processing technique to minimize mean square error.

APs connect to CPU through fronthaul connections and use more resources. Fronthaul signaling connected in topology to observe data signal which may connect locally, partially or fully. CF implement in Level 4 is fully centralized. APs forward all types of signal to central CPU. CPU estimates channel for MMSE. Here APs K send to CPU the received pilot signal

$$\left\{Z_{tk} : t = 1.......\tau_p, k = 1......K\right\}$$

The received signal

$$\{y_k = k = 1......, K\}$$

In this level, the required spectral efficiency of equipment l [1] is

$$S_{kn} \triangleq V_{nk}^{H} y_l = V_{nk}^{H} h_{nk} s_l + \sum_{i=1}^{N} V_{nk}^{H} h_{iK} \, S_i + V_{nk}^{H} + s_k \tag{3}$$

Level 3 estimates channels for data instead of sending to CPU. It computes local estimates and given to CPU. The local estimates of S_{kn} is

$$S_{kn} \triangleq V_{nk}^{H} y_l = V_{nk}^{H} h_{nk} s_l + \sum\nolimits_{i=1}^{N} V_{nk}^{H} h_{iK} S_i + V_{nk}^{H} + s_k \tag{4}$$

Level 3 is used for local MMSE. It gives better spectral efficiency. But it locally combines at every AP. So, with increase in AP and UE, complexity increases exponentially. It is also designed with RZF precoding. Level 2 precoding method is used to overcome drawback of Level 3.CPU creates average of estimated signals. Level 1 is used for small cells. In this method, signals from UE to AP is decoded at AP locally and estimate channels also locally. CF is compared with SC depending on large-scale-fading ß.

5 Result Analysis

We assume a cellular network with four square cells (1 × 1 km^2 area) and each BS equip with 100 antennas. A CF network is assumed in same 1 × 1 km^2 area. APs are placed 400 number with antenna each or 100 number with 4 antennas each. Product of APs and antennas are same in both cases. APs are placed at 10 m height and 50 m distance from each other. Forty UEs are connected to APs, and assigned the pilot signal to more UEs. Similarly, UE connects to BS with the assigned pilot signal in cellular configuration. All UEs uses 100 mW transmit power with 20 MHZ bandwidth. Configuration of CF is based on MICRO cell model in this environment [1].

$$Large - scale - fading\ \beta\ in\ dB = -30.5 - 36.7 \log_{10}\left(\frac{d_{kn}}{1m}\right) + F_{kn} \quad (5)$$

where d_{kn} *indicates* the distance between AP and UE, and

F_{kn} indicates the shadow fading.

Figure 2 shows variation of small cell signals in dB with respect to 40 UEs. It varies in a smaller range. It describes the change in signal level when UE change. It has less effect on signal.

Figure 3 shows variation of cell-free signals in dB with respect to 40 UEs. It varies more. It has more effect on signal in compare to small cells.

Figure 4 describes spectral efficiency in bits/s/Hz with variation of probability density function (PDF). We consider K = 400 APs with L = 1 antenna each and 40 UEs are randomly located. It compares CF with cellular by using MMSE processing in Level 4, L-MMSE (local-MMSE) processing in Level 3, L-MR (local-MR) processing in Level 2, and small cells in Level 1. Spectral efficiency is the variation data with change in bandwidth. By comparing all four levels with cellular, we observe that cell free and small cell are giving better spectral efficiency in compare to cellular network.

Figure 5 shows variation of spectral efficiency in bits/s/Hz for CF as well as cellular network with K = 400 APs, L = 1 antenna each and 40 UEs. It compares CF with cellular by using MMSE precoding in Level 4, RZF processing in Level 3,

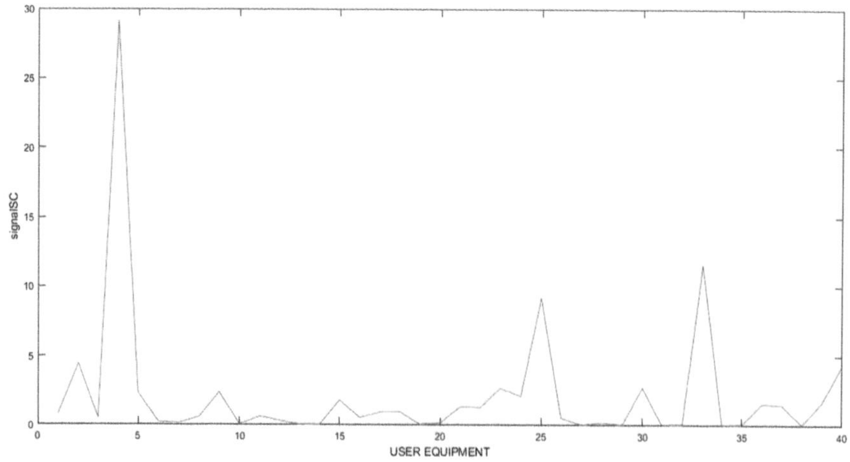

Fig. 2 Shows variation of small cell signals in dB with respect to 40 UEs

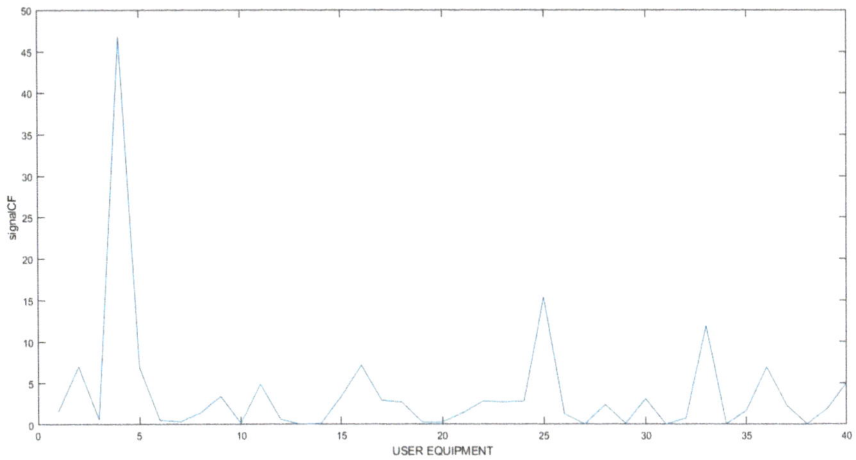

Fig. 3 Shows variation of cell-free signals in dB with respect to 40 UEs

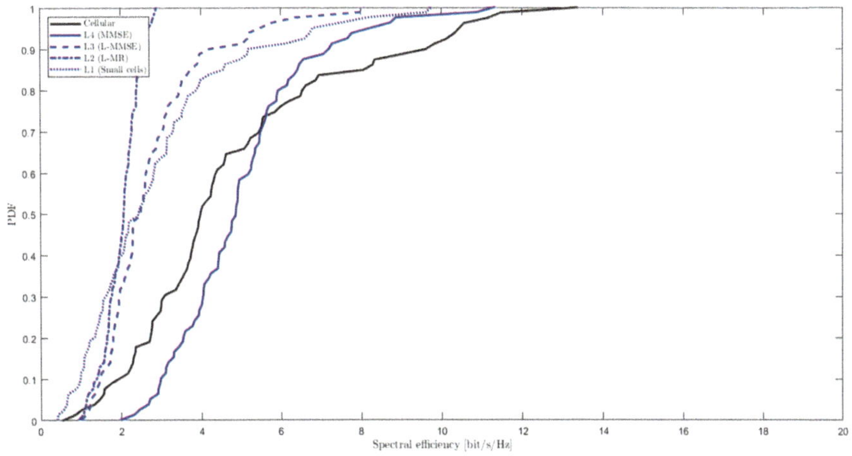

Fig. 4 Shows PDF versus spectral efficiency with K = 400, L = 1

L-MR (local-MR) processing in Level 2, and small cells in Level 1. We observe that Level 3 using RZF and Level 2 using L-MR are giving better spectral efficiency as compare to cellular networking environment.

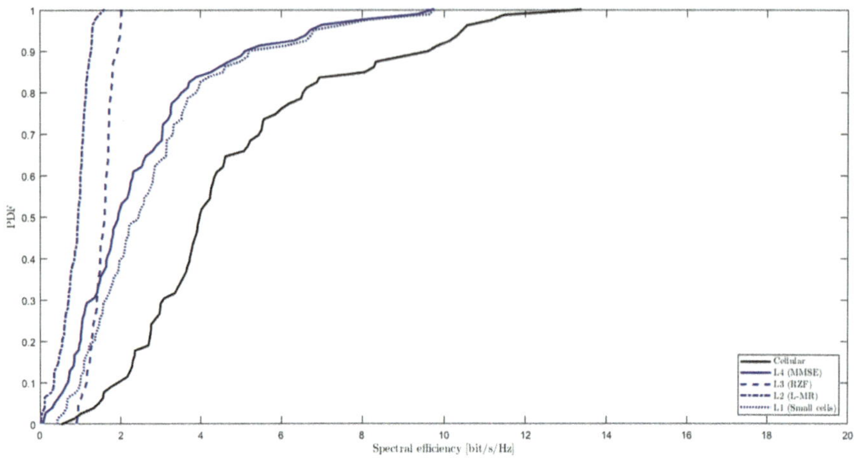

Fig. 5 Compare CF and cellular network spectral efficiency with K = 400, L = 1

6 Conclusion

We studied the level of performance in CF mMIMO by considering channel estimation and pilot assignment. We performed a comparison between CF-mMIMO and small-cell networks. This network can be implemented beyond-5G network system. In future, these levels of implementation can be applied to both centralized and distributed networks. Future network 6G can be designed by using this model. We discussed the variation of CF signal and variation of small-cell signal in dB with respect to UEs. It helps to enhance spectral efficiency by using different precoding method. Out of different precoding methods, CF-mMIMO is mostly preferable for RZF and MMSE. We studied the maximal variation of SE with PDF. We also observed that complexity of fronthaul capacity can be reduced.

References

1. Emil Björnson, F., Luca Sanguinetti, S.: Making cell-free massive MIMO competitive with MMSE processing and centralized implementation. IEEE. J. Wirel. Commun. **19**(1), 77–90 (2020)
2. Björnson, E., F. L., Sanguinetti, S.: Cell-free versus cellular massive MIMO: what processing is needed for cell-free to win?. In: Proceedings IEEE 20th International Workshop Signal Process Advances in Wireless Communication (SPAWC), pp. 1–5 (2019)
3. Dhalbisoi, S. F., Rout, A. S., Sahoo, R. K., T., Sethi, S., F.: A comparative analysis on 5G cell free massive mimo in next generation networking environment. In: ICICCSP(IEEE) Conference (2022)
4. Ahmadi, H., Farhang, A., Marchetti, S.N., Mackenzie, A.: A game theoretical approach for pilot contamination avoidance in massive MIMO. IEEE. Wireless. Commun. Lett. (2015)

5. Saleeb, B., F., Shehata, M., S., Mostafa, T., Fahmy, Y., F.: Performance evaluation of RZF precoding in multi-user MIMO systems. In: IEEE Conference (2019)
6. Ngo, H.Q., F., Tran, L.N., S., Duong, T.Q., T., Matthaiou, M., F., Larsson, E.G., F.: On the total energy efficiency of cell-free massive MIMO. IEEE. Trans. Green. Communication. Network. **2**(1):25–39 (2018)
7. Hong Yang, F., Marzetta, T.L.S.: Energy efficiency of massive MIMO: cell-free versus cellular. In: IEEE Conference on Wireless Communication (2018)
8. Riera-Palou, F., F., Femenias, G., S., Armada, A.G., T., Pérez-Neira, A., F.: Clustered cell-free massive MIMO. In: Proceedings IEEE Globecom Workshops (GC Wkshps), pp. 16 (2018)
9. Hien Quoc Ngo, F., Alexei Ashikhmin, S., Hong Yang, T., Erik, G., Larsson, F., Thomas, L., Marzetta, F.: Cell-free massive MIMO versus small cells. IEEE. J. Wirel. Commun. 1–17 (2017)

Machine Learning-Based Mental Stress Detection Through Cost-Benefit Analysis

Padmini Sethi, Ramesh K. Sahoo, Ashima Rout, M. Mufti, and Bichitra Mandal

Abstract Mental Stress has evolved as a major health issue that has negative impacts on humans mentally and physically. It causes significant physiological and psychological changes in the human body. Therefore, continuous observation of mental stress is highly essential for a human which increases the productivity of humans with sound health. A machine learning-based predictive model using Decision Tree (J48) algorithm in cross-validation mode has been proposed to detect the mental stress of any human with respect to different positions of the body and the behaviour of the person. Cost-benefit analysis approach has been proposed and implemented with the machine learning-based predictive model to enhance the accuracy of the model. Due to this, the accuracy of the model has been increased from 80 to 99.67%. Galvanic Skin Response (GSR), temperature, and airflow sensors have been used for the collection of health parameters from humans to have a vast dataset for analysis using machine learning-based predictive model with cost-benefit analysis.

Keywords Mental stress · Machine learning · Cost-benefit · Galvanic skin response · Physiological and psychological data

P. Sethi
RDW University, Bhubaneswar, India

R. K. Sahoo (✉) · A. Rout
IGIT Sarang, Dhenkanal, Odisha, India
e-mail: ramesh0986@gmail.com

M. Mufti
Nottingham University, Nottingham, UK
e-mail: mufti.mahmud@ntu.ac.uk

B. Mandal
OUTR, Bhubaneswar, India

S. Kumar Udgata et al. (eds.), *Intelligent Systems*, Lecture Notes in Networks and Systems 1149, https://doi.org/10.1007/978-981-97-8160-7_22

1 Introduction

In this modern era, human beings are making great achievements and developing in various fields for the betterment of society but all go through mental stress at some point in life which they reflect through many emotions like being fearful, anxious, angry, depressed, etc. Dealing with day-to-day life problems or any kind of situation human body generates stress. It results in releasing sweat in some specific body parts like palm, face, and sole [1]. Eccrine gland is one of the three types of sweat glands in the human body which can be found in the palm and it is also connected to anxiety and stress [2]. Sometimes stress helps for better performance but again sometimes stress due to certain physiological and psychological happenings in humans can have harmful or even an deadly effect on the human body physically and mentally. Thus, consistently observing stress level of people and analysing them is very much essential.

Physiological changes occurring in an individual's body have a huge impact on his/her performance, decision-making, thinking ability, behaviour, etc., which again has a strong impact on his/her mental health [3]. Sweating or increase in sweating is seen due to the above-mentioned processes [4]. Thus, Physiological data are important to consider for the analysis of human's stress level and his/her range properly. Stress level of the human body can be found by the help of skin conductance which authors have discussed experimentally in [5]. GSR plays an important role to monitor an individual's stress by using its two electrodes on the middle and index fingers. İt also helps in measuring the skin conductance of the body through sweat which reflects the electrical skin response because of psychological/physiological stimulation.

Using skin conductance GSR sensor-based technology is measured [6, 7], which is modeled as a wearable device for detecting stress. By using it, an individual's mental illness can also be detected in an early stage by measuring and analysing his/her stress levels, which will help in improving his/her quality of life. A device which is wearable has been developed [8] using e-Health sensor shield, Arduino Uno microcontroller, and GSR Sensor for the purpose of detecting the mental stress of an individual and is extended [9] by connecting remote data collection like GSM and Wi-Fi shields which is by Arduino-based devices. A discussion of the Data collection framework is done in [10]. Removal of all the added noise because of external factors can also be done for the enhancement of the model's accuracy [11]. Cost-benefit analysis is an approach that has been used to enhance the accuracy of the machine learning-based predictive model for ensuring data integrity in mobile crowdsensing environment [12, 13]. Analysis of an individual's mental stress is done through various machine learning algorithms [14] with 80% accuracy in Decision Tree (J48) algorithm. In the proposed work, the mental stress has been analysed in cross-validation mode using Decision Tree (J48) algorithm that is more robust from a quality point of view, and also to boost the accuracy level of the model, cost-benefit analysis approach has been implemented.

2 Methodology

2.1 Data Collection Framework

In the proposed work, body temperature is detected in degree centigrade using temperature sensor, and the breathing rate is detected by using airflow sensor to observe the low or high breathing rates which are abnormal conditions. GSR sensor is connected to the index and middle finger of the left hand, which will detect skin conductance of an individual which will reflect his/her stress level. Skin conductance is computed with the constant supply of 5-V by GSR sensor and the human nervous system always regulates a small amount of current flow throughout the body. When the stress level of an individual increases, his/her sweat gland activity also increases, which leads to an increase in the amount of flow of current throughout the body. Because the GSR sensor provides a constant voltage of 5-V voltage to an individual's body, the skin conductance of that individual will depend on the flow of current through the body. Thus, the Stress level of an individual is directly proportional to his/her skin conductance.

GSR, airflow, and temperature sensors are connected to the e-health sensor platform; it is a motherboard which is used for collecting vital health data from an individual's body and those data will be further analysed for the detection of actual health conditions. Next, it is connected with an Arduino Uno microcontroller which can be powered up and also programmed through any monitoring device like laptop/desktop to control various connected sensors and devices; also for analysis purpose, the data can be stored. Further, connecting hybrid network platforms like GSM Shield and Wi-Fi with Arduino Uno can be done for communicating data from Arduino to the cloud platform that is for storage and analysis. Wi-Fi shield will initially look for any Wi-Fi network that is available nearby, it will be automatically connected if available and the transmission of the data to the cloud platform will take place otherwise, the GSM shield will be used for the transmission of data to the cloud platform by using cellular network.

2.2 Working Principle

e-Health sensor platform connected with GSR, airflow, and Temperature sensors is considered for collecting vital health data from the human body, which is further connected to Arduino Uno microcontroller and also GSM shield and Wi-Fi for communicating data to cloud platform. Laptop/desktop is connected with Arduino Uno microcontroller for powering up the device and also uploading Arduino sketch. As per the program setup, all the sensors used will sense the individual's body to get all the vital health data which will be stored in the comma separated value (CSV) format locally in a micro SD card connected to Wi-Fi shield as shown in Fig. 1a and ultimately everything will get uploaded in the cloud platform for analysing and

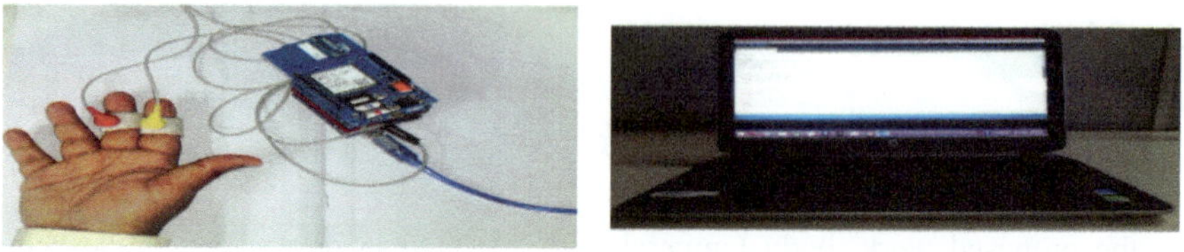

(a): Data acquisition Node (b): Observing Node

Fig. 1 Data acquisition and observing Node setup [9]

storing. This can be done either through the cellular network by using GSM shield or through the Wi-Fi by using Wi-Fi shield as per the network available as shown in Fig. 1b.

All the vital health data are collected from various individuals staying in different positions such as laying, sitting, and standing while being in different types of modes like exercising, being normal, and tensed. All these are considered to estimate mental stress of an individual with respect to different types of positions and modes. Ultimately analysis is done by using many Machine Learning (ML) algorithms which will eventually help in estimating the most accurate ML algorithm that can help in predicting mental stress.

2.3 *Analysis Through Machine Learning Algorithms*

In WEKA environment some well-known supervised and standard ML algorithms such as Decision Tree (J48) [15], Naive Bayes [15], Support Vector Machine (SVM) [15], and Random Forest [15] have been used for the analysis of the obtained dataset in Full training set mode [14] and 88% of accuracy in Decision Tree (J48) Algorithm. Also, the WEKA environment provides a space where ML algorithms are used for classification purpose in cross-validation, full Training set, and percentage split modes. In percentage split mode and Full training set, only one model is created by dividing the obtained dataset into testing and training set only once but in the case of Cross-validation mode, division of obtained dataset is done into testing and training set, and one multiple model is created for multiple times in as many as possible ways. Thus, the better and robust mode than the other specified modes is cross-validation mode but it is less accurate. Therefore, Cross-validation mode is used to analyse through classification in the proposed work instead of full training set mode used in [14]. Further, cost-benefit analysis approach has been used to enhance the accuracy level of the machine learning-based predictive model. Its efficiency has been judged through the parameters such as Cost, benefit, and Accuracy before and after virtual screening. Table 1 reflects the various parameters considered for the evaluation of ML algorithms and also for the cost-benefit analysis.

Table 1 Evaluation parameters and the symbols

Parameters	Symbols
Relative absolute error	RAE
Root relative squared error	RRSE
Mean absolute error	MAE
Root mean squared error	RMSE
True positive	TP
False positive	FP
True negative	TN
False negative	FN

2.4 Cost and Benefit Analysis

It is a cost sensitive approach in machine learning to enhance the accuracy level of the predictive model through proper minimization of cost and maximization of benefit. In machine learning approach, a data\record can be classified as True Positive (TP), True Negative (TN), False Positive (FP), and False Negative (FN). Total number of records classified as TPs and TNs represents the correctly classified data and total number of records classified as FPs and FNs represents the incorrectly classified data. In cost-benefit analysis, total number of incorrectly classified data reflects the cost and benefit is presented in terms of total number of correctly classified data. It can be represented mathematically through following equations.

$$C_{pM} = T_{fp} + T_{fn} \tag{1}$$

$$B_{pM} = T_{tp} + T_{tn} \tag{2}$$

In Eq. 1, Cost is reflected through C_{pM} computed using total numbers of FPs (T_{fp}) and total number of FNs (T_{fn}). Similarly in Eq. 2, B_{pM} reflects the benefits computed using total numbers of TPs (T_{tp}) and total number of TNs (T_{tn}).

The cost-benefit analysis is based on the conversion of FPs to TNs and FNs to TPs. Therefore, the cost will be minimized and the benefit will be maximized. This will enhance the accuracy of the predictive model through the minimization of classification error.

2.5 Virtual Screening

It is a set of computational approaches used for analysis and evaluation of huge datasets to have the required dataset for classification. It helps to choose the size and portion of the dataset that can be used as a training set to have the best classification

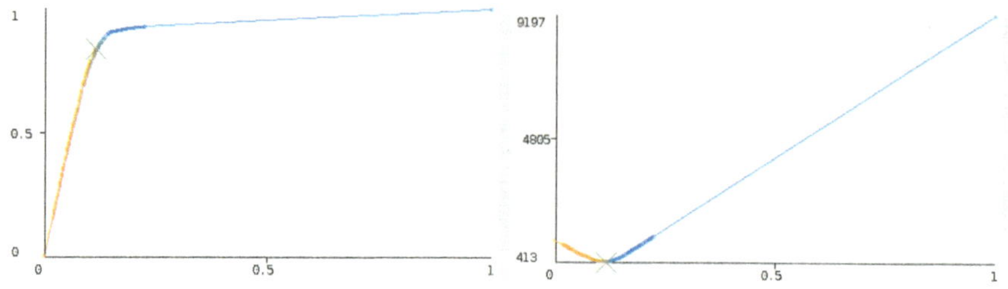

(a) Threshold Curve (b) Cost Benefit Curve

Fig. 2 Threshold and cost-benefit curve of Sit_tension position in Decision Tree (J48) model

pattern and it will be applied to the entire population for correct classification with minimal error. In the classification approach, a record can classified as TP, TN, FP, and FN.

The threshold curve is reflected in Fig. 2a in which X- and Y-axis reflect the Sample size and True Positive Rate, respectively. It reflects that 11.53% of the entire population has been considered for virtual screening (reflected through orange colour) and it converts FPs and FNs to TNs and TPs, respectively, to minimize the cost and achieve 83.59% of the target. It also represents the position at which the accuracy of the model is maximum through proper minimization of wrongly classified data using the symbol 'X'.

Cost-Benefit curve is reflected in Fig. 2b in which X- and Y-axis reflect the Sample size and the cost, respectively. It is observed that the cost before virtual screening is 1164 and that has been decreased to 413 by converting 751 number of FPs and FNs to TNs and TPs, respectively. Hence, the accuracy level of the model has been increased from 88.76 to 96.01%.

3 Result Analysis

Table 2 reflects a comparative study of machine learning-based predictive model using Decision Tree (J48) algorithm in Full Training set mode and Cross-validation mode. It is observed that due to the presence of higher classification errors in terms of RAE, RRSE, MAE, and RMSE, Accuracy in cross-validation mode is less compared to that of Full training set mode. Accuracy in Full Training set mode is 80.18% whereas it is 68.67% in Cross-validation mode. But cross-validation mode is the robust one and better in terms of quality [16]. Therefore, we have considered cross-validation mode for the proposed work. Further, Cost-benefit analysis has been used with virtual screening to minimize the classification error and enhance the accuracy of the machine learning-based predictive model using Decision Tree (J48) algorithm.

Tables 3 and 4 reflect data obtained for cost-benefit analysis using Decision Tree (J48) algorithm before and after the virtual screening in cross-validation mode. It is observed that cost-benefit analysis has no impact on the position Stand_Tension

Table 2 Comparative analysis of Decision Tree (J48) algorithm-based predictive model in full training set and cross-validation mode

Parameters	Full training set	Cross-validation
Accuracy (in %age)	80.1853	68.67
Relative Absolute Error (RAE) (in %age)	27.0698	34.72
Root Relative Squared Error (RRSE) (in %age)	52.0287	65.46
Mean Absolute Error (MAE)	0.0535	0.0686
Root mean squared error	0.1635	0.2057

and Stand_Normal. For Stand_Exercise Position, it is observed that before virtual screening (BVS), total number of FP is 0, FN is 1201, TP is 0, and TN is 9160 so the cost is 1201, but after virtual screening (AVS), TP and FP increased to 572 and 233, respectively, but FN and TN decreased to 629 and 8927, respectively. Hence, gradually total cost has decreased to 861. 7.81% of the entire population has been used for virtual screening, and it converts 47.87% of FP and FN records to TP)and TN; therefore, it achieves 28.31% of benefit. It can be observed in Fig. 3 that CostBVS (Blue Colour) represents the total cost before virtual screening, CostAVS (Red Colour) represents the total cost after virtual screening, and BenefitAVS (Yellow Colour) represents the total benefit after virtual screening. Due to minimization of classification error in terms of cost from 1201 to 861, the accuracy level of the model has been increased from 88.4 to 91.69% for Stand_Exercise position. It has been observed through Fig. 4 in which AccuracyBVS (Blue Colour) represents the accuracy level of the model before virtual screening and AccuracyAVS (Red Colour) represents the accuracy level of the model after virtual screening. Similarly, in all the body positions except Stand_Tension and Stand _Normal, the classification error in terms of Cost has been minimized after virtual screening and the benefit has been observed through enhanced accuracy level of the model in all body positions. Therefore, it is observed that cost-benefit analysis can be used to enhance the accuracy of the model through minimization of classification error. Since classification through cost-benefit analysis has been done in cross-validation mode, the minimum accuracy level observed in all categories can be considered as the overall accuracy level of the model. Therefore, accuracy level of the model using Decision Tree (J48) algorithm in cross-validation ranges from 88.72 to 99.67% after virtual screening which is better than 88.72–89.19% of accuracy observed in Decision Tree (J48) algorithm in cross-validation model before virtual screening. From the above analysis, it is observed that accuracy level of the Decision Tree (J48)-based predictive model in cross-validation mode without cost-benefit analysis has been increased to 99.67% from earlier 89.29% due to use of cost-benefit analysis. It is observed that Cost-benefit analysis is an approach that can be used to enhance the accuracy of the machine learning-based predictive model by minimizing the classification errors through virtual screening.

Table 3 Cost-benefit analysis before Virtual Screening (BVS) using Decision Tree (J48) algorithm

Class	Accuracy	TP	FP	FN	TN	Cost	Cost (%)
Stand_Tension	88.72	0	0	1168	9193	1168	11.27
Stand_Normal	88.72	0	0	1168	9193	1168	11.27
Stand_Exercise	88.4	0	0	1201	9160	1201	11.59
Sit_Tension	88.76	0	0	1164	9197	1164	11.23
Sit_Normal	89.4	0	0	1098	9263	1098	10.6
Sit_Exercise	89.09	0	0	1130	9231	1130	10.91
Lay_Tension	88.92	0	0	1148	9213	1148	11.08
Lay_Normal	89.19	0	0	1120	9241	1120	10.81
Lay_Exercise	88.76	0	0	1164	9197	1164	11.23

Table 4 Cost-benefit analysis after Virtual Screening (AVS) using Decision Tree (J48) algorithm

Class	Accuracy	TP	FP	FN	TN	Cost	Cost (%)	Pop (%)	Target (%)	Benefit	Benefit (%)
Stand_ Tension	88.72	0	0	1168	9193	1168	11.27	0	0	0	0
Stand_ Normal	88.72	0	0	1168	9193	1168	11.27	0	0	0	0
Stand_ Exercise	91.69	572	233	629	8927	861	8.31	7.81	47.87	340	28.31
Sit_ Tension	96.01	973	222	191	8975	413	3.99	11.53	83.59	751	64.52
Sit_ Normal	99.63	1085	25	13	9238	38	0.37	10.71	98.81	1060	96.54
Sit_ Exercise	97.08	1003	175	127	9056	302	2.91	11.36	88.76	828	73.27
Lay_ Tension	93.25	680	231	468	8982	699	6.75	8.79	59.23	449	39.11
Lay_ Normal	99.67	1094	20	26	9221	34	0.33	10.9	98.92	1086	96.96
Lay_ Exercise	94.06	927	378	237	8819	615	5.94	12.59	79.63	549	47.16

4 Conclusions and Future Scope

In the proposed work, vital health parameters have been collected using the sensors such as GSR, airflow, and temperature to estimate the stress level, breathing rate, and the body temperature in various positions such as Sitting, Standing, and laying and various moods of human such as Normal, tension, and exercise. Further, it has been analysed using machine learning-based predictive model to detect stress level of the human. Decision Tree (J48) algorithm in robust cross-validation mode has

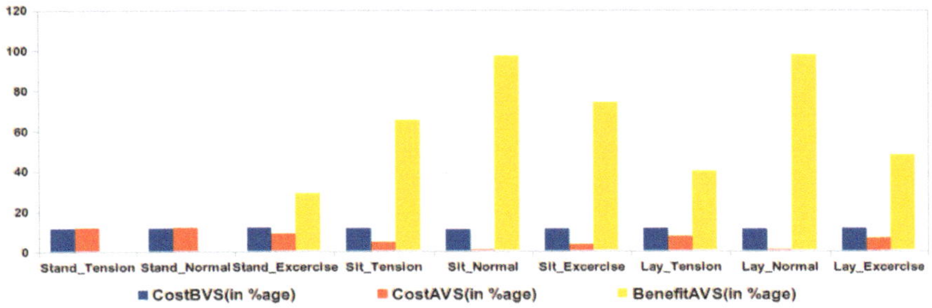

Fig. 3 Cost-benefit analysis using Decision Tree (J48) algorithm

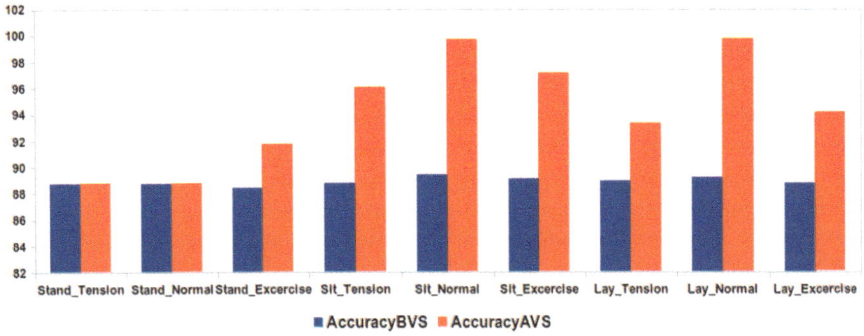

Fig. 4 Accuracy level before and after Virtual Screening using Decision Tree (J48) algorithm

been considered in the proposed work. Analysis has been done through cost-benefit analysis. Before virtual screening, accuracy level of the predictive model varies in between 88.72 and 89.19% for various categories whereas after virtual screening, due to use of cost-benefit approach accuracy level of the same predictive model has been increased and varies in between 88.72 and 99.67%. It is possible because of the reduction in cost in terms of classification errors available in the predictive model through cost-benefit approach.

Acknowledgements The authors are showing their gratitude to Prof. Srinivas Sethi, Principal Investigator, of SERB (DST) sponsored Project for carrying out the work related to Stress Analysis in wireless sensor network environment at CSEA Department, IGIT Sarang.

References

1. Harker, M.: Psychological sweating: a systematic review focused on aetiology and cutaneous response. Skin Pharmacol. Physiol. **26**, 92–100 (2013). https://doi.org/10.1159/000346930
2. Asahina, M., Poudel, A., Hirano, S.: Sweating on the palm and sole: "Physiological And Clinical Relevance". Clin. Auton. Res. **25**, 153–159 (2015). https://doi.org/10.1007/S10286-015-0282-1
3. Glanz, K., Schwartz, M.: "Stress, Coping, and Health Behavior", Health Behavior And Health Education: Theory, Research, and Practice, pp. 211–236 (2008)
4. Vijaya, P.A., Shivakumar, G.: Galvanic skin response: a physiological sensor system for affective computing. Int. J. Mach. Learn. Comput. **3**(1) (Feb 2013). Springer
5. Navea, R.F., Buenvenida, P.J., Cruz, C.D.: Stress detection using galvanic skin response: an android application. In: Journal of Physics: Conference Series, vol. 1372, p. 012001. Iop Publishing (2019). https://doi.org/10.1088/1742-6596/1372/1/012001
6. Sanchez-Comas, A., et al.: Correlation analysis of different measurement places of galvanic skin response in test groups facing pleasant and unpleasant stimuli. Sensors **21**(12), 4210 (2021). https://doi.org/10.3390/S21124210
7. Fenz, W.D., Epstein S.: Gradients of a physiological arousal of experienced and novice parachutists as a function of an approaching jump. Psychomatic Med. **29**, 33–51 (1967)
8. Sahoo, R., Sethi, S.: Functional analysis of mental stress based on physiological data of GSR sensor. In: Emerging ICT for Bridging the Future-Proceedings of the 49th Annual Convention of the Computer Society of India (CSI), vol. 1. Springer, Cham (2015)
9. Sahoo, R., Sethi, S.: Remotely functional-analysis of mental stress based on GSR sensor physiological data in wireless environment. In: Information Systems Design and Intelligent Applications, pp. 569–577. Springer, New Delhi (2015)
10. Sethi, S., Sahoo, R.K.: Design of WSN in real time application of health monitoring system. In: Virtual and Mobile Healthcare: Breakthroughs in Research and Practice, pp. 643–658. IGI Global (2020)
11. Sahoo, R.K., et al.: Mental stress detection using GSR sensor data with filtering methods. In: Intelligent Systems, pp. 537–548. Springer, Singapore (2022)
12. Sahoo, R.K., Pradhan, S., Sethi, S., Udgata, S.K.: Enhancing data integrity in mobile crowd sensing environment with machine learning and cost-benefit analysis. Int. J. Comput. Digit. Syst. **14**(1), 1–1 (2023)
13. Badajena, J.C., Sethi, S., Sahoo, R.K.: Data-driven approach to designing a BCI-integrated smart wheelchair through cost-benefit analysis. High-Confid. Comput. **3**(2), 100118 (2023)
14. Sethi, P., Sahoo, R.K., Rout, A., Mufti, M.: Galvanic skin response-based mental stress identification using machine learning. In: International Conference on Machine Learning, IoT and Big Data, pp. 561–570. Springer Nature Singapore, Singapore (Mar 2023)
15. Eligüzel, N., Çetinkaya, C., Dereli, T.: Comparison of different machine learning techniques on location extraction by utilizing geo-tagged tweets: a case study. Adv. Eng. Inform. **46**, 101151 (2020)
16. Badajena, J.C., Sethi, S., Sahoo, R.K.: Comparative analysis of full training set and cross-validation for machine learning approach to run smart wheelchair. In: Intelligent Systems and Applications: Select Proceedings of ICISA 2022, pp. 121–132. Springer Nature Singapore, Singapore (2023)

Odia Handwritten Character Recognition: A Comparative Study of EfficientNet-B0, ResNet-50, DenseNet-121, and Xception-71 by Leveraging GAN

Santosh Kumar Das and Swarupananda Bissoyi

Abstract Recognizing handwritten Odia characters is a complex challenge due to the vast number of possible combinations resulting from the large variety of vowels, consonants, conjuncts, and *matraas*, coupled with limited research and dataset availability. In this research, we offer a comparative examination of four deep learning models—EfficientNet-B0, ResNet-50, DenseNet-121, and Xception-71—in terms of their accuracy in Odia handwritten character recognition. We further investigate the improvements observed in these models when leveraging data augmentation techniques using Generative Adversarial Networks (GANs). Our experiments are conducted on the NITROHCSv1.0 dataset, which comprises 15,040 images of Odia characters. The results demonstrate that ResNet-50 and EfficientNet-B0 achieve the highest accuracy among the four models, with ResNet-50 attaining 100% accuracy and EfficientNet-B0 reaching 99.8% accuracy. Interestingly, the incorporation of GANs enhanced the performance of these models, with both ResNet-50 with GAN and EfficientNet-B0 with GAN achieving 100% accuracy. DenseNet-121 reported a respectable accuracy of 99.1%, while Xception-71 exhibited relatively lower performance, reaching 58.3% accuracy in 30 epochs. When augmented with GANs, DenseNet-121 and Xception-71 achieved 99.6% and 61.2% accuracy, respectively, in 30 epochs.

Keywords Generative Adversarial Network (GAN) · Generator · Discriminator · DCNN · EfficientNet-B0 · ResNet-50 · DenseNet-121 · Xception-71

S. K. Das · S. Bissoyi (✉)
Department of Computer Application, Maharaja Sriram Chandra Bhanja Deo University, Baripada, Odisha, India
e-mail: swarupananda.bissoyi@nou.nic.in

S. Kumar Udgata et al. (eds.), *Intelligent Systems*, Lecture Notes in Networks and Systems 1149, https://doi.org/10.1007/978-981-97-8160-7_23

1 Introduction

Odia, with its rich literary tradition and minimal influence from other languages, has been accorded the status of the sixth classical language in India. It is spoken by approximately 40 million people worldwide. Handwritten Character Recognition (HCR) is a significant area of research in Odia, given its varied applications in diverse fields such as document and manuscript scanning, text recognition, and image processing. Deep learning (DL) methodologies have been extensively employed for HCR tasks across various languages, including Odia. This paper endeavors to conduct a comparative performance analysis of four different DL models in the area of Odia HCR, namely EfficientNet-B0 [1], ResNet-50 [2], DenseNet-121 [3], and Xception-71 [4], for Odia HCR. A major obstacle in supervised learning is having enough labeled data to learn features that better describe the structure of the data. It's both time- and cost-consuming. Unsupervised learning is used for drawing conclusions from datasets comprising of input data and any labels. Generative Adversarial Networks (GANs) [5], a subfield of unsupervised learning, have been extensively utilized to enhance the performance of DL models by generating additional data. In this paper, we investigate the application of GANs to improve the performance of the four DL models discussed earlier.

1.1 Generative Adversarial Networks (GANs)

GANs are a machine learning technique that employs two competing neural networks, known as the Generator (G) and Discriminator (D), to enhance their predictive accuracy through DL methods. This unsupervised approach utilizes a cooperative zero-sum game framework to facilitate learning. In this section, we will explore the fundamental components of GANs, which can be leveraged to augment datasets with additional data.

1.1.1 Generator and Discriminator

A simple architecture of GAN, considered for our work, is depicted in Fig. 1. It has two components. The first component is called Generator (G) and the second component is called Discriminator (D). G's role is to generate images that eventually look as close as possible to the original ones without being aware of the original images, and D's role is to consume the generated image and classify it as original or artificially generated. The ones identified as original become part of our extended dataset whereas the ones identified as artificially generated are fed back to G. To address this challenge of maximizing the performance of G, we created a model composed of blocks of fully connected layers, with activation function as LeakyRelu, and additional dropout layers.

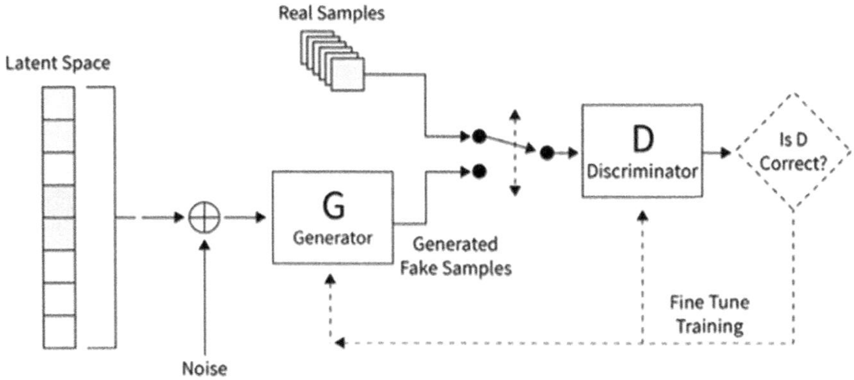

Fig. 1 GAN Architecture used for our study

1.1.2 Loss Function Computation

To make the G module's image generation process effective, it's fed with data, thereby increasing its effectiveness in creating real looking images. The D module examines the images generated by G and provides feedback to G with the probability of it being a real or artificially generated image. To effectively classify as real or generated, D receives images that are from the original dataset. If D receives and identifies an image as a generated image, it is not considered for data augmentation of the actual dataset. It's an attempt to maximize $D(x)$ and minimize $D(G(z))$. Based on this, G has the responsibility of getting positive feedback from D by generating images that look like real ones. The process attempts to ensure that when D gets an image from the generated dataset, it thinks that the image is coming from the original dataset. It's an attempt to maximize $D(G(z))$. To avoid the network feel overconfident about the wrong outputs by considering $D(G(z))$ as a loss function, $\log(D(G(z)))$ is used instead. In our implementation, we set up the G and D models using binary cross-entropy as the objective function to optimize.

The loss function utilized in our Generative Adversarial Network (GAN) is expressed as follows:

$$E_x[\log(D(x))] + E_z[\log(1 - D(G(z)))] \quad (1)$$

1.1.3 Optimizing Loss

For efficient training, we used a Stochastic Gradient Descent (SGD) optimization. D and G don't have the same losses. For effectively optimizing each other G and D are given different loss functions. After identifying appropriate functions, we trained the model to optimize the losses.

The following steps are used in our work:

1. Select an image from the input dataset and generate random noise of the same size.
2. Provide both images to D and get probability of original v/s generated.
3. Create a noise of the same size and provide it to G.
4. Train G for a few rounds.
5. Execute these steps for a predefined number of times until a reasonable image similar to input dataset images is generated.

1.2 Deep Convolutional Neural Network (DCNN) Models

Deep convolutional neural networks (DCNNs) are a specialized type of feed-forward neural network that excels at feature learning and image classification tasks. Unlike earlier neural networks, DCNNs overcome the challenges of vanishing and exploding gradients during backpropagation by utilizing regularized weights and fewer connections. For our study, we have used an original dataset along with the GAN augmented dataset with the following four DCNN variants.

1.2.1 ResNet-50

ResNet-50, a widely adopted DCNN architecture, comprises 50 layers and employs a unique residual learning framework to tackle the vanishing gradient problem. This architecture incorporates shortcut connections, enabling direct paths for gradient propagation during the training process. This innovative design helps to alleviate degradation issues, facilitating the training of deep networks. With 23.5 million parameters, ResNet-50 demonstrates exceptional performance in image classification tasks, achieving cutting-edge accuracy.

1.2.2 EfficientNet-B0

EfficientNet-B0 which is part of the EfficientNet CNN series is characterized by a balanced scaling of depth, width, and resolution. Introduced by Tan and Le in 2019, it efficiently balances model size and performance. With 5.3 million parameters, it achieves competitive accuracy on image classification tasks while being computationally lightweight. The compound scaling method used in EfficientNet-B0 allows for improved generalization and robustness, making it a popular choice for various applications, particularly in scenarios with limited computational resources.

1.2.3 DenseNet-101

DenseNet-101, a DL model belonging to the DenseNet (Densely Connected Convolutional Networks) family, is composed of 101 layers. Unlike traditional neural network architectures, DenseNet-101 establishes dense interconnections between its layers, allowing each layer to directly receive inputs from all preceding layers. This dense connectivity pattern encourages the reuse of features, improves parameter efficiency, and mitigates the vanishing gradient issue, which can hinder the training of deep networks. With a parameter count of 7 million, DenseNet-101 exhibits exceptional performance in image classification tasks, consistently achieving high accuracy. Its innovative architectural design facilitates seamless information flow and gradient propagation, leading to the development of robust and effective feature representations. DenseNet-101 represents a notable advancement in deep neural network architectures, setting new benchmarks in terms of performance and efficiency.

1.2.4 Xception-71

Xception-71 is an extension of the Xception neural network architecture, originally introduced by François Chollet in 2017. With 71 convolutional layers, Xception-71 strives to improve the performance of DL models. It utilizes depth-wise separable convolutions, a technique that decouples spatial and channel-wise convolutions, thereby reducing computational complexity and the number of parameters. This design choice enables Xception-71 to achieve top-tier performance while using fewer parameters compared to conventional architectures. The innovative use of depth-wise separable convolutions enhances feature learning and promotes better generalization, making Xception-71 well-suited for image classification tasks that prioritize computational efficiency and high accuracy.

1.3 Proposed Approach

Our proposed approach has three steps. The first step is pre-processing the input dataset by removing noise from data. This step includes image resizing and converting it to grayscale. In the second step, we create and train the GAN models (G and D) to generate new images of Odia handwritten characters. This will enable G to produce images looking like real ones. These images are later used in the deep convolutional neural network phase to compare accuracy with the augmented dataset. The third step is the deep convolutional neural network (DCNN) phase. This is the step where we build and train the deep conventional neural network on the new enlarged dataset of images that were newly generated by the GAN model and also the images from original dataset for comparison in our study. Implementing GANs helped us to avoid over-fitting issues.

2 Related Work

Despite significant research endeavors dedicated to optical character recognition in various languages, advancements in Odia, particularly in the realm of handwritten character recognition, have been limited. Dash et al. highlighted deficiencies and limitations in existing state-of-the-art approaches [6]. Das et al. discussed machine learning techniques used in the recognition of Odia character recognition [7]. Lecun et al. examined various gradient-based methods for recognizing handwritten characters through a standard test focused on recognizing handwritten digits. The analysis revealed that deep convolutional neural networks consistently surpassed alternative approaches, demonstrating superior performance in handling images with diverse variations [8]. Husnain et al. achieved accuracy rates of 92.7% and 95.6% in tenfold and eightfold cross-validation, respectively, when applying CNN for the recognition of Urdu handwritten characters [9]. Das et al. and Alwaqfi et al. studied the effect of data augmentation techniques on DCNN for Odia and Arabic characters, respectively [10, 11]. Das et al. demonstrated a 1% improvement in accuracy through the application of GANs to Odia handwritten characters, while Alwaqfi et al. observed a 3.5% enhancement in accuracy when employing GANs for Arabic handwritten characters. Jha et al. engaged in a study involving image augmentation using Generative Adversarial Networks (GANs) on an ISI-Kolkata dataset containing Devanagari, Bangla, Latin, and Odia scripts. GANs were utilized to generate synthetic sample images for the dataset without prior knowledge of potential variations between samples. Their methodology achieved a notable accuracy of 97.31% specifically for the Odia character set [12]. Mukherjee et al. studied the implementation of GAN using TensorFlow on the MNIST English digits dataset [13]. Chakraborty et al. enlarged the image samples within the Bangla Lekha-isolated character dataset, employing a DCNN model. They disclosed an accuracy of 91.81% for the original dataset alphabet characters and a heightened accuracy of 95.25% for an augmented dataset. The augmentation involved techniques such as position shifting, zoom, rotation, and shear, among others [14]. Among other recent studies, authors of [15–18] focused their studies on the identification of Odia alphabets and numerals. Authors of [19–21] employed neural networks and various DL methodologies to advance the domain of character classification. To the best of our knowledge, no comparative study of popular DL models has been done for the Odia Handwritten Character recognition task.

3 Methodology

Our proposed model depicted in Fig. 2 consists of three deep neural networks: One generator, one discriminator from GAN, and one of the DCNNs out of ResNet-50, EfficientNet-B0, Xception-71, and DenseNet-121 which is used to train the HCR model. In this section, we discuss in detail the different elements of the model we

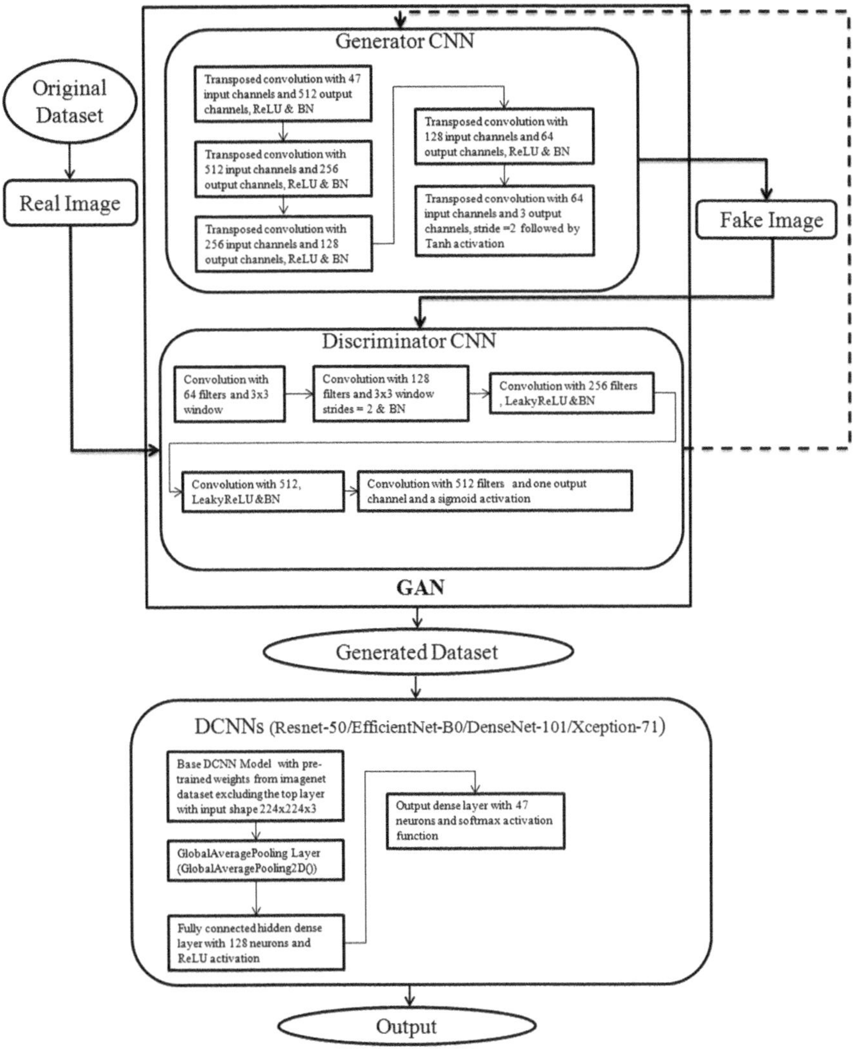

Fig. 2 The Proposed OCR Model Used for our Study

have used for our work and the methods followed to obtain the accuracies of each of the four models under study.

3.1 NITROHCSv1.0 Dataset

The dataset employed in our research is NITROHCSv1.0 [22], which includes 47 categories of handwritten Odia characters, with 320 images per category, totaling 15,040 sample images in bitmap format. Each image has a resolution of 81 × 81 pixels. This dataset consists of samples collected from individuals of different age groups at various time intervals. It is divided into training, validation, and test sets, containing 12,032, 1,504, and 1,504 images, respectively.

3.2 Configuring the Generator and Discriminator for GAN

In our research, we configured the models using Fully Connected layers and employed the ReLU activation function. Our generator consists of five layers. The initial layer is a transposed convolution with 47 input channels and 512 output channels, utilizing ReLU activation and batch normalization. The second layer is a transposed convolution with 512 input channels and 256 output channels, also employing ReLU activation and batch normalization. The third layer is a transposed convolution with 256 input channels and 128 output channels, with ReLU activation and batch normalization. The fourth layer is a transposed convolution with 128 input channels and 64 output channels, utilizing ReLU activation and batch normalization. The final layer is a transposed convolution with 64 input channels and 3 output channels, with a stride value of 2. To ensure the generated image aligns with the original NITROHCSv1.0 dataset, we applied the Tanh activation function for this specific layer of the generator.

In a similar manner, our Discriminator architecture consists of several layers. The initial layer is a convolutional layer with 64 filters and a 3 × 3 window. The second layer is a convolutional layer with 128 filters, a 3 × 3 window, and a stride value of 2, accompanied by batch normalization. The third layer is a convolutional layer with 256 filters, utilizing LeakyReLU activation and batch normalization. The fourth layer is a convolutional layer with 512 filters, again employing LeakyReLU activation and batch normalization. The final layer is a fully connected convolutional layer with 512 filters and a single output channel, concluding with a sigmoid activation function. This final sigmoid activation outputs the classification probability, which is the desired outcome of our experiment.

3.3 Generating Handwritten Odia Characters

We have used the training set of the NITROHCSV1.0 dataset to train the four DL models, and the validation set to fine-tune the hyperparameters of the models. The testing set is used to evaluate the performance of the models. We were able to generate handwritten characters from the NITROHCSV1.0 dataset. We created a method to store predictions from G for one batch of images. The module created random noise and passed them to G, processed it for display, and then saved it to a folder. We ran this helper function every 200 epochs. The GAN was trained for 4000 epochs with a batch size of 32. We saved the generated images every 100 epochs in the images folder. The generated images are added to our input dataset and the four models are trained to measure the improvement in accuracies.

For our study, we used each of the above DCNNs separately as a base model with pre-trained weights from imagenet dataset excluding the top layer with input shape 224 × 224 × 3. This was followed by a GlobalAveragePooling Layer, a fully connected hidden dense layer with 128 neurons and ReLU activation and finally output dense layer with 47 neurons and softmax activation function.

4 Results and Discussion

The real input images passed to the GAN and the images generated by the GAN are given in Fig. 3. The findings of the experiments are presented in Table 1.

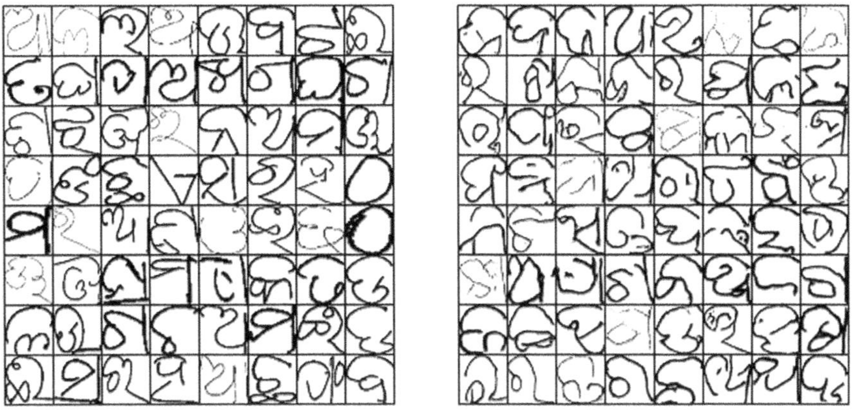

Fig. 3 Image generation using GAN, original (left) and generated (right)

Table 1 Accuracy of the four models with/without GAN

Model	Accuracy without GAN (%)	Accuracy with GAN (%)
ResNet-50	100	100
EfficientNet-B0	99.8	100
DenseNet-121	99.1	99.6
Xception-71	58.3	61.2 (70% with 75 epochs)

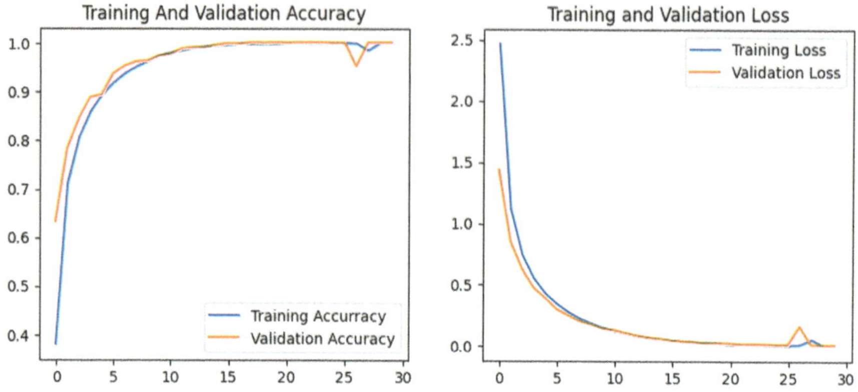

Fig. 4 Accuracy and loss metrics for ResNet-50 with GAN

The second column in Table 1 shows the accuracy of the four models without GAN, while the last column shows the accuracy of the models with input dataset augmented with GAN generated images. The results show that ResNet-50, EfficientNet-B0, and DenseNet-121 achieve high accuracy among the four models, with ResNet-50 achieving an accuracy of 100%, EfficientNet-B0 achieving 99.8%, and DenseNet-121 achieving 99.1% accuracy. Xception-71 has lower accuracy compared to the other three, achieving an accuracy of only 58.3% in the 30 epochs. When we add GAN generated dataset to the original dataset and generate the models, we can see that the accuracy of all four models improves. ResNet-50 using GAN, EfficientNet-B0 using GAN, and DenseNet-121 achieve the highest accuracy, with ResNet-50 using GAN achieving an accuracy of 100% in 21 epochs instead of 27, EfficientNet-B0 using GAN achieving an accuracy of 100%, and DenseNet-121 using GAN achieving an accuracy of 99.6%. Xception-71 using GAN also showed improvement in accuracy, achieving an accuracy of 61.2%. When the number of epochs was increased to 75, the accuracy of Xception-71 improved to 70%. The results prove that GANs improve the performance of DL models for image classification tasks. In particular, ResNet-50 and EfficientNet-B0, which are already powerful models, benefit the most from the addition of GANs, achieving state-of-the-art performance on the NITROHCSV1.0 dataset.

However, the performance of GANs can vary based on the specific implementation and training parameters used. Therefore, careful evaluation of the performance of GANs on a specific task and dataset is needed before using them in practice. Figures 4, 5, 6, 7, and 8 show the training and validation accuracies and losses of the four models. Figure 9 shows the test results of the four models with GAN.

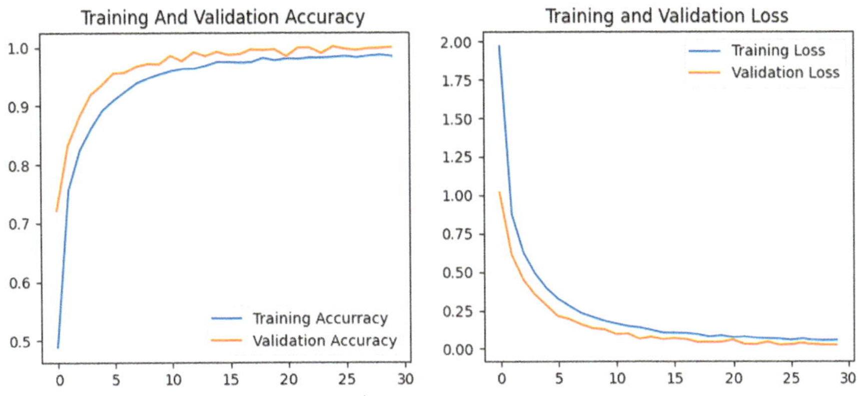

Fig. 5 Accuracy and loss metrics for in EfficientNet-B0 with GAN

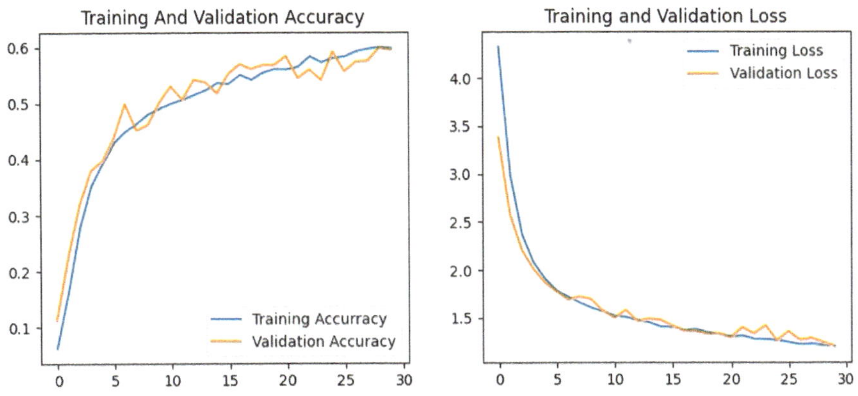

Fig. 6 Accuracy and loss metrics for training and validation in DenseNet-101 with GAN

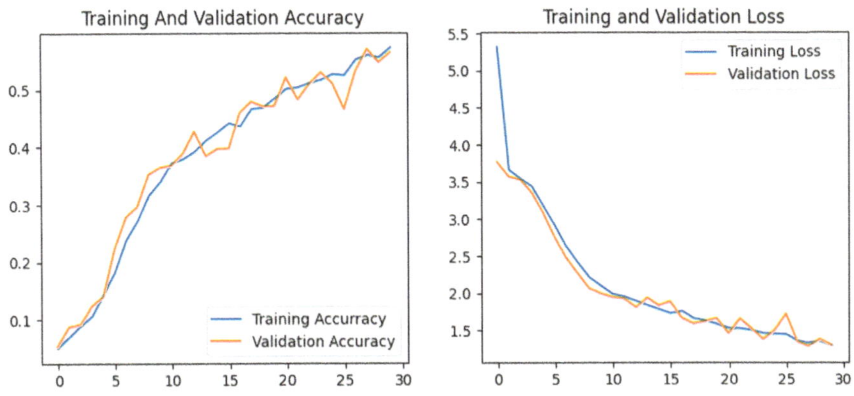

Fig. 7 Accuracy and loss metrics for Exception-71 with GAN

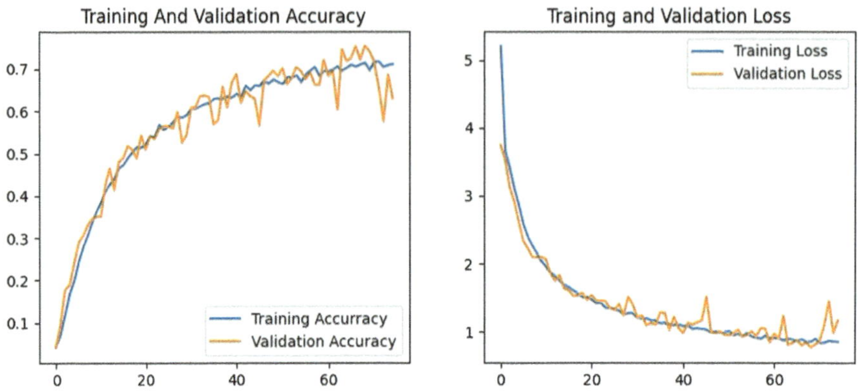

Fig. 8 Accuracy and loss metrics for Exception-71 (75 Epochs) with GAN

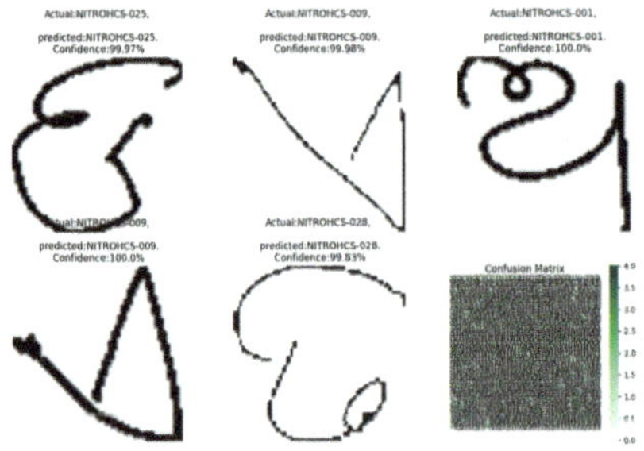

(a) Test Results for ResNet-50 with GAN

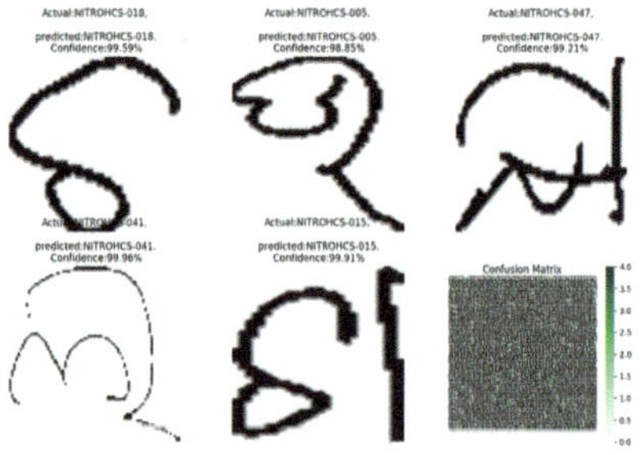

(b) Test Results for EfficientNet-B0 with GAN

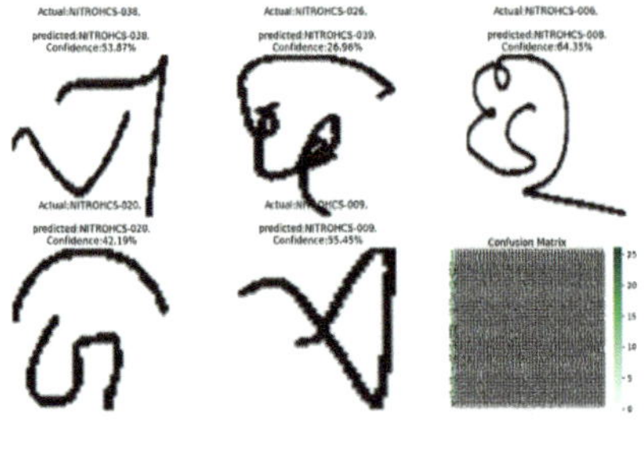

(c) Test Results for DenseNet-101 with GAN

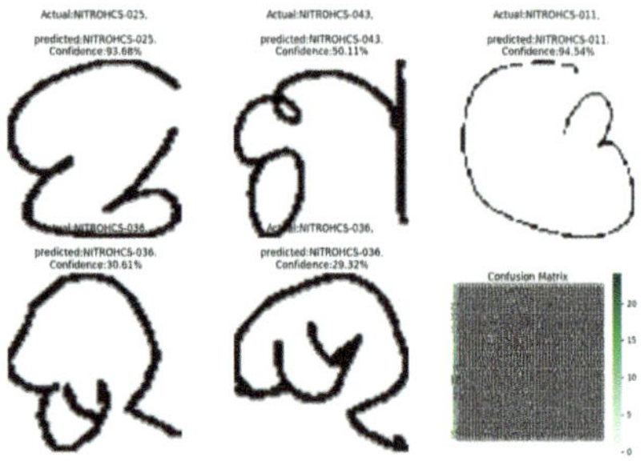

(d) Test Results for Xception-71 with GAN

Fig. 9 Test results

5 Conclusion

In this study, we used the NITROHCSv1.0 Odia handwritten character dataset and tested the performance of four different DL models ResNet-50, EfficientNet-B0, DenseNet-121, and Xception-71, with the default dataset and updated dataset with images augmented by GAN. The model was trained for 30 epochs in each case. However, due to poor accuracies reported by Xception-71, we ran it for 75 epochs but still the accuracy could not match with other models. ResNet-50 with 50 layers outperformed other DL models with a larger number of layers. The comparison of accuracies of these results of the experiments shows that GANs improved the performance of DL models for image classification tasks and that EfficientNet-B0 and ResNet-50 are models that can benefit the most from the addition of GANs for the Odia handwritten characters dataset.

References

1. Tan, M., Le, Q.: Efficientnet: Rethinking model scaling for convolutional neural networks. In: International Conference on Machine Learning, pp. 6105–6114. PMLR (2019)
2. He, K., Zhang, X., Ren, S., Sun, J.: Deep residual learning for image recognition. In: Proceedings of the IEEE Conference on Computer Vision and Pattern Recognition, pp. 770–778 (2016)
3. Huang, G., Liu, Z., Van Der Maaten, L., Weinberger, K.Q.: Densely connected convolutional networks. In: Proceedings of the IEEE Conference on Computer Vision and Pattern Recognition, pp. 4700–4708 (2017)
4. Chollet, F.: Xception: deep learning with depthwise separable convolutions. In: Proceedings of the IEEE Conference on Computer Vision and Pattern Recognition, pp. 1251–1258 (2017)
5. Goodfellow, I., Pouget-Abadie, J., Mirza, M., Xu, B., Warde-Farley, D., Ozair, S., Courville, A., Bengio, Y.: Generative adversarial nets. Adv. Neural Inf. Process. Syst. **27** (2014)
6. Dash, K.S., Puhan, N.B., Panda, G.: Odia character recognition: a directional review. Artif. Intell. Rev. **48**, 473–497 (2017). https://doi.org/10.1007/s10462-016-9507-5
7. Das, M., Panda, M., Dash, S.: 4 A comparative analysis of machine learning techniques for Odia character recognition. Mach. Learn. Appl., 65–90 (2020)
8. LeCun, Y., Bottou, L., Bengio, Y., Haffner, P.: Gradient-based learning applied to document recognition. Proc. IEEE **86**, 2278–2324 (1998)
9. Husnain, M., Saad Missen, M.M., Mumtaz, S., Jhanidr, M.Z., Coustaty, M., Muzzamil Luqman, M., Ogier, J.-M., Sang Choi, G.: Recognition of Urdu handwritten characters using convolutional neural network. Appl. Sci. **9**, 2758 (2019)
10. Das, M., Panda, M., Dash, S.: Enhancing the power of CNN using data augmentation techniques for Odia handwritten character recognition. Adv. Multimed. **2022**, 6180701 (2022). https://doi.org/10.1155/2022/6180701
11. Alwaqfi, Y.M., Mohamad, M., Al-Taani, A.T.: Generative adversarial network for an improved Arabic handwritten characters recognition. Int. J. Adv. Soft Comput. Appl. **14** (2022)
12. Jha, G., Cecotti, H.: Data augmentation for handwritten digit recognition using generative adversarial networks. Multimed. Tools Appl. **79**, 35055–35068 (2020). https://doi.org/10.1007/s11042-020-08883-w
13. Mukherjee, S.: Generating Images Using GANs in TensorFlow . https://www.scaler.com/topics/deep-learning/gans-image-generation/

14. Chakraborty, P., Islam, A., Abu Yousuf, M., Agarwal, R., Choudhury, T.: Bangla handwritten character recognition using convolutional neural network. In: Skala, V., Singh, T.P., Choudhury, T., Tomar, R., Abul Bashar, Md. (eds.) Machine Intelligence and Data Science Applications, pp. 721–731. Springer Nature Singapore, Singapore (2022). https://doi.org/10.1007/978-981-19-2347-0_56
15. Majhi, B., Satpathy, J., Rout, M.: Efficient recognition of Odiya numerals using low complexity neural classifier. In: 2011 International Conference on Energy, Automation and Signal, pp. 1–4. IEEE (2011)
16. Khedidja, D., Hayet, M.: Multiple classifiers and invariant features extraction for digit recognition. Int. J. Comput. Electr. Eng. **11**, 41–52 (2019)
17. Dash, K.S., Puhan, N.B., Panda, G.: On extraction of features for handwritten Odia numeral recognition in transformed domain. In: 2015 Eighth International Conference on Advances in Pattern Recognition (ICAPR), pp. 1–6. IEEE (2015)
18. Sarangi, P.K., Ahmed, P.: Recognition of handwritten Odia numerals using artificial intelligence techniques. Int. J. Comput. Sci. **2**, 35–38 (2013)
19. Bora, M.B., Daimary, D., Amitab, K., Kandar, D.: Handwritten character recognition from images using CNN-ECOC. Procedia Comput. Sci. **167**, 2403–2409 (2020)
20. Jana, R., Bhattacharyya, S., Das, S.: Handwritten digit recognition using convolutional neural networks. In: Bhattacharyya, S., Snasel, V., Hassanien, A.E., Saha, S., Tripathy, B.K. (eds.) Deep Learning: Research and Applications, pp. 51–68 (2020)
21. Chhajro, M.A., Khan, H., Khan, F., Kumar, K., Wagan, A.A., Solangi, S.: Handwritten Urdu character recognition via images using different machine learning and deep learning techniques. Indian J. Sci. Technol. **13**, 1746–1754 (2020)
22. NITROHCSv1.0 (Odia handwritten character set version 1.0). https://website.nitrkl.ac.in/docs/CS/Database/Windows/NITROHCSv1.0.rar

Security and Privacy for IoT-Based Blockchain Technology: Review, Solutions, and Challenges

Ahmed J. Obaid and Mohammed Ayad Alkhafaji

Abstract The Internet of Things (IoT) has extensively impacted our lifestyles. However, IoT networks' large size and widespread nature remain significant obstacles to IoT privacy. Furthermore, the decentralized nature of blockchain has solved numerous problems related to safety, authentication, and maintenance in IoT systems. A smart city as an intelligent society can be created in this digital age of rapid communication. It aims to review a wide range of literature to assess the safety challenges and problems associated with implementing blockchain technology for sustainability and smart governments. The convergence of blockchain technology and artificial intelligence is a unique catalyst for technological transformation in IoT applications that are intelligent and sustainable. Sustainable IOT applications could be advanced and developed through the use of blockchain. The discussion led us to develop a framework for processing and obtaining required information leveraging IoT devices, security, and privacy. Through digital analytics and storing results in decentralized cloud repositories, blockchain technology promotes a wide range of applications.

Keywords Security · Privacy · Blockchain · IOT

1 Introduction

Smart cities, smart homes, and other areas have benefited from the Internet of Things (IoT), among other things, farming [1], smart homes [2], wearable [3], smart cities [4], connected health [5], connected cars [6], and connected drones [7]. The Internet of Things allows communication, information sharing, and decision coordination. The Internet of Things transforms traditional objects into intelligent objects through

A. J. Obaid (✉)
Faculty of Computer Science and Mathematics, University of Kufa, Kufa, Iraq
e-mail: ahmedj.aljanaby@uokufa.edu.iq

M. A. Alkhafaji
College of Engineering, Medical Instruments Technology Engineering, National University of Science and Technology, Dhi Qar, Nasiriyah, Iraq

S. Kumar Udgata et al. (eds.), *Intelligent Systems*, Lecture Notes in Networks and Systems 1149, https://doi.org/10.1007/978-981-97-8160-7_24

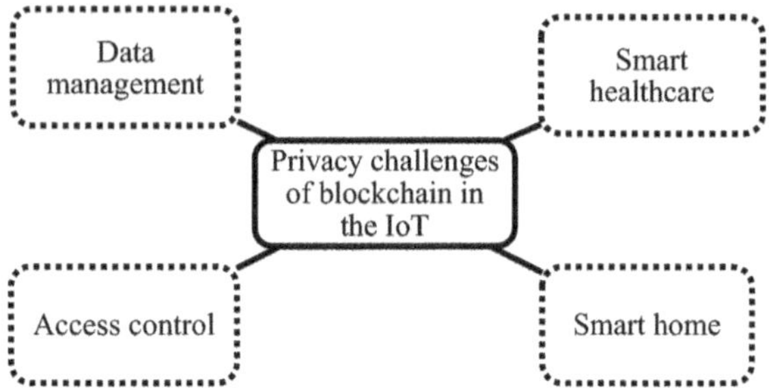

Fig. 1 Privacy challenges of IoT-based blockchain technology

communication technologies, Internet protocols, applications, and sensor networks [8]. Since IoT devices are developing exponentially, conserving their critical data is becoming increasingly important [9]. An IoT-cloud-based infrastructure housed essential IoT data in a third-party cloud service provider [10]. Because the cloud server has access to the entire data contained within the private IoT data, it can reveal it. Blockchain storage, however, has been viewed as a decentralized and distributed method to store data [11]. As IoT systems, networks, and things connect dynamically, a central server is justified as the central authentication point where all things or systems must relay. A decentralized and distributed storage method has been considered with blockchain technology. Numerous nodes worldwide share and maintain data through peer-to-peer networks, and replicated algorithms generate more copies [10].

Unlike any prior papers or surveys, our study focuses on the privacy concerns of blockchain-based IoT systems. In implementing IoT systems based on blockchain, we pay great attention to privacy concerns. Previous survey papers have dealt extensively with IoT integration and blockchain protection, and some have addressed privacy concerns related to blockchains. Despite that, no previous post has covered confidentiality protection techniques in a blockchain-based IoT platform from an application perspective, to the best of our knowledge. Several IoT technologies, such as smart homes, smart healthcare, access control, and real-time data management, were examined as examples of privacy protection techniques [12]. To resolve any privacy concerns raised by blockchain-operated IoT networks, Potential privacy policies will be developed based on the results of this study. The purpose of this article is explained in this section's continuation. Smart cities, smart medical devices, and management data are some topics covered in the articles. In Fig. 1, we illustrate some promising IoT-related applications.

1.1 Definition of Blockchain

Distributed ledger technology (DLT) or financial technology (FinTech) has been described as a type of blockchain. According to others, blockchain is a process that records transactional information, encrypts it, and governs its use through consensus mechanisms that surpass the classical financial ledger. Blockchain is defined differently by different disciplines from different perspectives [13], showing no single definition of blockchain [14]. According to the ICAEW, blockchain has the following characteristics: It is a protocol for recording transactions rather than a single technology. Unlike the Internet, where data is shared, blockchains enable ownership transfers between parties. Several reasons make blockchain an attractive model. It could, for example, save time reconciling disparate ledgers in markets with many participants. Due to its widespread distribution, it eliminates outages and costs associated with a central authority maintaining the ledger accuracy. Blockchain allows for increased transparency and self-auditing by participants in the ledger. ICAEW defines blockchain as an approach to recording transactions rather than a technology: blockchain is a protocol, not a single technology. Unlike the Internet, blockchains enable parties to transfer ownership between each other, as opposed to sharing data through the Internet. Several reasons make blockchain an attractive model. Multiple ledgers need not be reconciled if many parties transact in a market. As a result of being dispersed among all operators, it removes outages and the cost associated with paying for ledger accuracy to be maintained by a central authority. Participants can track past transactions using blockchain, enabling the system to self-audit and increase transparency. Accordingly, blockchain is described in [15] as an accounting technology that assures the constancy of financial records through the transfer of ownership and the maintenance of a ledger of accurate financial data [16].

1.2 Characteristics of a Smart City

A smart city can strengthen its amenities and facilities using smart technologies, software, and data. Water and electricity are required for that. We must ensure that affordable housing, better education, and healthcare are available and IT is accessible to all. Slums are unplanned and poorly planned, so smart cities can upgrade and build them in a manner that keeps communities safe and disaster-free. Fraud will be monitored with surveillance cameras, and women, children, and older adults will be secured with significant safety measures. Citizens will have easier access to a wide variety of public services. To minimize service charges and provide facilities without going to government offices, digital services will be delivered, especially through mobile devices [17]. The system will be more transparent, accountable, and engaging for citizens. Creating e-groups would facilitate citizen participation and suggestions, as well as tracking events and programs with the help of cyber tour sites. Parks and play areas need to be developed and preserved to mitigate urban impacts. Increasing populations will lead to more living spaces and better quality of life for residents. Enhancing public transit accessibility and improving traffic management, smart parking, and advanced modes of transportation. There should be fewer walking distances between essential administrative facilities in smart cities, making them more pedestrian and bike-friendly.

To make infrastructure more eco-friendly and efficient, waste production should be minimized, and natural resources should be used efficiently. To achieve success, smart cities leverage technology to transform and redevelop urban communities to maximize sustainability and living standards. Smart cities will use technology and data to upgrade their facilities and services. People from lower-middle-class backgrounds would benefit significantly, there would be more jobs, and wages would rise. As urban populations increase rapidly over the next few decades, sustainable and competitive cities will benefit from smart city infrastructure. According to BCC's research, the IT sector for smart cities will reach about $800 billion by 2022 [18]. In addition to advanced technology, smart cities contribute to economic growth, better living standards, and sustainability for their residents. Because of these factors, blockchain technology ensures data integrity in smart cities, promotes data exchange among organizations, including healthcare, schools, colleges, governments, and citizens, encourages collaborative decision-making, facilitates openness in city maintenance, and promotes the implementation of a smart city that is safe, reliable, transparent, and democratized [19]. In Fig. 2, blockchain technology is shown in action in smart cities. The paper also discusses the use of blockchain technology in government, transportation, management, trade, and finance, among other areas.

Fig. 2 Applications of smart cities

2 Literature Review

Using technologies such as anonymous authorization and anonymized data disclosure, IoT applications can enhance privacy through seven categories [20]. There are, therefore, several categories of security requirements [21]. Access control, confidentiality, non-repudiation, integrity, and authentication are the security types available in IoT applications. Green IoT-based agriculture security protocols should meet these requirements for security and privacy.

In a review [22], the author discussed the development of large-scale IoT-based pilot projects in agriculture. Using IoT for improved farming, the author presented a systematic survey [23]. IoT-based agriculture was examined in two recent surveys [24, 25] concerning its fundamental structures and how it impacts green agriculture. However, security and privacy research challenges are very limited in these surveys. As described in the paper [26], there are eight different frameworks for developing applications, each with security and privacy issues. It's difficult to develop IoT applications without security and privacy issues like authentication and data protection [27], but blockchains, fog computing, and machine learning can solve this problem (see Tables 1 and 2).

3 Blockchain Characteristics and Typology

In addition to being transparent, decentralized, immutable, tamper-resistant, and synchronized networks, blockchain technology also provides robust verification, coordinated networks, and agreement. It is, therefore, possible to transfer anything of value using blockchain technology, including intellectual property, health data, votes, and ideas. The blockchain technology has been classified according to its intended audience. For cryptocurrencies to be transacted digitally, Blockchain 10 is needed. Blockchain 2.0 uses smart contracts. Government, healthcare, and supply chains are among the areas in which blockchain 3.0 is applicable beyond cryptocurrency and finance. In Blockchain 4.0, artificial intelligence and blockchain will be combined [41].

Transaction validation and access rights are at the core of Blockchain technology, meaning different users can have different access rights and recording, updating, and validation rights. There are usually several characteristics that distinguish community and isolated blockchains: the scale, the permissions for joining, and the way transactions are validated. Consider control, data ownership, privacy, and access to help organizations determine the type of blockchain most suitable for their needs. A public or permissionless blockchain is an open ledger that anyone can access, has many nodes, and everyone can participate as a miner or validator. A consensus is costly and time-consuming because the distributed network is vast. Unlike public blockchains, private blockchains (like those developed by consortia) have very few participants, and users must obtain permission to participate in mining and validating

Table 1 IoT-based agriculture utilizing privacy-oriented blockchain-based solutions

Solution	Network model	Privacy-preserving model	Consensus algorithm	Countermeasures
[28]	IOT forwarding devices, Shelter modules, OrchApps, and controllers are all part of the network model	Controls that preserve privacy	N/A	Modules for OrchApp and Shelter
[29]	Local blockchain, local storage, and home mining represent the network model	Controls that preserve privacy	Smart home miner	Hashing using a generalized Diffie-Hellman key
[30]	There were six components in the network model: management hubs, blockchain networks, smart contracts, agent nodes, managers, and wireless sensor networks	Controls that preserve privacy	Proof-of-Work	Restrictions on access
[31]	The leader's device consists of the following components:	Controls that preserve privacy	Practical Byzantine fault tolerance	A secure hash algorithm based on elliptic curves for digital signatures
[32]	One device is designed as the master of the bubble and is composed of IoT devices	A reputation for trust and reliability	Proof-of-Work Proof-of-Stake	Cryptographic hash algorithm using elliptic curves
[33]	The network model comprises providers, distributors, retailers, processors, producers, and consumers	A traceable system	Proof-of-Work	Ethereum Sawtooth Hyperledger
[34]	An ecosystem that integrates data-driven IoT platforms, IoT data providers, and IoT devices	Privacy-preserving machine learning training Training machines with privacy in mind	Proof-of-Work	The hash value for the Pailler cryptosystem
[35]	Devices, fog, and clouds compose the network model	Controls that preserve privacy	Proof-of-Work	Cryptographic hashes with symmetrical and asymmetric properties

(continued)

Table 1 (continued)

Solution	Network model	Privacy-preserving model	Consensus algorithm	Countermeasures
[36]	This network model comprises three layers: a blockchain, an application, and a data layer	An organization's reputation and trustworthiness	Reputation-based adaptive block validation	Hash algorithm
[37]	Attribute authorities and IoT devices make up the network model	Controls that preserve privacy	Practical Byzantine fault tolerance	Cryptographic system based on elliptic curves Authentication and key agreement protocol based on identity hash algorithms-
[38]	The application, transport, and sensing layers are the network model's components	Data protection for blockchains	Practical Byzantine fault tolerance	Using double-chains for tamper-proofing Blind signature algorithms Hash algorithms
[39]	Four components comprise the system model: Virtual switches, SDN controllers, IP networks, and private clouds	Data protection for blockchains	N/A	Learning random subspaces based on nearest neighbors

transactions. With known identities, private blockchains require less blind trust and are faster and cheaper. Data privacy and confidentiality can be protected with this design.

Blockchain applications are also transforming due to smart contracts [42]. A digital contract is a self-enforcing and tamperproof contract defined by technical codes and automatically enforced and executed [43]. All participants within a network benefit from smart contracts by ensuring transparency and reducing contracting costs [44].

3.1 An Integration Model Between Blockchain and IoT

IoT (Internet of Things) is a key component of Industry 4.0 [45]. The IoT should rely more heavily on sensing devices, big data volumes, and various network patterns in Industry 5.0 [95]. IoT 2.0 is the name we need to give it [46]. Rather than focusing on devices and data, IoT 2.0 has moved to the world of actionable intelligence

Table 2 Blockchain-based consensus algorithms adapted to IoT-based agriculture

Consensus algorithm	Blockchain-based solution	Type	Performance	Limitations
Proof-of-Work (PoW)	Bitcoin	Consensus among competitors	DDoS attack resistance Defeat spam attacks with ease	A double-spend attack uses a lot of computational resources
Proof-of-Stake (PoS)	Peercoin Nxt	Consensus among competitors	Using less energy Attacking is more expensive	Easily influenced by a no-stakes problem
Delegated Proof-of-Stake (DPoS)	BitShares Sreemit	Consensus among competitors	Increasing node interaction efficiency Faster validation and more transactions	The limited number of delegates makes it vulnerable to centralization
Delegated Proof-of-Work (DPoW)	Komodo	Consensus among competitors	Security and resistance to attacks: 51% Based on a PoW or PoS blockchain	Easily hacked, eclipsed, and double-spent
Proof-of-Capacity (P0C)	Burstcoin	Consensus among competitors	Hardware is not dedicated Hard disk drives do not need to be constantly upgraded	Malware attacks are possible
Proof-of-Stake Velocity (PoSV)	Reddcoin	Consensus among competitors	Mining waste reduction Low energy or computing requirements	Vulnerable against attack 51% Invulnerable to 51% of attacks
Proof-of-Authority (PoA)	POA network VeChain	Consensus among competitors	Scalability should be provided Reducing energy consumption	Blockchains that are private and permissioned can use this technology
Proof-of-History (PoH)	Solana	Consensus among competitors	Streamlining data management Shadeless	Attacks involving double-spending are vulnerable
Proof-of-Activity (PoAC)	Decred	Consensus among competitors	Consumption of less energy The topology of the network has been improved	Reducing the incentive is necessary because it is difficult to reward

(continued)

Table 2 (continued)

Consensus algorithm	Blockchain-based solution	Type	Performance	Limitations
Proof-of-Weight (PoW)	Alorand	Consensus among competitors	Attacks by Sybil can be prevented Develop a scalable solution	The only nodes that have a chance of becoming miners are those willing to burn more money
Proof-of-Burn (PoB)	Slimcoin	Consensus among competitors	Attacks by Sybil can be prevented Scale the system	Public blockchain networks are not suitable for use
Proof-of-Elapsed Time (PoET)	Hyperledger sawtooth	Consensus among competitors	Costs associated with participation have been reduced	Problematic with nothing-at-all
Proof-of-Importance (PoI)	NEM	Consensus among competitors	Hardware is not required Reducing energy consumption	Participant costs are reduced
Stellar Consensus Protocol (SCP)	Stellar consensus	Consensus among competitors	A decentralized control system that is low latency and flexible in terms of trust	Due to the lack of block rewards, an incentive is required
Proof-of-Reputation (PoR)	GoChain	Consensus among competitors	Permission-based and private adaptations Reducing energy consumption	A private or permissioned blockchain can use this protocol
Delegated Byzantine Fault Tolerance (dBFT)	Neo	Consensus among competitors	Consumption of less energy Hardware that is not powerful enough to mine	Bookkeepers use real identities

[47]. Due to the integration of IoT 2.0 with other technologies such as AI, Cloud Computing, and Data Science, the digital transformation of IoT 2.0 moves into the Internet of Transformation [48]. As a result of blockchain technology integrating with the Internet of Things, various domains are predicted to undergo a digital transformation, which is predicted to revolutionize the world's economy [49]. Among the most promising applications of blockchain and IoT by 2030 will be global supply chain management and tracking. The IoT and Blockchain will have an even greater impact on smart grids and digital energy by 2030 [50]. In the following sections, you will find more details about blockchain design patterns for IoT, major blockchain solutions to IoT difficulties, and recent blockchain protocols relevant to IoT (see Fig. 3).

Fig. 3 Top five predictions for integrating blockchain and IoT by 2030

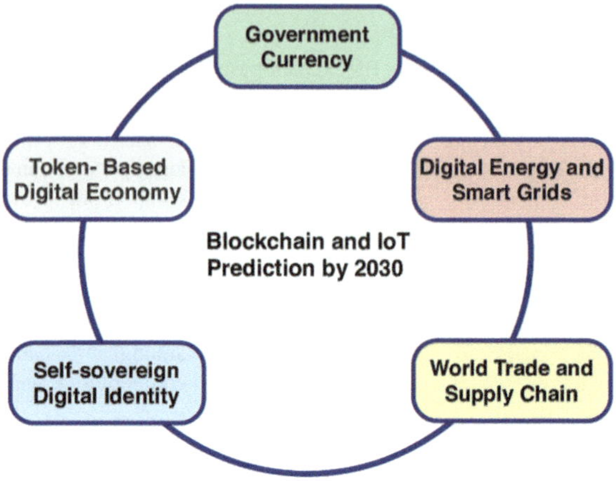

4 Security Challenges in IOT Smart Cities

There are many security concerns and challenges associated with IoT-based smart cities. The city's implementation and safety are adversely affected by some challenges. Table 3 outlines five cryptographic primitives preserved by blockchains (BC): authenticity, integrity, confidentiality, availability, and non-repudiation.

A. **Integrity**: An entity that does not alter data without authorization ensures its integrity. The integrity of the data is at risk. An attacker attempts to alter the data and configurations of IoT devices by gaining unauthorized access. The blockchain mechanism is used to manage the values associated with the entities and the access control rules on the system so that the system's trust level and information integrity are maintained. Another type of integrity attack is a forgery attack, a Trojan horse attack, a fabrication attack, or an attack on a biometric template. Provide a three-step framework for ensuring data integrity in smart cities. Firstly, the user must determine the subject (user) and object (task) parameters in the system. Secondly, it must enact logical and non-contradictional access policies. To conclude, the blockchain must

Table 3 Security features

Security issues	Proposed solution
Confidentiality	All transactions are encrypted with a symmetric key
Integrity	Hashes are used in all transactions
Availability	It ensures that all system functions are available to authorized users when they request them over the local or canopy network
Authentication	Access headers and shared keys are used to accomplish this
Non-repudiation	For non-repudiation, the transaction generator signs all local and canopy transactions. A transaction cannot be committed by both requester and requested without their collaboration

incorporate system-sensitive values, and smart contracts must incorporate access policies. [51] A framework for data integrity service has been proposed to create a reliable system that checks data integrity without third-party intervention. Moreover, [52] provided a crowdsourcing system, which distributes tasks to workers and receives them from requesters. The system has been tested on a decentralized network and has proven to be scalable.

B. **Availability**: Authentic users are ensured to have access to the system's services. As a result of this method, IoT equipment will become inoperable. Denial of Service attacks occur when attackers inject fake data into IoT software or flood a server with illegal data. Availability attacks include botnet attacks.

C. **Authentication**: The authentication process ensures that only authentic and authorized users can access data. Devices and nodes are accessed by hackers using identity forgery. When requesting to access a given ICT platform's data or functions, user and machine identities are authenticated by authenticating their attributes. Information about identities is stored and managed by Identity Management Systems (IdM) [53]. The authorization process refers to the decision to allow or deny the use of a resource according to its owners' access control policies. Authentication and Authorization Service (AAS) implements it. It is possible to represent such policies in various ways, such as using roles, attributes, and so on [54], and exchange protocols and data formats, such as the Security Assertion Markup Language (SAML).

D. **Confidentiality**: Even if an attacker intercepts data, he cannot understand it. Smart systems can be undermined by unregistered intruders using access points at the sensor layer to listen in on IoT-based devices and networks. This is a passive attack as long as the attacker doesn't cause any disruption. However, it can potentially escalate into advanced attacks, where the attacker manipulates the victim into making harmful decisions and committing harmful acts. Three techniques are employed when an adversary attacks a computer using identity-based attacks. The security of public blockchains depends heavily on data privacy. The nodes can view blockchain transactions. The blockchain network will, therefore, be protected by private or consortium blockchains that permit only certified organizations to participate.

E. **Privacy**: Information will only be disclosed if it has been permitted. Intercepting or tracing information about an individual's identity or behavior is never good. Attackers may capture and manipulate agro-perception devices to gain access to consumer information. Using public key cryptography in blockchain transactions helps maintain a high level of privacy by concealing the identity of users. Due to public access to cryptographic keys and transactions, blockchain cannot guarantee transactional anonymity. Tracking the transactions allows for recognizing the user.

F. **Non-repudiation**: Using it ensures that a message cannot be retrieved later. There is no way for a system element, even a storage server, to reject receiving data or controlling commands. Private keys require that the owner know/own the key and that the owner cannot contradict the signature, as the signature cannot confirm effectively that the owner did not sign the document.

Table 4 Open challenges and guidelines

Challenge	Cause	Guidelines
Privacy	A distributed and open technology like blockchain; (b) the leakage of personal data; (c) the 51% attack	In public blockchains, personal information should be stored at a minimum and kept anonymous
Legislation	(a) Technology like blockchain that is distributed and open; (b) data leakage; (c) 51% attacks	There is a tradeoff between the quality of services and privacy leakage in the context of new international mutual legislation (a);
Acceptance of blockchain technology	(a) A lack of user trust; (b) a complex operation; (c) a privacy concern	Several factors impact user trust: (a) new trust models, (b) user-friendly interfaces (ease of use), and (c) strong privacy protection
Robust blockchain architecture	(a) As a result of high costs; (b) as a result of high processing, storage, and communication demands; (c) as a result of the lack of scalability; (d) as a result of the complexity of operations	An architecture that relies on smarts, reliability, speed, scalability, and resource efficiency

5 Research Challenges

Using blockchain technology to secure smart cities presents many challenges (see Table 4).

6 Conclusion

Using IoT-based blockchain technology, we explored existing security and privacy solutions. This paper proposes a blockchain-based security framework as part of a smart city's data communication framework. Blockchain is resilient to many threats, which is its main advantage. The system possesses several unique features. These include increased reliability, improved fault tolerance, faster operation, and scalability. Integrating blockchain technology with devices in a smart city will allow communication securely in a distributed environment. Future work will focus on designing a system-level model to examine the interoperability and scalability of different smart city platforms.

References

1. Ahmed, N., De, D., Hussain, I.: Internet of Things (IoT) for smart precision agriculture and farming in rural areas. IEEE Internet Things J. **5**(6), 4890–4899 (2018)
2. Song, T., Li, R., Mei, B., Yu, J., Xing, X., Cheng, X.: A privacy preserving communication protocol for IoT applications in smart homes. IEEE Internet Things J. **4**(6), 1844–1852 (2017)
3. Ometov, A., Bezzateev, S.V., Kannisto, J., Harju, J., Andreev, S., Koucheryavy, Y.: Facilitating the delegation of use for private devices in the era of the internet of wearable things. IEEE Internet Things J. **4**(4), 843–854 (2016)
4. Rani, P., Sharma, R.: Intelligent transportation system performance analysis of indoor and outdoor internet of vehicle (iov) applications towards 5g. Tsinghua Sci. Technol. (2023)
5. Al-Hamadi, H., Chen, R.: Trust-based decision making for health IoT systems. IEEE Internet Things J. **4**(5), 1408–1419 (2017)
6. Rani, P., Sharma, R.: Intelligent transportation system for internet of vehicles based vehicular networks for smart cities. Comput. Electr. Eng. Electr. Eng. **105**, 108543 (2023). https://doi.org/10.1016/j.compeleceng.2022.108543
7. Chen, Y.-J., Wang, L.-C.: Privacy protection for internet of drones: a network coding approach. IEEE Internet Things J. **6**(2), 1719–1730 (2018)
8. Kumar, N., Rani, P., Kumar, V., Verma, P.K., Koundal, D.: TEEECH: three-tier extended energy efficient clustering hierarchy protocol for heterogeneous wireless sensor network. Expert Syst. Appl. **216**, 119448 (2023)
9. Augustine, D.P., Raj, P.: Blockchain and IoT security. Blockchain Technol. Appl., 51–64 (2020)
10. Ge, C., Liu, Z., Fang, L.: A blockchain based decentralized data security mechanism for the Internet of Things. J. Parallel Distrib. Comput. **141**, 1–9 (2020)
11. Hussain, N., Rani, P., Kumar, N., Chaudhary, M.G.: A deep comprehensive research architecture, characteristics, challenges, issues, and benefits of routing protocol for vehicular ad-hoc networks. Int. J. Distrib. Syst. Technol. IJDST **13**(8), 1–23 (2022)
12. Bhola, B., et al.: Quality-enabled decentralized dynamic IoT platform with scalable resources integration. IET Commun., cmu2.12514 (Oct 2022). https://doi.org/10.1049/cmu2.12514
13. Xu, M., Chen, X., Kou, G.: A systematic review of blockchain. Financ. Innov. **5**(1), 1–14 (2019)
14. Sheldon, M.D.: A primer for information technology general control considerations on a private and permissioned blockchain audit. Curr. Issues Audit. Issues Audit. **13**(1), A15–A29 (2019). https://doi.org/10.2308/ciia-52356
15. Khan, A.H., et al.: Blockchain and 6G: the future of secure and ubiquitous communication. IEEE Wirel. Commun.Wirel. Commun. **29**(1), 194–201 (2022). https://doi.org/10.1109/MWC.001.2100255
16. Rani, P., Verma, S., Yadav, S.P., Rai, B.K., Naruka, M.S., Kumar, D.: Simulation of the lightweight blockchain technique based on privacy and security for healthcare data for the cloud system. Int. J. E-Health Med. Commun. IJEHMC **13**(4), 1–15 (2022)
17. Jararweh, Y., Otoum, S., Al Ridhawi, I.: Trustworthy and sustainable smart city services at the edge. Sustain. Cities Soc. **62**, 102394 (2020)
18. BCC Research: Global Markets Reports and Industry Analysis. https://www.bccresearch.com/. Accessed 24 Sep 2023
19. Aloqaily, M., Bouachir, O., Al Ridhawi, I.: Blockchain and fl-based network resource management for interactive immersive services. In: 2021 IEEE Global Communications Conference (GLOBECOM), pp. 01–06. IEEE (2021)
20. Cha, S.-C., Hsu, T.-Y., Xiang, Y., Yeh, K.-H.: Privacy enhancing technologies in the Internet of Things: perspectives and challenges. IEEE Internet Things J. **6**(2), 2159–2187 (2018)
21. Ferrag, M.A., Maglaras, L.A., Janicke, H., Jiang, J., Shu, L.: Authentication protocols for internet of things: a comprehensive survey. Secur. Commun. Netw. **2017** (2017)
22. Brewster, C., Roussaki, I., Kalatzis, N., Doolin, K., Ellis, K.: IoT in agriculture: designing a Europe-wide large-scale pilot. IEEE Commun. Mag.Commun. Mag. **55**(9), 26–33 (2017)
23. Ray, P.P.: Internet of things for smart agriculture: Technologies, practices and future direction. J. Ambient Intell. Smart Environ. **9**(4), 395–420 (2017)

24. Khanna, A., Kaur, S.: Evolution of Internet of Things (IoT) and its significant impact in the field of Precision Agriculture. Comput. Electron. Agric. Electron. Agric. **157**, 218–231 (2019)
25. Ruan, J., et al.: A life cycle framework of green IoT-based agriculture and its finance, operation, and management issues. IEEE Commun. Mag.Commun. Mag. **57**(3), 90–96 (2019)
26. Ammar, M., Russello, G., Crispo, B.: Internet of Things: a survey on the security of IoT frameworks. J. Inf. Secur. Appl. **38**, 8–27 (2018). https://doi.org/10.1016/j.jisa.2017.11.002
27. Hassija, V., Chamola, V., Saxena, V., Jain, D., Goyal, P., Sikdar, B.: A Survey on IoT security: application areas, security threats, and solution architectures. IEEE Access **7**, 82721–82743 (2019). https://doi.org/10.1109/ACCESS.2019.2924045
28. Sharma, P.K., Singh, S., Jeong, Y.-S., Park, J.H.: Distblocknet: a distributed blockchains-based secure sdn architecture for iot networks. IEEE Commun. Mag.Commun. Mag. **55**(9), 78–85 (2017)
29. Dorri, A., Kanhere, S.S., Jurdak, R., Gauravaram, P.: Blockchain for IoT security and privacy: the case study of a smart home. In: 2017 IEEE international conference on pervasive computing and communications workshops (PerCom workshops), , pp. 618–623. IEEE (2017)
30. Novo, O.: Blockchain meets IoT: an architecture for scalable access management in IoT. IEEE Internet Things J. **5**(2), 1184–1195 (2018)
31. Zhou, L., Wang, L., Sun, Y., Lv, P.: Beekeeper: a blockchain-based iot system with secure storage and homomorphic computation. IEEE Access **6**, 43472–43488 (2018)
32. Hammi, M.T., Hammi, B., Bellot, P., Serhrouchni, A.: Bubbles of Trust: a decentralized blockchain-based authentication system for IoT. Comput. Secur. Secur. **78**, 126–142 (2018). https://doi.org/10.1016/j.cose.2018.06.004
33. Caro, M.P., Ali, M.S., Vecchio, M., Giaffreda, R.: Blockchain-based traceability in Agri-Food supply chain management: a practical implementation. In: 2018 IoT Vertical and Topical Summit on Agriculture-Tuscany (IOT Tuscany), pp. 1–4. IEEE (2018)
34. Shen, M., Tang, X., Zhu, L., Du, X., Guizani, M.: Privacy-preserving support vector machine training over blockchain-based encrypted IoT data in smart cities. IEEE Internet Things J. **6**(5), 7702–7712 (2019)
35. Ma, M., Shi, G., Li, F.: Privacy-oriented blockchain-based distributed key management architecture for hierarchical access control in the IoT scenario. IEEE Access **7**, 34045–34059 (2019)
36. Dedeoglu, V., Jurdak, R., Putra, G.D., Dorri, A., Kanhere, S.S.: A trust architecture for blockchain in IoT. In: Proceedings of the 16th EAI International Conference on Mobile and Ubiquitous Systems: Computing, Networking and Services, pp. 190–199. ACM, Houston Texas USA (Nov 2019). https://doi.org/10.1145/3360774.3360822
37. Ding, S., Cao, J., Li, C., Fan, K., Li, H.: A novel attribute-based access control scheme using blockchain for IoT. IEEE Access **7**, 38431–38441 (2019)
38. Si, H., Sun, C., Li, Y., Qiao, H., Shi, L.: IoT information sharing security mechanism based on blockchain technology. Future Gener. Comput. Syst. **101**, 1028–1040 (2019). https://doi.org/10.1016/j.future.2019.07.036
39. Derhab, A., et al.: Blockchain and random subspace learning-based IDS for SDN-enabled industrial IoT security. Sensors **19**(14), 3119 (2019). https://doi.org/10.3390/s19143119
40. Casino, F., Dasaklis, T.K., Patsakis, C.: A systematic literature review of blockchain-based applications: Current status, classification and open issues. Telemat. Inform. **36**, 55–81 (2019). https://doi.org/10.1016/j.tele.2018.11.006
41. Angelis, J., Da Silva, E.R.: Blockchain adoption: a value driver perspective. Bus. Horiz.Horiz. **62**(3), 307–314 (2019)
42. Iansiti, M., Lakhani, K.R., Mohamed, H.: It will take years to transform business, but the journey begins now. Harv. Bus. Rev. Bus. Rev. **95**(1), 118–128 (2017)
43. Cong, L.W., He, Z.: Blockchain disruption and smart contracts. Rev. Financ. Stud. **32**(5), 1754–1797 (2019). https://doi.org/10.1093/rfs/hhz007
44. Moll, J., Yigitbasioglu, O.: The role of internet-related technologies in shaping the work of accountants: new directions for accounting research. Br. Account. Rev. **51**(6), 100833 (2019)

45. Lu, Y.: Industry 4.0: A survey on technologies, applications and open research issues. J. Ind. Inf. Integr.Integr. **6**, 1–10 (2017)
46. Sheth, A.: Internet of things to smart iot through semantic, cognitive, and perceptual computing. IEEE Intell. Syst.Intell. Syst. **31**(2), 108–112 (2016)
47. "Unstructured data: turning data into actionable intelligence," i-SCOOP. https://www.i-scoop.eu/big-data-action-value-context/unstructured-data/. Accessed 24 Sep 2023
48. Darwish, A., Hassanien, A.E., Elhoseny, M., Sangaiah, A.K., Muhammad, K.: The impact of the hybrid platform of internet of things and cloud computing on healthcare systems: opportunities, challenges, and open problems. J. Ambient. Intell. Humaniz. Comput.Intell. Humaniz. Comput. **10**, 4151–4166 (2019)
49. Makhdoom, I., Abolhasan, M., Abbas, H., Ni, W.: Blockchain's adoption in IoT: the challenges, and a way forward. J. Netw. Comput. Appl.Netw. Comput. Appl. **125**, 251–279 (2019)
50. Yang, T., et al.: Applying blockchain technology to decentralized operation in future energy internet. In: 2017 IEEE Conference on Energy Internet and Energy System Integration (EI2), pp. 1–5. IEEE (2017)
51. Liu, B., Yu, X.L., Chen, S., Xu, X., Zhu, L.: Blockchain based data integrity service framework for IoT data. In: 2017 IEEE International Conference on Web Services (ICWS), pp. 468–475. IEEE (2017)
52. Li, M., et al.: CrowdBC: a blockchain-based decentralized framework for crowdsourcing. IEEE Trans. Parallel Distrib. Syst.Distrib. Syst. **30**(6), 1251–1266 (2018)
53. Miyata, T., et al.: A survey on identity management protocols and standards. IEICE Trans. Inf. Syst. **89**(1), 112–123 (2006)
54. Tourani, R., Misra, S., Mick, T., Panwar, G.: Security, privacy, and access control in information-centric networking: a survey. IEEE Commun. Surv. Tutor. **20**(1), 566–600 (2017)

Analysis Reliable and Link Stability-Based Energy-Aware Multipath Routing for Mobile Ad Hoc Networks Under Random Waypoint Mobility

Mohammed Ayad Alkhafaji, Muhammet Tahir Güneşer, and Cihat Şeker

Abstract It is possible to set up and launch mobile ad hoc networks (MANETs) quickly and easily. Routing protocols that are reliable and energy-efficient pose one of the biggest challenges in these networks. It is necessary to implement a multicast routing protocol that is efficient and reliable in MANETs due to their limited resources. An energy-aware multipath routing protocol is proposed in this paper based on link stability and reliability. From the source node to a group of destination nodes, the proposed scheme performs better than ODMRP and finds energy-efficient multicast routes. Compared with ODMRP, EODMRP, and LSMRP routing algorithms, the proposed scheme outperforms them.

Keywords MANET · ODMRP · EODMRP · LSMRP · Internet of Things (IoT)

1 Introduction

A proliferation of mobile devices has led to an exponential growth in MANETs. There is no centralized administration, fixed base station, or centralized system of manually exchanging data between these wireless networks. Using MANETs in rescue, emergency operations, and battlefield communication is very straightforward because of their self-organizing features. However, MANETs modify their topology randomly due to mobility and self-organization. Data transmission is typically done by neighboring mobile nodes when the mobile node's transmission range is limited. Thus,

M. A. Alkhafaji (✉)
College of Technical Engineering, National University of Science and Technology, DhiQar, Nasiriyah 64001, Iraq
e-mail: mohammed.alkhafaji@nust.edu.iq

M. T. Güneşer
Electronics and Communication Engineering Department, Faculty of Electrical and Electronics, Istanbul Technical University, Istanbul, Turkey

C. Şeker
Electric Electronic Department, Collage of Engineering, Karabuk University, Karabuk, Turkey

S. Kumar Udgata et al. (eds.), *Intelligent Systems*, Lecture Notes in Networks and Systems 1149, https://doi.org/10.1007/978-981-97-8160-7_25

MANET performance is largely determined by node-to-node routing reliability [1]. Due to rapid advancements in wireless communication technologies, the Internet of Things (IoT) is gaining momentum for a variety of reasons, including for business purposes, for smart homes, and for smart cities [2–6]. There are millions of physical devices on the Internet of Things that communicate, see, and hear each other. In the era of ad hoc networks, embedded devices, internet protocols, and sensor networks, physical devices are becoming more 'smart' by utilizing a plethora of ad hoc networks, embedded devices, and internet protocols. Physical, software, network, and encryption attacks are among the many security threats IoT systems face [7]. In IoT, data integrity and security provision have become increasingly crucial [8, 9]. A MANET has no central authority or infrastructure [10, 11]. Generally, MANETs rely on unreliable connections to establish communication between nodes.

Because mobile nodes are free to move between networks, new nodes can join or leave MANETs anytime. For various vision-based applications, MANETs offer topological flexibility, robustness, fault tolerance, self-healing, and rapid deployment among their advantages [12, 13]. Node-to-node packet exchanges within MANETs are governed by routing protocols, which define the parameters involved [12, 14]. MANET challenges include dynamic topology changes, limited bandwidth, node mobility-related link failures, low power consumption on mobile nodes, and routing computation-induced power consumption [15]. Various routing protocol algorithms have been developed to address these challenges.

1.1 Objective and Contributions

In real-time communications scenarios, reactive routing protocols are more suitable since they can maintain a reasonable convergence time. No fixed infrastructure or centralized ancillaries are required to coordinate and control MANETs. To establish a connection between end users, mobile nodes act as both hosts and routers. Depending on the status of intermediate nodes, a route should be determined between mobile nodes in MANETs. Using MANETs, reliable and stable routes can be established between sources and destinations. Despite continued improvements in channel link quality, mobile nodes are still constrained by energy constraints, mobility, and channel noise. Among these challenges are the following:

2 Background and Related Work

Links, paths, and other methods for predicting mobility are described in this section. Further, this section discusses selecting stable links to achieve optimal PDR and minimum E2E delay. The article also discusses link estimation methods and existing mobility protocols. The lifespan of nodes and links is assessed through the energy depletion rate, mobility evaluation, and the lifespan of nodes and links [16, 17].

The recovery of DSR competence also depends on two processes. A node in an adjacent neighborhood and a node in a path of loss are the two primary mobility constraints [18]. Signal-to-noise ratio is used to calculate link losses. Stable paths between source and destination were constructed using a trustworthy node based on BEB computation. A relative and local stability judgment was shown to be important when it came to multicast routing [19]. The strength of the node can be estimated based on two decision factors. An effective multicast tree with a spontaneous method will be formed by choosing a stable path. Link Stability creates a QoS routing protocol by combining the mobility degree between different nodes [20]. With Long Life Time Multicast Routing Protocols, route discovery mechanisms will be modified during a network failure to maintain a constant route map [21]. There has been reduced control overhead, delays, and an extension of route lifetimes [22].

Wireless technologies have become more accessible thanks to recent developments. The Internet of Things (IoT) [23], Cognitive Radio (CR) [24], Massive Multiple Input Multiple Output (MIMO) [25] Fog Computing [26], MANET [27, 28], Device-to-Device Communication (D2D) [29], Coordinated Multipoint (CoMP) [30], and Vehicular Ad hoc Networks (VANET) [31] are some of these technologies. Several existing technologies will be incorporated into 5G, including power optimization mechanisms and scheduling algorithms that will meet the needs of users [21, 32–34]. A swift investigation is required to determine the overall extent of the damage and provide the rescue team with an estimate of how many people need to evacuate during an emergency. This process requires instantaneous communication, so MANETs can easily deliver this kind of instant communication [35]. MANETs are an innovative form of wireless networking that is able to generate itself and self-organize without a centralized infrastructure [36–38]. As a result, they are widely used in disaster recovery and rescue operations in various situations to establish emergency communications.

The authors [39] describe a multicast scheduling protocol based on scheduled links. A load-balanced approach considers dependability and quality of service requests when planning SLMRP paths. The best path for each origin-target pair is determined by several loop-free and disjoint paths. Path Timeout Timer (PTT) and Multicast Routing Activation Timer (MRAT) monitor whether these paths are being activated or deactivated. An alternative route discovery protocol based on localization was proposed in [40] when a network connection fails. In telecommunications, the EOD-LLR protocol is known as EOD-LLR (Efficient on-demand link failure multicast local recovery routing protocol). Control packets can be delayed and overhead reduced using the proposed method, resulting in an increase in throughput. Various multicast routing protocols are efficient, including ODMRP (On-Demand Multicast Routing Protocol) [41]. Multicast paths between a multicast origin and multicast target are created with nodes with maximum Cost Stability Link (LSF) functionality.

3 Proposed Methodology

The underlying concepts will be used to validate our simulation results in this mathematical model. An N-node mobile network is composed of N mobile nodes connected by a set of links (those that are within the range of communication). There are graph $G(V, E)$ with $V = \{v_1, v_2, v_3, ..., v_n\}$ nodes and $E = \{e_1, e_2, e_3, ..., e_m\}$, edges connecting the network's vertices, which looks like a graph problem. Communication between two nodes is possible if they are in range, there is an edge or a link between them. The route R_i of a multicast MANET There are a number of j nodes ($j \leq n$) and k edges ($k \leq m$) for sources and multicast group destinations (MGD). In this example, Edges e_x are intermediate links between sources and destinations along a specific route, R_i. The link e_x in selected route R_i may fail due to mobility or unstable and unreliable links.

Stable links are used in this estimate to overcome the problem of link failures. Our approach does not require global positioning systems (GPS) or location-based methodologies to predict node movements. An estimation of stable links is made as part of the mobility prediction process by estimating the maximum amount of time that the requesting node will spend in the communication range of the computing node.

3.1 *Mathematical Proof of the Concept*

Vertices v_a in the set V are considered trustworthy nodes for sending data packets or for determining routes for destinations. v_a^{CR} is the communication range for the node v_a. In the network, let's take a look at the node v_b. Node v_b is moving into the range of communication of node v_a from anywhere in the network. In addition to the threshold communication range (TCR) (where $v_a^{\mathrm{TCR}} \leq v_a^{CR}$) we can also define a communication range for node VA (where VATCRVACR) that can be used to predict the movement of a node v_a.

Using this approach, we can determine the maximum traveling time between two nodes within their communication range. This allows us to predict the optimal route based on different route requirements between source and destination. As node v_b moves toward node v_a with speed V, it moves towards its communication range. Taking the chord AB and the node v_a as an example, we can calculate the total travel time.v_b travel time is represented by Eq. (3).

$$t^{\mathrm{TCR}} \leftarrow \frac{\text{distance}\,(D)}{\text{speed}\,(V)} \tag{1}$$

Three scenarios are presented in this network, Node travel time is calculated as well v_b, we will calculate the total travel time in the communication range of node v_a.

Figure 1 illustrates the first case where a node enters a v_a^{CR} communication range but does not enter a v_a^{TCR} communication range. Nodes of this kind cannot be predicted (estimated) in advance. An evaluation of the minimum amount of time (≥0) a task takes to reach the threshold communication range can be made using a minimum amount of time (t_{TCR}). Despite having a minimum link activation time, we cannot use this information to stabilize our route since we cannot predict it.

According to Fig. 2, Node v_b enters the communication range of node v_a^{CR}. just as it touches the boundary of the threshold communication range v_a^{TCR}. In order for multimedia traffic to travel in a reliable manner, it is necessary to have a minimum link activation time. Unless we are unable to find another link with an upper activation time, we can accept this link. A total distance is shown in Eq. 2. On chord *AB* created by node v_b. We have placed a perpendicular L to calculate the total distance. Mathematically, the chord is divided into two equal parts of a perpendicular line drawn from the circle's centre [42, 43].

In other words, *BL* and *AL* add up to a total distance equal to $2 * AL$.

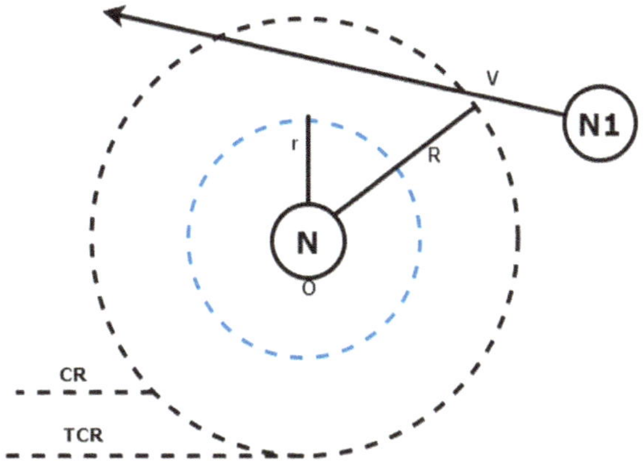

Fig. 1 Mobility prediction model outside of communication range

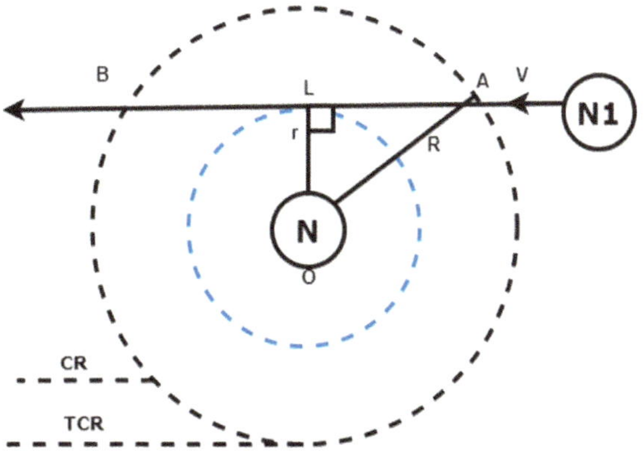

Fig. 2 TCR boundary just touched by node inside

$$D \leftarrow AB \leftarrow AL + BL \tag{2}$$

$$D \leftarrow 2 * AL \because AL = BL \tag{3}$$

Our next step is to calculate *AL*, the base of a right-angled triangle. In this case, using the perpendicular on the chord *AB* leads to a right-angled triangle. Pythagoras' theorem has been used to calculate *AL*.

$$(OL)^2 - (AO)^2 \rightarrow (AL)^2 \tag{4}$$

Circle radii are represented by the other edges of the triangle. Perpendicular lines (*OL*) define a small circle's radius, and hypotenuse lines (*AO*) define a large circle's radius.

$$(AL)^2 \leftarrow R^2 - r^2 \tag{5}$$

Now you can calculate the total traveling time in a node's communication range V_a^{CR} based on the distance (D) between the node and the home node. During a communication range, this time is required for the data transmission to be completed.

According to Fig. 3, Node v_b enters the communication ranges of nodes v_a^{CR} and v_a^{TCR}, as shown in the third case. The minimum existence time is calculated in the last case after crossing the threshold. When the node is alive, we can calculate when the node is in communication range. The maximum time a link has existed can, therefore, be determined.

Assume node v_a is in communication range with node v_b as *AB*. Total distance travelled by node v_b. The points *A* and *B* represent the points at which node v_b enters and exits the node v_a communication range, respectively.

$$AB = AA' + A'L + LB' + B'B \tag{6}$$

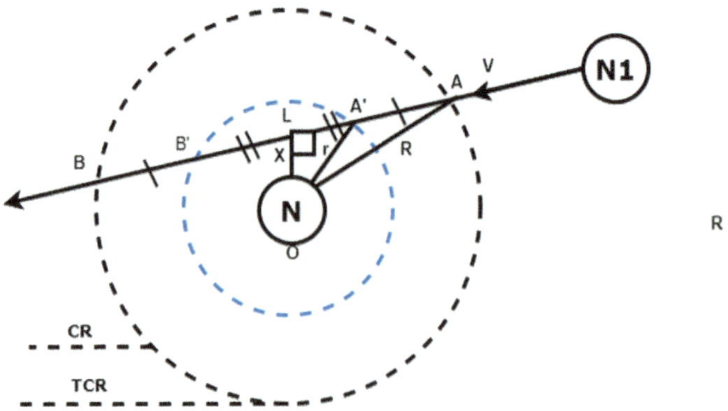

Fig. 3 Node coming inside communication range

Our distance calculation consists of four points: $AA^{'}$, $A^{'}L$, $LB^{'}$, $B^{'}B$. To determine node v_b total travel time, we must calculate all parts of AB. The inner circle is intersected by A' and B' when node v_b moves toward node v_a. Right-angled triangles can be calculated by projecting perpendicular L onto chord A. A chord divided into two equal half by OL is always divided into two halves by OL due to its perpendicularity to $A^{'}B^{'}$. In this case, $A^{'}L$ and LB' are equivalent.

$$\therefore A'L = LB'$$

Additionally, OL bisects AB since it is perpendicular to it. A common chord on concentric circles can be seen to have the same length for two small line segments $AA^{'}$ and BB'.

$$\therefore AA' = BB'$$

We can reduce Eq. 6 by using these two statements. The exact distance of chord AB can now be calculated with only two terms.

$$\therefore AB = 2\left(AA' + A'L\right) \tag{7}$$

$A^{'}LO$ and $\measuredangle ALO$ are the right-angled triangles formed after projecting perpendicular. A triangle's edges can now be calculated. Pythagoras' theorem is applied to the first triangle to arrive at Eq. 8:

$$\left(OA'\right)^2 - (OL)^2 = \left(A'L\right)^2 \tag{8}$$

A reliable data transmission requires a minimum amount of time between the source and destination, assumed as 'r' between node v_a and a specific point. Our temporary OL parameter is x.

$$\left(A'L\right)^2 = r^2 - x^2 \tag{9}$$

$\measuredangle ALO$ triangle can be found by applying Pythagoras' theorem as follows:

$$(AL)^2 = \left(OA'\right)^2 - (OL)^2 \tag{10}$$

Due to the fact that we have already calculated the maximum range of communication, R can be substituted for $OA^{'}$ in Eq. 10.

$$(AL)^2 = R^2 - x^2 \tag{11}$$

AA' and $A'L$ are added together to form AL. Therefore, we can break Eq. 11 using Eq. 12

$$\because AA' + A'L = (AL) \tag{12}$$

$$\therefore (A'L + AA')^2 = R^2 - x^2 \tag{13}$$

We can estimate $A'L$ as follows using Eqs. 9 and 13:

$$A'L = \frac{R^2 - r^2 - (AA')^2}{2 * (AA')} \tag{14}$$

Because we already know R^2 and r^2, respectively, we can calculate $A'L$ in Eq. 14. The length of AA'' must be computed. Our node v_b can be configured with a timer that calculates the total elapsed time (t_e) for reaching the communicative range of node v_a.

$$AA'' = t_e * v \tag{15}$$

Equation 15 Based on speed V and time t_e, it is possible to compute AA' distance. Using speed (V) and t_e time, Eq. 13 can be used to calculate AA' distance.

4 Experimental Setup and Simulation Parameter

ODMRP, EODMRP, LSMRP, and proposed model packet delivery ratios are illustrated in Fig. 4. There is an increment step of 2 km/h in the mobility of nodes between 2 and 18 km/h. A decrease in PDR has been observed with increased node mobility in both protocols. Nodes move randomly and rapidly, which makes links more likely to break.

Our proposed protocol and ODMRP illustrate how end-to-end delay varies as node mobility increases. Our proposed protocol offers a lower end-to-end delay than ODMRP, as evident from the results. As node mobility increases, the protocol ignores issues related to link stability, leading to more chances of link failure. Whenever a link fails, it increases the probability of data packets being retransmitted since alternate routes are switched over. When using ODMRP, the delay is also affected by the time elapsed during route reconfiguration and data retransmission. Retransmission and delay are also problems faced by EODMRP and LSMRP. It tries to reduce some link failures by selecting stable links, so LSMRP protocol delay is less than ODMRP and EODMRP (see Fig. 5).

As shown in Fig. 6, the multicast routing protocol provides a very low prediction overhead. During prediction, a node's total time travel is estimated from the communications range of the estimating node and is an additional overhead. During prediction, a node's total time travel is estimated from the communications range of the estimating node and is an additional overhead. When evaluating prediction overhead, the ODMRP protocol is not taken into account. ODMRP is not as effective

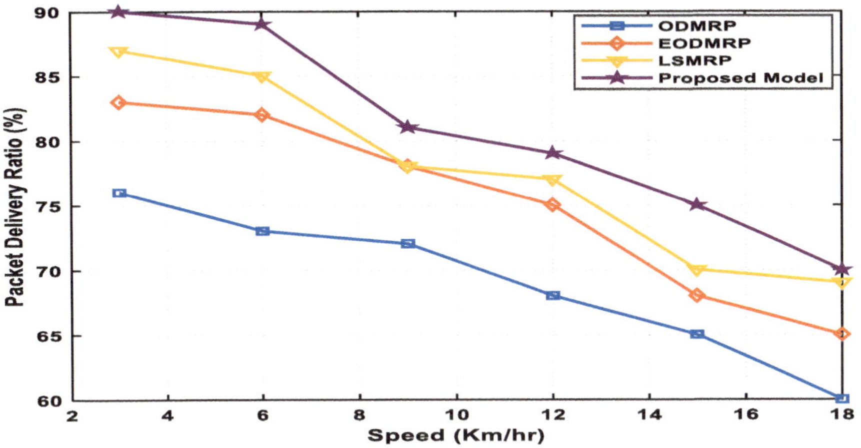

Fig. 4 Packet delivery ratio PDR (%) versus mobility speed (km/h)

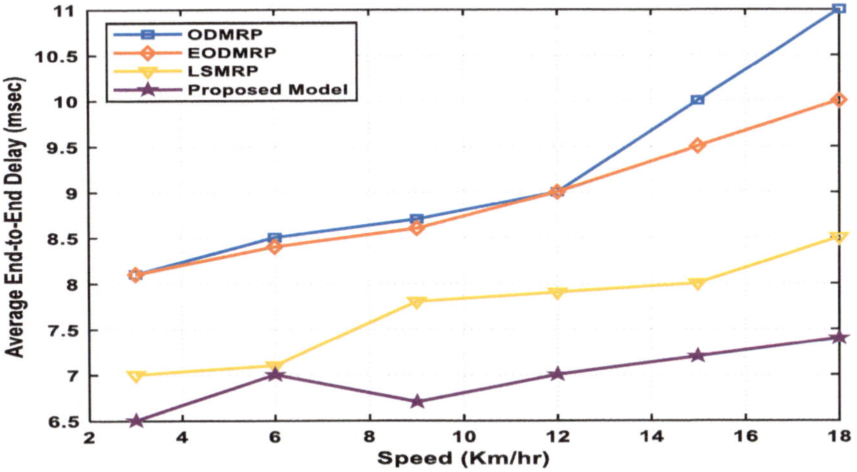

Fig. 5 Average end-to-end delay (msec) versus mobility speed (km/h)

as the other programs in this comparison. The ODMRP protocol has a very low overhead. Since LSMRP does not send any extra packets for route establishment, it gives less routing overhead than all other protocols. With regard to network overhead, the proposed model is more efficient than the existing model.

Data rates are shown in Fig. 7 according to packet delivery ratio (PDR). A simulation environment with various data rates was tested for streaming video data. Results are shown on 128 kbps, 512 kbps, and 1024 kbps. Comparing the proposed protocol to existing ones, the proposed protocol shows a higher PDR. Our simulator supports

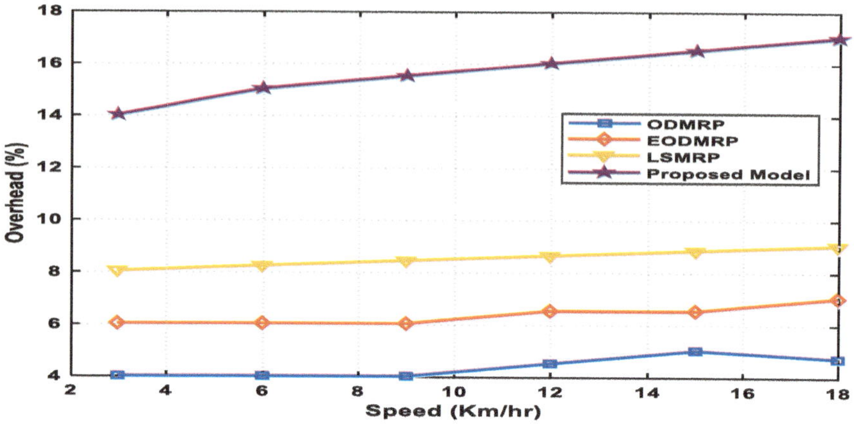

Fig. 6 Overhead (%) versus mobility speed (km/h)

video data communication at a maximum data rate according to the measurement shown in this figure.

In Fig. 8, various routing protocols are compared on different data rates to show their end-to-end delays. There is already a discussion of the reason for the ups and downs in the performance metric section. Using this data rate, video data will be transmitted with a lower end-to-end delay and a higher proportional delay reduction. Having a data rate of (128 × 512) kbps ensures flawless communication because there is no congestion. Our protocol has been tested in comparison with the results obtained with minimum and maximum contention counts. A node with the lowest contention count was selected as the forwarding node in the first case, whereas a node with the highest contention count was selected in the second case.

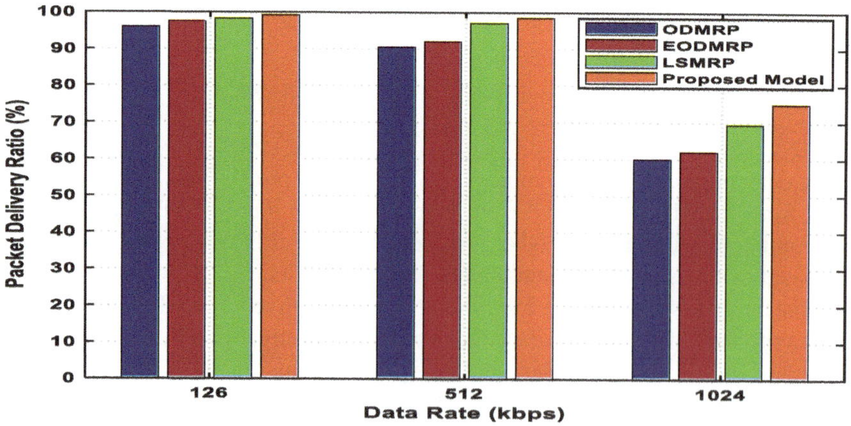

Fig. 7 Packet delivery ratio (%) versus data rate (kbps)

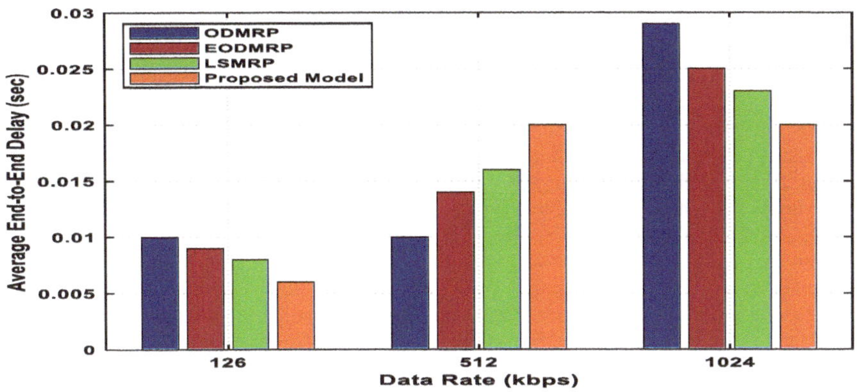

Fig. 8 Average end-to-end delay (s) versus data rate (kbps)

5 Conclusion

The current state of mesh-based multi-path routing is discussed using the example of edge-based ad hoc networks. As a result of our earlier discussions, choosing the right QoS metric for each domain is important, especially in multi-path routing. The "goodness" of a routing solution can be assessed using a suitable QoS metric. The proposed model is designed with the help of the Quality of Service and routing modifications with multi-path routing. Protocols are classified by how they are modified to improve packet delivery, packet loss, control overhead, and end-to-end delay, according to the paper. Despite their advantages and limitations, existing multi-path routing systems are critically reviewed. There have also been discussions about multicast routing protocols in MANETs.

References

1. Rani, P., Sharma, R.: IMFOCA-IOV: intelligent moth flame optimization based clustering algorithm for internet of vehicle. In: 2023 14th International Conference on Computing Communication and Networking Technologies (ICCCNT), Delhi, India, pp. 1–6. IEEE (Jul 2023). https://doi.org/10.1109/ICCCNT56998.2023.10307646
2. Al-Fuqaha, A., Guizani, M., Mohammadi, M., Aledhari, M., Ayyash, M.: Internet of things: a survey on enabling technologies, protocols, and applications. IEEE Commun. Surv. Tutor. **17**(4), 2347–2376 (2015)
3. Ray, P.P.: A survey on Internet of Things architectures. J. King Saud Univ.-Comput. Inf. Sci. **30**(3), 291–319 (2018)
4. Lin, J., Yu, W., Zhang, N., Yang, X., Zhang, H., Zhao, W.: A survey on internet of things: architecture, enabling technologies, security and privacy, and applications. IEEE Internet Things J. **4**(5), 1125–1142 (2017)
5. Rani, P., Sharma, R.: Intelligent transportation system for internet of vehicles based vehicular networks for smart cities. Comput. Electr. Eng. **105**, 108543 (2023)

6. Rani, P., Sharma, R.: An experimental study of IEEE 802.11n devices for vehicular networks with various propagation loss models. In: Dubey, A.K., Sugumaran, V., Chong, P.H.J. (eds.) Advanced IoT Sensors, Networks and Systems. Lecture Notes in Electrical Engineering, vol. 1027, pp. 125–135. Springer Nature Singapore, Singapore (2023). https://doi.org/10.1007/978-981-99-1312-1_11
7. Yang, Y., Wu, L., Yin, G., Li, L., Zhao, H.: A survey on security and privacy issues in Internet-of-Things. IEEE Internet Things J. **4**(5), 1250–1258 (2017)
8. Mosenia, A., Jha, N.K.: A comprehensive study of security of internet-of-things. IEEE Trans. Emerg. Top. Comput. **5**(4), 586–602 (2016)
9. Bhola, B., et al.: Quality-enabled decentralized dynamic IoT platform with scalable resources integration. IET Commun., cmu2.12514 (Oct 2022). https://doi.org/10.1049/cmu2.12514
10. Leite, J.E., Ursini, E.L., Martins, P.S.: Simulation of AdHoc networks including clustering and mobility. In: Ad-hoc, Mobile, and Wireless Networks: 16th International Conference on Ad Hoc Networks and Wireless, ADHOC-NOW 2017, Messina, Italy, September 20–22, 2017, Proceedings 16, pp. 199–209. Springer (2017)
11. Vinoth Kumar, V., Ramamoorthy, S.: Secure adhoc on-demand multipath distance vector routing in MANET. In: Proceedings of the International Conference on Computing and Communication Systems: I3CS 2016, NEHU, Shillong, India, pp. 49–63. Springer (2018)
12. Inamdar, S.R., Yadahalli, R.M.: Paradigm shift in routing approaches for high speed MANET applications. Eur. J. Eng. Technol. Res. **2**(1), 59–64 (2017)
13. Vijayalakshmi, D., ECE, G., Rao, D.S., ECE, J.: Efficient multicast routing protocol for manets. Int. J. Comput. Netw. Wirel. Commun. IJCNWC **6**, 56–61 (2016)
14. Hussain, N., Rani, P., Kumar, N., Chaudhary, M.G.: A deep comprehensive research architecture, characteristics, challenges, issues, and benefits of routing protocol for vehicular ad-hoc networks. Int. J. Distrib. Syst. Technol. IJDST **13**(8), 1–23 (2022)
15. Kumar, N., Rani, P., Kumar, V., Athawale, S.V., Koundal, D.: THWSN: enhanced energy-efficient clustering approach for three-tier heterogeneous wireless sensor networks. IEEE Sens. J. **22**(20), 20053–20062 (2022)
16. De Rango, F., Guerriero, F., Fazio, P.: Link-stability and energy aware routing protocol in distributed wireless networks. IEEE Trans. Parallel Distrib. Syst. **23**(4), 713–726 (2010)
17. Kumar, N., Rani, P., Kumar, V., Verma, P.K., Koundal, D.: TEEECH: three-tier extended energy efficient clustering hierarchy protocol for heterogeneous wireless sensor network. Expert Syst. Appl. **216**, 119448 (2023)
18. Abbas, N.I., Ilkan, M., Ozen, E.: Fuzzy approach to improving route stability of the AODV routing protocol. EURASIP J. Wirel. Commun. Netw. **2015**, 1–11 (2015)
19. Xia, H., Xia, S., Yu, J., Jia, Z., Sha, E.H.-M.: Applying link stability estimation mechanism to multicast routing in MANETs. J. Syst. Archit. **60**(5), 467–480 (2014)
20. Moussaoui, A., Semchedine, F., Boukerram, A.: A link-state QoS routing protocol based on link stability for Mobile Ad hoc Networks. J. Netw. Comput. Appl. **39**, 117–125 (2014)
21. Al-Hemyari, A., Ismail, M., Hassan, R., Saeed, S.: Improving link stability of multicasting routing protocol in MANETs. J. Theor. Appl. Inf. Technol. **55**(1) (2013)
22. Sridhar, K.N., Chan, M.C.: Stability and hop-count based approach for route computation in MANET. In: Proceedings. 14th International Conference on Computer Communications and Networks, 2005. ICCCN 2005, pp. 25–31. IEEE (2005)
23. Shikh-Bahaei, M., Choi, Y.-S., Hong, D.: Full-duplex and cognitive radio networking for the emerging 5G systems **2018** (2018). Hindawi
24. Elijah, O., Rahman, T.A., Orikumhi, I., Leow, C.Y., Hindia, M.N.: An overview of Internet of Things (IoT) and data analytics in agriculture: Benefits and challenges. IEEE Internet Things J. **5**(5), 3758–3773 (2018)
25. Zhang, J., Dai, L., Li, X., Liu, Y., Hanzo, L.: On low-resolution ADCs in practical 5G millimeter-wave massive MIMO systems. IEEE Commun. Mag. **56**(7), 205–211 (2018)
26. Peng, M., Yan, S., Zhang, K., Wang, C.: Fog-computing-based radio access networks: Issues and challenges. Ieee Netw. **30**(4), 46–53 (2016)

27. Conti, M., Giordano, S.: Mobile ad hoc networking: milestones, challenges, and new research directions. IEEE Commun. Mag. **52**(1), 85–96 (2014)
28. Tilwari, V., Dimyati, K., Hindia, M.N., Fattouh, A., Amiri, I.S.: Mobility, residual energy, and link quality aware multipath routing in MANETs with Q-learning algorithm. Appl. Sci. **9**(8), 1582 (2019)
29. Tehrani, M.N., Uysal, M., Yanikomeroglu, H.: Device-to-device communication in 5G cellular networks: challenges, solutions, and future directions. IEEE Commun. Mag. **52**(5), 86–92 (2014)
30. Qamar, F., Dimyati, K.B., Hindia, M.N., Noordin, K.A.B., Al-Samman, A.M.: A comprehensive review on coordinated multi-point operation for LTE-A. Comput. Netw. **123**, 19–37 (2017)
31. Lu, Z., Qu, G., Liu, Z.: A survey on recent advances in vehicular network security, trust, and privacy. IEEE Trans. Intell. Transp. Syst. **20**(2), 760–776 (2018)
32. Qamar, F., et al.: Investigation of future 5G-IoT millimeter-wave network performance at 38 GHz for urban microcell outdoor environment. Electronics **8**(5), 495 (2019)
33. Qamar, F., Dimyati, K., Hindia, M.N., Noordin, K.A., Amiri, I.S.: A stochastically geometrical poisson point process approach for the future 5G D2D enabled cooperative cellular network. IEEE Access **7**, 60465–60485 (2019)
34. Qamar, F., Hindia, M.N., Dimyati, K., Noordin, K.A., Amiri, I.S.: Interference management issues for the future 5G network: a review. Telecommun. Syst. **71**, 627–643 (2019)
35. Rosati, S., Krużelecki, K., Heitz, G., Floreano, D., Rimoldi, B.: Dynamic routing for flying ad hoc networks. IEEE Trans. Veh. Technol. **65**(3), 1690–1700 (2015)
36. Maheswar, R., Jayarajan, P., Vigneswaran, D., Udaiyakumar, R., Theepak, C.G., Amiri, I.S.: VSMART—A simulation tool for performance analysis of wireless sensor node using queue threshold. In: 2018 International Conference on Communication and Signal Processing (ICCSP), pp. 0234–0237. IEEE (2018)
37. Udaiyakumar, R., Joseph, S., Sundararajan, T.V.P., Vigneswaran, D., Maheswar, R., Amiri, I.S.: Performance analysis in digital circuits for process corner variations, slew-rate and load capacitance. Wirel. Pers. Commun. **103**, 99–115 (2018)
38. Hindia, M.N., Fadoul, M.M., Abdul Rahman, T., Amiri, I.S.: A stochastic geometry approach to full-duplex MIMO relay network. Wirel. Commun. Mob. Comput. **2018** (2018)
39. Singh, S., Koslia, M., Poonia, R.C.: A GA-QMR: genetic algorithm oriented MANET QoS multicast routing. Recent Pat. Comput. Sci. **11**(4), 268–275 (2018)
40. Vodnala, D., Phani Kumar, S., Aluvala, S.: An efficient on-demand link failure local recovery multicast routing protocol. In: Proceedings of 2nd International Conference on Intelligent Computing and Applications: ICICA 2015, pp. 603–616. Springer (2017)
41. Singal, G., et al.: Multi-constraints link stable multicast routing protocol in MANETs. Ad Hoc Netw. **63**, 115–128 (2017)
42. Obaid, A.J.: (2021) Critical research on the novel progressive, JOKER an opportunistic routing protocol technology for enhancing the network performance for multimedia communications. In: Kumar, R., Quang, N.H., Kumar Solanki, V., Cardona, M., Pattnaik, P.K. (eds.) Research in Intelligent and Computing in Engineering. Advances in Intelligent Systems and Computing, vol. 1254. Springer, Singapore. https://doi.org/10.1007/978-981-15-7527-3_36
43. Obaid, A.J.: Wireless Sensor Network (W.S.N.) routing optimization via the implementation of fuzzy ant colony (F.A.C.O.) algorithm: towards enhanced energy conservation. In: Kumar, R., Mishra, B.K., Pattnaik, P.K. (eds.) Next Generation of Internet of Things. Lecture Notes in Networks and Systems, vol. 201. Springer, Singapore (2021). https://doi.org/10.1007/978-981-16-0666-3_33

Automatic Facial Expression Detection Using Genetic Algorithm with SVM and Haar Wavelet Features

Hassanain K. Alrammahi and Kadim A. Jabbar

Abstract Face recognition technology has been developed as a result of the development of computer technology. In the present day, computer technology and network technology have made it possible to apply face recognition technology to a wide range of fields. We have designed and developed an automated learning-free facial landmark detection technique that performs similarly to other state-of-the-art landmark detection methods, yet consumes significantly less time. A consistent expression recognition solution has been developed in low resolution images using the proposed method, which has been found to work well in different resolutions. In experiments, the proposed system was shown to be effective on the CK+ and JAFFE facial expression databases.

Keywords CK+ · JAFFE · SVM and haar wavelet features

1 Introduction

Human affective behavior is largely assessed based on facial expressions as visual cues. There are still challenges in this field despite the various techniques reported. In previous research, facial expressions were primarily recognized using static images, but not their temporal information. A temporal dynamics analysis of a dynamic process, however, is a powerful tool for improving real-world affect analysis systems. Statistics (such as mean and standard deviation) have been utilized in previous literature [1, 2] to reduce the number of Gabor features. As an example, we used Genetic Algorithms (GA) to select filters that discriminate different emotions more effectively [3]. In other words, GA reduces the length of a feature vector by selecting Gabor filters based on frequency and orientation. Gabor features can reduce feature

H. K. Alrammahi (✉)
General Directorate of Education in Najaf, Najaf, Iraq
e-mail: hassanaink.alrammahi@uokufa.edu.iq

K. A. Jabbar
College of Engineering, Medical Instruments Technology Engineering, National University of Science and Technology, Dhi Qar, Nasiriyah, Iraq

S. Kumar Udgata et al. (eds.), *Intelligent Systems*, Lecture Notes in Networks and Systems 1149, https://doi.org/10.1007/978-981-97-8160-7_26

lengths and preserve the unique information in high-dimensional feature vectors, unlike mean and standard deviation.

Using Gabor features, we interpret affective facial expressions from video sequences using a dynamic approach. Also, by using GA, we select the most relevant Gabor filters rather than using the entire set of Gabor filters. FER has been solved using many existing approaches. As an example, PCA and SVM were combined for [4, 5]. The support vector machine (SVM), the higher-order spectral transform (HOS), and the radon transform (RT) were later employed by [6]. Using pyramid histograms of oriented gradients (PHOG) and genetic algorithms was suggested by the author [7]. LDA was used as the final step in the analysis. Orthogonal wavelet entropy (BWE) was used by the author [8].

Recognition of faces is most successful when facial expressions are detected and extracted. Feature extraction and classification in this study is carried out using wavelet transforms (HWTs) based on radial basis functions.

1. Generate Fuzzy-C-Means from Genetic Algorithms and Face Images.
2. CK+ and JAFFE databases were used in the experiment.
3. CMU, BioID, Long Distance, and FEI are the other databases used for face detection.
4. Using fdlibmex algorithm, this algorithm detects faces. Using contrast enhancement, however, we can improve this algorithm's limitations. Using low contrast images and long distances, we will solve the problem.
5. Noise can be removed from an image using median filtering during the preprocessing stage. Feature extraction is improved during this stage. A pattern recognition task often involves finding an image based on the image components.

2 Literature Review

There are two types of facial recognition systems: target-oriented and gesture-oriented [9]. One key frame from a sequence of images, involving the peak facial expression, is selected by target-oriented systems. A gesture-based approach leverages the temporal information contained in a video sequence by tracking certain facial features across all frames. Viola and Jones proposed a boosting algorithm that selects Haar-like features [10, 11]. Based on comparisons of classification accuracy and computational time, they evaluated the Haar+Adaboost approach against the Gabor+SVM approach [12]. There is no difference in recognition rate between the two approaches, but the first approach takes less time to compute. According to Author [13], the hair features can be used to detect emotion in humans since Viola-Jones rectangles are better suited to classifying faces, [13, 14]. The work presented here proposes new rectangle features that differentiate expressions from one subject from one another. The features of human facial expressions were used to define some expression-specific features.

Based on skin color, the Gabor filter, and neural networks, the author proposed a facial detection algorithm [15]. It is already possible to recognize faces using Gabor filters and neural networks. Nevertheless, the proposed paper's main objective is to reduce computation time by implementing skin color selection before Gabor filters. The detection quality is poor and high computation time is observed from this method by perfuming experimental search. Arithmetically subtraction of nearest frames is used to detect the edges of faces [16, 17]. Gaussian filtering is then used to remove unwanted edges and noise. Detecting edges corresponding to face videos is then achieved by logically combining the two previous output frames with the noiseless face contour frame. Complex backgrounds are not compatible with this proposal. Another [18] face recognition technique is proposed by us and it depends on multi-scale histograms. A speed of more than 10 times as fast can be achieved when this proposed strategy is legitimately applied when the square size is set to more than 10 times that of the LBP-based technique. The author [19] skin shading location, UCS space is utilized to section pictures rapidly. At that point to identify human face in light of sectioned skin shading region Adaboost calculation is utilized. The aftereffect of the test plainly demonstrates this proposed speed of face discovery calculation is high. Other than that, false location rate is lower and missing discovery rate than Adaboost calculation. Reference for the outline of real face location framework to some degree can be given by the proposed confronts identification strategy. The author's [20–22] paper concentrates on the significance of face discovery and its element parts. Viola–Jones calculation was actualized in this technique. The critical piece of this paper is to highlight extraction and the calculation utilized for the reason for existing is adjusted neighborhood paired examples calculation. For most extreme ages, this work is wasteful. The exploration of the Weighted Least Square sifting gave a concise similar investigation of the programmed Face Expression Recognition by invalidating the impact of age on the acknowledgment procedure. The channels Gabor and Log Gabor are for the most part helpful to concentrate facial components. The SVM classifier is intended to first know the information pictures and afterward characterize the obscure info pictures. It is excessively tedious to highlight the extraction handle [23].

3 Proposed Methodology

Figure 1 illustrates the proposed methodology. Because there are unwanted things present in the image, this section injects an image with emotion into the preprocessing step. The Input Face Expression Image undergoes the preprocessing operation in two stages namely, noise removal and contrast enhancement.

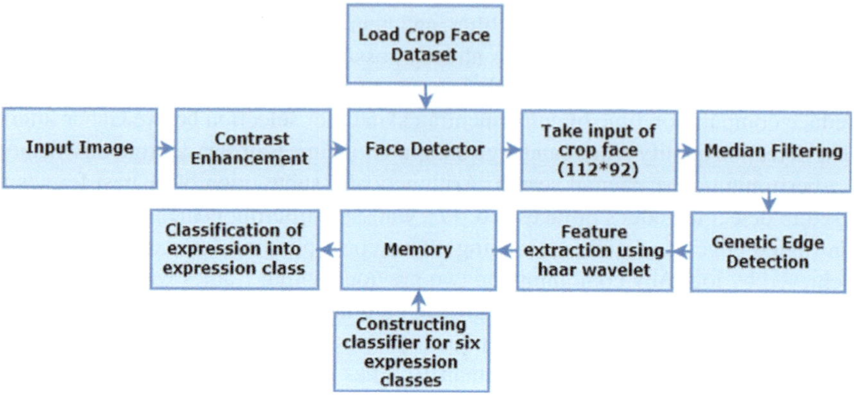

Fig. 1 Block diagram of proposed architecture

Noise removal: In this section, uneven illumination, light intensity blur, occlusion problem, and noise present in the image are removed by using a novel Adaptive Mode Guided Filter (AMGF). A guided Filter is an edge-preserving smoothening light filter that retains the sharp edges while filtering out the noise. Due to the limitations of the existing Guided Filter, it ignores structural inconsistencies between the reference and target images, such as color and infrared images acquired under different conditions. Sometimes, it causes over smoothing and blurriness in an image; to overcome that the adaptive mode function is used to calculate the Cost Function instead of the mean.

Let the input image be (F) and the filtered output can be obtained by considering the guided image (G) which can be the input image itself or a different image. The guided image filters the input image by imprinting its structure over the original input image. The filtration is the linear transformation of a guided image in a window (w_k) centered at the pixel (k). Hence the filtering output at a ith pixel of the image can be expressed as,

$$F_i^{\text{filter}} = G_i l_k + B_k \tag{1}$$

where F_i^{filter} denotes the filtering output at the ith pixel of the input image, l_k and B_k are the linear coefficient and bias assumed in the w_k, G_i is the guided image at the ith pixel. Input images are transformed linearly during the filtration process. In order to determine the linear coefficients and bias, we calculate the minimum cost function by using the following formula:

$$C = \sum_{i \ni w_k} ((G_i l_k + B_k - F_i)^2 + \ni l_k^2 \tag{2}$$

where,

$$l_k = \frac{1/l \sum_{i \ni w_k} F_i G_i + u_k F_{\dot{k}}}{\Delta_k^2 + \ni} \tag{3}$$

$$B_k = F_{\dot{k}} - l_k u_k \tag{4}$$

where, $\ni$ denotes the regularization, u_k and Δ_k^2 are the adaptive mode and variance of the guided image at w_k, $F_{\dot{k}}$ is the adaptive mode of F in the window. Here, the adaptive coefficient is calculated for the guided image, which is represented as,

$$U_k = h_i + \frac{(Y_p - Y_{\text{pre}}) * > H}{(Y_p - Y_{\text{pre}}) + (Y_p - Y_{\text{suc}})} \tag{5}$$

where, h_i denotes the lower pixel value, Y_p is the frequency of pixel, Y_{pre} is the frequency of preceding pixels, Y_{suc} is the succeeding pixel, H is the interval between the pixels. Similarly, an adaptive mode of input image has been calculated. Through the above process, the output of the filtered image filter F^{filter} can be obtained.

Contrast Enhancement: The filtered image F^{filter} is enhanced with CLAHE. To enhance the contrast of an image, it is a variant of adaptive histogram equalization. Here the contrast amplification can be mitigated by a clip limit. Initially, an equal number of pixels is assigned to each contextual region of the filtered facial image. Tiles are another name for it. Based on the pixels in the image, we calculate the histograms for each tile. Taking the gray level R_{avg} as an example, the average pixel is calculated as follows.

$$R_{\text{avg}} = \frac{N_s - N_t}{N_g} \tag{6}$$

where, an indication of how many pixels are in the contextual region in the x direction is given by N_s, N_t is the number of pixels in the contextual region in the y direction, Gray levels are represented by N_g. In order to calculate the clip limit, the average gray level should be taken into account, as follows:

$$F_{CL} = r_{\text{avg}} \times F^{\text{filter}}_{\text{Max(avg)}} \tag{7}$$

In this case, $F^{\text{filter}}_{\text{Max(avg)}}$ refers to the maximum number of pixels in each gray level of the contextual region, while F_{CL} refers to the clip limit. Pixels exceeding the clip point limit are re-distributed for each gray level once they reach the clip point limit. Utilizing bilinear interpolation, the input image is combined after equalization to eliminate boundaries. This can therefore be referred to as F^{COM} for the image after contrast enhancement.

Haar Wavelet Features Extraction: Sine and cosine waves can be used to represent any signal in Fourier theory. Fourier representations lack information about time; they only represent frequency resolution. With Wavelet transforms [24] frequency information is provided at precise time, thus overcoming the shortcoming of Fourier transforms. There are many applications of wavelets in pattern recognition, signal processing, and machine vision. In a multi-resolution setting, wavelets provide time and frequency localization of information. Due to its separability, linearity, and completeness, 2D wavelets are faster than 3D wavelets [25]. Recognition tasks tend to be dominated by coarser information, as demonstrated in previous studies. In a discrete wavelet transform, input signals are divided into four subbands: coefficients of approximation (LL), vertical (HL), diagonal (HH), and horizontal (LH). Discretized wavelet transforms are produced with a bank of low-pass and high-pass filters [24]. Diagonal features correlate closely with the details and texture of the pattern, while approximation features correlate closely with the input shape. The low-pass filter is used to calculate approximation features and the high-pass filter is used to derive detailed features. In Fig. 2, you can see how 2D DWT is performed.

The Haar wavelet is composed of square signals rescaled to form the wavelet base [26]. Fourier transforms are similar to Haar transforms, but the Haar transform is more intuitive and useful. Haar can be calculated very easily. As defined below, $\psi(t)$ is the mother wavelet function.

$$\psi(t) = \begin{cases} 1 & 0 \leq t < 0.5 \\ -1 & 0.5 \leq t < 1 \\ 0 & \text{otherwise} \end{cases} \tag{8}$$

In order to scale the wavelet (father wavelet), we need to use the following equation,

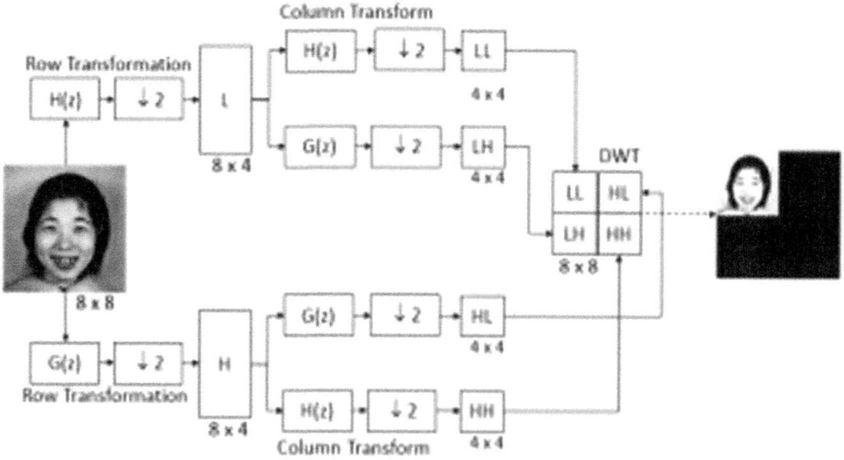

Fig. 2 Digital signal transformed into discrete wavelets in two dimensions

$$\varnothing(t) = \begin{cases} 1 & 0 \le t < 1 \\ 0 & \text{otherwise} \end{cases} \tag{9}$$

In Haar wavelet analysis, the original signal is divided into trends (running averages) and fluctuations (running differences). A half-size signal is used for both parts. In comparison with the original signal, the trend signal is compressed, preserving most of its original energy. It is possible to decompose the first trend signal into trend and fluctuation components for the second level of haar decomposition. For the $(i + 1)$st level of haar decomposition, the above steps are repeated for the ith trend signal.

Most of the energy of a signal can be captured by Haar trend coefficients. We can determine trend and fluctuation signals for rows and columns separately. Many signal processing applications can benefit from this property of haar. Literature provides detailed information about haar [27].

Feature selection optimization using GA and Classification: Since there are 53,050 features in a JAFFE face, PCA is used to reduce the number of features. 99% of the information is retained in this fitted PCA model. Using the PCA model, 53,050 features in JAFFE images are reduced to 212 with the PCA model. We then optimize both SVM hyperparameters and PCA features simultaneously to maximize model accuracy. The selection of features and the exhaustive value of C and y would allow evaluation of numerous combinations possible ($2^{n_y+n_c+n_f}$ combinations), where n_f measures how many PCA features there are, n_c In hyperparameters, bits represent parameters C, and n_y In hyperparameters, bits represent parameters Y.

As $n_y+n_c+n_f$ value grows, this process becomes impractical due to the amount of computer work required. It is for this reason that feature selection and hyperparameter optimization are important in practice. Genetic Algorithms (GA) are widely used to solve problems like optimizing hyperparameters in SVMs and selecting features. Also, the chromosome size in our case is not too large compared to other problems, so we use the GA instead of other metaheuristic methods. Our population must contain a few chromosomes, so we need a few chromosomes. We aren't faced with that problem here, since population size can cause an exponential effect on computation costs. This GA choice should be reviewed if other GAs, such as Gabors, need to be included for large datasets like FER-2013.

The simple GA maintains constant numbers of candidate solutions P throughout its iterative process. For each iteration step (generation), the fitness value is calculated using the cost function and the three genetic operators (reproduction, mutation, and crossover). A new population of candidate solutions is formed based on these genetic operators and evaluations. Our system's feature selection is an important step. In order to improve the system, we will choose a group of features from PCA. The use of binary coding, where one indicates inclusion in the subset of features, and zero indicates exclusion, is sufficient and better than real coding for this purpose. Each chromosome represents a prediction model in our case. There are two genes that represent binary values of the SVM hyperparameters (C and Y) and one gene for feature binary position that represents whether features should be included (1) or not (0). Figure 3 illustrates the process of optimizing hyperparameters and selecting features for SVM.

(a) **Initialization and chromosome design**

First, chromosomes must be created and initialized in the population. Due to GA's stochastic nature, genes on chromosomes are usually initialized at random, so three different parts of the chromosome are involved: C, Y, and the feature binary position. Chromosomes were represented using binary coding. Our model represents binary chromosomes in Fig. 6 [28]. As represented by $g_C^1 \sim g_C^{n_c}$ and $g_y^1 \sim g_y^{n_y}$, the hyperparameter binary values of C and y are represented by $g_f^1 \sim g_f^{n_f}$; where n_C is the number of bits corresponding to the hyperparameter C, n_y is corresponding to the hyperparameter y and n_f represents corresponding to the feature [chromosome design inspired by [29]. Equation (2) converts genotype into phenotype based on

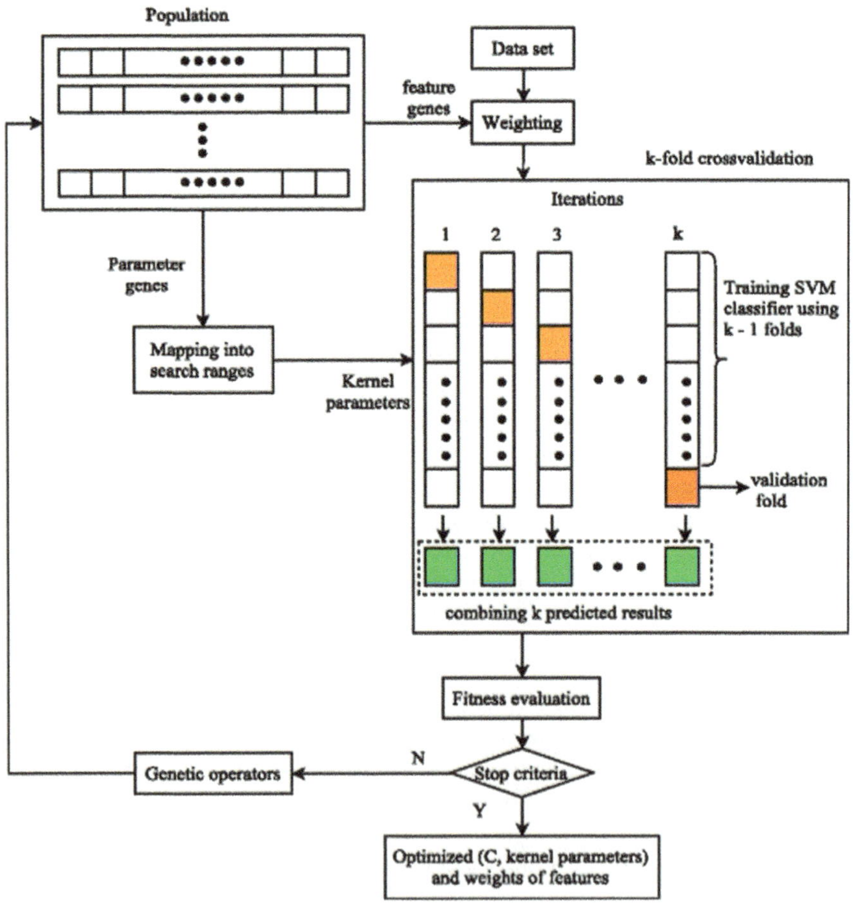

Fig. 3 Feature selection and hyperparameter optimization for SVMs based on GA

bitstrings representing C and Y hyperparameters [29]. In order to display hyperparameters accurately, a bit string (n_c and n_y) has a maximum and minimum length, and the user specifies the values of these parameters. Chromosomes with values 1 and 0 represent features, respectively, and bit 1 represents the selected feature.

$$p = \min_p + \frac{\max_p - \min_p}{2^l - 1} \times d \tag{10}$$

It consists of P, $\min_p$, $\max_p$, d is a decimal; l is a length; P is the phenotype of bit string; P is $\min_p$; P is $\max_p$; P is the minimum value; P is the maximum value. Following several experiments, we found that: $n_y = 20$, $n_c = 20$, $\min_c = 10^{-2}$, $\max_c = 100$, $\min_y = 10^-15$, $\max_y = 10^{-3}$ and for feature length n_f, PCA features determine whether it works or not. (Section 5.2 validates the choice of $N = 200$ as the population size).

(b) **Fitness assignment**

Rank-based fitness assignment is the most common method for assigning fitness [30]. All chromosome errors are sorted out with this method. In determining a chromosome's fitness, the position in the ranking of those chromosomes is the only factor considered, not the actual error. According to the Rank-based method, each chromosome will have the following fitness value:

$$\varnothing(i) = w * R(i), i = 1 \ldots..N \tag{11}$$

The value of w is between 1 and 2, and it is referred to as selective pressure [30]. A higher level of selective pressure will result in a greater likelihood of crossover between the fittest chromosomes. A chromosome's rank is represented by the parameter $R(i)$. Fitness values are calculated using this value $w = 1.5$.

(c) **Selection**

N determines how many chromosomes should be selected based on the population size $N/2$. (half the number of people) is selected. Roulette wheel is a popular selection method. In accordance with this, the fitness level of each chromosome is reflected in the area of its chromosome on a roulette wheel. A roulette wheel is then used to select chromosomes at random. Chromosomes are crossed over [31].

(d) **Crossover**

When the selection operator has chosen half of the population, the chromosomes are recombined in the crossover operation. In uniform crossover, two chromosomes are jointly combined at random until the new population equals the old population [32].

(e) **Mutation**

There can be very strong similarities between the parents and the offspring created by the crossover operator. This can lead to a lack of diversity in the next generation.

By changing specific bits randomly in the offspring, the mutation operator solves this problem. A random number between 0 and 1 determines whether a bit will be mutated, and if it is less than the mutation rate [33], that variable is flipped.

When the stopping criterion is met, the best chromosome is obtained. Utilizing Gabor filters to extract the features, we use the GA's subset feature vectors to select the best PCA features. A feature subset will only be selected if it has a value of 1. The SVM model is trained using GA parameters and the hyperparameters are optimized to predict facial emotion from the feature vector [38–40].

4 Result Analysis and Discussion

The MATLAB programming language supports matrix operations as well as matrices, making it a versatile tool for data analysis and visualization. As well as its own powerful programming language, MATLAB has excellent graphics capabilities. One of the reasons MATLAB has become so popular is because it can be used to support specific tasks. A toolbox is a set of programs, and the image processing and image acquisition toolbox are of particular interest to us. Cohn-Kanade (CK+) and JAFFE are two widely used facial expression databases.

CK+: Based on CK+ [34], which contains 837 images under seven class labels, the RFER-EADL technique was experimentally justified using Table 1. A sample image is shown in Fig. 4. In the FER-2013 dataset, 28,000 tags have been assigned to images. A development set consists of 3500 tagged images, and a test set consists of 3500. There are seven different emotions represented in FER-2013, from joy and sadness to anger, terror, astonishment, disgust, and neutrality.

JAFFE: The proposed method is robust to various facial expressions when using the JAFFE [35]. The database contains 213 pictures of 10 Japanese female models with seven facial expressions (happy, angry, fearful, sad, disgusted, and surprised). A total of 60 Japanese subjects rate each image according to six emotion adjectives. As shown in Fig. 5, there are a few samples from the JAFFE dataset.

Table 1 CK+ dataset details

Label	Description	No. of image
An	Anger	45
Co	Contempt	18
Di	Disgust	59
Fe	Fear	25
Ha	Happy	69
Nu	Neutral	593
Sa	Sad	28
Total number of image		837

Fig. 4 Sample images

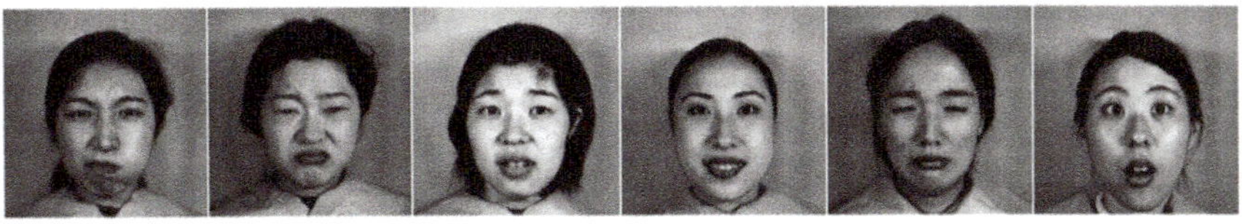

Fig. 5 Samples of JAFFE database

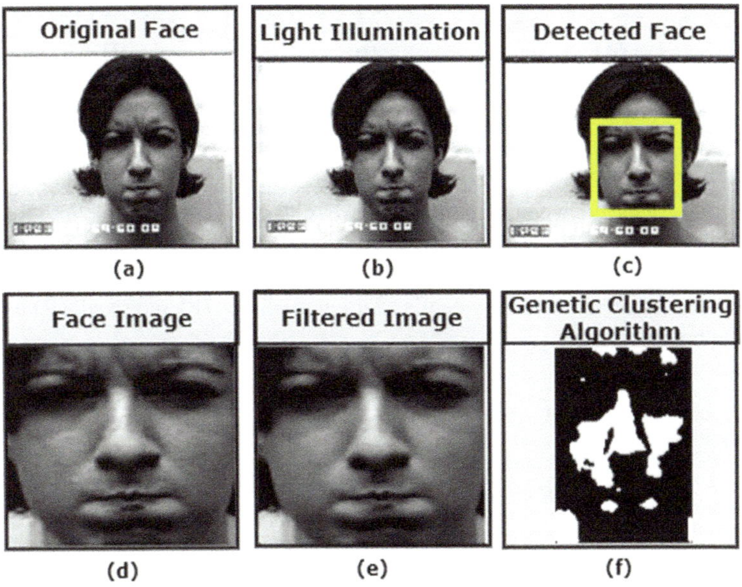

Fig. 6 Framework for proposed system on CK+ dataset, **a** Original Face, **b** Enhanced Image, **c** Face Detection, **d** Crop Face Image, **e** Noise Removed Image **f** Face Segmentation using GA

5 Experimental Results

We will utilize tenfold cross validation to evaluate the proposed method's performance. The faces were then scaled to a common resolution using face detection and scaling, as discussed earlier. A face image was analyzed to detect facial landmarks

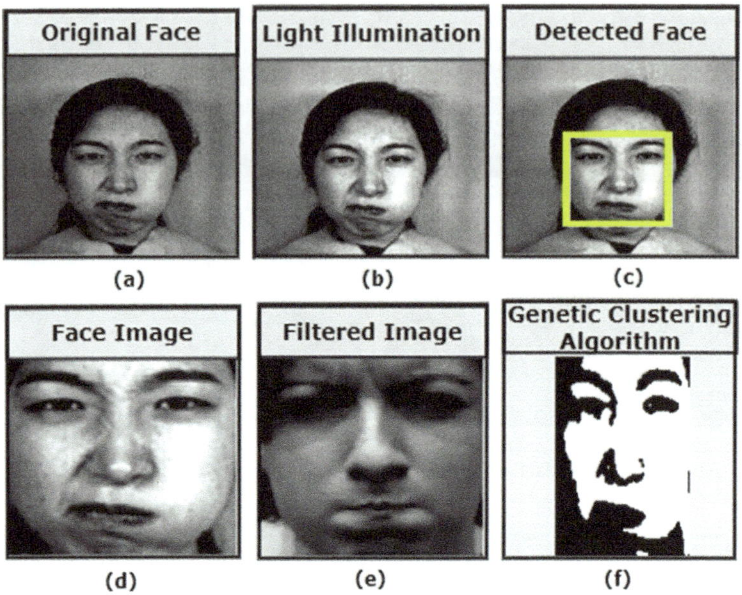

Fig. 7 Framework for proposed system on JAFFEE dataset, **a** Original Face, **b** Enhanced Image, **c** Face Detection, **d** Crop Face Image, **e** Noise Removed Image **f** Face Segmentation using GA

and extract edges. Each pair of expressions was classified by an SVM classifier during the training stage. Concatenation of features between expression classes was carried out using the edges containing discriminative characteristics. Analyze the test set's performance. Calculate your system's average precision (P) and recall (R) using the formula below:

$$P = \frac{\text{No. of relevant images matched}}{\text{Total number of image matched}}$$

$$R = \frac{\text{No. of relevant image matched}}{\text{number of image in the datebese}}$$

1. Using the formula below, calculate the system's matching rate (MR):

$$\text{MR} = 100 \times ((\text{TP} + \text{TN})/N)$$

N is the size of the dataset, where TP and TN are true positives and negatives, respectively (Figs. 6, 7, 8, 9, 10, 11 and 12 and Tables 2 and 3).

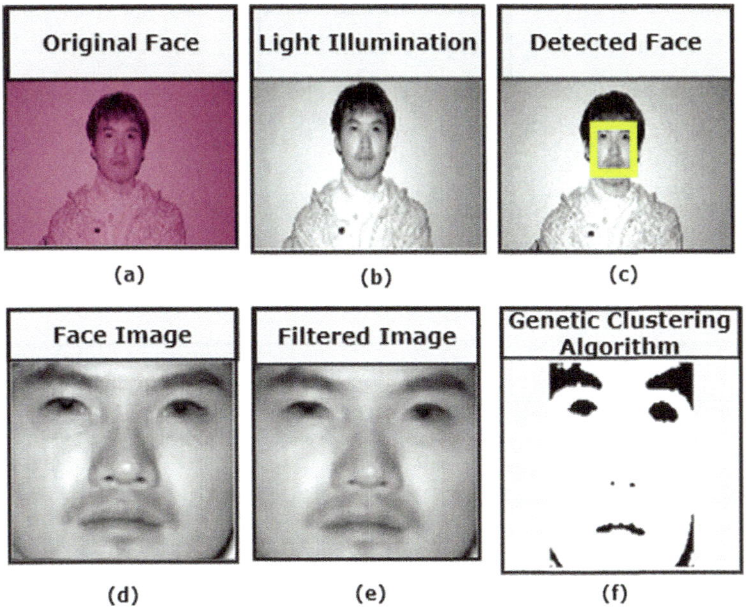

Fig. 8 Framework for proposed system on long distance dataset for 1 m distance, **a** Original Face, **b** Enhanced Image, **c** Face Detection, **d** Crop Face Image, **e** Noise Removed Image **f** Face Segmentation using GA

Based on the CK+ dataset, Proposed model performs better than Ref. [37] model as shown in Fig. 13. When the best values are selected among the results, we can attain a 100% proposed classification rate, while a 94% reference classification rate can be reached (see Figs. 14 and 15 and Tables 4 and 5).

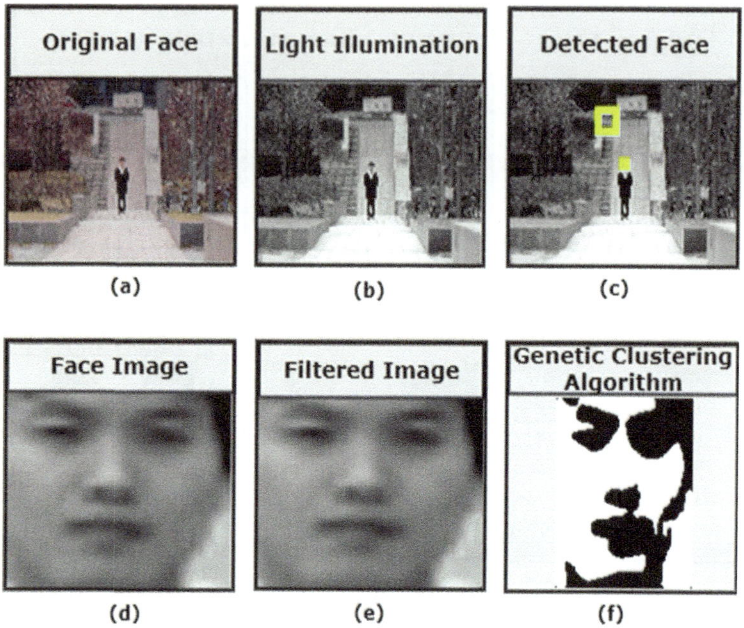

Fig. 9 Framework for proposed system on long distance dataset for 150 m distance, **a** Original Face, **b** Enhanced Image, **c** Face Detection, **d** Crop Face Image, **e** Noise Removed Image **f** Face Segmentation using GA

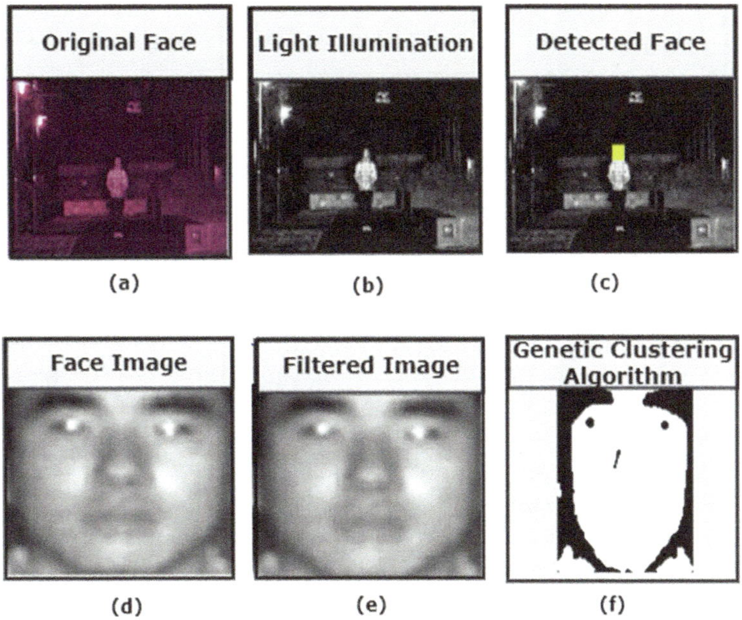

Fig. 10 Framework for proposed system on long distance dataset for 100 m distance, **a** Original Face, **b** Enhanced Image, **c** Face Detection, **d** Crop Face Image, **e** Noise Removed Image **f** Face Segmentation using GA

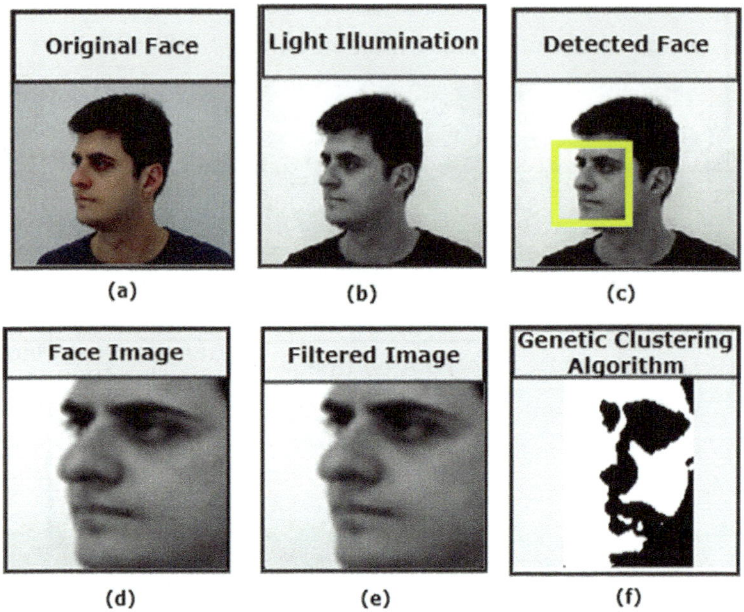

Fig. 11 Framework for Proposed System FEI dataset at 45 degrees, **a** Original Face, **b** Enhanced Image, **c** Face Detection, **d** Crop Face Image, **e** Noise Removed Image **f** Face Segmentation using GA

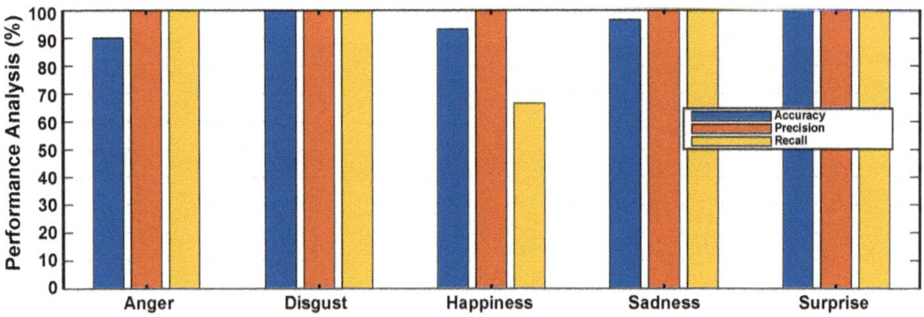

Fig. 12 The performance analysis (%) of the proposed model on CK+ dataset

Table 2 The performance analysis (%) of the proposed model on CK+ dataset

Expression	Accuracy	Precision	Recall
Anger	90	100	100
Disgust	100	100	100
Happiness	93.33	100	66.66
Sadness	96.66	100	100
Surprise	100	100	100

Table 3 The performance comparison of proposed model with existing model on CK+ dataset

Expression	S.L. Happy [36]	Accuracy
Anger	87.8	90
Disgust	93.33	100
Happiness	94.2	93.33
Sadness	96.42	96.66
Surprise	98.46	100

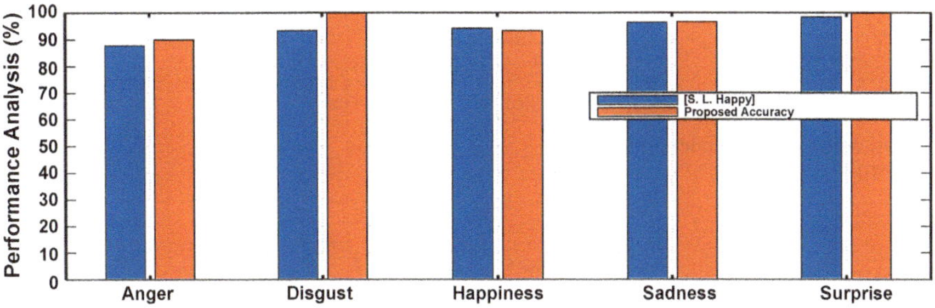

Fig. 13 The performance comparison of proposed model with existing model on CK+ dataset

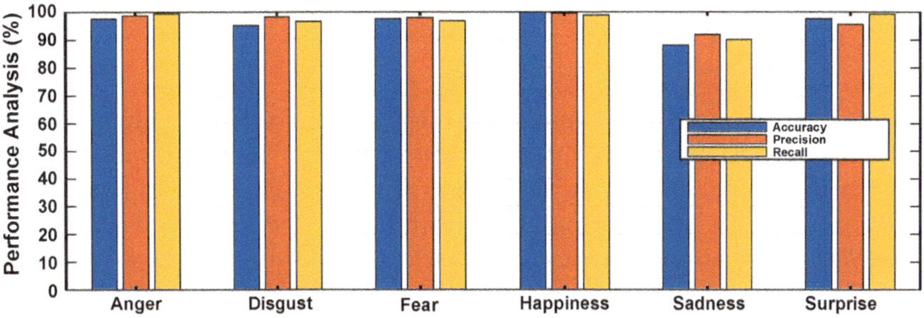

Fig. 14 The performance analysis (%) of the proposed model on JAFFEE dataset

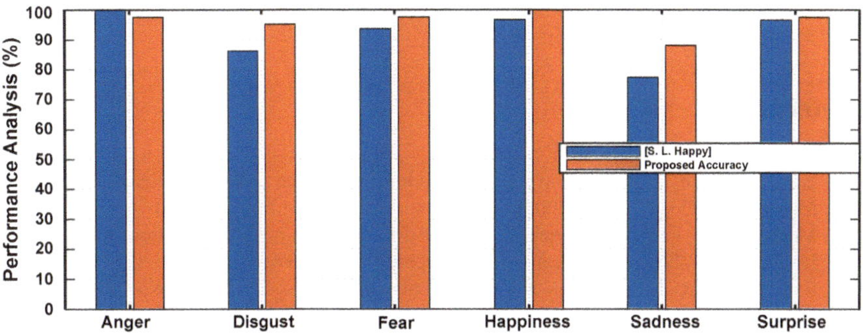

Fig. 15 The performance comparison of proposed model with existing model on JAFFEE dataset

Table 4 The performance analysis (%) of the proposed model on JAFFEE dataset

Expression	Accuracy	Precision	Recall
Anger	97.61	98.72	99.52
Disgust	95.23	98.26	96.64
Fear	97.61	98.0	96.87
Happiness	100	99.62	98.88
Sadness	88.09	91.89	90.10
Surprise	97.61	95.62	99.32

Table 5 The performance comparison of proposed model with existing model on JAFFEE dataset

Expression	S.L. Happy [36]	Accuracy
Anger	100	97.61
Disgust	86.2	95.23
Fear	93.75	97.61
Happiness	96.77	100
Sadness	77.41	88.09
Surprise	96.66	97.61

6 Conclusion

Identity authentication is becoming increasingly necessary with the advancement of information technology. Authentication methods that use face recognition are convenient and accurate. In academic and industrial settings, facial recognition plays an important role. An innovative facial recognition method was proposed in this paper. Wavelet coefficients were extracted from facial images using discrete wavelet transform. Secondly, the features were reduced through principal component analysis. Our final and most important innovation is genetic optimization for training the weights and biases of the classifier. The proposed approach has better overall accuracy as compared to existing state-of-the-art approaches.

References

1. Wang, Y., Guan, L.: Recognizing human emotional state from audiovisual signals. IEEE Trans. Multimed. **10**(5), 936–946 (2008)
2. Meghjani, M., Ferrie, F., Dudek, G.: Bimodal information analysis for emotion recognition. In: 2009 Workshop on Applications of Computer Vision (WACV), pp. 1–6. IEEE (2009)
3. Rani, P., Sharma, R.: IMFOCA-IOV: intelligent moth flame optimization based clustering algorithm for internet of vehicle. In: 2023 14th International Conference on Computing Communication and Networking Technologies (ICCCNT), Delhi, India, pp. 1–6. IEEE (Jul 2023). https://doi.org/10.1109/ICCCNT56998.2023.10307646

4. Drume, D., Jalal, A.S.: A multi-level classification approach for facial emotion recognition. In: 2012 IEEE International Conference on Computational Intelligence and Computing Research, pp. 1–5. IEEE (2012)
5. Ansari, G., Rani, P., Kumar, V.: A novel technique of mixed gas identification based on the group method of data handling (GMDH) on time-dependent MOX gas sensor data. In: Mahapatra, R.P., Peddoju, S.K., Roy, S., Parwekar, P. (eds.) Proceedings of International Conference on Recent Trends in Computing. Lecture Notes in Networks and Systems, vol. 600, pp. 641–654. Springer Nature Singapore, Singapore (2023). https://doi.org/10.1007/978-981-19-8825-7_55
6. Ali, H., Hariharan, M., Yaacob, S., Adom, A.H.: Facial emotion recognition based on higher-order spectra using support vector machines. J. Med. Imaging Health Inform. **5**(6), 1272–1277 (2015)
7. Boubenna, H., Lee, D.: Feature selection for facial emotion recognition based on genetic algorithm. In: 2016 12th International Conference on Natural Computation, Fuzzy Systems and Knowledge Discovery (ICNC-FSKD), pp. 511–517. IEEE (2016)
8. Zhang, Y.-D., et al.: Facial emotion recognition based on biorthogonal wavelet entropy, fuzzy support vector machine, and stratified cross validation. IEEE Access **4**, 8375–8385 (2016)
9. Cowie, R., et al.: Emotion recognition in human-computer interaction. IEEE Signal Process. Mag. **18**(1), 32–80 (2001)
10. Whitehill, J., Omlin, C.W.: Haar features for FACS AU recognition. In: 7th International Conference on Automatic Face and Gesture Recognition (FGR06), pp. 5–101. IEEE (2006)
11. Viola, P., Jones, M.: Rapid object detection using a boosted cascade of simple features. In: Proceedings of the 2001 IEEE Computer Society Conference on Computer Vision and Pattern Recognition. CVPR 2001, pp. I–I. Ieee (2001)
12. Rani, P., Sharma, R.: Intelligent transportation system for internet of vehicles based vehicular networks for smart cities. Comput. Electr. Eng. **105**, 108543 (2023)
13. Lu, K., Zhang, X.: Facial expression recognition from image sequences based on feature points and canonical correlations. In: 2010 International Conference on Artificial Intelligence and Computational Intelligence, pp. 219–223. IEEE (2010)
14. Yadav, S.P., Jindal, M., Rani, P., De Albuquerque, V.H.C., Dos Santos Nascimento, C., Kumar, M.: An improved deep learning-based optimal object detection system from images. Multimed. Tools Appl. (Sep 2023). https://doi.org/10.1007/s11042-023-16736-5
15. Chihaoui, M., Elkefi, A., Bellil, W., Amar, C.B.: Implementation of skin color selection prior to Gabor filter and neural network to reduce execution time of face detection. In: 2015 15th International Conference on Intelligent Systems Design and Applications (ISDA), pp. 341–346. IEEE (2015)
16. Dey, A.: A contour based procedure for face detection and tracking from video. In: 2016 3rd International Conference on Recent Advances in Information Technology (RAIT), pp. 483–488. IEEE (2016)
17. Verma, S., et al.: An automated face mask detection system using transfer learning based neural network to preventing viral infection. Expert Syst., e13507 (Nov 2023). https://doi.org/10.1111/exsy.13507
18. Lin, C.-Y., Fu, J.T., Wang, S.H., Huang, C.-L.: New face detection method based on multi-scale histograms. In: 2016 IEEE Second International Conference on Multimedia Big Data (BigMM), pp. 229–232. IEEE (2016)
19. Liuliu, W., Mingyang, L.: Multi-pose face detection research based on AdaBoost. In: 2016 Eighth International Conference on Measuring Technology and Mechatronics Automation (ICMTMA), pp. 409–412. IEEE (2016)
20. Muttu, Y., Virani, H.G.: Effective face detection, feature extraction & neural network based approaches for facial expression recognition. In: 2015 International Conference on Information Processing (ICIP), pp. 102–107. IEEE (2015)
21. Kumar, N., Rani, P., Kumar, V., Athawale, S.V., Koundal, D.: THWSN: enhanced energy-efficient clustering approach for three-tier heterogeneous wireless sensor networks. IEEE Sens. J. **22**(20), 20053–20062 (2022)

22. Kumar, N., Rani, P., Kumar, V., Verma, P.K., Koundal, D.: TEEECH: three-tier extended energy efficient clustering hierarchy protocol for heterogeneous wireless sensor network. Expert Syst. Appl. **216**, 119448 (2023)
23. Kulkarni, K.R., Bagal, S.B.: Facial expression recognition. In: 2015 Annual IEEE India Conference (INDICON), pp. 1–5. IEEE (2015)
24. Lim, J., Kim, Y., Paik, J.: Comparative analysis of wavelet-based scale-invariant feature extraction using different wavelet bases. In: Signal Processing, Image Processing and Pattern Recognition: International Conference, SIP 2009, Held as Part of the Future Generation Information Technology Conference, FGIT 2009, Jeju Island, Korea, December 10–12, 2009. Proceedings, pp. 297–303. Springer (2009)
25. Papageorgiou, C., Poggio, T.: A trainable system for object detection. Int. J. Comput. Vis. **38**, 15–33 (2000)
26. Stankovic, R.S., Falkowski, B.J.: The Haar wavelet transform: its status and achievements. Comput. Electr. Eng. **29**(1), 25–44 (2003)
27. Mallat, S.G.: A theory for multiresolution signal decomposition: the wavelet representation. IEEE Trans. Pattern Anal. Mach. Intell. **11**(7), 674–693 (1989)
28. Yu, J., Bhanu, B.: Evolutionary feature synthesis for facial expression recognition. Pattern Recognit. Lett. **27**(11), 1289–1298 (2006)
29. Huang, C.-L., Wang, C.-J.: A GA-based feature selection and parameters optimizationfor support vector machines. Expert Syst. Appl. **31**(2), 231–240 (2006)
30. Whitley, L.D.: The GENITOR algorithm and selection pressure: why rank-based allocation of reproductive trials is best. Colorado State University, Department of Computer Science (1989)
31. Lipowski, A., Lipowska, D.: Roulette-wheel selection via stochastic acceptance. Phys. Stat. Mech. Its Appl. **391**(6), 2193–2196 (2012)
32. Spears, W.M., Kenneth, A.: De Jong. On the virtues of parametrized uniform crossover. In: Proceedings of the Fourth International Conference on Genetic Algorithms, pp. 230–236 (1995)
33. Janson: training product unit neural networks with...—Google Scholar. Accessed 11 Dec 2023
34. Sunitha, G., Geetha, K., Neelakandan, S., Pundir, A.K.S., Hemalatha, S., Kumar, V.: Intelligent deep learning based ethnicity recognition and classification using facial images. Image Vis. Comput. **121**, 104404 (2022). https://doi.org/10.1016/j.imavis.2022.104404
35. Cortes, C., Vapnik, V.: Support-vector networks. Mach. Learn. **20**, 273–297 (1995)
36. Happy, S.L., Routray, A.: Automatic facial expression recognition using features of salient facial patches. IEEE Trans. Affect. Comput. **6**(1) (2014). Art. no. 1
37. Viola, P., Jones, M.J.: Robust real-time face detection. Int. J. Comput. Vis. **57**, 137–154 (2004)
38. Yamsani, N., Ahammad, S.H., Obaid, A.J., Saikumar, K., Alshathr, A.H., Ali, Z.S.M.: Analysis on exposition of speech type video using S.S.D. and CNN techniques for face detection. In: Khanna, A., Polkowski, Z., Castillo, O. (eds.) Proceedings of Data Analytics and Management. Lecture Notes in Networks and Systems, vol. 572. Springer, Singapore (2023). https://doi.org/10.1007/978-981-19-7615-5_71
39. Alrammahi, H.K., Obaid, A.J.: Human face recognition and facial attribute analysis using data analytics techniques in smart grid using image processing. In: Kumar Sharma, D., Sharma, R., Jeon, G., Kumar, R. (eds.) Data Analytics for Smart Grids Applications—A Key to Smart City Development. Intelligent Systems Reference Library, vol. 247. Springer, Cham (2023). https://doi.org/10.1007/978-3-031-46092-0_21
40. Abdulbaqi, A.S., Turki, N.A., Obaid, A.J., Dutta, S., Panessai, I.Y.: Spoof attacks detection based on authentication of multimodal biometrics face-ECG signals. In: Agarwal, P., Khanna, K., Elngar, A.A., Obaid, A.J., Polkowski, Z. (eds.) Artificial Intelligence for Smart Healthcare. E.A.I./Springer Innovations in Communication and Computing. Springer, Cham (2023). https://doi.org/10.1007/978-3-031-23602-0_30

A Real-Time Aerial Semantic Segmentation System Based on U-Net Deep Learning Using Drone Images

Muhammet Tahir Güneşer, Chihat Şeker, and Mohammed Ayad Alkhafaji

Abstract Image segmentation at the pixel level is a time-consuming and difficult task in computer vision and image processing. Aerial (satellite/drone) photo segmentation is considered in this paper. Data from high-resolution remote sensing has enabled new applications such as more detailed per-pixel object classification. U-Net with VGG16 has made segmentation and categorization of images much more efficient and intelligent. U-Net models with pre-trained VGG16 backbones perform best across all tested scenarios. Adding the near-infrared band improves prediction results slightly compared with using RGB image bands alone. The ability to transfer images between sensors, especially between satellites and aerial images, could be improved through train-time enhancement and contrast enhancement. Further improving performance could be achieved by adding noisy training data from free online resources.

Keywords U-Net · Deep learning · Drone images

1 Introduction

Data extraction from databases (images, videos, point clouds, etc.) has seen significant progress in photogrammetry and remote sensing in recent years [1, 2]. Data semantic segmentation has been a major challenge in photogrammetry and remote sensing over the last three decades [3]. Using semantic segmentation, scenes can be classified by their segment labels (e.g., cars, trees, vegetation, roads, skies, or

M. T. Güneşer
Faculty of Electrical and Electronics, Istanbul Technical University, Istanbul, Turkey

C. Şeker
Collage of Engineering, Department of Electrical Electronic Engineering, Karabuk University, Karabuk, Turkey

M. A. Alkhafaji (✉)
College of Technical Engineering, National University of Science and Technology, Dhi Qar 64001, Iraq
e-mail: mohammed.alkhafaji@nust.edu.iq

S. Kumar Udgata et al. (eds.), *Intelligent Systems*, Lecture Notes in Networks and Systems 1149, https://doi.org/10.1007/978-981-97-8160-7_27

people). Semantic segmentation is a technique for labeling each pixel of an image in a predefined class by dividing it into meaningful semantic objects. The use of segmentation for urban planning [4], land cover mapping [5], extraction of buildings and roads from aerial images [6], and autonomous vehicle driving are among many examples [7, 8]. Semantic segmentation has evolved into an essential and, simultaneously, challenging aspect of urban aerial scene understanding with the progress of deep learning networks [9].

Multitemporal data can be obtained using UAVs that support a controllable flight path without revisiting interval limitations. An automatic warning system and traffic analysis are some of the tasks that require fast response times [10, 11]. A UAV camera can also capture multiview remote sensing images through oblique projection imaging. To map structural damage to buildings, it can provide a great deal of information [12]. Research on semantic segmentation has emerged as an emerging and challenging approach to acquiring image information for UAV remote sensing images. A semantic segmentation predicts pixel-level labels using the semantic information in pixels within a picture. Recently, convolutional neural networks (CNNs) have been extensively used to improve semantic segmentation tasks [13]. Remote sensing uses semantic segmentation to determine the terrain type, analyze objects on the ground, and identify roads.

Computer vision has shifted to deep learning in recent years. A multilayer computing system can learn and analyze data using deep learning at multiple abstraction levels. DNNs use convolution as their main mathematical operation [14]. CNNs have reached a high level of performance since LeCun introduced them in the early 1990s. When recognizing images, CNNs perform better than any other technique. Classifying, recognizing, and segmenting images using a convolutional neural network is possible by compressing the input images into a vector representation. Vectors representing the objects in an image are called feature vectors. A higher-order function is recognized by combining features extracted from successive layers [15]. There are three layers in a CNN: input, hidden, and output [16].

As encoders, well-known classification architectures were used to derive the real-time efficient object segmentation model. The VGG-16 encoders were used to implement U-Net in this work. An encoder that is reflected as a decoder is always used. Object segmentation details are presented in Fig. 1 as follows.

A deep learning VGG-16 architecture is used to normalize high-resolution RGB images. A preprocessing step is performed, followed by a normalization step and an image resizing step. RGB images are created from the output. Label mask information is matched with each segmented label of the pixel of an object class. As a result, a segmented object or objects are obtained with color values assigned.

2 Literature Review

Transfer learning and deep nets have recently been used to be successful at classifying images [17, 18]. A hybrid proposal classifier model was used to demonstrate transfer for visual recognition, detection, as well as instance and semantic segmentation [19].

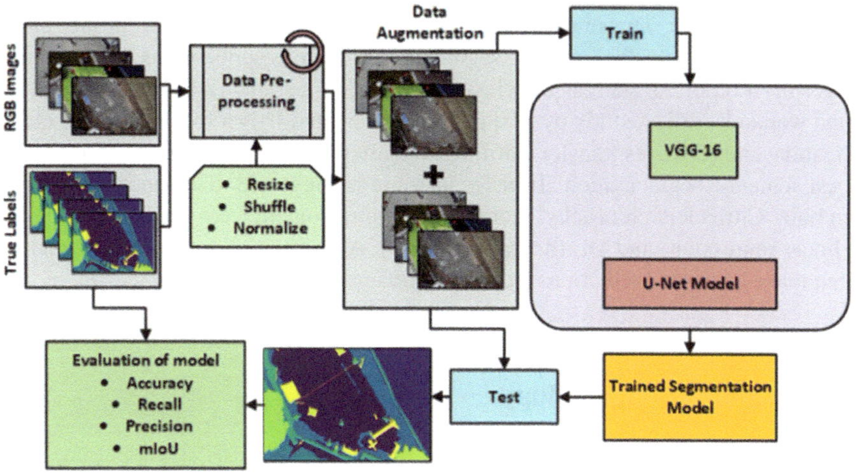

Fig. 1 The proposed system model based on the deep learning model

Using new architectures and fine-tunings, we predict semantic segmentation directly and densely [20].

The goal of developing more accurate approaches for determining damage from images taken by UAVs or robots is to develop approaches with higher accuracy; in addition to VGG16 [21], GoogLeNet [17], AlexNet [22], DenseNet-121 [23], and ResNet [24], multiple layer structured CNNs are also being used in machine learning. In one experiment, deep learning was used to classify images of concrete cracks [25] and a road with cracks [26], and the steel may also have defects [27]. The traditional classification methods rely on CNN models to detect damage locations from images based only on the categories of damages available to them.

In [28], the author presents a model for extracting target features from sonar images using CNN and a Support Vector Machine to classify sonar onboard an unmanned underwater vehicle, which improves semantic segmentation efficiency and automates target recognition. According to [29], underwater scene semantic segmentation can be achieved with encoder-decoder networks. This paper describes an underwater enhancement architecture and an encoder based on a pre-trained VGG-16 network for feature extraction. As a result of the author's [30] proposal, underwater obstacle detection can be improved by using semantic segmentation based on monocular images. This method based on encoder-decoder-based U-Nets is proposed by [31] to monitor high quality and high temporal coverage of specific regions.

Drone data can be analyzed using a variety of techniques. DL and ML are the two most advanced image analysis techniques for drones. Drone images were classified into roads, vineyards, asphalt, and roofs using machine learning techniques [32]. Ensemble decision trees are used in the study based on an object-oriented approach. According to the study, the segmentation of agricultural fields was carried out using multi-resolution object-based segmentation. Numerous other pixel-based studies have used support vector machines (SVM) to map agriculture and monitor reefs with drones. The study used DL to segment concrete cracks in drone images [33]. CNNs are used for semantic segmentation. Various applications of this method have been reported, such as the classification of urban land [34], forest cover [35], and wetlands [36]. A study by [37] uses ResNet50 and U-Net for tree species classification, and [38] uses transfer learning to segment aerial images AeroScapes for the best semantic segmentation. Transfer learning is suggested to enhance the analysis in both. Citrus leaf chlorophyll content and water content were predicted using ML (linear regression) and DL (neural networks). According to the study, ML and DL can make accurate predictions from UAV images.

3 Proposed Methodology

Convolutional networks have become increasingly popular in research due to their proven success in various fields. A brief description of U-Net and VGG-16 is provided in this section. Following that, a structured proposal is presented.

3.1 VGG-16

Base networks are based on VGG-16 architectures [21]. With the Image Net dataset, this is a convolutional neural network architecture [39]. Various classification tasks are handled with this vision model architecture, which is considered one of the best. Instead of producing many hyperparameters, VGG-16 used a filtered 3×3 convolution layer with a stride of 1 and a padding layer with a filter size of 2×2 and a stride of 2 [21]. Convolution layers and max-pooling layers are supported continuously throughout the architecture. An output classification layer is added to the end with two fully connected layers. In architecture, there are 16 weighted layers in total. 138 million parameters make up this network (see Fig. 2).

3.2 U-Net-Based Segmentation Model

An approach to object segmentation in aerial images was based on the U-Net model. This method is built around fully convolutional networks, as introduced by [40]. Rather than dense layers, it consists of convolutional layers. As a result, it accepts images of any size. According to Figure, the general architecture consists of two main sections: encoders and decoders. The encoder uses layers of convolutional and maximum pooling to obtain the image's context. The decoder section uses reversed or transposed convolutions to enable precise localization to create symmetric expanding paths. This section, also known as down-sampling, uses several deep learning classification architectures as the backbone. A 2×2 max-pooling layer is used with each

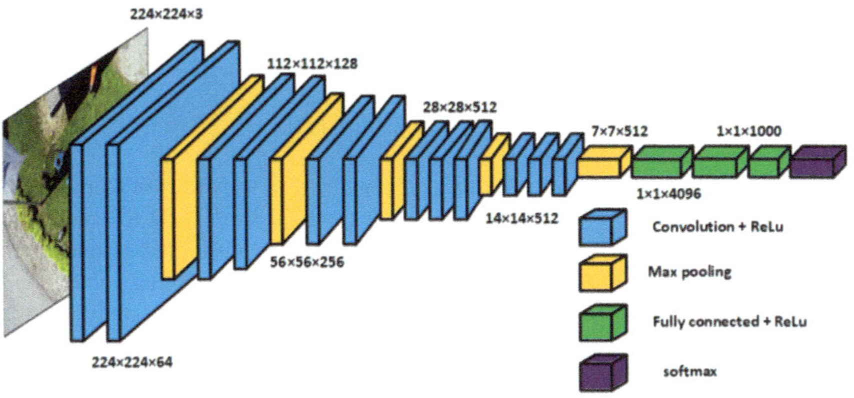

Fig. 2 A visual representation of the architecture of the VGG-16

step based on an in-depth analysis. Each step initially utilizes two 3 × 3 convolution layers with batch normalization. In the horizontal bottleneck, there are two layers of convolutions of 3 by 3 and two layers of up-convolutions of 2 by 2. Additionally, there are four stages to the up-sampling path: a decoder with three convolutional layers and two up-sampling layers. Feature maps are halved at each step.

Figure 3 shows how the architecture provides local and global information during up-sampling by skipping path connections. Feature maps represent how many segments the segmented map should contain, and the final convolutional layer constructs it.

To segment the input images, the architecture receives the preprocessed RGB images. It is determined that the input images have the shape of $(H, N, W, 3)$. Now, N indicates the number of input channels; channels are represented by 3, height and width by H and W. The map's energy function is determined using a pixel-wise

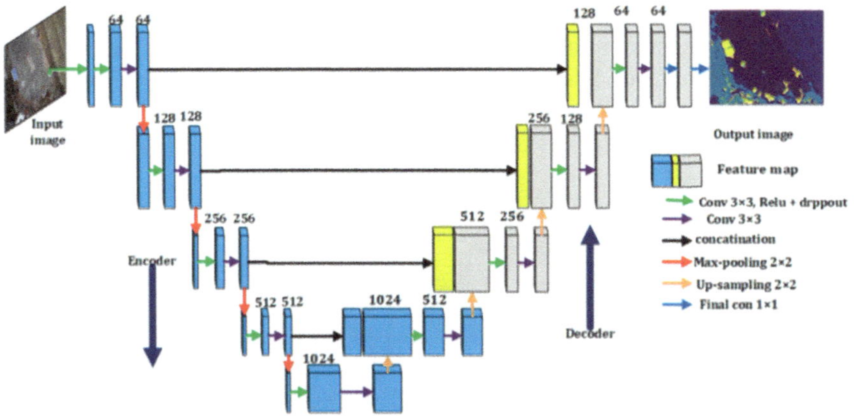

Fig. 3 Architecture visualization for U-Nets

softmax [40]. The description is as follows:

$$P_k(x) = \frac{\exp(a_k(x))}{\sum_{k'=1}^{K} \exp\left(a'_k(x)\right)} \quad (1)$$

The rectified linear unit (ReLu) in Eq. (1) symbolizes activation in feature maps or a_k. An object class's total number, K, is expressed by K, and the maximum approximation is expressed by $p_k(x)$. Assuming $a_k(x)$ at position k is maximally activated, $p_k(x)$ is approximately equal to 1. When it is set to another value, it will be considered $p_k(x) \approx 0$. A penalty is then applied to the cross entropy function at each location, as follows: [40]:

$$E = \sum_{w(x)\epsilon\Omega} \log\left(p_{l(x)}(x)\right) \quad (2)$$

Each pixel's true label is described by $l : \Omega \rightarrow 1,2,\ldots\ldots.K$ in Eq. (2). As $w : \Omega \rightarrow IR$, the weight map that emphasizes pixels during training is provided [40]. Pixels of various frequencies are segmented pre-computed. For some classes, we used morphological operations to generate weight maps.

$$w(x) = w_c(x) + w_0.\exp$$

As a result, $w_c : \Omega \rightarrow IR$ stands for weight map applied to class frequencies to balance them. There are two distances, d_1 and d_2, to the adjacent edges. $w_0 = 10$ and $\sigma \approx 5$ are both values [40].

4 Result Analysis and Discussion

The proposed model is evaluated using drone image datasets. It is accomplished by classifying every pixel in an image into the appropriate class, such as red for all people, purple for roads, blue for vehicles, yellow for street signs, etc. In Table 1, we display the initial deep learning hyperparameters (see Fig. 4).

Table 1 Initial hyperparameters

Hyper	Parameter
Batch size	8
Steps per epoch	200.0
Validation steps	10.0
Input shape	(512,512,3)
Initial learning rate	0.0001
Number of epochs	20

Fig. 4 Sample images from aerial image dataset

4.1 Dataset Overview

The semantic understanding of urban scenes can be improved by using this dataset to make autonomous drone flights and landing procedures safer. From a nadir (bird's eye) perspective, the imagery depicts more than 20 houses. The images were captured using a high-resolution camera with a resolution of 6000 × 4000 pixels. As shown in Fig. 5, the semantic segmentation mask of the original image was extracted. 400 images from the training set are publicly available, while 200 are private. The training and test sets for person detection include annotations of bounding boxes. The same training and test sets were annotated with pixel-accurate information. A dataset with 20 classes is considered complex.

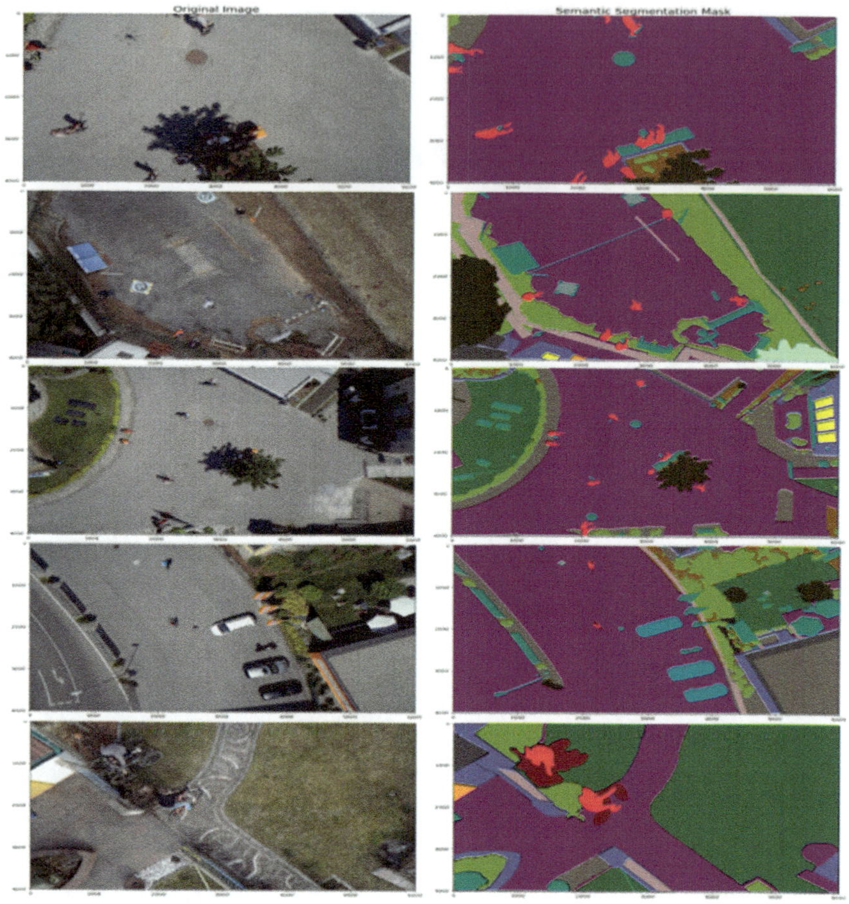

Fig. 5 Original image for the semantic segmentation

4.2 *Experimental Results*

Data augmentation is essential to teach a network's invariance and robustness properties. The invariance of shift and rotation is essential for microscopical images and robustness to variations in gray values and deformations. Figures 6, 7, and 8 illustrate how an annotated image segmentation network can be trained utilizing random elastic deformations of training samples. With a coarse grid of 3 by 3, we generate smooth deformations. A Gaussian distribution with a standard deviation of 10 pixels is used to sample displacements. In the next step, bicubic interpolation is used to compute per-pixel displacements. After the contracting path is completed, drop-out layers are used to augment implicit data further.

400 images were available in the dataset, from which 320 were used as training images (80%) and 80 as validation images (20%). Data augmentation is preferred

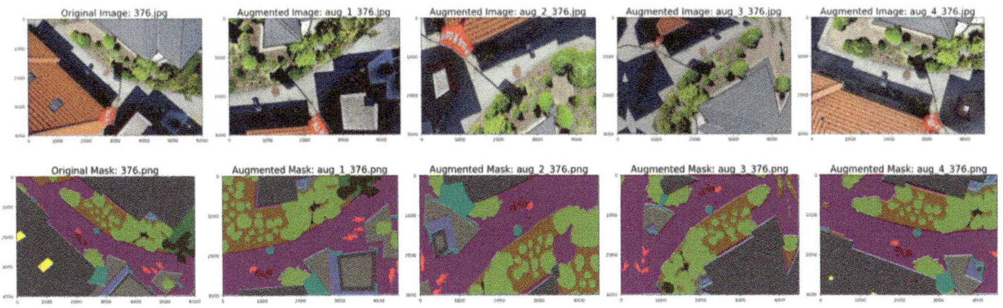

Fig. 6 Original image for augmented images and masks of the dataset

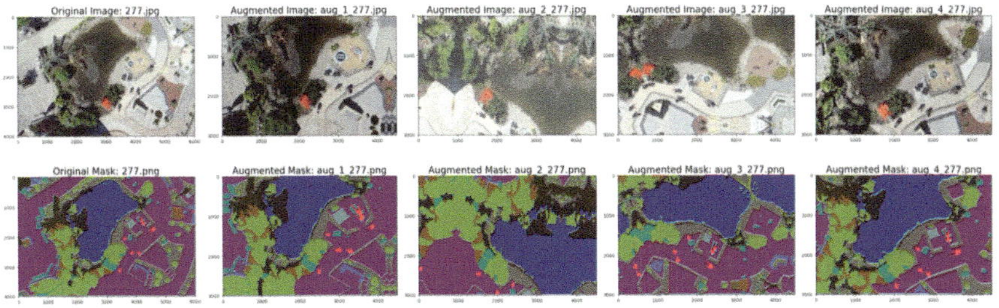

Fig. 7 Original image for augmented images and masks of the dataset

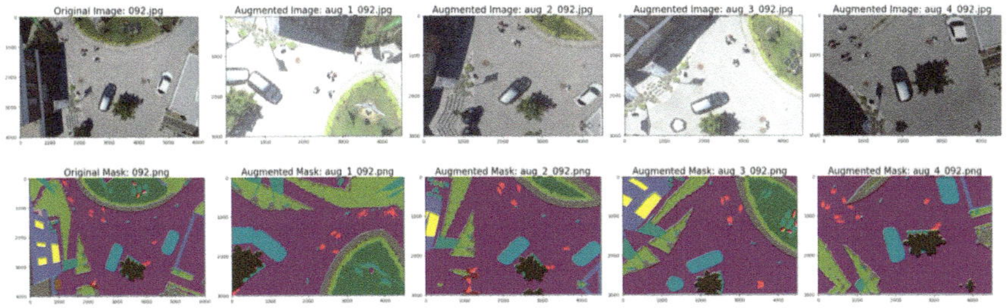

Fig. 8 Original image for augmented images and masks of the dataset

to avoid overfitting since there is a relatively small amount of data. In what follows, 1600 images are included in the training dataset and 80 images in the validation dataset after data augmentation.

In the proposed model, we increased the data by performing deformation on the images with batch size from 1 to 8 and changing the Learning Rate (LR) from the default (10^{-3}) to LR of (10^{-4}) (see Table 2).

After the model has been trained for 100 epochs, Figs. 9 and 10 show the loss and accuracy curves for the training phase. With an increasing number of epochs, the loss decreases nonlinearly. Additionally, IoU loss is very effective in segmenting small and large areas. The proposed model is illustrated in Figs. 11, 12, and 13.

Table 2 The results of training

Model	Epochs	Train dice coefficient	Train loss	Vol. dice coefficient	Vol loss	Max (initial) LR	Min LR	Total training time
VGG16-U-Net	20 (best weights at 18th epoch)	0.8781	0.2599	0.8702	0.29959	1.000×10^{-4}	1.122×10^{-5}	23569 s (06:32:49)

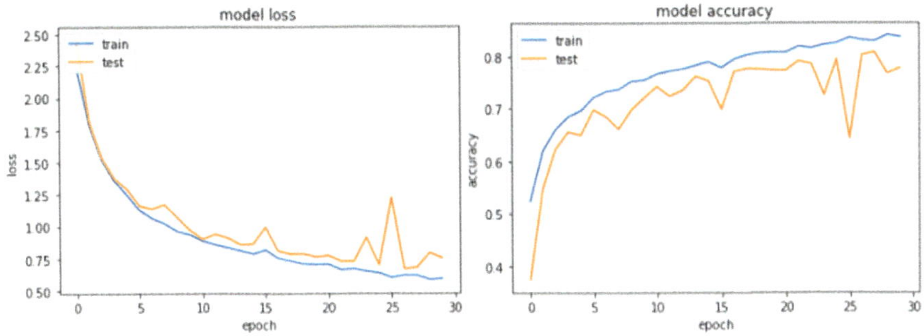

Fig. 9 Model training and loss of the augmented data

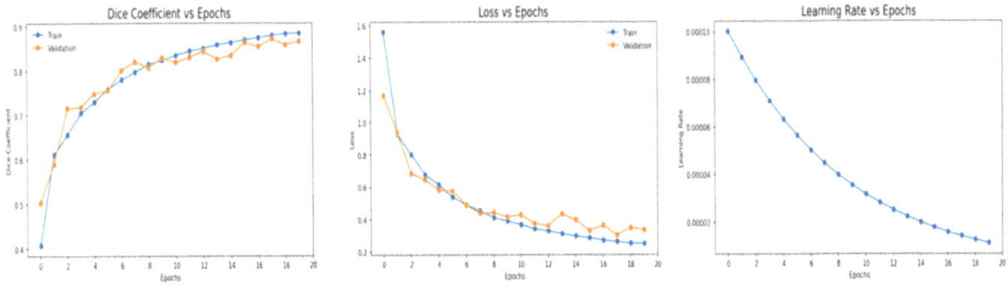

Fig. 10 Model performance concerning dice coefficient, loss, and learning rate for epochs

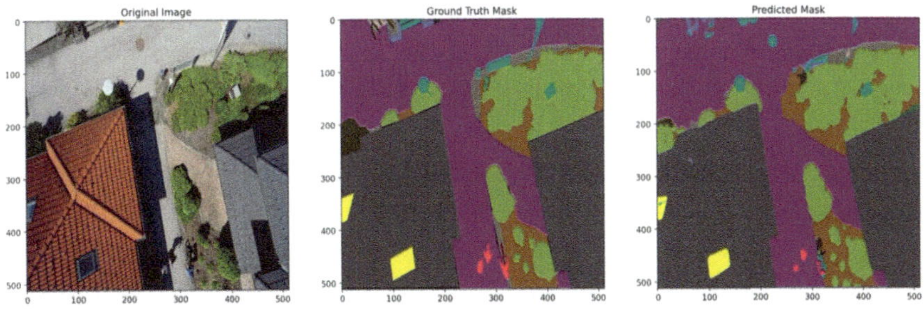

Fig. 11 The model prediction results based on a ground truth mask of the validation set

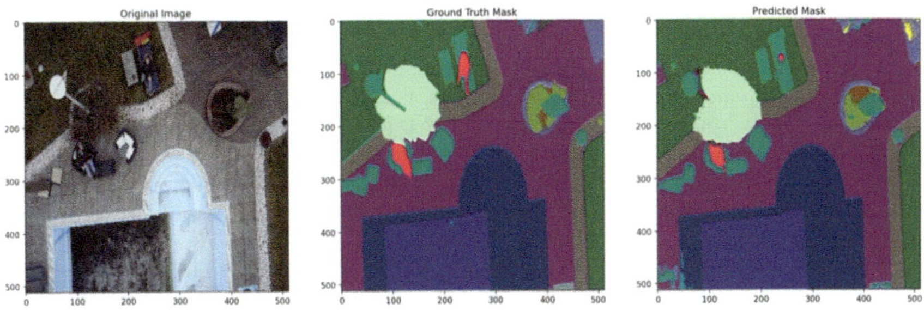

Fig. 12 Validation set with ground truth mask for model prediction

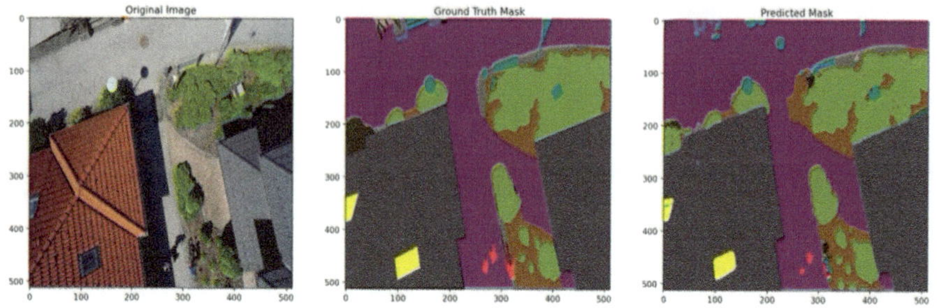

Fig. 13 Predictions based on the ground truth mask on the validation set

5 Conclusion

Semantic segmentation of aerial images has been extensively used to monitor environmental changes, plan cities, and manage disasters. The accuracy of the image semantic segmentation algorithms determines the reliability of these decision support systems. The U-Net architecture performs very well on a wide range of biomedical segmentation applications. U-Net is recommended to segment pixels and train them to predict classes. The segmentation results of this network can be improved through further research and optimization. We obtained a matrix with the same dimensions as the input image, so U-Net was applied to the aerial images, and the Pytorch library performed accurate and fast pixel prediction through the U-Net. Our future work might include other deep learning segmentation models. A fine-tuned model is also expected to improve the model's performance on a variety of benchmark datasets.

References

1. Yadav, S.P., Jindal, M., Rani, P., De Albuquerque, V.H.C., Dos Santos Nascimento, C., Kumar, M.: An improved deep learning-based optimal object detection system from images. Multimed. Tools Appl. (Sep 2023). https://doi.org/10.1007/s11042-023-16736-5
2. Verma, S., et al.: An automated face mask detection system using transfer learning based neural network to preventing viral infection. Expert Syst., e13507 (Nov 2023). https://doi.org/10.1111/exsy.13507
3. Yuan, X., Shi, J., Gu, L.: A review of deep learning methods for semantic segmentation of remote sensing imagery. Expert Syst. Appl. **169**, 114417 (2021)
4. Yao, H., Qin, R., Chen, X.: Unmanned aerial vehicle for remote sensing applications—a review. Remote Sens. **11**(12), 1443 (2019)
5. Matikainen, L., Karila, K.: Segment-based land cover mapping of a suburban area—comparison of high-resolution remotely sensed datasets using classification trees and test field points. Remote Sens. **3**(8), 1777–1804 (2011)
6. Rani, P., Sharma, R.: An experimental study of IEEE 802.11n devices for vehicular networks with various propagation loss models. In: Dubey, A.K., Sugumaran, V., Chong, P.H.J. (eds.) Advanced IoT Sensors, Networks and Systems. Lecture Notes in Electrical Engineering, vol. 1027, pp. 125–135. Springer Nature Singapore, Singapore (2023). https://doi.org/10.1007/978-981-99-1312-1_11

7. Wang, L., Li, R., Wang, D., Duan, C., Wang, T., Meng, X.: Transformer meets convolution: a bilateral awareness network for semantic segmentation of very fine resolution urban scene images. Remote Sens. **13**(16), 3065 (2021)
8. Rani, P., Sharma, R.: IMFOCA-IOV: intelligent moth flame optimization based clustering algorithm for internet of vehicle. In: 2023 14th International Conference on Computing Communication and Networking Technologies (ICCCNT), Delhi, India, pp. 1–6. IEEE (Jul 2023). https://doi.org/10.1109/ICCCNT56998.2023.10307646
9. Minaee, S., Boykov, Y., Porikli, F., Plaza, A., Kehtarnavaz, N., Terzopoulos, D.: Image segmentation using deep learning: a survey. IEEE Trans. Pattern Anal. Mach. Intell. **44**(7), 3523–3542 (2021)
10. Khan, M.A., Ectors, W., Bellemans, T., Janssens, D., Wets, G.: Unmanned aerial vehicle-based traffic analysis: a case study for shockwave identification and flow parameters estimation at signalized intersections. Remote Sens. **10**(3), 458 (2018)
11. Ke, R., Li, Z., Tang, J., Pan, Z., Wang, Y.: Real-time traffic flow parameter estimation from UAV video based on ensemble classifier and optical flow. IEEE Trans. Intell. Transp. Syst.Intell. Transp. Syst. **20**(1), 54–64 (2018)
12. Nex, F., Duarte, D., Steenbeek, A., Kerle, N.: Towards real-time building damage mapping with low-cost UAV solutions. Remote Sens. **11**(3), 287 (2019)
13. Chen, L.-C., Zhu, Y., Papandreou, G., Schroff, F., Adam, H.: Encoder-decoder with atrous separable convolution for semantic image segmentation. In: Proceedings of the European Conference on Computer Vision (ECCV), pp. 801–818 (2018)
14. Rani, P., Verma, S., Yadav, S.P., Rai, B.K., Naruka, M.S., Kumar, D.: Simulation of the lightweight blockchain technique based on privacy and security for healthcare data for the cloud system. Int. J. E-Health Med. Commun. IJEHMC **13**(4), 1–15 (2022)
15. Becquet, C., Blachon, S., Jeudy, B., Boulicaut, J.-F., Gandrillon, O.: Strong-association-rule mining for large-scale gene-expression data analysis: a case study on human SAGE data. Genome Biol. **3**(12), 1–16 (2002)
16. Creighton, C., Hanash, S.: Mining gene expression databases for association rules. Bioinformatics **19**(1), 79–86 (2003). https://doi.org/10.1093/bioinformatics/19.1.79
17. Szegedy, C., et al.: Going deeper with convolutions. In: Proceedings of the IEEE Conference on Computer Vision and Pattern Recognition, pp. 1–9 (2015)
18. Donahue, J., et al.: Decaf: a deep convolutional activation feature for generic visual recognition. In: International Conference on Machine Learning, PMLR, pp. 647–655 (2014)
19. Rani, P., Sharma, R.: Intelligent transportation system performance analysis of indoor and outdoor internet of vehicle (iov) applications towards 5g. Tsinghua Sci. Technol. (2023)
20. Rani, P., Singh, P.N., Verma, S., Ali, N., Shukla, P.K., Alhassan, M.: An implementation of modified blowfish technique with honey bee behavior optimization for load balancing in cloud system environment. Wirel. Commun. Mob. Comput. **2022**, 1–14 (2022)
21. Simonyan, K., Zisserman, A.: Very deep convolutional networks for large-scale image recognition (2014). ArXiv14091556
22. Krizhevsky, A., Sutskever, I., Hinton, G.E.: Imagenet classification with deep convolutional neural networks. Adv. Neural Inf. Process. Syst. **25** (2012)
23. Huang, G., Liu, Z., Van Der Maaten, L., Weinberger, K.Q.: Densely connected convolutional networks. In: Proceedings of the IEEE Conference on Computer Vision and Pattern Recognition, pp. 4700–4708 (2017)
24. He, K., Zhang, X., Ren, S., Sun, J.: Deep residual learning for image recognition. In: Proceedings of the IEEE Conference on Computer Vision and Pattern Recognition, pp. 770–778 (2016)
25. Cha, Y., Choi, W., Büyüköztürk, O.: Deep learning-based crack damage detection using convolutional neural networks. Comput.-Aided Civ. Infrastruct. Eng. **32**(5), 361–378 (2017). https://doi.org/10.1111/mice.12263
26. Zhang, L., Yang, F., Zhang, Y.D., Zhu, Y.J.: Road crack detection using deep convolutional neural network. In: 2016 IEEE international conference on image processing (ICIP), pp. 3708–3712. IEEE (2016)

27. Soukup, D., Huber-Mörk, R.: Convolutional neural networks for steel surface defect detection from photometric stereo images. In: International Symposium on Visual Computing, pp. 668–677. Springer (2014)
28. Zhu, P., Isaacs, J., Fu, B., Ferrari, S.: Deep learning feature extraction for target recognition and classification in underwater sonar images. In: 2017 IEEE 56th annual conference on decision and control (CDC), pp. 2724–2731. IEEE (2017)
29. Zhou, Y., Wang, J., Li, B., Meng, Q., Rocco, E., Saiani, A.: Underwater scene segmentation by deep neural network (2019)
30. Arain, B., McCool, C., Rigby, P., Cagara, D., Dunbabin, M.: Improving underwater obstacle detection using semantic image segmentation. In 2019 International Conference on Robotics and Automation (ICRA), pp. 9271–9277. IEEE (2019)
31. Möller, T., Nilssen, I., Nattkemper, T.W.: Tracking sponge size and behaviour with fixed underwater observatories. In: Pattern Recognition and Information Forensics: ICPR 2018 International Workshops, CVAUI, IWCF, and MIPPSNA, Beijing, China, August 20–24, 2018, Revised Selected Papers 24, pp. 45–54. Springer (2019)
32. Treboux, J., Genoud, D.: Improved machine learning methodology for high precision agriculture. In: 2018 Global Internet of Things Summit (GIoTS), pp. 1–6. IEEE (2018)
33. Lee, D., Kim, J., Lee, D.: Robust concrete crack detection using deep learning-based semantic segmentation. Int. J. Aeronaut. Space Sci. **20**(1), 287–299 (2019). https://doi.org/10.1007/s42405-018-0120-5
34. Majidizadeh, A., Hasani, H., Jafari, M.: Semantic segmentation of UAV images based on U-Net in urban area. ISPRS Ann. Photogramm. Remote Sens. Spat. Inf. Sci. **X-4/W1-2022**, 451–457 (2023). https://doi.org/10.5194/isprs-annals-X-4-W1-2022-451-2023
35. Dechesne, C., Mallet, C., Le Bris, A., Gouet-Brunet, V.: Semantic segmentation of forest stands of pure species combining airborne lidar data and very high resolution multispectral imagery. ISPRS J. Photogramm. Remote Sens.Photogramm. Remote Sens. **126**, 129–145 (2017). https://doi.org/10.1016/j.isprsjprs.2017.02.011
36. Cui, B., Zhang, Y., Li, X., Wu, J., Lu, Y.: WetlandNet: semantic segmentation for remote sensing images of coastal wetlands via improved UNet with deconvolution. In: Genetic and Evolutionary Computing: Proceedings of the Thirteenth International Conference on Genetic and Evolutionary Computing, November 1–3, 2019, Qingdao, China 13, pp. 281–292. Springer (2020)
37. Kentsch, S., Lopez Caceres, M.L., Serrano, D., Roure, F., Diez, Y.: Computer vision and deep learning techniques for the analysis of drone-acquired forest images, a transfer learning study. Remote Sens. **12**(8), 1287 (2020). https://doi.org/10.3390/rs12081287
38. Do, D., Pham, F., Raheja, A., Bhandari, S.: Machine learning techniques for the assessment of citrus plant health using UAV-based digital images. In: Autonomous Air and Ground Sensing Systems for Agricultural Optimization and Phenotyping III, SPIE, pp. 189–200 (May 2018). https://doi.org/10.1117/12.2303989
39. Deng, J., Dong, W., Socher, R., Li, L.-J., Li, K., Fei-Fei, L.: Imagenet: a large-scale hierarchical image database. In: 2009 IEEE conference on computer vision and pattern recognition, pp. 248–255. IEEE (2009)
40. Ronneberger, O., Fischer, P., Brox, T.: U-net: convolutional networks for biomedical image segmentation. In: Medical Image Computing and Computer-Assisted Intervention–MICCAI 2015: 18th International Conference, Munich, Germany, October 5–9, 2015, Proceedings, Part III 18, pp. 234–241. Springer (2015)

Multispectral Satellite Image Classification Using Symbolic Machine Learning for Land Cover Area and Land Cover

Mohammed Ridha Hammoodi and Ali Ashoor Issa

Abstract For sustainable urban development globally, it is essential to monitor human-induced changes and develop reliable and methodologically consistent urban area maps. In GHSL, multispectral satellite images are segmented using symbolic machine learning (SML) to produce reliable and automated maps of built-up areas. Experiments conducted on the benchmark Landsat-8 dataset demonstrated SML's accurate classification of large satellite images. Large datasets were measured with high accuracy using both simple and quantum-entanglement-free kernels.

Keywords Multispectral satellite image · Machine learning · Land cover area

1 Introduction

A wide range of geographic scales can be obtained from satellite data due to their temporally consistent nature over large areas [1]. Similar sub-regions and low-level features in different satellite image classes make classification a challenging task [2–15]. Assigning a land use category to an image is the main objective of classification. A subfield of artificial intelligence, machine learning algorithms, is used to extract information automatically from data using classification techniques. The features of the images, however, must first be extracted, so that fewer resources are required for the description of a large dataset, and then ML algorithms like SVM are used to learn the features of the images [16, 17], RF [6, 18, 19], etc. There are two types of algorithms: supervised [20–23] and unsupervised procedures [6, 24], which select pixels from an image that represent specific classes and then classify the rest of the image using those training sites as references. Data collected from the LANDSAT-8

M. R. Hammoodi (✉)
Faculty of Computer Science and Mathematics, University of Kufa, Kufa, Iraq
e-mail: mridha.hamoodi@uokufa.edu.iq

A. A. Issa
College of Engineering, Medical Instruments Technology Engineering, National University of Science and Technology, Dhi Qar, Nasiriyah, Iraq

S. Kumar Udgata et al. (eds.), *Intelligent Systems*, Lecture Notes in Networks and Systems 1149, https://doi.org/10.1007/978-981-97-8160-7_28

satellite were used to perform supervised classification using ML algorithms in the Moroccan territory [22]. In this paper, we describe six supervised algorithms for mapping land cover in the study area, including SVMs [16], RF [18], CART [15], MD [25], DT [15], and Gradient Tree Boost [26].

Storage devices have greatly improved over the past decades thanks to improvements in digital technology. In addition to texts, videos, and images comprise this large amount of data. In cases where there are multiple images to manage, such as satellite photos, it's crucial to set up data processing systems that help consumers find relevant content. Satellite classification and analysis can be used in various fields, including disaster recovery, geography, forestry, and monitoring damage from natural disasters. Additionally, satellite images are classified to map pixels, objects, and other features. Researchers have benefited from remote sensing in various research fields, including academia, society, policy, and other interested citizens. The development of satellite image processing has increased interest in remote sensing. Large areas cannot be analyzed by the human eye alone. An incorrect labeling of the image may lead to incorrect feedback for decision-making in the case of strategic development, and the image information is mistakenly labeled water instead of soil. Highly specific techniques are required to classify images accurately. The correct classification of images, therefore, requires the application of efficient techniques. Even so, machine learning can generate approaches that enable the collation of images. This paper presents a framework for automatically classifying satellite images using machine learning techniques. Using different datasets, we will also examine how different machine learning techniques perform [27, 28].

Through multiple processing layers, ML models represent data at different abstraction levels [29]. A combination of CNNs and graphics processing units (GPUs) has yielded astonishing results in the detection and classification of objects [30]. CNN-based algorithms have dominated object detection and classification in photographs since 2012 [31]. CNN-based products and services have already been deployed by Google, Facebook, and Microsoft [29].

Satellite imagery is valuable for finding important objects, facilities, and events. Law enforcement, mudslides, and flooding seek out unlicensed quarrying processes, illegal spinning pitchers are mapped by disaster response teams, and financial investors monitor agricultural development and oil wells. It is necessary to automate the search process because of the massive geographic area to cover and the limited number of analysts who can conduct the searches. Despite their inaccuracy and inefficiency, traditional object detection and classification algorithms can be used to help solve this problem. Objects and facilities must be identified and labeled automatically with deep learning.

2 Related Work

The author considered handcrafted features as part of the CNN architecture [32]. Image classification problems can be solved using this architecture. Their accuracy scores were 99.90% for handcrafted features and 99.84% for CNN feature maps constructed using this framework. Convolutional neural networks can be fed with an appropriate scale by developing a technique [33]. To determine the best output, CNNs were initially trained with scales chosen for a dummy time. The CNN model undergoes a full training phase after selecting the optimal scale with the highest precision. This step can be made more precise by embedding an adaptive scale mechanism. An unsupervised ordered clustering algorithm for deep satellite image analysis has been proposed using a deep satellite image examination system [34]. Despite the promise to examine the images with feeble socio-economic and demographic foundations, the main disadvantage of their method was the high requirement for computing power. Using k-means optimization, [35] presents a bare-bone explosives procedure for image classification based on bare-bones fireworks [35]. Compared with the original k-means algorithm, the model performs better for image classification.

There have recently been several annotations of imagery and activities related to detection and classification. Land cover is used to classify building and non-building area detection for the most prevalent use of deep learning (DL) on remotely sensed imagery. For example, a survey from the U.S. Geological Survey is included in the UC Merced Land-use Dataset [31]. Each pixel in the images is 0.3 m across and is 256×256 pixels in size. Agribusiness, roads, water, storage tanks, tennis courts, and other facilities are included in the 21 classes. UC Merced images were classified into land cover types using VGG, ResNet, and Inception CNNs [36–38], with one study reporting 98.5% accuracy in classification [38]. While this dataset is quite large and has a wide variety of classes and regions, it is limited in its size and type of classes.

The author [39] states that this study aims to estimate the performance of ML models under unpredictable conditions to enhance their use for recognizing urban patterns based on object-based image analysis. This study has selected and implemented four machine learning concepts (tree-based, probabilistic, function-based, and nearest neighbor) on a free and open-source source. These techniques use various methods, including Normal Bayes, KNN, RT, and SVM. A particular focus is assessing how well machine learning algorithms generalize to different image types and scenes and their transferability [41, 42].

3 Methodology: Symbolic Machine Learning (SML)

This experiment describes a new supervised classification technique introduced by [40]. This technique has proven scalable for large and complex multi-scene satellite data processing scenarios. This technique involves two main steps: (i) the quantization sequencing of the image and (ii) the association rules analysis.

$m \times n = mn$ spatial samples or pixels with F features or descriptors are included in $D_{m \times n \times F}$. In this example, $X_{mn \times F}$ is a 2D image matrix, $X = [x_1, x_2, \ldots.., x_i, \ldots..x_F]$, where F means the number of features used, and $x_i \in Z_{+^{mn}}$ represents the number of features not used. A set of unique instances of X can be defined as $\widehat{X}$.

For the data quantization and sequencing phase, an equal quantization approach was implemented for each epoch feature. Quantized data instances $X(q) = \left[\text{round}\left(\frac{x}{q}\right)\right]_{mn \times F}$ are those with $q = \frac{\max(x)}{s}, x \in X_i$.

Throughout the diverse data epochs, the numerator q, the symbols s, and the attributes F are shown in Table 1. Using $\widehat{X}$ estimate of $p_\mu = |X|/\left|\widehat{X}\right|$ and selecting a value in the range of 10^2 to 10^3 as p_μ, the q value was determined empirically. Using noisy training sets with noise characteristics related to those investigated here, these p_μ orders of magnitude were satisfactory for classification exercises [40]. It is determined by the sensor characteristic in a particular epoch how many spectral attributes are available at X, while the number of symbols available for each spectral attribute is determined by combining *DN* encoding, sensor data *D*, and the q parameter. The new Quality Assessment $(QA)^{14}$ band was also included in the image data sequence for the epoch 2014 containing *L8* data. Depending on the surface, sensor conditions, and atmosphere, image elements (pixels) can collect radiometric information. SML's symbolic nature allows seamless integration into the input data attributes for classification. QA attribute quantization was not applied because it encodes a classification output.

Evidence-based Normalized Differential Index (ENDI)Φ_E formulation does not require mutual exclusivity between Y^{+,Y^-} training sets and is thus suitable for other confidence measures $Y^{+\cap Y^{-\geq\varnothing}}$. As a result of this property, partially overlapping or inconsistent training sets (by scale and by theme) can be exploited, such as the ones used in this study.

In the Φ_E, values are assumed between [−1, 1]. Data instances and target classes or foreground information are estimated to be associated with them. Data instances with values close to 1 have strong correlations to the presence of foreground information (buildings, for example). In contrast, values close to −1 represent data instances with a strong correlation to the presence of the complement (background, not built-up) as indicated by the negative evidence. When Φ_E values are close to 0, and a comparable amount of evidence can support both the foreground and background information hypothesis. Thus, they fall within the decision space's boundaries.

Table 1 Analyzes image data for various epochs by showing the parameters applied

Year	2014	2000	1990	1975
Q	1024	8	4	2
S	16	32	64	128
F	7(+1)	6	6	4
Max x	65,535	255	255	255

3.1 Computational Complexity

A data instance/example/observation is denoted by m, a feature is denoted by n, a variable is denoted by v, and Y classes are denoted by x_i rather than j. For brevity, we assume that $v = v_j, j = 1, \ldots, n$ and $V = \prod_{j=1}^{n} v_j$ are the variables. The data taxonomy and data frequencies' computation contribute to the total time complexity of SML training, which is $O(m.n)$. An upper bound of order $O(m^2.n)$ can be found using the brute force approach. An ov^n-scale time complexity can be achieved by exploiting bit arrays and hash tables. An extra structure of size $(m.y)$ contains instances of every unique data instance observed in each class when all the data instances of the initial dataset are unique, creating a worst-case memory requirement of $O(m.(n + Y))$. Pragmatic space complexity of OY, v^n was estimated from the previous analysis. The class label of a new n-datum is determined by its frequency in each class and the threshold applied to it, resulting in $O(Y)$ at the classification stage. In addition to the $O(V.n)$ cost to match the n-datum to an element of the unique data instance matrix, there is also a cost of $o(1)$ when hashing is applied; the space complexity remains the same, $O(Y.v^n)$.

4 Result Analysis and Discussion

Human settlement is assessed using the Global Human Settlement Layer (GHSL). A built-up map, a population map, and a settlement map are among the types of global spatial information that it provides. A machine learning algorithm generates this information by using evidence-based analytics. This layer is generated using Earth Observation Satellites and Census data. The development pipeline includes three stages: BUILT UP → POP → SMOD.

Dataset: In this paper, two different types of datasets are used.

Sentinel-2:

- The European Space Agency operates a satellite for Earth observation
- There are 13 bands in the multispectral data
- Three spatial resolutions: 10, 20, and 60 m

Landsat-8:

- Earth observation satellite operated by the United States Geological Survey (USGS)
- With 11 bands, multispectral data is available
- Resolutions of 15, 30, and 100 m

Symbolic machine learning (SML)

Step 1—Data Collection

- *X*: Sentinel-2/Landsat-8 multispectral satellite images were collected
- *Y*: Corine Land Cover (CLC) with land cover labels (Water bodies, Urban region, etc.) as categorical reference data

Step 2—Data Reduction

- Taxonomy: meaning quantization of numerical values
- Sequencing: construct sequences of tuples $X \rightarrow Y$
- Unique sequences: identify unique sequences of $X \rightarrow Y$.

Example representation of each pixel: (Band1, Band2, Band3) → (Built)

Step 3—Association Analysis

Interestingness measure:

Data sequences in *X* are scored based on the occurrence of their sequences in each of the reference classes in *Y* using an Evidence Normalized Differential (ENDI) measure.

$$EndI_A = f_{\text{pos}} - \frac{f_{\neg}}{f_{\text{pos}} + f_{\neg}} \quad ENDI = \frac{ENDI_A + ENDI_B}{2}$$

$$ENDI^B = p_{\text{pos}} - \frac{p_{\neg}}{p_{\text{pos}} + p_{\neg}} \quad ENDI \text{ range} = [-1, +1].$$

A positive reference instance and a negative reference instance are observed to occur more than once in *X* data instances. It is known as f_{pos} and $f_{\neg}$.

p_{pos} and $p_{\neg}$ represent the probabilities.

Classification: argmaxof *ENDI* measures of each class. These results in Built-Up or Non-Built-up classification of an individual pixel in an input image based on the accuracy, F1 score, Precision, Recall, MeanIoU, and Macro average.

GHS BUILT-UP: The input is considered the Multispectral satellite imagery. Symbolic Machine Learning is used for the built-up classification. The output of the built-up classification on the image (0 = Built up, 1 = No built BUILT UP (0/1) classification and BUILT UP (confidence 0 100) as shown in Fig. 1.

GHS-POP: The Input Gridded Population of the World (GPW v4) is considered. It models the distribution of the human population (counts and densities) on a raster

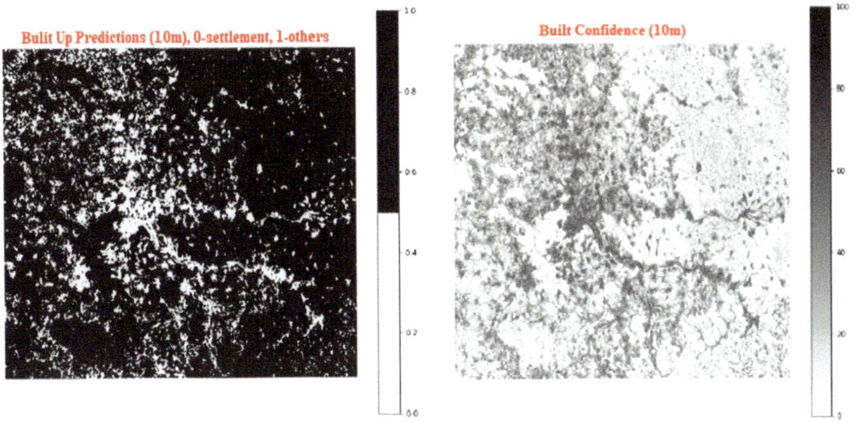

Fig. 1 The result of the build-up model with the prediction and confidence imagery

surface. This system produces population grids that depict the density and distribution of the population in terms of people per cell.

Step 1 Divide the region into sub-regions with different classes of population (0–6) with 0 = no_population to 6 = very_high_population

Step 2 Identify the types of population and calculate their densities

Step 3 Each population type needs to be calculated based on its proportions

Step 4 Assign each region with population based on the evidence of GHS BUILT, distribute by proportion

Setting different thresholds for population distribution can lead to different results. Density and Count-based POP generation do not work well with the same asymmetric mapping technique. Experimented to generate high resolution (10 m) POP layer as shown in Fig. 2.

GHS-SMOD: SMOD results for GHS-BUILT and GHS_POP models must be input for GHS-SMOD analysis. Based on the number of people living per cell, divide the total area into urban_centres, urban_clusters, rural_areas, and no_populations. The three SMOD rules are used for the classification.

- Urban centre: If ((GHS POP > 1500) or (GHS BUILT UP > 50%)) and (4 connected cells GHS POP > 50,000)
- Urban cluster: If ((GHS POP > 300) and (GHS BUILT UP > 3%)) and (4 connected cells GHS POP > 5000)
- Rural area: If ((GHS POP > 1) and (single or connected cells GHS) POP < 5000)

Categorizing areas into urban centres, clusters and rural works differently in pixels around water, mountains, etc. Experimented to generate high resolution (10 m) SMOD layer. Results are influenced by BUILT UP and POP, as shown in Fig. 3.

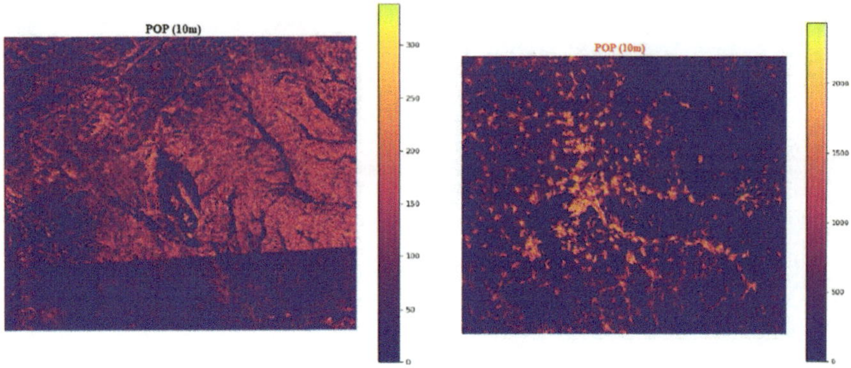

Fig. 2 POP result of the population representation imagery with (Yellow = high population, Blue = low)

Fig. 3 SMOD result imagery with (Red = urban centre, Yellow = urban cluster, Green = rural region, White = no)

Figure 4 shows that the attention weights for some interest classes differ for all annotated pixels, illustrating the attention operator's interest in different classes. The decision-making process exploits all bands. These weights closely match the ones obtained at the validation and test scales. The class 'Water' is an exception. Considering channels or the temporal dimension, we can easily merge attention weights. However, implementing the same logic for a space dimension is more challenging.

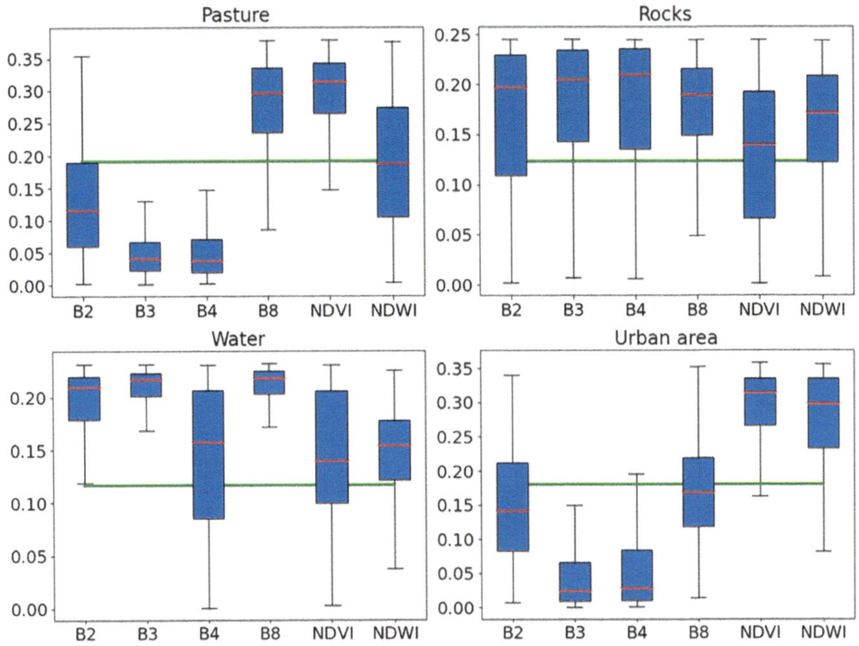

Fig. 4 A box plot compares channel weight attention for four classes normalized by class sums for each annotated pixel. An activation threshold is represented by horizontal lines (0.5 in the case of a sigmoid)

5 Conclusion

This manuscript presents a novel framework based on spatial analysis of multi-temporal very high resolution (VHR) satellite images for firm spatial classification of a geographic area. Machine learning is a powerful approach to finding patterns in voluminous data, commonly known as big data. Some environmental problems can be solved using remote sensing and ML concepts. LANDSAT sensors can capture raw spatial data, drawing meaningful insights according to particular arena requirements. The purpose of this paper is to introduce SML for the detection of clouds in multispectral satellite images with the use of quantum kernels. Experimental studies were conducted using widely used Landsat-8 imagery to examine such quantum-kernel models with SML. SML has used the image segmentation process in machine learning to segment the image to estimate the land cover area and get better accuracy.

References

1. Abburu, S., Babu Golla, S.: Satellite image classification methods and techniques: a review. Int. J. Comput. Appl. **119**(8), 20–25 (2015). https://doi.org/10.5120/21088-3779
2. Ouchra, H., Belangour, A., Erraissi, A.: A comprehensive study of using remote sensing and geographical information systems for urban planning (2022)
3. Ouchra, H., Belangour, A., Erraissi, A.: Satellite data analysis and geographic information system for urban planning: a systematic review. In: 2022 International Conference on Data Analytics for Business and Industry (ICDABI), pp. 558–564. IEEE (2022)
4. Ouchra, H., Belangour, A., Erraissi, A.: Spatial Data Mining technology for GIS: a review. In: 2022 International Conference on Data Analytics for Business and Industry (ICDABI), pp. 655–659. IEEE (2022)
5. Ouchra, H., Belangour, A., Erraissi, A.: Machine learning for satellite image classification: a comprehensive review. In: 2022 International Conference on Data Analytics for Business and Industry (ICDABI), pp. 1–5. IEEE (2022)
6. Ouchra, H., Belangour, A.: Satellite image classification methods and techniques: a survey. In: 2021 IEEE International Conference on Imaging Systems and Techniques (IST), pp. 1–6. IEEE (2021)
7. Ouchra, H., Belangour, A.: Object detection approaches in images: a survey. In: Jiang, X., Fujita, H. (eds.) Thirteenth International Conference on Digital Image Processing (ICDIP 2021), p. 85. Singapore, Singapore: SPIE (Jun 2021). https://doi.org/10.1117/12.2601452
8. Sharifi, A.: Estimation of biophysical parameters in wheat crops in Golestan province using ultra-high resolution images. Remote Sens. Lett. **9**(6), 559–568 (2018). https://doi.org/10.1080/2150704X.2018.1452058
9. Sharifi, A.: Flood mapping using relevance vector machine and SAR data: a case study from Aqqala, Iran. J. Indian Soc. Remote Sens. **48**(9), 1289–1296 (2020). https://doi.org/10.1007/s12524-020-01155-y
10. Sharifi, A., Amini, J., Tateishi, R.: Estimation of forest biomass using multivariate relevance vector regression. Photogramm. Eng. Remote Sens. **82**(1), 41–49 (2016). https://doi.org/10.14358/PERS.83.1.41
11. Jalayer, S., Sharifi, A., Abbasi-Moghadam, D., Tariq, A., Qin, S.: Modeling and predicting land use land cover spatiotemporal changes: a case study in chalus watershed, Iran. IEEE J. Sel. Top. Appl. Earth Obs. Remote Sens. **15**, 5496–5513 (2022)
12. Esmaeili, M., Abbasi-Moghadam, D., Sharifi, A., Tariq, A., Li, Q.: Hyperspectral image band selection based on CNN embedded GA (CNNeGA). IEEE J. Sel. Top. Appl. Earth Obs. Remote Sens. **16**, 1927–1950 (2023)
13. Ghaderizadeh, S., Abbasi-Moghadam, D., Sharifi, A., Tariq, A., Qin, S.: Multiscale dual-branch residual spectral–spatial network with attention for hyperspectral image classification. IEEE J. Sel. Top. Appl. Earth Obs. Remote Sens. **15**, 5455–5467 (2022)
14. Farmonov, N., et al.: Crop type classification by DESIS hyperspectral imagery and machine learning algorithms. IEEE J. Sel. Top. Appl. Earth Obs. Remote Sens. **16**, 1576–1588 (2023). https://doi.org/10.1109/JSTARS.2023.3239756
15. Hu, Y., Dong, Y., Batunacun: An automatic approach for land-change detection and land updates based on integrated NDVI timing analysis and the CVAPS method with GEE support. ISPRS J. Photogramm. Remote Sens. **146**, 347–359 (2018). https://doi.org/10.1016/j.isprsjprs.2018.10.008
16. Awad, M.: Google Earth Engine (GEE) cloud computing based crop classification using radar, optical images and Support Vector Machine Algorithm (SVM). In: 2021 IEEE 3rd International Multidisciplinary Conference on Engineering Technology (IMCET), pp. 71–76. IEEE (2021)
17. Rani, P., Sharma, R.: Intelligent transportation system for internet of vehicles based vehicular networks for smart cities. Comput. Electr. Eng. **105**, 108543 (2023)
18. Magidi, J., Nhamo, L., Mpandeli, S., Mabhaudhi, T.: Application of the random forest classifier to map irrigated areas using google earth engine. Remote Sens. **13**(5), 876 (2021). https://doi.org/10.3390/rs13050876

19. Ouchra, H., Belangour, A., Erraissi, A.: A comparative study on pixel-based classification and object-oriented classification of satellite image. Int. J. Eng. Trends Technol. **70**(8), 206–215 (2022). https://doi.org/10.14445/22315381/IJETT-V70I8P221
20. Dhingra, S., Kumar, D.: A review of remotely sensed satellite image classification. Int. J. Electr. Comput. Eng. IJECE **9**(3), 1720 (2019). https://doi.org/10.11591/ijece.v9i3.pp1720-1731
21. Usama, M., et al.: Unsupervised machine learning for networking: techniques, applications and research challenges. IEEE Access **7**, 65579–65615 (2019)
22. Shafaey, M.A., Salem, M.A.-M., Ebied, H.M., Al-Berry, M.N., Tolba, M.F.: Deep learning for satellite image classification. In: Hassanien, A.E., Tolba, M.F., Shaalan, K., Azar, A.T. (eds.) Proceedings of the International Conference on Advanced Intelligent Systems and Informatics 2018. Advances in Intelligent Systems and Computing, vol. 845, pp. 383–391. Springer International Publishing, Cham (2019). https://doi.org/10.1007/978-3-319-99010-1_35
23. Ansari, G., Rani, P., Kumar, V.: A novel technique of mixed gas identification based on the group method of data handling (GMDH) on time-dependent MOX gas sensor data. In: Mahapatra, R.P., Peddoju, S.K., Roy, S., Parwekar, P. (eds.) Proceedings of International Conference on Recent Trends in Computing. Lecture Notes in Networks and Systems, vol. 600, pp. 641–654. Springer Nature Singapore, Singapore (2023). https://doi.org/10.1007/978-981-19-8825-7_55
24. Rani, P., Verma, S., Yadav, S.P., Rai, B.K., Naruka, M.S., Kumar, D.: Simulation of the lightweight blockchain technique based on privacy and security for healthcare data for the cloud system. Int. J. E-Health Med. Commun. IJEHMC **13**(4), 1–15 (2022)
25. Borra, S., Thanki, R., Dey, N.: Satellite image analysis: clustering and classification. In: Springer Briefs in Applied Sciences and Technology. Springer Singapore, Singapore (2019). https://doi.org/10.1007/978-981-13-6424-2
26. Natekin, A., Knoll, A.: Gradient boosting machines, a tutorial. Front. Neurorobotics **7** (2013). https://doi.org/10.3389/fnbot.2013.00021
27. Rani, P., et al.: Federated learning-based misbehaviour detection for the 5G-enabled internet of vehicles. IEEE Trans. Consum. Electron., 1–1 (2023). https://doi.org/10.1109/TCE.2023.3328020
28. Hussain, N., Rani, P.: Comparative studied based on attack resilient and efficient protocol with intrusion detection system based on deep neural network for vehicular system security. In: Distributed Artificial Intelligence, pp. 217–236. CRC Press (2020). https://www.taylorfrancis.com/chapters/edit/10.1201/9781003038467-13/comparative-studied-based-attack-resilient-efficient-protocol-intrusion-detection-system-based-deep-neural-network-vehicular-system-security-naziya-hussain-preeti-rani. Accessed 19 Oct 2023
29. LeCun, Y., Bengio, Y., Hinton, G.: Deep learning. Nature **521**(7553), 436–444 (2015)
30. Kumar Agrawal, N., et al.: TFL-IHOA: three-layer federated learning based intelligent hybrid optimization algorithm for internet of vehicle. IEEE Trans. Consum. Electron. (2023)
31. Challenge, L.S.V.R.: Large scale visual recognition challenge (ILSVRC)–ImageNet. LSVRC
32. Liu, Q., et al.: Deepsat v2: feature augmented convolutional neural nets for satellite image classification. Remote Sens. Lett. **11**(2), 156–165 (2020)
33. Laban, N., Abdellatif, B., Ebied, H.M., Shedeed, H.A., Tolba, M.F.: Multiscale satellite image classification using deep learning approach. Mach. Learn. Data Min. Aerosp. Technol., 165–186 (2020)
34. Block, J., et al.: An unsupervised deep learning approach for satellite image analysis with applications in demographic analysis. In: 2017 IEEE 13th International Conference on e-Science (e-Science), pp. 9–18. IEEE (2017)
35. Tuba, E., Jovanovic, R., Tuba, M.: Multispectral satellite image classification based on bare bone fireworks algorithm. In: Information and Communication Technology for Sustainable Development: Proceedings of ICT4SD 2018, pp. 305–313. Springer (2020)
36. Liang, Y., Monteiro, S.T., Saber, E.S.: Transfer learning for high resolution aerial image classification. In: 2016 IEEE Applied Imagery Pattern Recognition Workshop (AIPR), pp. 1–8. IEEE (2016)
37. Castelluccio, M., Poggi, G., Sansone, C., Verdoliva, L.: Land use classification in remote sensing images by convolutional neural networks (2015). ArXiv150800092

38. Scott, G.J., England, M.R., Starms, W.A., Marcum, R.A., Davis, C.H.: Training deep convolutional neural networks for land–cover classification of high-resolution imagery. IEEE Geosci. Remote Sens. Lett. **14**(4), 549–553 (2017)
39. Wieland, M., Pittore, M.: Performance evaluation of machine learning algorithms for urban pattern recognition from multispectral satellite images. Remote Sens. **6**(4), 2912–2939 (2014)
40. Pesaresi, M., Syrris, V., Julea, A.M.: Benchmarking of the Symbolic Machine Learning classifier with state of the art image classification methods-application to remote sensing imagery (2015)
41. Obaid, A.J., Suman Rajest, S., Silvia Priscila, S., Shynu, T., Ettyem, S.A.: Dense convolution neural network for lung cancer classification and staging of the diseases using N.S.C.L.C. images. In: Swaroop, A., Polkowski, Z., Correia, S.D., Virdee, B. (eds.) Proceedings of Data Analytics and Management. I.C.D.A.M. 2023. Lecture Notes in Networks and Systems, vol. 788. Springer, Singapore (2023). https://doi.org/10.1007/978-981-99-6553-3_28
42. Lavanya, K., Obaid, A.J., Sumaiya Thaseen, I., Abhishek, K., Saboo, K., Paturkar, R.: Terrain mapping of LandSat8 images using M.N.F. and classifying soil properties using ensemble modelling. Int. J. Nonlinear Anal. Appl. **11**(Special Issue), 527–541 (2020). https://doi.org/10.22075/ijnaa.2020.4750

Multimodality Fusion Technique Using Hybrid Pulse-Coupled Neural Network with Supervised Learning Classifier for Medical Images Diagnosis

Khulood Abdel-Khaliq Al-Salim, Mohammed Ayad Alkhafaji, and Ahmed J. Obaid

Abstract Multimodal image fusion improves the quality of fused images considerably. Preserving all important information from the source images using image fusion techniques without introducing additional distortions or flaws is possible. The multimodal fusion of medical images provides accurate clinical analysis and surgical development by providing an informative and qualitative image. The principal component analysis (PCA) with discrete wavelet transform (DWT) is the most common and effective approach for medical image fusion. Based on the prominent information in the input images, the DWT extracts the principal component rather than only taking average values of low-frequency components. A hybrid-PCNN and KNN method is also proposed for selecting which combination of modalities results in the best discrimination between groups. An MRI and PET fusion study of glioma diseases illustrates the fusion process. A preliminary clinical validation is achieved, and the results are presented and evaluated.

1 Introduction

Data fusion is a technique that facilitates and fetches additional information from scientific images by merging them without prohibiting them. Fusion is derived from data fusion. An important amount of irrelevant information can be minimized with this technique [1]. As a result of image fusion, uncertainty can be reduced, and output redundancy is minimized. Human perception, information theory, and feature

K. A.-K. Al-Salim
Dejla University College, Baghdad, Iraq

M. A. Alkhafaji
College of Engineering, Medical Instruments Technology Engineering, National University of Science and Technology, Dhi Qar, Iraq

A. J. Obaid (✉)
Faculty of Computer Science and Mathematics, University of Kufa, Kufa, Iraq
e-mail: ahmedj.aljanaby@uokufa.edu.iq

S. Kumar Udgata et al. (eds.), *Intelligent Systems*, Lecture Notes in Networks and Systems 1149, https://doi.org/10.1007/978-981-97-8160-7_29

extraction use the fusion methodology. Previous developers used PCA, intensity, hue, saturation, high-pass filters, discrete wavelet transforms, and continuous wavelet transform methods. The discrete wavelet transform is most commonly used. Medical image fusion is a promising new research field widely used to capture the information and features of medical images. Medical images can be analyzed using magnetic resonance imaging (MRI), computed tomography (CT), positron emission tomography (PET), and electromagnetic radiation [2].

The fusion approach employs a massive collection of tools focusing on medical outcomes based on images of human bodies, cells, and organs [3]. Combining images reduces the data and time required for transmissions, removes artefacts, and improves reliability [4]. The registration process determines that all the points correspond by aligning two images of the same object. A perfect image fusion requires first registering the initial images acquired from different sensors and then applying the fusion process to the registered input images. Spatial and transform domains are the two basic domains of fusion methods. Pixel values are directly manipulated by spatial domain methods to achieve the desired results [5, 6]. A transform method involves transforming the images into a different domain, like a temporal domain, fusing them, and applying an inverse transform to produce the desired result. Digital image processing can be performed using pulse-coupled neural networks (PCNNs), as illustrated in Fig. 1. The author [7] discusses the connected models and their applications. It was discussed in detail how to apply PCNN models to image processing, including interconnecting cortical models (ICM), PCNN models, and multichannel models. In addition to object detection, image registration, and compression, salient regions can be detected for various processes. A new cortical PCNN model is described in this article that enables parallel processing at different stages and behaves like a primary visual system.

This article contributes the following main contributions:

1. It is possible to accomplish the fusion task effectively with an image fusion structure built on boundary-measured PCNNs.
2. Most qualitative and quantitative applications of hybrid-PCNN are superior to those using other state-of-the-art image fusion techniques.

2 Literature Review

Fusing partially blurred images is crucial to obtaining a more explanatory image. The image fusion techniques combine partial blurry images into other single images based on the information they contain [8]. Partially blurred images contain more information than fused images [9]. Multi-focus image fusion is widely used in surveillance systems, restoration of hazy [10] and foggy [11] images, image recognition, pan sharpening, etc. Researchers have recently developed image fusion techniques for diagnosing patients' diseases [12].

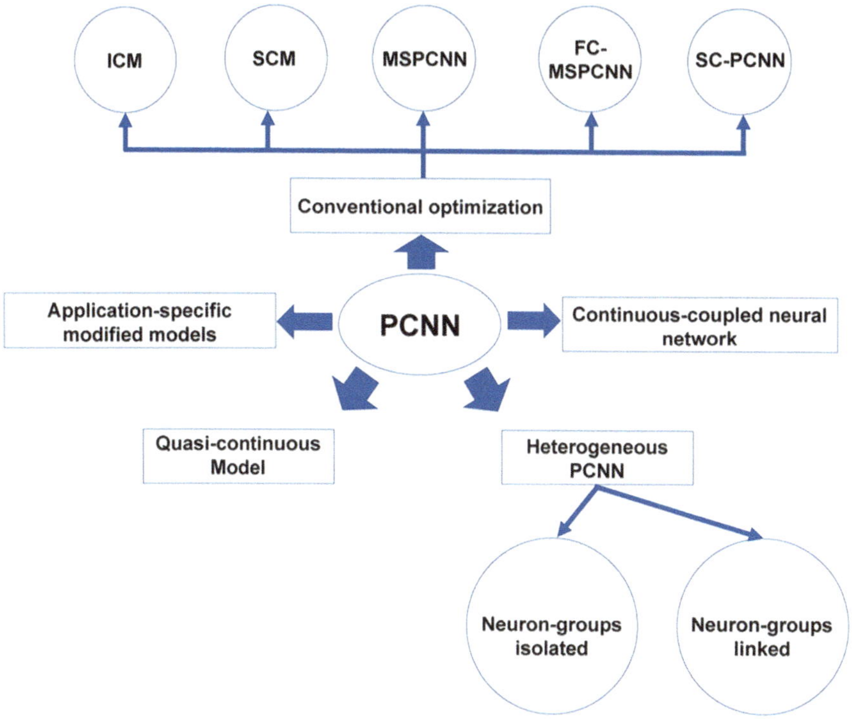

Fig. 1 Framework of PCNN's development

An independent component analysis (ICA) [13] and PCA [14] both use dimensionality reduction techniques to reduce dimensions. It is common for these dimensionality reduction techniques to be combined with methods such as wavelet-based ones for feature processing [15]. A volumetric image fusion of the first order is accessible in [16]. Multimodal image fusion based on PCA with intensity-hue-saturation (IHS) transforms spatial features without colour distortion [14]. Several mathematical transformations can be applied to features in order to enhance image fusion performance. Image fusion can be achieved using complex contourlet transforms combined with wavelets [17]. In addition, transform-based methods are used to diagnose liver disease, combine risk factors, predict multifactorial diseases, perform parametric classification, analyze local images [16], and fuse multimodal images [18]. Possible clustering methods are superior to fuzzy c-means clustering in image fusion registration stages. The best way to fuse images is at the pixel level [19]. Author [20] developed pixel and region-based techniques for fusion using complex wavelets. The image is decomposed into numerous features using a DCWT, and a region map is also calculated. To fuse the images using the DCWT domain, region-based techniques are considered for computing the features of each region [20]. The study showed that deep learning (DL) enhanced fusion rates. Images can be successfully fused using deep learning-based fusion [21].

There have been successful uses of multimodal deep learning models outside of medicine, such as the internet of vehicles [22, 23] and video classification [2, 24]. Medical imaging leveraging fusion strategies is primarily motivated by the need to incorporate complementary contextual information and overcome the limitations of image-only models, in addition to the justifications for these efforts in medical applications.

3 Proposed Hybrid Image Fusion Methods

Figure 2 shows the hybrid ML fusion method for classifying images into two groups, usual or irregular, for functionality abstraction. MRI and PET scans are used for pre-treatment during the planned fusion process. In addition to its ability to analyze images at various resolution levels, DWT is an efficient tool for extracting features from images. Traditional combinations require the capability of obtaining widespread excellence images. Combining both equations suggests that pulse-coupled neural networks (PCNNs) with K-nearest neighbour Networks (KNNs) be used in the proposed plan. It is more difficult to calculate this technique due to the massive volume required. In order to decrease the measurements, another method was used. This PCA reduced the vector aspect of the function and increased its descriptive power. Traditional brain tumour detection methods are not as accurate or faster as PCA.

KNN is an uncertain and stable classification method that performs well in complex and simple applications. Data in a dataset is categorized by KNN using a reserve of functions. The collection is formed when there is a close aloofness between the data, and the many groups are formed when the data is extended. KNNs are widely used in MRI and PET research to classify proteins. A KNN was developed to distinguish epileptic and normal brain functions based on MRI and PET images. KNNs are considered efficient and robust classification methods and are suitable even for complex applications. KNNs categorize stories based on aloof function sets in the information set. Various groups can be formed when there is a close gap between the data and a long distance between them. MRI and PET images were classified using KNN as epileptic and typical brain activity.

3.1 The PCNN Model

PCNN is used to extract visual features from images as signatures. Our technique uses a PCNN model, as shown in Fig. 3. There are two main components in this model: connecting and feeding components. In addition to calculating thresholds, this model also determines the final value of neurons via two minor modules.

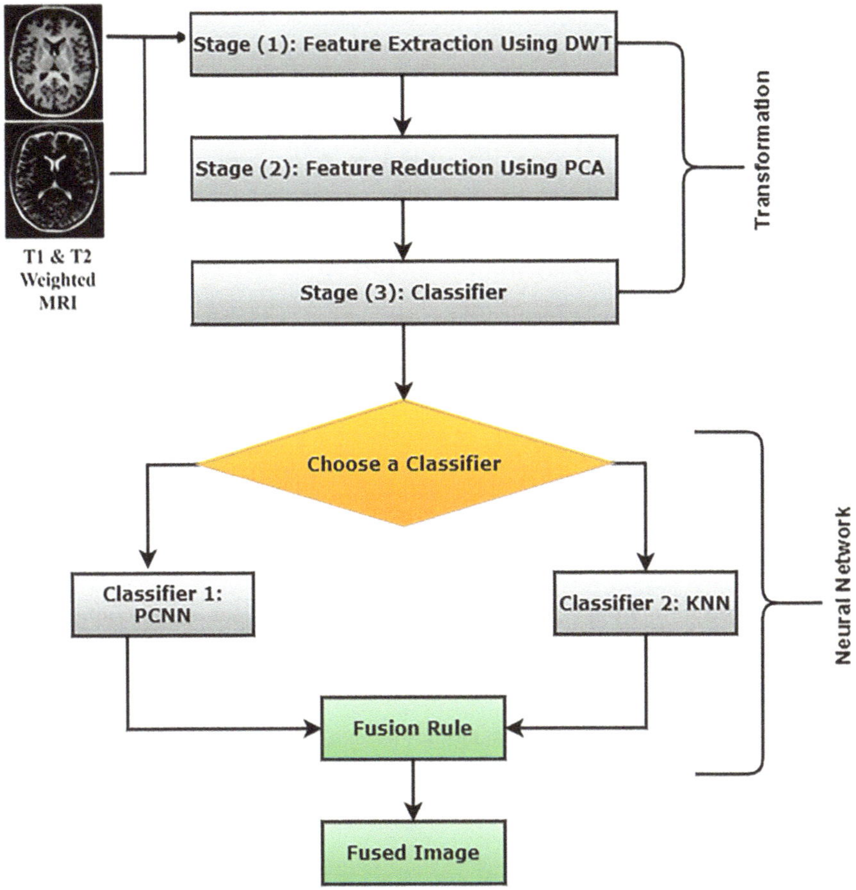

Fig. 2 Hybrid image fusion for medical imaging

There are two layers of neurons in the PCNN, each corresponding to a pixel in the image. The feeding component receives and processes the pixel value following Eq. (1). Additionally, the linking component operates through Eq. (2).

$$F_{ij}(n) = F_{ij}(n-1) * e^{-\frac{1}{AF}} + VF^{*}Y_{kl}(n-1) + A_{ij} \tag{1}$$

$$L_{ij}(n) = L_{ij}(n-1) * e^{-\frac{1}{AL}} + VL^{*}Y_{mn}(n-1) + A_{ij} \tag{2}$$

VF, AF, VL, and AL are constants and F_{ij} and L_{ij} represent the feeding and linking values of neurons (i, j), respectively. Also, A_{ij} is the amount of image pixel $(i, j)Y_{kl}$ and Y_{mn} are the outcome values of neurons from prior iterations, and n is the iteration number. Contrary to the original PCNN model described in [25], synaptic weights

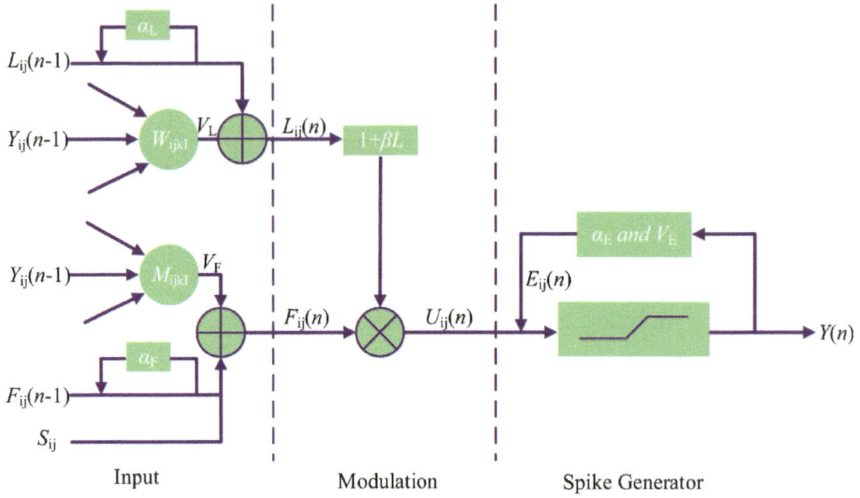

Fig. 3 The structure of the PCNN

are considered for the linking and feeding components. Equation (3) combines the outputs from the feeding and linking components to calculate the internal value of the neuron.

$$U_{ij}(n) = F_{ij}(n) + \left(1 + \beta * L_{ij}(n)\right) \tag{3}$$

where β is the constant value. Using Eq. (5), the binary function calculates the neuron's value internally.

$$Y_{ij} = \begin{cases} 1 \; if \; U_{ij} > T_{ij} \\ 0 \; Otherwise \end{cases} \tag{4}$$

where T_{ij} j is a dynamic threshold value. Binomial functions use threshold values to determine whether neuronal firing occurs. Calculation and updating of threshold values are illustrated in Eq. (5).

$$T_{ij}(n) = T_{ij}(n-1) * e^{-\frac{1}{AT}} + VT^{*}Y_{ij}(n-1) \tag{5}$$

where Y_{ij} is the output value of neuron (i, j). The values of these parameters determine a signature's quality. A parameter optimization technique thus improved the generated signature.

3.2 KNN Classifier

The KNN is one of the most widely used and simple classifiers. The technique has also been applied to other areas of research, including machine learning (ML) and pattern recognition [26]. KNN is used to measure the distance between input image signatures and image signatures in different database classes. This image is classified according to the lowest distances between the image and the innovative signature. In addition to Euclidean, Hamming, Manhattan, and Minkowski distances, the KNN classifier can also use some distance functions [27]. Considering that Euclidean distance is widely used, we selected Eq. (6) to determine distances [28].

$$eucl(x, y) = \sqrt{\sum_{i=1}^{k} (X_i - Y_i)^2} \tag{6}$$

In this case, X, Y are the original and target image signatures, respectively, and k is the length of the signature.

4 Experimental Result and Analysis

We conducted experiments on datasets representing glioma diseases in order to authorize the proposed system. A comparison was conducted between the proposed system and methods currently used in image fusion. The performance of the software was assessed qualitatively and quantitatively.

4.1 Datasets

An algorithm for multimodal medical images using MRI-PET images was tested. Whole Brain Atlas makes the image pairs available for download [29]. The images are 256 × 256 for the correctly registered. Figure 4 shows an example image from the dataset.

4.2 Evaluation Metrics

A quantitative analysis of the proposed algorithm is performed by using five metrics such as standard deviation (SD), entropy (EN), structure similarity (SS), and normalized mutual information (NMI). Both SD and EN can measure the knowledge of a

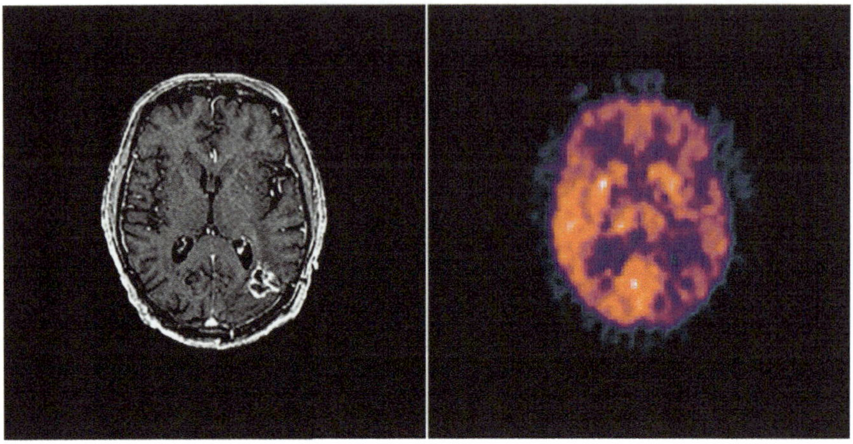

Fig. 4 Fusion dataset for MRI-PET

fused image. Source images and fused images are measured based on NMI. The references for each fusion method provide more information about these evaluation metrics.

Entropy (H): The Entropy (H) of an image measures its amount of information. A grey level with the same frequency over the whole range produces the highest entropy. A higher entropy value of the fused image indicates that it contains more information than the parent image.

$$H = -\sum_{g=0}^{L-1} p(g) log_2 p(q)$$

Peak Signal to Noise Ratio (PSNR): As a result of fusion, the PSNR of the final result is defined as follows: PSNR represents the ratio between the maximum possible power of the signal and the maximum power of corrupting noise which affects its fidelity.

$$PSNR = 10Xlog \frac{(f_{max})^2}{(RMSE)^2}$$

The higher the PSNR value, the better the fusion algorithm. f_{max} stands for the maximum grey scale value of the pixels in the fused image.

$$RMSE = \sqrt{\sum_{n=1}^{M} \sum_{n=1}^{N} (R(m, n) - F(m, n))^2}$$

It is usually better for the fusion algorithm to have a smaller RMSE than a larger RMSE. $R(m, n)$ and $F(m, n)$ are the reference (MRI) images and the fused images, respectively.

Normalized Mutual Information (NMI): Based on mutual information (MI), one variable is measured by the amount of information it contains about the other variables based on their statistical dependence. Here, mutual information refers to how the intensity distributions of the corresponding images are similar. Let A and B be the random variables with marginal probability distributions. $P_A(a)$ and $P_B(b)$ and joint probability distribution $P_{AB}(a, b)$.

$$I_{AB}(a, b) = \sum_{x,y} P_{AB}(a, b) log \frac{P_{AB}(a, b)}{P_A(a) P_B(b)}$$

The equation above can be used to calculate how much information a fused image F contains about two input images A and B

$$I_{FA}(f, a) = \sum_{x,y} P_{FA}(f, a) log \frac{P_{FA}(f, a)}{P_F(f) P_A(a)}$$

$$I_{FB}(f, b) = \sum_{x,y} P_{FB}(f, b) log \frac{P_{FB}(f, b)}{P_F(f) P_B(b)}$$

Thus, the MI is given by

$$M_F^{AB} = I_{FA}(f, a) + I_{FB}(f, b)$$

4.3 Experimental Results

Medical image fusion results cannot be fully determined by evaluating visual effects. If the feature information is not lost and the medical diagnosis is not misjudged due to losing the information, the visual effect will be acceptable. Figure 5 illustrates a graphical representation of the hybrid model.

As shown in Table 1, the proposed hybrid-PCNN model has been analyzed based on various performance metrics.

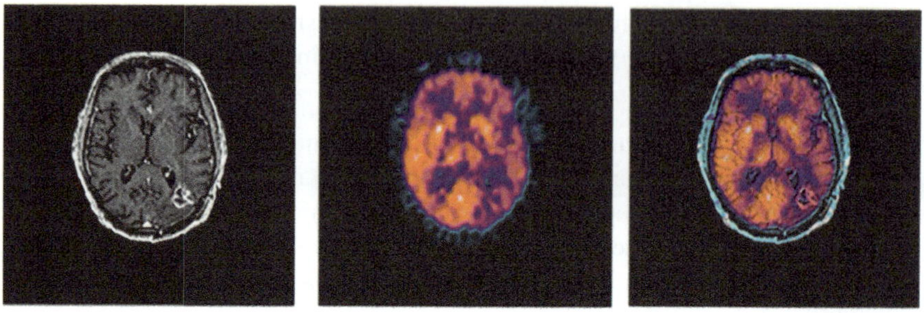

Fig. 5 Compares MRI and PET images with the proposed method (hybrid-PCNN)

Table 1 Performance values for the proposed hybrid-PCNN

Performance parameter	Performance values
EN	15.52
PSNR	105.412
SD	93.62
NMI	0.301
SS	0.742

4.4 Comparison Methods

Hybrid techniques based on PCNN algorithms are compared with seven existing fusion methods. The methods include convolutional neural networks (CNNs), convolutional sparsity-based morphological component analyses (CSMCAs), information of interest in Laplacian filtering domains (LLF-IOIs), neuro-fuzzy approaches (NFAs), and parameter-adaptive PCNNs (PAPCNNs). In Fig. 6, the feature information does not seem to be lost despite different visual effects. Further quantitative evaluation requires objective evaluation indicators.

Table 2 and Figs. 7 and 8 summarize the mean values for each metric of the different fusion methods. A majority of cases perform better with the proposed method. The NMI of the normal ageing case is still the second strongest, even if it is not the highest value in a particular column.

4.5 Conclusion

A medical imaging fusion method registers and combines multiple imaging modalities to improve image quality and reduce randomness, thereby improving the clinical utility of medical images. The fusion of multimodal medical images has demonstrated notable improvements in the clinical accuracy of medical image-based decisions. We

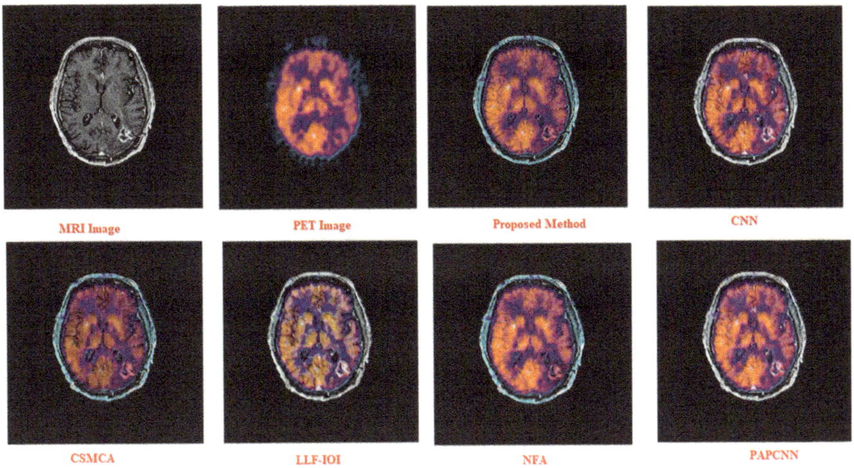

Fig. 6 The Comparative graphical analysis of two commonly used medical MRI and PET images

Table 2 Performance value analysis of various methods

Method	EN	SD	NMI	SS
CNN	3.54	53.96	0.753	0.732
CSMCA	3.41	54.96	0.731	0.734
LLF-IOI	3.33	50.12	0.684	0.693
NFA	3.41	53.63	0.701	0.742
PAPCNN	3.40	53.99	0.702	0.753
Hybrid-PCNN	15.21	93.62	0.301	0.742

propose a method for the fusion of multimodal medical images based on PCNN and KNN. Having both fusion strategies fit different scales is the main advantage of the hybrid-PCNN method. Wavelet transforms can be used to decompose images into different scales, thus enabling two fusion strategies to be fully utilized. Based on publicly available datasets, hybrid-PCNN has demonstrated its state-of-the-art performance.

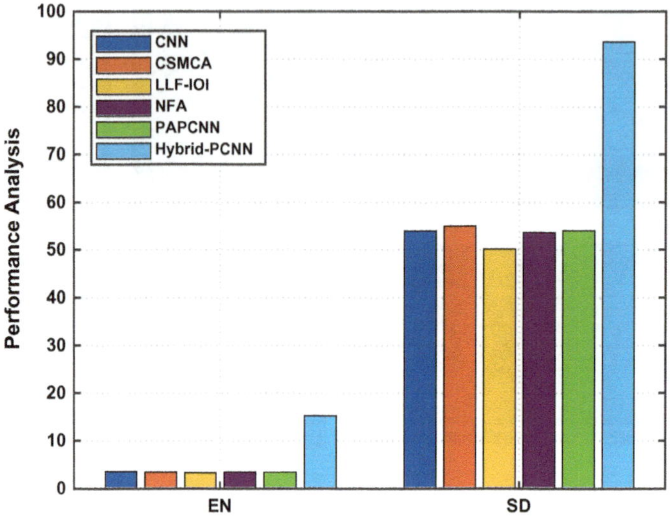

Fig. 7 The comparative performance analysis of EN and SD

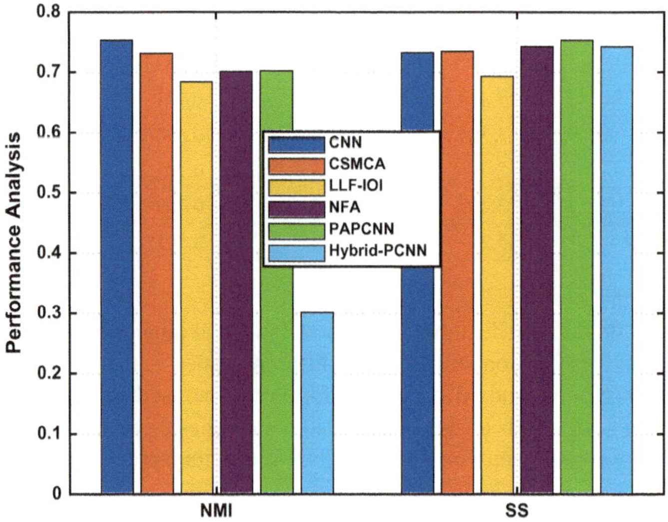

Fig. 8 The comparative performance analysis in terms of NMI and SS

References

1. Pavithra, C., Bhargavi, S.: Fusion of two images based on wavelet transform. Int. J. Innovat. Res. Sci., Engin. Technol. **2**(5), 1814–1819 (2013)
2. Yadav, S.P., Jindal, M., Rani, P., De Albuquerque, V.H.C., Dos Santos Nascimento, C., Kumar, M.: An improved deep learning-based optimal object detection system from images. Multimed. Tools Appl. https://doi.org/10.1007/s11042-023-16736-5
3. James, A.P., Dasarathy, B.V.: Medical image fusion: a survey of the state of the art. Inf. Fus. **19**, 4–19 (2014)
4. Masood, S., Sharif, M., Yasmin, M., Shahid, M.A., Rehman, A.: Image fusion methods: a survey. J. Engin. Sci. Technol. Rev. **10**(6) (2017)
5. Morris, C., Rajesh, R.S.: Two stage spatial domain image fusion techniques. ICTACT J. Image Video Process.: Special Issue Video Process. Multimedia Syst. **5**(01) (2014)
6. Rani, P., Sharma, R.: Intelligent transportation system performance analysis of indoor and outdoor internet of vehicle (iov) applications towards 5g. Tsinghua Sci. Technol. (2023)
7. Johnson, J.L., Padgett, M.L.: PCNN models and applications. IEEE Trans. Neural Netw. **10**(3), 480–498 (1999)
8. Ouerghi, H., Mourali, O., Zagrouba, E.: Non-subsampled shearlet transform based MRI and PET brain image fusion using simplified pulse coupled neural network and weight local features in YIQ colour space. IET Image Proc. **12**(10), 1873–1880 (2018)
9. Lee, S.L., Zare, M.R., Muller, H.: Late fusion of deep learning and handcrafted visual features for biomedical image modality classification. IET Image Proc. **13**(2), 382–391 (2019)
10. Singh, D., Garg, D., Singh Pannu, H.: Efficient landsat image fusion using fuzzy and stationary discrete wavelet transform. Imaging Sci. J. **65**(2), 108–114 (2017)
11. Singh, D., Kumar, V.: Dehazing of remote sensing images using improved restoration model based dark channel prior. Imaging Sci. J. **65**(5), 282–292 (2017)
12. Venkatrao, P.H., Damodar, S.S.: HWFusion: Holoentropy and SP-Whale optimisation-based fusion model for magnetic resonance imaging multimodal image fusion. IET Image Proc. **12**(4), 572–581 (2018)
13. Calhoun, V.D., Adali, T.: Feature-based fusion of medical imaging data. IEEE Trans. Inf. Technol. Biomed. **13**(5), 711–720 (2008)
14. He, C., Liu, Q., Li, H., Wang, H.: Multimodal medical image fusion based on IHS and PCA. Proc. Engin. **7**, 280–285 (2010)
15. Cui, Z., Zhang, G., Wu, J.: Medical image fusion based on wavelet transform and independent component analysis. In: 2009 International Joint Conference on Artificial Intelligence, pp. 480–483. IEEE (2009)
16. Wang, C., Ye, Z.: First-order fusion of volumetric medical imagery. IEE Proc.-Vis., Image Signal Process. **153**(2), 191–198 (2006)
17. Al-Azzawi, N.A., Sakim, H.A.M., Abdullah, A.K.W., An efficient medical image fusion method using contourlet transform based on PCM. In: IEEE Symposium on Industrial Electronics and Applications. IEEE **2009**, 11–14 (2009)
18. Yang, L., Guo, B.L., Ni, W.: Multimodality medical image fusion based on multiscale geometric analysis of contourlet transform. Neurocomputing **72**(1–3), 203–211 (2008)
19. Chang, L., Feng, X., Zhu, X., Zhang, R., He, R., Xu, C.: CT and MRI image fusion based on multiscale decomposition method and hybrid approach. IET Image Proc. **13**(1), 83–88 (2019)
20. Lewis, J.J., O'Callaghan, R.J., Nikolov, S.G., Bull, D.R., Canagarajah, N.: Pixel-and region-based image fusion with complex wavelets. Inf. Fus. **8**(2), 119–130 (2007)
21. Liu, Y., Chen, X., Wang, Z., Wang, Z.J., Ward, R.K., Wang, X.: Deep learning for pixel-level image fusion: recent advances and future prospects. Inf. Fus. **42**, 158–173 (2018)
22. Rani, P., Hussain, N., Khan, R.A.H., Sharma, Y., Shukla, P.K.: Vehicular intelligence system: time-based vehicle next location prediction in software-defined internet of vehicles (SDN-IOV) for the smart cities. In: Al-Turjman, F., Nayyar, A., Devi, A., Shukla, P.K., (eds.), Intelligence of Things: AI-IoT Based Critical-Applications and Innovations, pp. 35–54. Springer International Publishing, Cham (2021). https://doi.org/10.1007/978-3-030-82800-4_2.

23. Hussain, N., Rani, P.: Comparative studied based on attack resilient and efficient protocol with intrusion detection system based on deep neural network for vehicular system security. In: Distributed Artificial Intelligence, pp. 217–236. CRC Press (2020). Accessed: Oct. 19, 2023.
24. Verma, S. et al.: An automated face mask detection system using transfer learning based neural network to preventing viral infection. Expert Syst. e13507 (2023). https://doi.org/10.1111/exsy.13507
25. Johnson, J.L., Ranganath, H., Kuntimad, G., Caulfield, H.J.: Pulse-coupled neural networks. In: Neural Networks and Pattern Recognition, pp. 1–56. Elsevier (1998)
26. Bermejo, S., Cabestany, J.: Adaptive soft k-nearest-neighbour classifiers. Pattern Recogn. **33**(12), 1999–2005 (2000)
27. Mafteiu-Scai, L.O.: A new dissimilarity measure between feature-vectors. Int. J. Comput. Appl. **64**(17) (2013)
28. Souza, R., Rittner, L., Lotufo, R.: A comparison between k-optimum path forest and k-nearest neighbors supervised classifiers. Pattern Recogn. Lett. **39**, 2–10 (2014)
29. The Whole Brain Atlas. Accessed: Dec. 25, 2023. https://www.med.harvard.edu/aanlib/

Semantic Segmentation Using Global–Local Networks (GLNets) from Ultra-High-Resolution Images

Israa Falih Muslm, Shaymaa Shnain, and Najlaa Nsrulaah Faris

Abstract A high-resolution (HR) image provides more detailed information about structural conditions than a low-resolution (LR) image. Increasing interest and rapid progress in Ultra-High-Resolution (UHR) segmentation have created the need for a large-scale benchmark with dense fine-grained annotations covering a wide range of scenes. The modified convolutional neural network architecture based on Global–Local Networks was used to implement the recognition algorithm. Various image segmentation applications have been found, including urban planning, forest management, climate modelling, etc. The algorithm's performance has been demonstrated on DeepGlobe and Inria Aerial datasets. The proposed method outperforms existing methods on two benchmark datasets with significant improvements.

Keywords Semantic segmentation · GLNets images · UHR

1 Introduction

There have been decades of studies on semantic segmentation. Segmentation requires large-scale input images compared to other vision tasks [1–3]. In remote sensing, ultra-high-resolution images of pixels are crucial [4–6], autonomous driving [7–9] and medical imagery applications [10–13]. Although general segmentation systems can hold ultra-high-resolution images, they have memory and computational limitations. Currently, segmentation procedures primarily focus on designing neural networks for regular resolution images but overlook larger scale inputs. In order

I. F. Muslm (✉)
Department of Performance Evaluation, University of Babylon, Babil Govermorate, Iraq
e-mail: israa.muslim@uobabylon.edu.iq

S. Shnain
College of Basic Education, University of Babylon, Babil Govermorate, Iraq
e-mail: shaima.almorshedy3@uobabylon.edu.iq

N. N. Faris
College of Engineering, Medical Instruments Technology Engineering, National University of Science and Technology, Dhi Qar, Iraq

S. Kumar Udgata et al. (eds.), *Intelligent Systems*, Lecture Notes in Networks and Systems 1149, https://doi.org/10.1007/978-981-97-8160-7_30

to encounter the recall and speed requirements of a deep and complicated model, some detail is discarded during downsampling, which results in poor performance. It is possible to adopt a shallow and lightweight model to handle larger scale inputs [14]. Because ultra-high-resolution images require unaffordable computing resources, robotic systems with limited computational power will have difficulty handling them [15, 16]. It is typically the case that ultra-high-resolution images are down-scaled before being segmented, but this strategy often leaves out smaller, thinner and curvier regions. Furthermore, as input image resolutions increase, GPU memory requirements increase significantly, making this method unsuitable for systems with limited GPU memory capacities. With limited GPU memory, is semantic segmentation possible for arbitrarily large images. A significant challenge with global–local processing is maintaining consistency throughout the restored image. To restore ultra-high-resolution images stepwise, we propose a Global–Local Stepwise Generative Network (GLSGN) using two restoring pathways. Using a local pathway, fine-grained images can be restored over patches of high resolution in a local area. The coarse restoration of an intact, scaled-down image is carried out with a global pathway, which captures noise patterns and semantic information. The global pathway provides global cues that can be used to achieve successively a clean background using three local paths.

Under GPU memory constraints, the paper considers a new ultra-high-resolution segmentation problem. It is appropriate to separate local patches instead of combining all local segmentation results before merging them. This approach, however, often degrades performance, so to address this concern, additional context beyond the local patch must be introduced. As a consequence of introducing additional context, GPU memory usage increases as well. In this paper, we propose a novel method for segmenting local patches without considering the context beyond them. In this manner, GPU-based ultra-high-resolution semantic segmentation frameworks can be created that are memory-efficient and effective.

2 Related Work

Global–local integration in GLNet combines the local and global contexts to improve segmentation results [17]. A classification branch is presented by PPN based on the GLNet to choose significant local patches for fusing using overall images [18]. Furthermore, the Author [19] refines coarse global and local segmentation maps with a general cascade structure. A multiscale architecture has also been introduced in MagNet [20]. On the coarsest scale, the rough output will be gradually refined. A recent study found that FCtL [21] was able to improve local segmentation by combining local-context information with adaptive feature fusion. These ultra-high-resolution (UHR) segmentation approaches, however, include satellite imagery into patches that fit into short GPU memory, causing redundant calculations and very slow inference speeds. We integrate shallow and deep networks directly, without cropped patches, so the inference speed is significantly accelerated without cropped patches.

The global–local reliability modelling strategy is a standard modelling technique that has been confirmed in a variety of computer vision applications. Learning global–local consistency can be done in many ways [6], for example, by extracting features from various sizes of convolutional kernels and aggregating them [22]. A deep pyramid network derived from the image pyramid is introduced to restore blurry images from different levels of degradation that take advantage of cross-scale similarity and self-similarity within the corrupted image [23]. As another method of knowledge global–local reliability [24], different networks are employed to separate the global image from the local patches before they are aggregated. Author [31] planned a multi-patch hierarchical system to restore blurred images coarse-to-finely. According to the author [28], a global–local filtering module based on global dependencies and dynamically exploiting neighbouring pixels improves performance significantly. The author proposed a innovative local to global dual-stream system for remote sensing images [25]. High-resolution imaging has yet to fully exploit global–local consistency in modelling intact noise patterns. Feature-sharing between the two branches is improved by implementing a dual-stream architecture. To test our model, Inria's aerial image labeling data and DeepGlobe's building detection data were used.

The Coupled Global–local (CGL) system can effectively and efficiently address the abovementioned problems. Local features from downscaled global images are simultaneously extracted with local features from local images using our parallel feature extraction network. The adaptive dimensionality reduction convolution (ADRConv) scheme is suggested to improve multiscale feature presentation, followed by a multiscale fusion module. A small computation is needed to distinguish large substances in the global detection branch. Existing architectures often produce global losses twice as the limited branch, frequently dominating experiments' final losses. Details are seriously lost if the local branch information loss is considered in addition to the global loss. Furthermore, the global outlet magnifies performance problems in two ways. The global branch reduces the size of small objects. Smaller objects have a harder time extracting features and converge to a minor enough damage. The location accuracy of local objects is higher than that of global objects. A proposal filter reduces global and local losses, while a nonlocal suppression algorithm (NLS) suppresses global branch confidence for more accurate object locations. Common detection models can easily incorporate the CGL network. According to the experiments, the proposed method provides considerably better detection, with only a slight increase in memory demand [26, 27].

3 Proposed Methodology

Figure 1 shows our GLNet architecture. Images are downsampled and cropped on the global and local branches, respectively. Collaboration across global and local feature maps is reinforced by sharing and regularizing feature maps.

A high-level feature map is aggregated from two branches to generate the final segmentation. From images and segmentations of N ultra-high-resolution data $D = \{I_i, S_i\}_{i=1}^{N}$ where $I_i, S_i \in R^{H\times W}$, Low-resolution images are taken by the global branch G, $D^{lr} = \{I_i^{lr}, S_i^{lr}\}_{i=1}^{N}$ and patches with the same resolution are cropped from D to the local branch $D^{hr} = \{I_i^{hr}, S_i^{hr}\}_{i=1}^{N}$, where I_i and S_i in D contains η_i covers. To facilitate training and inference, both I_i and S_i were cropped (rather than randomly). $I_i^{lr}, S_i^{lr} \in R^{H_1\times W_1}$ & $I_i^{hr}, S_i^{hr} \in R^{H_2\times W_2}$, where H_1, H_2, as well as w_1, w_2. Both G and L have the same backbone, which consists of layers $1L$ (Fig. 2).

A deep sharing of feature maps occurs during the segmentation of each branch (Sect. 3.1). A branch aggregation layer f_{agg} is then applied to produce the absolute segmentation cover (Sect. 3.2). Local branches are also trained under the influence of weakly coupled regularizations to limit the two outlets and become constant in their training [46].

3.1 Deep Feature Map Sharing

First, we crop the feature map from the global branch, aligning it with the local patch's spatial location. Next, we upsample to match the features in the local branch. As an extra channel in the same layer, the local feature maps are concatenated. Before feeding the local branch feature maps into the next layer, we need to concatenate the local branch feature maps with the global branch feature maps. In Fig. 2, all layers except the last branch layer show how deep feature maps are shared layer-by-layer. Maps of shared features can be unidirectional, for example, the feature of the overall branch to the feature of the local branch ($G \rightarrow L$), as well as bidirectional ($G \leftrightarrows L$,

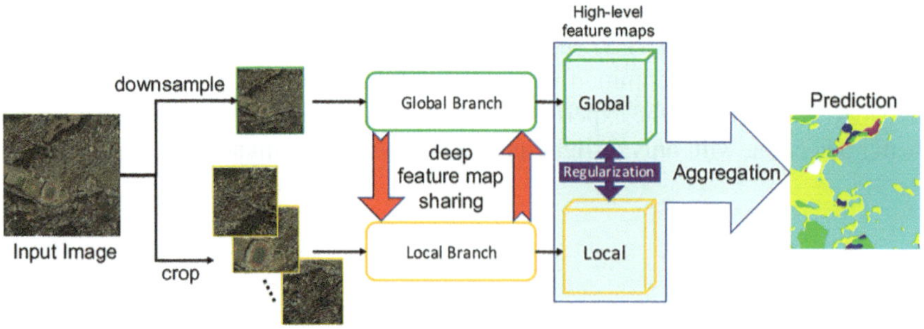

Fig. 1 The proposed GLNet architecture

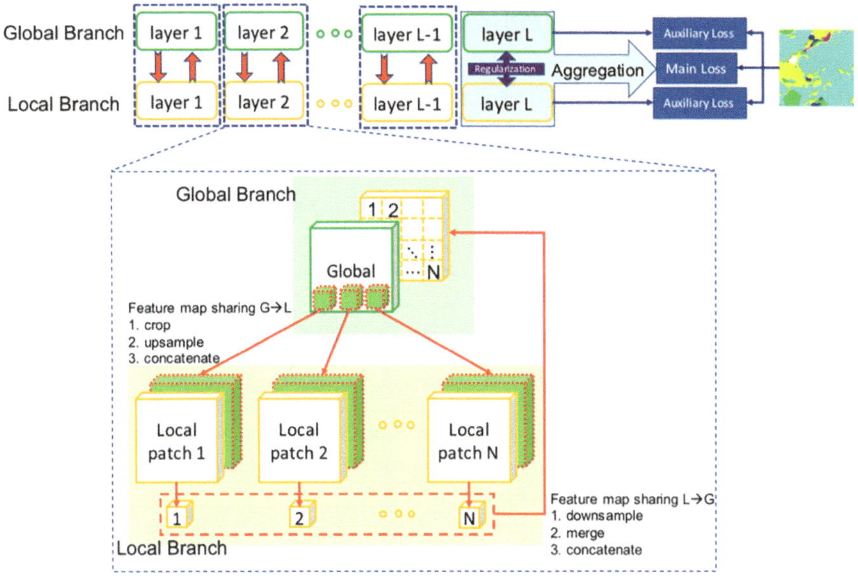

Fig. 2 Local and global branches share deep feature maps

which combines the fine structure and location features). Local fine essential features are joined with global contextual features at every layer.

3.2 *Branch Combination with Regularization*

These global and local branches will be combined by a convolutional layer 3×3 filter called the aggregation layer. f_{agg}. The feature maps are with the local branch's Lth layer $\widehat{X}_L^{LOC}$, are concatenated along the channel with the ones from the global branch $\widehat{X}_L^{Glb}$. Final segmentation output $\widehat{S}^{Agg}$ will be generated by f_{agg}. The segmentation maps of our local branch $\widehat{S}^{Loc}$ and global branch $\widehat{S}^{Gib}$ are accumulated to segmentation loss required on $\widehat{S}^{Agg}$ (local patch/global downsampled). It stabilizes the training by ensuring the segmentation outputs are close to the corresponding segmentation maps. Local branches tend to overfit and "override" global branches in practice. Our weakly coupled regularization is added to both branches' last layers to prevent the local branch from learning more rapidly. The Euclidean distance $\lambda \|\widehat{X}_L^{LOC} - \widehat{X}_L^{Gib}\|_2$ is applied to discourage large relative changes between $\widehat{X}_L^{LOC}$ and $\widehat{X}_L^{Gib}$ in our work, where L has been empirically fixed at 0.15. In this regularization process, the parameters are only updated in local branches in order to "slow down" the local branch training and make it more synchronized with the global branch learning.

3.3 Jointly Supervised Parameter Learning

End-to-end optimization is performed on the parameters of our proposed method.

Multi-stage Pixel Reconstruction Loss: Each pathway is reconstructed pixel-by-pixel using the L_1 loss:

$$L_{pixel} = \alpha_1 \sum_{i=1}^{N} L_1\left(B.\widehat{B_i}|\right) + \beta_1 L_1\left(B, \hat{B}_{out}\right) \tag{1}$$

$$L_{pixel} = \alpha_1 \sum\nolimits_{i=1}^{N} L_1\left(B.\widehat{B_i}|\right) + \beta_1 L_1\left(B, \hat{B}_{out}\right)()$$

where B_i indicates the repaired image by i^{th} and $\widehat{B}_{out}$ indicates the repaired image of the Laplacian Pyramid. Interpolation also provides ground truth B at various scales. $\alpha_1 = 0.1$ and $\beta_1 = 0.5$ are set empirically.

Loss of perception in multiple stages: Each pathway of the perceptual loss is supervised to learn the stability of semantic information [28]:

$$L_{perc} = \alpha_2 \sum_{i=1}^{N} L_{VGG}\left(B.\widehat{B_i}|\right) + \beta_2 L_{VGG}\left(B, \hat{B}_{out}\right) \tag{2}$$

According to a pre-trained VGG-19 [29] and $\alpha_2 = 0.1$ and $\beta_2 = 0.5$, L_{VGG} indicates the perceptional distance between two images.

Loss due to adverse conditions: The final repaired image $\widehat{B}_{out}$ is encouraged as accurate as the ground-truth image B. The adversarial learning process is stabilized by spectral normalization [30]:

$$L_{adv} = -E_{BP_{GLSGN}}$$

Each convolution layer uses the discriminator D^{sn} with spectral normalization. This leads to the following definition of the total loss function:

$$L = \lambda_1.L_{pixel} + \lambda_2.L_{perc} + \lambda_3.L_{adv} \tag{3}$$

To obtain equilibrium among various loss functions, the hyper-parameters $\lambda_1 = 1$ and $\lambda_2 = \lambda_3 = 0.01$ are used.

4 Result Analysis and Discussion

This section aims to estimate our model's capability over two public benchmarks. The proposed model is first associated with the previous state-of-the-art techniques performed to understand its structure.

A. Datasets

Inria Aerial [31]: Five cities are represented in this dataset, including dense metropolitan districts as well as alpine resorts. Building areas are separated from non-building extents in the package with a binary (0–1) mask. A training image, a validation image and a testing image were selected.

DeepGlobe [31]: The dataset consists of 803 images with a 2448x2448 pixel size, each of them. There are seven categories of landscape regions, with agriculture, urban, rangeland, water, forest, unknown and barren regions not considered in the challenge. All images were separated: 454 were used as training, 207 as validation and 142 as testing images.

B. Experimental Result

Research pertaining to the semantic segmentation of very high-resolution imageries has been sparse. It is known that high resolution details or spatial context lost through typical ad hoc techniques such as down sampling or patch cropping. Such imagery can be obtained from a 6 M-pixel ultra-high-resolution imagery [32]. According to Fig. 3, these methods have a high accuracy rate and use a lot of GPU memory. The accuracy of recent fast segmentation methods such as ICNet [34], whose memory usage is greatly reduced, drops. Additional studies (Figs. 3, 4 and 5) show that typical models are unable to balance accuracy and GPU memory consumption when local patches and global images are used at various sizes.

As a test of our model's capability under large GPU memory limits, GLNet and ISDNet are compared regarding mIoU, GPU Memory and FPS. Tables 1 and 2 show our comparison results following most published studies. According to our observations, the proposed model is as good as the state-of-the-art (SOTA) ISDNet on DeepGlobe, but it improves by at least 3% on Inria Aerial. In addition to performance and GPU memory, our approach allows us to trade off speed and speed and GPU memory.

5 Conclusion

The increasing availability of very high-resolution imagery in complex scenes presents challenges for extraction methods due to the diversity of appearances, varying measures and complex structures of buildings. Numerous works have shown that context-aware deep learning methods can provide strong recognition as well as detection of objects by capturing spatial relation cues. It is still challenging to

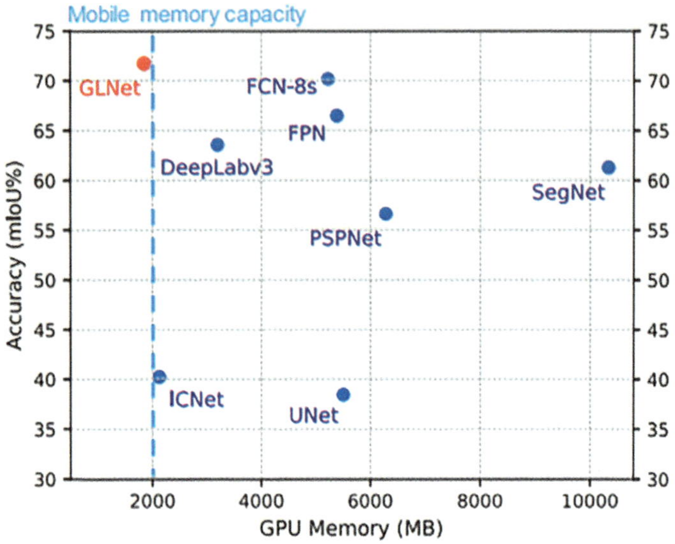

Fig. 3 Best performance achievable accuracy (mIOU) versus GPU memory (MB) on the DeepGlobe dataset

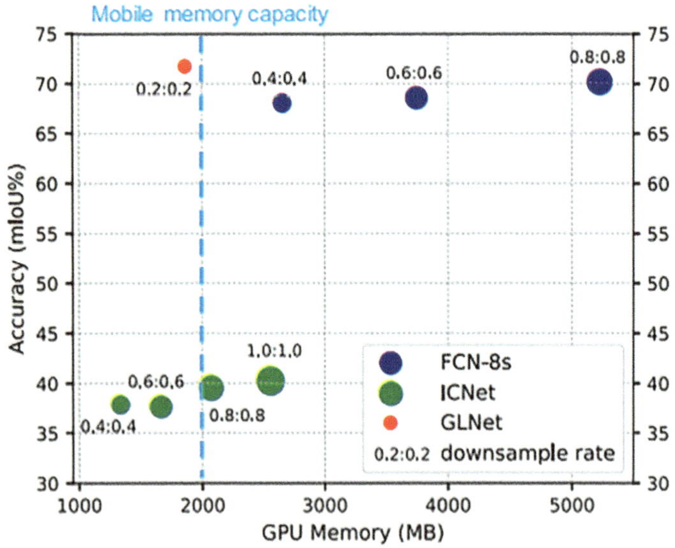

Fig. 4 Performance trained on global image accuracy (mIOU) versus GPU memory (MB) on the DeepGlobe dataset

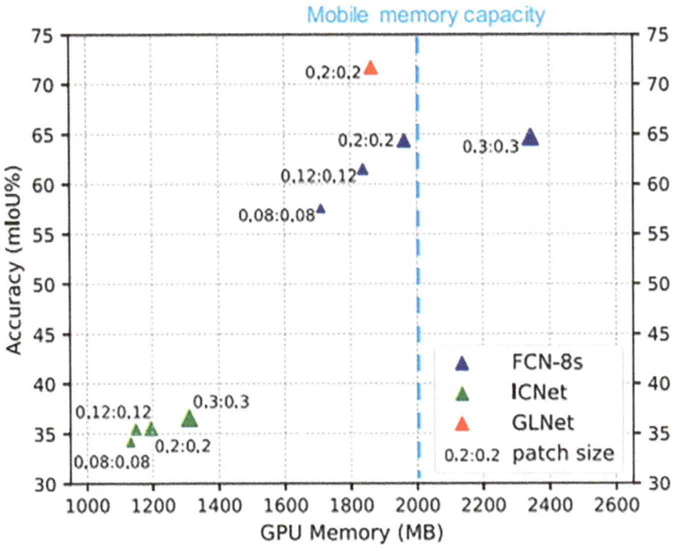

Fig. 5 Performance trained on local patches accuracy (mIOU) versus GPU memory (MB) on the DeepGlobe dataset

Table 1 On the Inria aerial dataset, a comparison analysis was conducted using large GPU memory limits

Model	mIoU	Memory	FPS
GLNet	71.20	2663	0.05
ISDNet	74.23	4680	6.90
Proposed method	78.72	4904	5.63

Table 2 DeepGlobe memory limits in comparison with large GPUs

Model	mIoU	Memory	FPS
U-Net	28.53	3511	9.34
GLNet	71.60	1865	0.17
ISDNet	73.30	1948	27.70
Proposed method	74.80	1986	25.44

distinguish building classes from other artificial classes in public datasets using various convolutional neural network (CNN) architectures. The proposed model is efficient and accurate due to its contextual weighting and close coupling of single-scale patches. Contextual weights and single-scale patches are coupled to achieve competitive mIoU performance compared to state-of-the-art methods.

References

1. Chen, L.-C., Papandreou, G., Schroff, F., Adam, H.: Rethinking atrous convolution for semantic image segmentation. arXiv, Dec. 05, 2017. https://doi.org/10.48550/arXiv.1706.05587
2. Ding, X., Zhang, X., Han, J., Ding, G.: Diverse branch block: building a convolution as an inception-like unit. In: Presented at the Proceedings of the IEEE/CVF Conference on Computer Vision and Pattern Recognition, 2021, pp. 10886–10895 (2021). Accessed: Dec. 23, 2023. https://openaccess.thecvf.com/content/CVPR2021/html/Ding_Diverse_Branch_Block_Building_a_Convolution_as_an_Inception-Like_Unit_CVPR_2021_paper.html
3. Hu, J., Shen, L., Sun, G.: Squeeze-and-excitation networks. In: Presented at the Proceedings of the IEEE Conference on Computer Vision and Pattern Recognition, 2018, pp. 7132–7141 (2018). Accessed: Dec. 23, 2023. https://openaccess.thecvf.com/content_cvpr_2018/html/Hu_Squeeze-and-Excitation_Networks_CVPR_2018_paper.html
4. Wu, X., Huang, T.-Z., Deng, L.-J., Zhang, T.-J.: Dynamic cross feature fusion for remote sensing pansharpening. In: Presented at the Proceedings of the IEEE/CVF International Conference on Computer Vision, 2021, pp. 14687–14696 (2021). Accessed: Dec. 23, 2023. https://openaccess.thecvf.com/content/ICCV2021/html/Wu_Dynamic_Cross_Feature_Fusion_for_Remote_Sensing_Pansharpening_ICCV_2021_paper.html
5. Zheng, Z., Ma, A., Zhang, L., Zhong, Y.: Change is everywhere: single-temporal supervised object change detection in remote sensing imagery. In: Presented at the Proceedings of the IEEE/CVF International Conference on Computer Vision, 2021, pp. 15193–15202 (2021). Accessed: Dec. 23, 2023. https://openaccess.thecvf.com/content/ICCV2021/html/Zheng_Change_Is_Everywhere_Single-Temporal_Supervised_Object_Change_Detection_in_Remote_ICCV_2021_paper.html
6. Zheng, Z., Zhong, Y., Wang, J., Ma, A.: Foreground-aware relation network for geospatial object segmentation in high spatial resolution remote sensing imagery. In: Presented at the Proceedings of the IEEE/CVF Conference on Computer Vision and Pattern Recognition, 2020, pp. 4096–4105 (2020). Accessed: Dec. 23, 2023. https://openaccess.thecvf.com/content_CVPR_2020/html/Zheng_Foreground-Aware_Relation_Network_for_Geospatial_Object_Segmentation_in_High_Spatial_CVPR_2020_paper.html
7. Fu, J., et al.: Dual attention network for scene segmentation. In: Presented at the Proceedings of the IEEE/CVF Conference on Computer Vision and Pattern Recognition, 2019, pp. 3146–3154 (2019). Accessed: Dec. 23, 2023. https://openaccess.thecvf.com/content_CVPR_2019/html/Fu_Dual_Attention_Network_for_Scene_Segmentation_CVPR_2019_paper.html
8. Pan, X., Shi, J., Luo, P., Wang, X., Tang, X.: Spatial as deep: spatial CNN for traffic scene understanding. In: Proceedings of the AAAI Conference Artificial Intelligent, vol. 32, no. 1 (2018). https://doi.org/10.1609/aaai.v32i1.12301
9. Rani, P., Sharma, R.: Intelligent transportation system for internet of vehicles based vehicular networks for smart cities. Comput. Electr. Eng.. Electr. Eng. **105**, 108543 (2023)
10. Reiss, S., Seibold, C., Freytag, A., Rodner, E., Stiefelhagen, R.: Every annotation counts: multi-label deep supervision for medical image segmentation. In: Presented at the Proceedings of the IEEE/CVF Conference on Computer Vision and Pattern Recognition, 2021, pp. 9532–9542 (2023). Accessed: Dec. 23, 2023. https://openaccess.thecvf.com/content/CVPR2021/html/Reiss_Every_Annotation_Counts_Multi-Label_Deep_Supervision_for_Medical_Image_Segmentation_CVPR_2021_paper.html
11. He, Y., Yang, D., Roth, H., Zhao, C., Xu, D.: DiNTS: differentiable neural network topology search for 3D medical image segmentation. In: Presented at the Proceedings of the IEEE/CVF Conference on Computer Vision and Pattern Recognition, 2021, pp. 5841–5850 (2021). Accessed: Dec. 23, 2023. https://openaccess.thecvf.com/content/CVPR2021/html/He_DiNTS_Differentiable_Neural_Network_Topology_Search_for_3D_Medical_Image_CVPR_2021_paper.html
12. Ji, W., et al.: Learning calibrated medical image segmentation via multi-rater agreement modelling. In: Presented at the Proceedings of the IEEE/CVF Conference on Computer

Vision and Pattern Recognition, 2021, pp. 12341–12351 (2021). Accessed: Dec. 23, 2023. https://openaccess.thecvf.com/content/CVPR2021/html/Ji_Learning_Calibrated_Medical_Image_Segmentation_via_Multi-Rater_Agreement_Modeling_CVPR_2021_paper.html
13. Verma, S., et al.: An automated face mask detection system using transfer learning based neural network to preventing viral infection. Expert Syst. e13507 (2023). https://doi.org/10.1111/exsy.13507
14. Yadav, S.P., Jindal, M., Rani, P., De Albuquerque, V.H.C., Dos Santos Nascimento, C., Kumar, M.: An improved deep learning-based optimal object detection system from images. Multimed. Tools Appl. (2023). https://doi.org/10.1007/s11042-023-16736-5
15. Ferrarini, B., Milford, M.J., McDonald-Maier, K.D., Ehsan, S.: Binary neural networks for memory-efficient and effective visual place recognition in changing environments. IEEE Trans. Robot. **38**(4), 2617–2631 (2022). https://doi.org/10.1109/TRO.2022.3148908
16. Grossman, L., Plancher, B.: Just round: quantized observation spaces enable memory efficient learning of dynamic locomotion. In: 2023 IEEE International Conference on Robotics and Automation (ICRA), May 2023, pp. 3002–3007 (2023). https://doi.org/10.1109/ICRA48891.2023.10160293
17. Chen, W., Jiang, Z., Wang, Z., Cui, K., Qian, X.: Collaborative global-local networks for memory-efficient segmentation of ultra-high resolution images. In: presented at the Proceedings of the IEEE/CVF Conference on Computer Vision and Pattern Recognition, 2019, pp. 8924–8933 (2019). Accessed: Dec. 23, 2023. https://openaccess.thecvf.com/content_CVPR_2019/html/Chen_Collaborative_Global-Local_Networks_for_Memory-Efficient_Segmentation_of_Ultra-High_Resolution_Images_CVPR_2019_paper.html
18. Wu, T., Lei, Z., Lin, B., Li, C., Qu, Y., Xie, Y.: Patch proposal network for fast semantic segmentation of high-resolution images. Proc. AAAI Conf. Artif. Intell. **34**(07), 12402–12409 (2020). https://doi.org/10.1609/aaai.v34i07.6926
19. Cheng, H.K., Chung, J., Tai, Y.-W., Tang, C.-K.: CascadePSP: toward class-agnostic and very high-resolution segmentation via global and local refinement. In: Presented at the Proceedings of the IEEE/CVF Conference on Computer Vision and Pattern Recognition, 2020, pp. 8890–8899 (2020). Accessed: Dec. 23, 2023. https://openaccess.thecvf.com/content_CVPR_2020/html/Cheng_CascadePSP_Toward_Class-Agnostic_and_Very_High-Resolution_Segmentation_via_Global_and_CVPR_2020_paper.html
20. Huynh, C., Tran, A.T., Luu, K., Hoai, M.: Progressive semantic segmentation. In: Presented at the Proceedings of the IEEE/CVF Conference on Computer Vision and Pattern Recognition, 2021, pp. 16755–16764 (2021). Accessed: Dec. 23, 2023. https://openaccess.thecvf.com/content/CVPR2021/html/Huynh_Progressive_Semantic_Segmentation_CVPR_2021_paper.html
21. Li, Q., Yang, W., Liu, W., Yu, Y., He, S.: From contexts to locality: ultra-high resolution image segmentation via locality-aware contextual correlation. In: Presented at the Proceedings of the IEEE/CVF International Conference on Computer Vision, 2021, pp. 7252–7261 (2021). Accessed: Dec. 23, 2023. https://openaccess.thecvf.com/content/ICCV2021/html/Li_From_Contexts_to_Locality_Ultra-High_Resolution_Image_Segmentation_via_Locality-Aware_ICCV_2021_paper.html
22. Szegedy, C., et al.: Going deeper with convolutions. In: Proceedings of the IEEE Conference on Computer Vision and Pattern Recognition, pp. 1–9 (2015)
23. Rani, P., Verma, S., Yadav, S.P., Rai, B.K., Naruka, M.S., Kumar, D.: Simulation of the lightweight blockchain technique based on privacy and security for healthcare data for the cloud system. Int. J. E-Health Med. Commun. IJEHMC **13**(4), 1–15 (2022)
24. Zamir, S.W., et al.: Multi-stage progressive image restoration. In: Presented at the Proceedings of the IEEE/CVF Conference on Computer Vision and Pattern Recognition, 2021, pp. 14821–14831 (2021). Accessed: Dec. 23, 2023. https://openaccess.thecvf.com/content/CVPR2021/html/Zamir_Multi-Stage_Progressive_Image_Restoration_CVPR_2021_paper.html
25. Rani, P., Singh, P.N., Verma, S., Ali, N., Shukla, P.K., Alhassan, M.: An implementation of modified blowfish technique with honey bee behavior optimization for load balancing in cloud system environment. Wirel. Commun. Mob. Comput.. Commun. Mob. Comput. **2022**, 1–14 (2022)

26. Bhola, B., et al.: Quality-enabled decentralized dynamic IoT platform with scalable resources integration. IET Commun. cmu2.12514 (2022). https://doi.org/10.1049/cmu2.12514
27. Hussain, N., Rani, P., Kumar, N., Chaudhary, M.G.: A deep comprehensive research architecture, characteristics, challenges, issues, and benefits of routing protocol for vehicular ad-hoc networks. Int. J. Distrib. Syst. Technol. IJDST **13**(8), 1–23 (2022)
28. Johnson, J., Alahi, A., Fei-Fei, L.: Perceptual losses for real-time style transfer and super-resolution. In: Computer Vision–ECCV 2016: 14th European Conference, Amsterdam, The Netherlands, October 11–14, 2016, Proceedings, Part II 14, pp. 694–711. Springer (2016)
29. Simonyan, K., Zisserman, A.: Very deep convolutional networks for large-scale image recognition (2014). arXiv Prepr. arXiv:14091556
30. Miyato, T., Kataoka, T., Koyama, M., Yoshida, Y.: Spectral normalization for generative adversarial networks (2018). ArXiv Prepr. arXiv:180205957
31. Maggiori, E., Tarabalka, Y., Charpiat, G., Alliez, P.: Can semantic labeling methods generalize to any city? The inria aerial image labeling benchmark. In: 2017 IEEE International Geoscience and Remote Sensing Symposium (IGARSS), pp. 3226–3229. IEEE (2017)
32. Long, J., Shelhamer, E., Darrell, T.: Fully convolutional networks for semantic segmentation. In: Proceedings of the IEEE Conference on Computer Vision and Pattern Recognition, 2015, pp. 3431–3440 (2015)
33. Badrinarayanan, V., Kendall, A., Cipolla, R.: Segnet: a deep convolutional encoder-decoder architecture for image segmentation. IEEE Trans. Pattern Anal. Mach. Intell.Intell. **39**(12), 2481–2495 (2017)
34. Zhao, H., Qi, X., Shen, X., Shi, J., Jia, J.: Icnet for real-time semantic segmentation on high-resolution images. In: Proceedings of the European Conference on Computer Vision (ECCV), 2018, pp. 405–420 (2018)

Quality Image Enhancement Using Convolutional Neural Network from Low-Resolution Camera Image

Hassanain K. Alrammahi and Mohammed Brayyich

Abstract Low-resolution (LR) images can only be obtained by long-distance shooting with low light because of the limitations of physical devices. A high-quality detached image is best obtained with optical lenses, which are fairly expensive and bulky. Enhancing low-light Night photography, video surveillance, and remote sensing using Super-Resolution (SR) is essential. We propose an architecture using DenseNet with Skip-connections to transform images from low to high resolution. A dataset containing images captured from various cameras is used for the analysis. In this paper, we used the DPED (DSLR-Photo Enhancement Dataset). Convolution neural networks (CNN) generate results that can be applied to any camera model. As a result of our architecture, we achieved good PSNR and SSIM results.

Keywords Quality image · Convolutional neural network · Low-resolution camera image

1 Introduction

Due to the simplicity and portability of mobile phones, more and more people are taking pictures with them nowadays. A mobile phone camera and a high-end digital single-lens reflex (DSLR) camera typically do not take photos. Smartphones are not the only mobile devices that can take photos; drones, tablets, and sports cameras can also take photos. Therefore, improving the perceived quality of mobile images is very important. Details, contrast, brightness, noise reduction, other characteristics, and current image enhancement procedures [1–3] enhance low-quality images. Mobile devices pose new computational challenges when it comes to enhancing

H. K. Alrammahi (✉)
General Directorate of Education in Najaf, Najaf, Iraq
e-mail: hassanaink.alrammahi@uokufa.edu.iq

M. Brayyich
College of Engineering, Medical Instruments Technology Engineering, National University of Science and Technology, Dhi Qar, Iraq

S. Kumar Udgata et al. (eds.), *Intelligent Systems*, Lecture Notes in Networks and Systems 1149, https://doi.org/10.1007/978-981-97-8160-7_31

perceptual images. Despite its processing speed and memory consumption, it has recently achieved impressive results in perceptual image enhancement [4].

However, the problem of automatic image quality enhancement remains unsolved despite the enormous improvement in deep learning techniques [5] for various subtasks. There are several current methods for enhancing photo quality, including increasing resolution [6], dehazing [7] and adjusting photos [8]. Each of these tasks involves converting an input image to an output image. As camera sensors have improved, as have post-processing tools and advanced software, mobile photography has experienced a significant improvement over the last decade. The aesthetic quality of photos produced by mobile devices is still inferior to those produced by DSLRs. In addition to greater resolution and colour rendition, larger sensors and higher aperture optics can produce low noise due to physical changes. A low-quality image would create significant barriers to follow-up work in autonomous driving [9, 10] or surveillance systems [11].

2 Related Work

Low resolution and motion blur are two of the most common attributes of image quality enhancement methods. The author presents a state-of-the-art (SOTA) method for transferring styles and photo-realistic super-resolutions based on [12, 13]. All these techniques are used for common image superiority improvement, for example, to create better visual results than competitive products. We are most interested in the work of [2]. Translating phone camera images into high-quality DSLR images was an attempt to convert low-quality phone images into high-quality DSLR images. A convolution layer of three kernels with 64 feature maps sits on top of an input layer, along with some similar residual blocks. According to [13], their image enhancement networks followed the architectural guidelines and employed techniques similar to those proposed in [6]. The researchers incorporated loss of content, colour, and texture into a composite error function. Many low-level computer vision tasks are being largely solved by convolutional neural networks (CNNs) [14, 15], including super-resolutions [12, 16–18], image enhancement [19] deblurring [20], denoising [21], and dehazing [7]. To determine the local colour transform coefficients of an image input and enhanced using a neural network, it has been proposed to use a neural network [19]. The author used a perceptual loss to generate realistic image textures. The author proposed a novel neural network architecture based on bilateral grid processing and local affine colour transforms. The architecture was capable of processing real-time 1080p video on smartphones.

In this study, the author [2] used a residual CNN to improve the colour rendition and sharpness of images taken with DSLRs compared to ordinary photos. We describe an improved mobile image enhancement method based on deep learning that performs better than previous methods. The author developed an approach to the regularized image upsampling problem based on models of the acquisition and display processes [22]. Using edge-directed filtering as the basis for our analysis, we provide a new analytic perspective defining the requirements of total-variation regularization. A regularization method that combines data fidelity with total variation can be created by applying this technique. A level-set motion method minimizes this objective function by interacting two types of motion simultaneously.

The author [17] proposes an end-to-end trainable system called DehazeNet for medium transmission estimation. An input hazy image is processed by DehazeNet using a scattering model based on atmospheric scattering. The DehazeNet algorithm uses a CNN-based architecture, whose layers incorporate established image dehazing assumptions and priors. Therefore, the outcome is used almost always for extracting haze-relevant features. High-resolution images (HRI) are created by converting low-resolution images into visually appealing high-resolution images. A three-layer convolutional neural network, SRCNN [12], approximates the complex non-linear mapping between LR and HR images. Transposed convolutions to accelerate SRCNNs (FSRCNNs) are proposed [23]. Using extremely deep super-resolution networks (VDSRs), the author [16] achieves exceptional SR performance with a larger model capacity. According to [24], an image reconstruction network using a Laplacian pyramid can reconstruct the residual sub-bands of high-resolution images. The author presents a parameter-sharing method based on DRRNs [19]. Furthermore, this paper proposes a new architecture for CNNs that addresses multiple long-term dependency problems using a very deep persistent memory network (MemNet). VGG-based loss functions and adversarial networks are used to achieve photo-realistic super-resolution task results.

3 Proposed Photo Quality Matching Scheme

With our image translation framework, we aim to make low-quality photos look as good as DSLRs. Using an I_s image of low quality and an I_t Image of DSLR quality, we attempt to learn a mapping between them. As a result of the training set $\left\{I_s^j, I_t^j\right\}_{j=1}^{N}$, we train a deep residual CNN with weight W to determine the translation function

$$W = argmin_w \frac{1}{N} \sum_{j=1}^{N} L \tag{1}$$

The loss function L represents a multi-component perceptual loss function. This figure illustrates the first step in defining our quality enhancement network architecture.

3.1 Quality Matching Network

Model architectures are described in this section. In Fig. 2, you can see the model architecture. The main differences between our network and other image translation work [2] are the revised residual blocks, the absence of a discriminator, and the concatenation of pre-adjusted images as input.

Computer vision problems can be solved effectively with residual networks [25]. Many take a typical architecture known as the Resblock [26] for granted. In their super-resolution analysis, [4] presented how batch normalization layers can discard useful information because they normalize the feature by eliminating range flexibility from the network [26]. Following their paper on super-resolution, the batch normalization layer is removed [17]. As a result, revisions to residual blocks can significantly reduce GPU memory consumption when layer and feature channels are increased. Several image generation problems can be solved with GANs by achieving photo-realistic results [6]. Taking advantage of this advantage may not be desirable

(a)

(b)

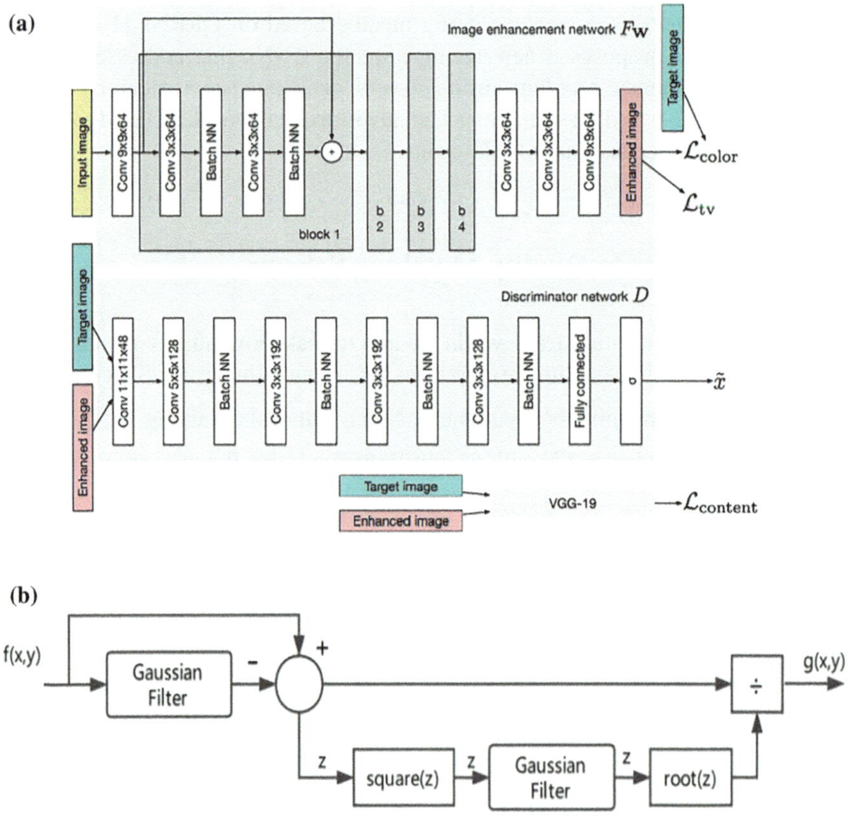

Fig. 1 **a** Proposed framework. **b** Local contrast normalized layer

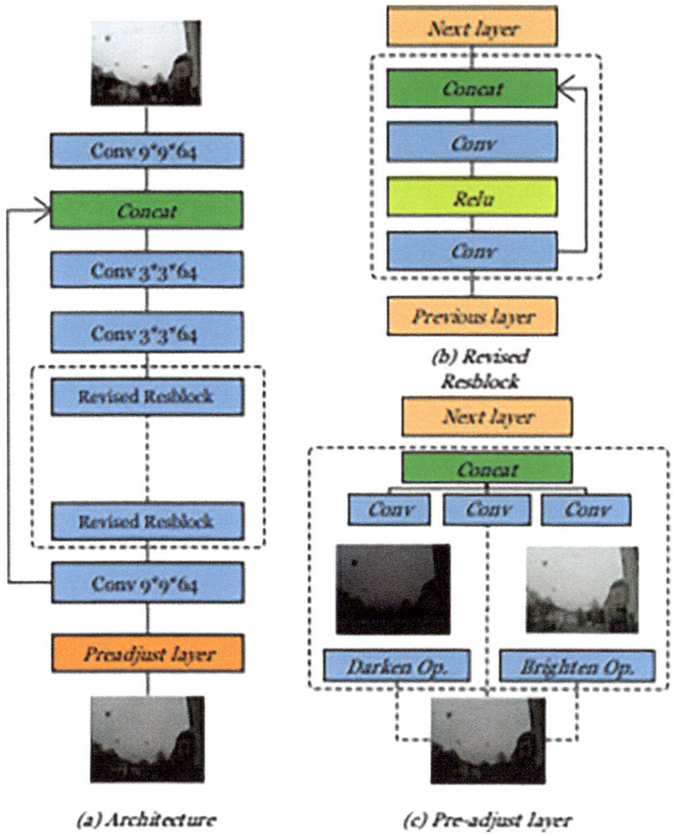

Fig. 2 This network is designed to match image quality based on its architecture

because your task does not require adding new content. As an alternative to the original image, we use traditional image processing methods to modify the brightness and exposure, and then we concatenate these three images together. Our training procedure is also stabilized by residual scaling.

3.2 *A Multi-Component Loss Function for Perceptual Loss*

Color Loss: Prediction errors can be measured using L_2 regression.

$$L(X, Y) = \|X_Y\|_2^2 \tag{2}$$

However, colour distributions with multimodality cannot be handled by regression [27]. As a result, the grey colour tends to be favoured with a loss of L_1 or L_2 [28].

Instead, we calculate distributions over bins of colored pixels, as described in [29]:

$$L(X, Y) = D_{KL}(X \| Y), \tag{3}$$

A histogram over bins is described by $\gamma = [0,1]^K$, and a divergence over K is described by D_{KL}. Lab colour space provides more perpetual results for its underlying metric, which expresses colour coherence. As a result, *L*, *a* and *b* marginal distributions are encoded individually for each lab axis and binned by Gaussian quantiles ($\mu = 0, \sigma = 25$). Combining both colour losses improved performance in our experiments.

Texture Loss: The texture difference between enhanced and target images can be measured with a texture quality metric and a texture difference metric. It is similar to determining the sharpness of an object to determine how much of the high-frequency content is attenuated by its edges in a space. Alternatively, we can use "local contrast normalization and perceptual similarity" (referred to as *lcn2lcn*) to define our first texture loss.

$$L \tag{4}$$

After *j*th a VGG-19 network with a convolutional layer, $\check{F}_W(I_s), \check{I}_t$, an image processing process that enhances grayscale images and normalizes local contrast in target images, and $\psi\left(\check{F}_W(I_s)\right) - \psi\left(\check{I}_t\right)$ refers to feature maps from their corresponding grayscale-enhanced images [30]. The $relu_{11}$ layer produced features that we specially used since CNN learns edges primarily in the first layer. Since CNN learns edges primarily from the first layer, we used the $relu_{11}$ layer to produce features we particularly liked.

Applying local non-linear operations to normalize local variance and remove local mean displacements can accentuate the decorrelating effect of local contrast normalizing in spatial image analysis [31]. In Fig. 3, the normalization procedure is shown in action. Early human vision can be modelled by using such an operation.

Based on the formulations in [27, 28], we can formulate the local normalization as follows:

$$\bar{I}(i,j) = \frac{I(i,j) - \mu(i,j)}{\sigma(i,j) + C} \tag{5}$$

In which place

$$\mu(i,j) = \frac{1}{(1P+1)(2Q+1)} \sum_{p=-P}^{p=P} \sum_{q=-Q}^{q=Q} I(i+p, j+q) \tag{6}$$

As well as

(a) Original image. (b)Normalized image.

Fig. 3 A normalization procedure has an effect

$$\sigma^2(i,j) = \sum_{p=-P}^{p=P} \sum_{q=-Q}^{q=Q} (I(i+p,j+q) - \mu(i,j))^2 \tag{7}$$

Its normalized value is $\mu(i,j)$, and its output value is (i,j), $\overline{I}(i,j)$ is its normalized value, $\mu(i,j)$ is its mean value, $\sigma(i,j)$ is its mean value, SD value ($C = 1$), and constant. The window size is $[2P + 1{,}2Q + 1]$, where $P = Q = 3$.

Additionally, our second texture loss (grad2grad) is inspired by our first texture loss (grad2grad) [32].

$$L(X, Y) = \|\nabla_x F_W(I_s) - \nabla_x I_t\| + \|\nabla_y F_w\left(I_s^j\right) - \nabla_y I_t\|| \tag{8}$$

Content Loss: Perceptual similarity defines our content loss [13].

$$L(X, Y) = \frac{1}{C_j H_j W_j} \|\psi(X) - \psi(Y)\| \tag{9}$$

VGG-19 network with jth convolutional layers, where J is its feature map. The $relu_{54}$ layer was specially used to produce features.

Total Variation Loss: An image with unrestrained detail, which might be spurious, has a higher total variation, which results in a loss of total variation. Removing unwanted noise whilst maintaining spatial smoothness and preserving details like edges is possible when the total variation of the image is reduced. A gradient's absolute integral is defined as:

$$L = \frac{1}{C_j H_j W_j} \|\nabla F_W\left(I_s^j\right) - \nabla F_W\left(I_s^j\right)\| \tag{10}$$

Fig. 4 DPED test images from our method

This image $F_W\left(I_s^j\right)$ has dimensions C, H and W. Spatial smoothness is greatly encouraged in generated images using it.

Total Loss. All of these losses are weighted to determine our final loss:

$$L_{Total} = L_{Content} + 0.1L_{Tecture} + 400L_{TV} \tag{11}$$

Preliminary experiments were used to determine coefficients.

4 Result Analysis & Discussion

A large-scale dataset for photo quality enhancement, DPED [2], is used to evaluate our method in this section. Three phones and one digital single lens reflex camera were used simultaneously to take 4549 images taken by Sony, 5727 images taken by iPhone, and 6015 images taken by Canon and Blackberry (Fig. 4). Tests were performed on around 100 original images, while training and validation were performed using patches of size 100 by 100. A low-quality image is used as training input and high-quality image patches are used as training input. We then augment

Fig. 5 BlackBerry DPED test images based on our method

the data by swapping colours and rotating and flipping geometrical coordinates to avoid overfitting. Training patches are multiplied by about 10 in this process (Figs. 5 and 6).

This section aims to discuss several terms of the proposed loss function. We consider four loss combinations here: (1) the proposed loss, (2) the loss combined with content and texture, (3) the loss coupled with MSE and texture, and (4) the loss combined with MSE alone. On DPED's test subset, a CNN was trained for each target loss combination. In Fig. 7, we show the visual results; in Figs. 8, 9, we present the experimental results. There is less sensitivity to image mismatches with content loss in this case. The resulting images are further improved by adding an extra colour term, which enhances the colour saturation and brings the colours closer to the target. With natural colour rendition, the PSNR and SSIM values are evident, but there is a strong artifacting effect caused by a single MSE. SSIM scores were highest with our loss, while visual results were also very good.

In order to evaluate our results, we conduct qualitative and quantitative comparisons with quality improvement techniques from the literature. PSNR and structural similarity index measure (SSIM) measure image quality. According to Table 1, our method has a significant margin on PSNR compared to the state-of-the-art SSIM, and Figs. 10 and 11 show its competitive PSNR. Compared to the conventional method, our method significantly reduces those undesired effects [12, 13].

Fig. 6 Sony DPED test images based on our method

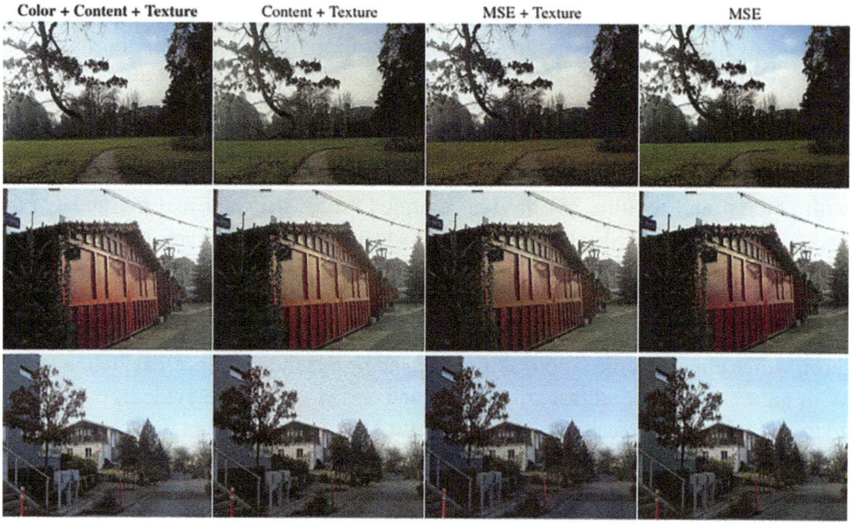

Fig. 7 Image results for four target loss functions using the iPhone camera

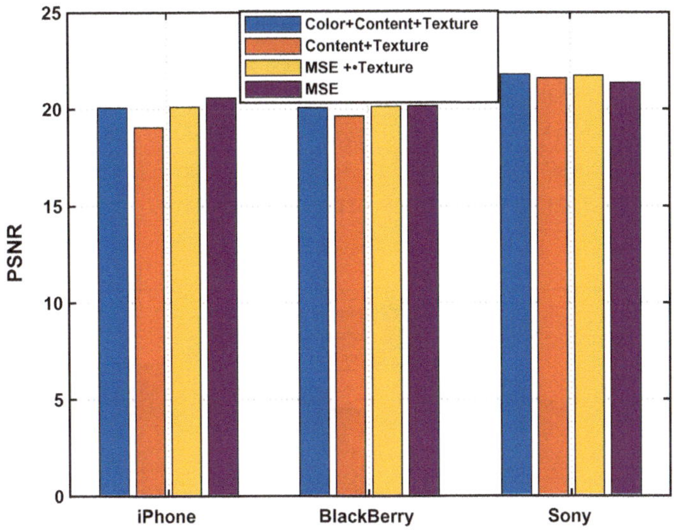

Fig. 8 We analyzed our method's PSNR performance

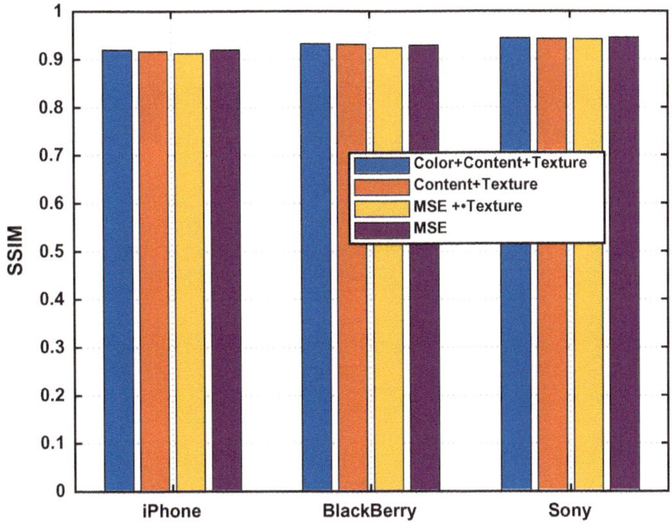

Fig. 9 Analysis of our method's performance using SSIM

Table 1 DPED test image PSNR/SSIM averages

Phone	Dong et al.,		Johnson et al.,		Ours Method	
	PSNR	SSIM	PSNR	SSIM	PSNR	SSIM
iPhone	19.37	0.8992	20.32	0.9161	20.08	0.9201
BlackBerry	18.89	0.9134	20.11	0.9298	20.07	0.9328
Sony	21.21	0.9382	21.33	0.9434	21.81	0.9437

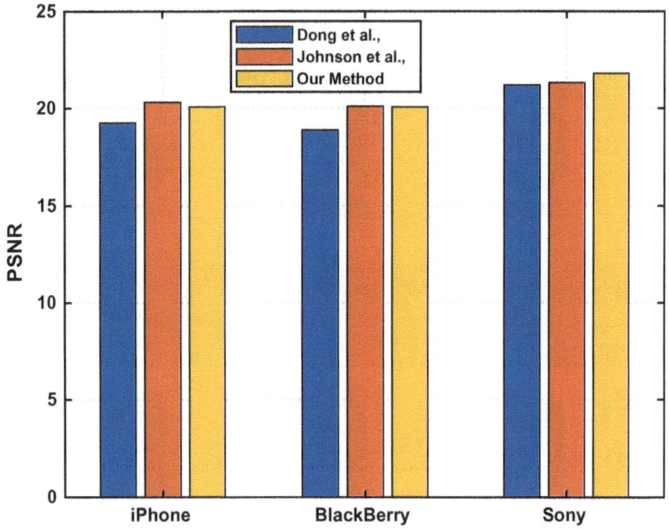

Fig. 10 Analyzing the PSNR's performance

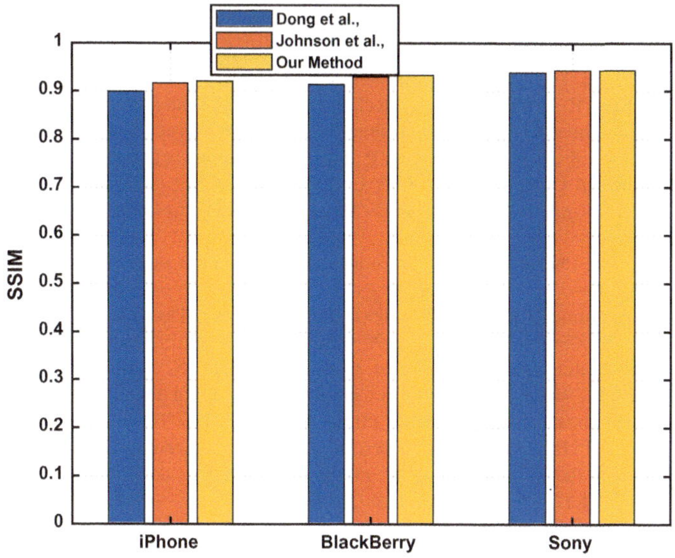

Fig. 11 Analyzing the SSIM's performance

5 Conclusion

Smartphone cameras are improving in configuration, but the quality of their photos still cannot match those of DSLRs due to physical space, hardware, and cost limitations. This work presents a fast and accurate approach for improving smartphone photos through generative adversarial networks. This paper proposes a lightweight convolutional network that maps ordinary photos to DSLR-quality images. PSNR and SSIM performance parameters are effectively studied with the constructed DPED Dataset, and experiments demonstrate the effectiveness of the proposed learning strategies.

References

1. Chen, Y.-S., Wang, Y.-C., Kao, M.-H., Chuang, Y.-Y.: Deep photo enhancer: unpaired learning for image enhancement from photographs with GANs. In: 2018 IEEE/CVF Conference on Computer Vision and Pattern Recognition, Jun. 2018, pp. 6306–6314 (2018). https://doi.org/10.1109/CVPR.2018.00660
2. Ignatov, A., Kobyshev, N., Timofte, R., Vanhoey, K., Van Gool, L.: DSLR-quality photos on mobile devices with deep convolutional networks. In: Presented at the Proceedings of the IEEE International Conference on Computer Vision, 2017, pp. 3277–3285 (2017). Accessed: Dec. 23, 2023
3. Rani, P., Verma, S., Yadav, S.P., Rai, B.K., Naruka, M.S., Kumar, D.: Simulation of the lightweight blockchain technique based on privacy and security for healthcare data for the cloud system. Int. J. E-Health Med. Commun. IJEHMC **13**(4), 1–15 (2022)

4. Yadav, S.P., Jindal, M., Rani, P., De Albuquerque, V.H.C., Dos Santos Nascimento, C., Kumar, M.: An improved deep learning-based optimal object detection system from images. Multimed. Tools Appl. (2023). https://doi.org/10.1007/s11042-023-16736-5
5. Ansari, G., Rani, P., Kumar, V.: A novel technique of mixed gas identification based on the group method of data handling (GMDH) on time-dependent MOX gas sensor data. In: Mahapatra, R.P., Peddoju, S.K., Roy, S., Parwekar, P. (eds.), Proceedings of International Conference on Recent Trends in Computing, vol. 600, Lecture Notes in Networks and Systems, vol. 600, pp. 641–654. Springer Nature Singapore, Singapore (2023). https://doi.org/10.1007/978-981-19-8825-7_55
6. Ledig, C., et al.: Photo-realistic single image super-resolution using a generative adversarial network. In: Presented at the Proceedings of the IEEE Conference on Computer Vision and Pattern Recognition, 2017, pp. 4681–4690 (2017). Accessed: Dec. 23, 2023
7. Ren, W., Liu, S., Zhang, H., Pan, J., Cao, X., Yang, M.-H.: Single image dehazing via multi-scale convolutional neural networks. In: Leibe, B., Matas, J., Sebe, N., Welling M (eds.), Computer Vision – ECCV 2016. Lecture Notes in Computer Science, vol. 9906, pp. 154–169. Springer International Publishing, Cham (2016). https://doi.org/10.1007/978-3-319-46475-6_10
8. Ignatov, A., et al.: PIRM challenge on perceptual image enhancement on smartphones: report. In: presented at the Proceedings of the European Conference on Computer Vision (ECCV) Workshops, 2018 (2018). Accessed: Dec. 23, 2023
9. Rani, P., Sharma, R.: Intelligent transportation system performance analysis of indoor and outdoor internet of vehicle (iov) applications towards 5g. Tsinghua Sci. Technol. (2023)
10. Rani, P., Sharma, R., IMFOCA-IOV: intelligent moth flame optimization based clustering algorithm for internet of vehicle. In: 2023 14th International Conference on Computing Communication and Networking Technologies (ICCCNT), pp. 1–6. IEEE, Delhi (2023). https://doi.org/10.1109/ICCCNT56998.2023.10307646
11. Koshimizu, T., Toriyama, T., Babaguchi, N.: Factors on the sense of privacy in video surveillance. In: Proceedings of the 3rd ACM workshop on Continuous archival and retrival of personal experences, in CARPE '06 , pp. 35–44. Association for Computing Machinery, New York (2006). https://doi.org/10.1145/1178657.1178665
12. Dong, C., Loy, C.C., He, K., Tang, X.: Learning a deep convolutional network for image super-resolution. In: Fleet, D., Pajdla, T., Schiele, B., Tuytelaars, T., (eds.), Computer Vision – ECCV 2014, Lecture Notes in Computer Science, vol. 8692, pp. 184–199. Springer International Publishing, Cham (2014). https://doi.org/10.1007/978-3-319-10593-2_13
13. Johnson, J., Alahi, A., Fei-Fei, L.: Perceptual losses for real-time style transfer and super-resolution. In: Leibe, B., Matas, J., Sebe, N., Welling, M., (eds.), Computer Vision – ECCV 2016. Lecture Notes in Computer Science, pp. 694–711. Springer International Publishing, Cham (2016). https://doi.org/10.1007/978-3-319-46475-6_43
14. Verma, S., et al.: An automated face mask detection system using transfer learning based neural network to preventing viral infection. Expert Syst. e13507 (2023). https://doi.org/10.1111/exsy.13507
15. Rani, P., Singh, P.N., Verma, S., Ali, N., Shukla, P.K., Alhassan, M.: "An implementation of modified blowfish technique with honey bee behavior optimization for load balancing in cloud system environment. Wirel. Commun. Mob. Comput.. Commun. Mob. Comput. **2022**, 1–14 (2022)
16. Kim, J., Lee, J.K., Lee, K.M.: Accurate image super-resolution using very deep convolutional networks. In: Proceedings of the IEEE Conference on Computer Vision and Pattern Recognition, 2016, pp. 1646–1654 (2016)
17. Lim, B., Son, S., Kim, H., Nah, S., Mu Lee, K.: Enhanced deep residual networks for single image super-resolution. In: Presented at the Proceedings of the IEEE Conference on Computer Vision and Pattern Recognition Workshops, 2017, pp. 136–144 (2017). Accessed: Dec. 23, 2023
18. Kligvasser, I., Shaham, T.R., Michaeli, T.: xunit: Learning a spatial activation function for efficient image restoration. In: Proceedings of the IEEE Conference on Computer Vision and Pattern Recognition, 2018, pp. 2433–2442 (2018)

19. Yan, Z., Zhang, H., Wang, B., Paris, S., Yu, Y.: Automatic photo adjustment using deep neural networks. ACM Trans. Graph. **35**(2):11:1–11:15 (2016). https://doi.org/10.1145/2790296
20. Noroozi, M., Chandramouli, P., Favaro, P.: Motion deblurring in the wild. In: Roth, V., Vetter, T., (eds.), Pattern Recognition. Lecture Notes in Computer Science, vol. 10496, pp. 65–77. Springer International Publishing, Cham (2017). https://doi.org/10.1007/978-3-319-66709-6_6.
21. Zhang, K., Zuo, W., Chen, Y., Meng, D., Zhang, L.: Beyond a Gaussian denoiser: residual learning of deep CNN for image denoising. IEEE Trans. Image Process. **26**(7), 3142–3155 (2017). https://doi.org/10.1109/TIP.2017.2662206
22. Aly, H.A., Dubois, E.: "Image up-sampling using total-variation regularization with a new observation model. IEEE Trans. Image Process. **14**(10), 1647–1659 (2005). https://doi.org/10.1109/TIP.2005.851684
23. Dong, C., Loy, C.C., Tang, X.: Accelerating the super-resolution convolutional neural network. In: Leibe, B., Matas, J., Sebe, N., Welling, M., (eds.), Computer Vision – ECCV 2016. Lecture Notes in Computer Science, vol. 9906, , pp. 391–407. Springer International Publishing, Cham (2016). https://doi.org/10.1007/978-3-319-46475-6_25
24. Tai, Y., Yang, J., Liu, X.: Image super-resolution via deep recursive residual network. In: Proceedings of the IEEE Conference on Computer Vision and Pattern Recognition, 2017, pp. 3147–3155 (2017)
25. He, K., Zhang, X., Ren, S. and Sun, J.: Deep residual learning for image recognition. In: Proceedings of the IEEE Conference on Computer Vision and Pattern Recognition, 2016, pp. 770–778 (2016)
26. Nah, S., Hyun Kim, T., Mu Lee, K.: Deep multi-scale convolutional neural network for dynamic scene deblurring. In: Proceedings of the IEEE Conference on Computer Vision and Pattern Recognition, 2017, pp. 3883–3891 (2017)
27. Leibe, B.: Computer vision - ECCV 2016: 14th European conference, Amsterdam, The Netherlands, October 11–14, 2016 Proceedings Part 2. in Lecture Notes in Computer Science, no. 2 (2016)
28. Isola, P., Zhu, J.Y., Zhou, T., Efros, A.A.: Image-to-image translation with conditional adversarial networks. In: Proceedings of the IEEE Conference on Computer Vision and Pattern Recognition, 2017, pp. 1125–1134 (2017)
29. Larsson, G., Maire, M., Shakhnarovich, G.: Learning representations for automatic colorization. In: Leibe, B., Matas, J., Sebe, N., Welling, M., (eds.), Computer Vision – ECCV 2016. Lecture Notes in Computer Science, vol. 9908, pp. 577–593. Springer International Publishing, Cham (2016). https://doi.org/10.1007/978-3-319-46493-0_35
30. Simonyan, K., Zisserman, A.: Very deep convolutional networks for large-scale image recognition (2014). ArXiv Prepr. arXiv:14091556
31. Mittal, A., Moorthy, A.K., Bovik, A.C.: No-reference image quality assessment in the spatial domain. IEEE Trans. Image Process. Image Process. **21**(12), 4695–4708 (2012). https://doi.org/10.1109/TIP.2012.2214050
32. Zhu, J.: Image gradient-based joint direct visual odometry for stereo camera. In: IJCAI, 2017, pp. 4558–4564 (2017)
33. Lv, Y., Jiang, G., Yu, M., Xu, H., Shao, F., Liu, S.: Difference of Gaussian statistical features based blind image quality assessment: A deep learning approach. In: 2015 IEEE International Conference on Image Processing (ICIP), pp. 2344–2348. IEEE (2015)

An Experimental Evaluation of Deepfake Recognition and Detection Using Deep Learning Model

Ahmed J. Obaid and Huda Qasim Owaied

Abstract The development of powerful image generation models has accelerated the development of face manipulation techniques. Deep Fake recognition for Expression Reenactment and Phasing Face Recognition attract much attention for their flexibility and ease of generating high-quality synthesis results. Recently, both subjects have been the subject of active research. However, most existing methods ignore the similarities between the two tasks. Face recognition is significantly impacted by Deepfake recognition. This paper develops a state-of-the-art method for classifying and detecting deepfake and deepswap. Evaluation results demonstrate the generalization capability and performance of the algorithm on different deepfake datasets with extensive experiments on the challenging FaceForensic+ + datasets.

1 Introduction

The development of deep learning has contributed greatly to the rapid improvement of computer vision over the past few years, particularly for manipulating images of people. The development of deep learning has contributed greatly to the rapid improvement of computer vision over the past few years, particularly for manipulating images of people. To enhance the quality of face swapping, deep learning has been incorporated with generative models [1–3]. Although the task is challenging, especially when the source image is just one frame and targets multiple scenarios, the task is still challenging. Two essential challenges in blending the swapped face seamlessly into the target while maintaining implicit attributes: explicitly capturing identification information and seamlessly blending the swapped face into the target.

A face swap, as illustrated in Fig. 1, involves swapping facial features while maintaining other attributes, such as poses, expressions, and backgrounds. Fake content

A. J. Obaid (✉)
Faculty of Computer Science and Mathematics, University of Kufa, Kufa, Iraq
e-mail: ahmedj.aljanaby@uokufa.edu.iq

H. Q. Owaied
College of Engineering, Medical Instruments Technology Engineering, National University of Science and Technology, Dhi Quar, Iraq

S. Kumar Udgata et al. (eds.), *Intelligent Systems*, Lecture Notes in Networks and Systems 1149, https://doi.org/10.1007/978-981-97-8160-7_32

Fig. 1 DeepFakeLab (DFL) results for face swapping

is generated by swapping faces between sources and destinations while maintaining facial movements and expressive deformations at both locations. The development of deep learning has contributed greatly to the rapid improvement of computer vision over the past few years, particularly for manipulating images of people.

Many applications of face swapping have attracted great attention [4, 5], including entertainment, filmmaking, creating virtual humans, and protecting privacy. In addition, attributes such as pose, expression, lighting, and background are added to the target frame or video to transfer a person's facial identity [6, 7]. In a nutshell, GANs (Generative Adversarial Networks) are the basic motivation for face manipulation techniques [8]. As StyleGAN [9] and StyleGAN2 [10] synthesize more and more faces, the human brain becomes incapable of determining the difference between them. GAN-based face-swapping methods are widespread in spoof videos published on YouTube and other video websites. ZAO2 and FaceApp3, two commercially available applications that allow netizens to swap images and videos easily, significantly contribute to the growth of these deepfakes [11].

Two approaches have been taken by previous studies to address the above problems. Long ago, graphics-based methods [3] and [4] used landmarks and 3D models [12] to perform face swapping using prior knowledge of intermediate structures. Using GANs [13], researchers have been able to extract identity and expressions from this information [14, 15]. The inaccuracy of structural information would greatly affect the stability and coherence of generated results, especially in videos. Various face-generating tasks have been investigated using StyleGAN architectures [16–18], including editing face attributes [19, 20], enhancing faces [21], and even reenacting facial features [22, 23]. The strong expressibility and latent space manipulation advantages of style-based generators are to blame. Nevertheless, face-swapping architectures [24, 25] have not been sufficiently explored. As a result of the fixed generator's limited distribution, the lighting conditions in [24] are severely compromised. StyleSwap, in particular, enables face swapping through a style-based generator. Compared to previous methods, it provides better robustness, fidelity, and identity similarity (i.e., it does not produce visible artifacts). Additionally, it is simple to implement and easy to train on. StyleGAN2 architecture must be adapted through simple modifications to handle swapping data flows, and the generator must take advantage of identity optimization.

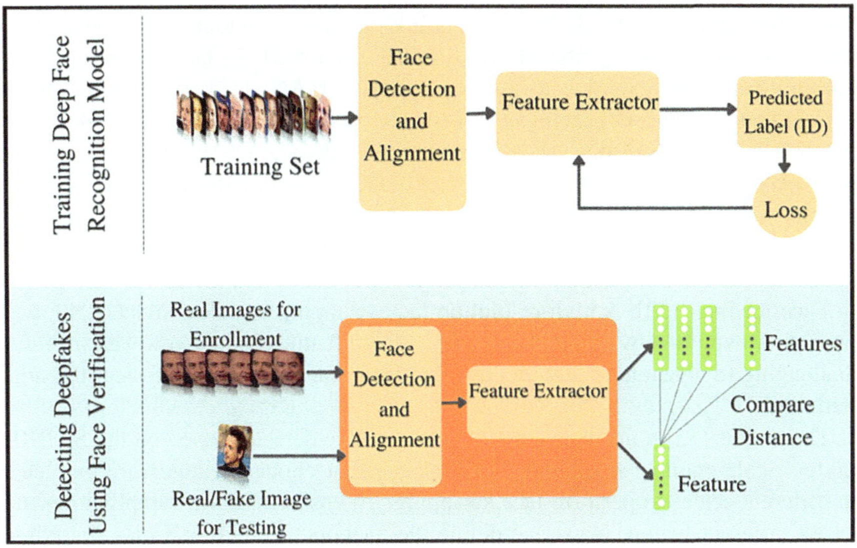

Fig. 2 Deep fake recognition is used to detect deep fakes

This study aims to compare the effectiveness of a deep facial recognition system to identify deepfakes generated by various methods using various loss functions. It will be possible to determine the best loss function to use when detecting deepfake generation techniques using face recognition technology. Model training on fake images is not needed in this work. However, deepfake detection requires identity labels when comparing template and query images. Deep fake recognition is used in Fig. 2 to detect deep fakes.

2 Related Work

The well-trained GAN model reconstructs the original image by using the latent code of the input image. The latent vector is generally optimized directly to minimize the errors [19, 20]. The time and effort required to use these methods could result in high-quality reconstructions. In subsequent methods, the latent space is mapped end-to-end via an encoder. A StyleGAN2 generator is pre-trained as part of the pSp and GHF [15, 26] procedures. The encoder developed by generates a base style code with an offset vector corresponding to the final style code derived from the base code [27, 28].

As part of face swapping, other aspects of the face are kept the same, but the facial identity changes. In the early stages, 3D-based methods were used to preserve the expression of the target, but they required manual interaction. 3D morphable models (3DMMs) are used in Face2Face [29, 30] to fit the source and target faces.

The author [26] combines 3DMM with face segmentation in unprecedented circumstances. Furthermore, significant progress has been made in face-swapping using learning-based methods due to the popularity of GANs [12]. Despite its ability to train encoder–decoder architectures for two distinct identities, DeepFakes [14] cannot generalize its results. The disentanglement paradigm is used in some works. In IPGAN [31], it disentangles the attributes of the source face from those of the target face and recombines them to preserve the source face's identity. FaceShifter [10] can integrate identity embeddings and attribute embeddings in an attentional manner in an adaptive way. A three-dimensional prior, texture code and identity code are used by FaceInpainter [32]. A high-resolution face-swapping decoder, StyleGAN2, was previously exploited by MegaFS [24]. The global feature fusion, however, makes it challenging to generate highly identity-consistent faces using the above referential methods.

The author [33] trained on a scale-invariant face detector based on the SAEHD model. Scale-equitable-face detection ensures that enough features are extracted on different scales to perform face swapping. To ensure uniform sampling density across different feature maps, we define the anchors at different scales according to the effective receptive field and employ the proportional sampling method. With the real-time speed of DeepFakeLab, we can achieve relatively small source and destination losses by replacing the face with a frame.

Face swapping based on deep learning has improved significantly. In the DeepFakeLab [14], an encoder–decoder is trained to swap faces automatically based on the situation, and impressive results have been achieved. In this method, however, only two specific identities can be swapped. Subject-agnostic methods have recently been made more convenient by introducing more convenient methods. It is generally possible to divide these methods into source- and target-oriented categories.

The images generated by a neural network trained for face swapping generally have obvious artefacts, making it difficult to train the network. With knowledge distillation, we can achieve a more stable training process and better results when we move from face-swapping to paired training tasks. Therefore, we train our lightweight network by employing a well-trained network as the teacher. It is possible, however, for the teacher model to produce some failures, which can result in an image that looks quite different from the source image or a low degree of identity similarity. This failure case information can lead to suboptimal results for the student model. The loss reweighting module proposed in this paper alleviates this problem. Additionally, the weighting of distillation loss is adaptively adjusted based on the quality of teacher outputs. Thus, students can benefit from teachers' better outputs and generate better-generated results with fewer artefacts and similar identities (Table 1).

Table 1 Analysis of deepfakes in comparison

Study	Method	Classifiers	Best performance	Databases
[34]	GAN-Pipeline Features	SVM	Area Under The Curve (AUC) = 70.0%	NIST MFC2018
[35]	GAN-Pipeline Features	SVM	Accuracy (Acc) = 84.7%	Own (InterFaceGAN, StyleGAN)
[36]	GAN-Pipeline Features	KNN, LDA SVM,	Acc. = 99.81%	Own (StyleGAN2,AttGAN, StarGAN, GDWCT, StyleGAN,)
[37]	Steganalysis Features	CNN	Equal Error Rate (EER) = 12.3% [38]	100 K-Faces (StyleGAN)
[39]	Deep Learning Features	CNN	Acc. = 99.5%	Own (ProGAN, CramerGAN, SNGAN, MMDGAN)
[40]	Deep Learning Features	CNN + Incremental Learning	Acc. = 99.3%	Own (CycleGAN, ProGAN, Glow, StarGAN, StyleGAN)
[41]	Deep Learning Features	CNN + Incremental Learning	AUC = 100% EER = 0.1%	DFFD (StyleGAN, ProGAN,)
[38]	Deep Learning Features	CNN	EER = 0.3%	100 K-Faces (StyleGAN)
			EER = 4.5%	iFakeFaceDB
[42]	Deep Learning Features	CNN, AE	Acc. = 99.8%	Own (Glow, StarGAN, StyleGAN ProGAN,)

3 The Proposed Approach

In most state-of-the-art (SoTA) architectures, there are advantages and disadvantages. Particularly when switching between single images and videos, face edge errors and inconsistent eye gazes pose the greatest disadvantages. The FaceShifter [10] model was used to implement several new steps in order to improve the quality and stability of the output results.

From the source image, X_s is the cropped face, while for the target image, X_t is the cropped face. We should generate a new face called $\widehat{Y}_{s,t}$. In terms of dimensions, X_s, X_t is while $\widehat{Y}_{s,t}$ is $256 \times 256 \times 3$. As shown in Fig. 3, the architecture is composed of the following components:

(1) A vector zid encoder is a trained ArcFace model that determines a person's identity from the source image using the vector zid of size 1 by 512.
(2) U-Net-based attribute encoders extract features from target images [43]: $X_t - z_{att}^1, z_{att}^2, \ldots, z_{att}^n$.

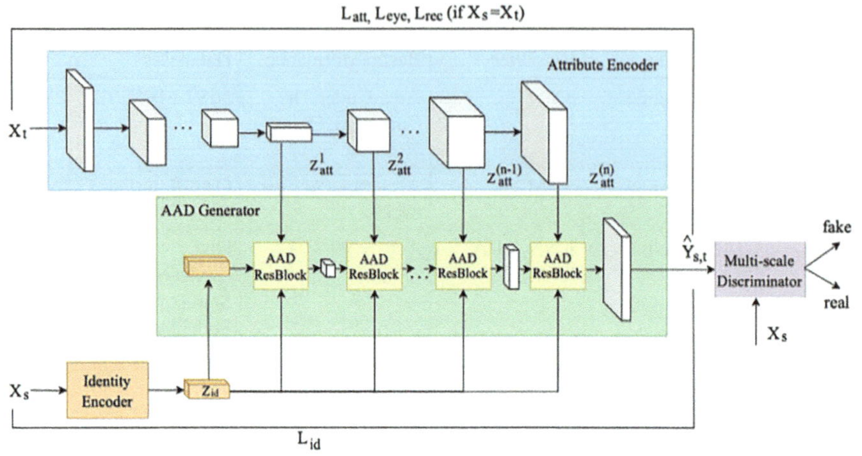

Fig. 3 Faceshifter-based model architecture

(3) With AAD ResBlocks, a new face $\widehat{Y}_{s,t}$ is generated using source identity and target attribute vectors from X_t and X_s, respectively.

(4) As the multiscale discriminator improves the output synthesis quality when comparing real and fake images, it is based on comparing real and fake images[44] improves the output synthesis quality when comparing real and fake images, it is based on comparing real and fake images.

The loss function we developed combines features that improve stability and increase swap accuracy over the baseline approach. We have improved the quality of the output due to these improvements. The experiments section describes various experiments we conducted with the architecture itself and loss improvements.

In general, the architecture part of AEI-Net consists of the following:

(1) Reconstruction loss is represented by L_{rec}. $X_s = X_t$ is randomly used as the model input, and $\widehat{Y}_{s,t} = X_t$ is the expected output.

$$L_{rec} = \begin{cases} \|\widehat{Y}_{s,t} - X_t\|_2^2 & X_s = X_t \\ 0 & otherwise \end{cases} \tag{1}$$

(2) Reconstruction loss is represented by L_{att}. It is important that $\widehat{Y}_{s,t}$ and X_t $z_{att}^1, z_{att}^2, \ldots, z_{att}^n$ att features are close to each other:

$$L_{att} = \sum_{k=1}^{n} \|z_{att}^k\left(\widehat{Y}_{s,t}\right) - z_{att}^k(X_t)\|_2^2 \tag{2}$$

(3) Reconstruction loss is represented byL_{id}. For $\widehat{Y}_{s,t}$ and X_t, we assume their cosine similarity outputs are the same:

$$L_{id} = 1 - \cos\left(Z_{id}\left(\widehat{Y}_{s,t}\right), Z_{id}(X_s)\right) \quad (3)$$

(4) Using a multi-scale discriminator, L_{adv} represents the loss associated with GAN (adversarial loss).

Here are the modifications we need to make to our loss. We have adapted the SimSwap model idea to modify the reconstruction loss [9]. FaceShifter's original architecture was based on the idea that two identical images of a person would yield the same picture. As far as $P\prime \in P$ is concerned, X_s and X_t are not all members of the same set. This loss modification was possible because we used datasets with multiple photos per person:

$$L_{rec} = \begin{cases} \|\widehat{Y}_{s,t} - X_t\|_2^2 & if X_s = P\prime and X_t \in P^i \\ 0 & otherwise \end{cases} \quad (4)$$

Face swap results can also be perceived visually based on gaze direction, especially when using image-to-video swaps. A realistic perception can be achieved by representing every frame's direction of sight. Therefore, we decided to correct the gaze direction by adding a special eye loss. Based on the face key point model, L_{eye} compared X_t and $\widehat{Y}_{s,t}$ eye heatmaps $hm(.)$ as follows:

$$L_{eye} = \|hm(\widehat{Y}_{s,t}) - hm(X_t)\|_2^2 \quad (5)$$

Loss as a whole is expressed as follows:

$$L = \lambda_{adv}L_{adv} + \lambda_{att}L_{att} + \lambda_{rec}L_{rec} + \lambda_{id}L_{id} + \lambda_{eye}L_{eye} \quad (6)$$

4 Result and Analysis

Our proposed method is compared with FaceShifter [10], MegaFS [24], and SimSwap [9] on FaceForensics + + . Figure 4 shows artifacts caused by large differences in source and target faces. In our study, MegaFS seems to be able to address these challenges, but MegaFS do not preserve the skin color. In addition to this, our results in rows 4 and 5 are much closer to the source face's eye color than those of other methods. A comparison is also made with the methods of FaceShifter and FaceInpainter [32] for wild face images.

The distance between the corresponding target and swapped faces is shown in Fig. 5. Figures 6, 7, and 8 show the comparison results. SimSwap maintains more target face attributes while retaining less consistency regarding identity. FaceShifter

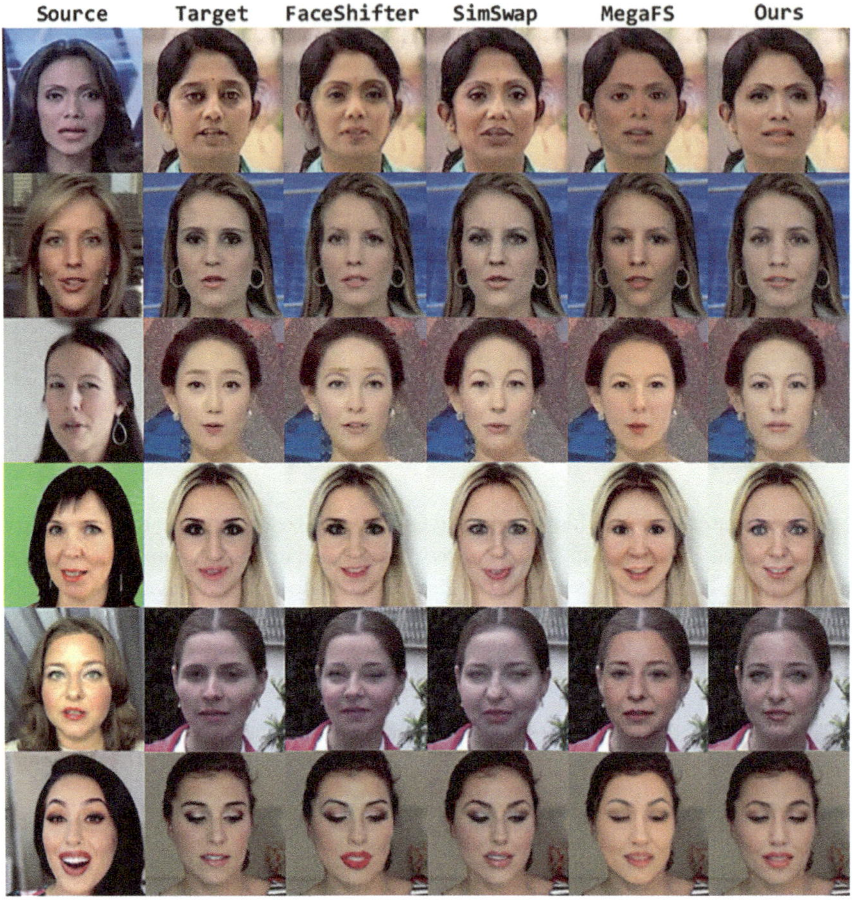

Fig. 4 Analyzing FaceForensics+ + data in comparison with state of the art

produces pose and expression errors comparable to our method, outperforming MegaFS by a wide margin.

FaceSwap and Deepfakes-based fake creation techniques are visualized using the t-SNE technique (two components for each user). FaceSwap clearly distinguishes between genuine and deepfake facial features. In addition to identity swapping, other methods of identity swapping were observed to have a similar effect. Expression-swapping approaches its the term for the process or exchange of expressions among faces, however its correct.

As shown in Fig. 9, the distribution of scores for genuine and deepfake templates was determined by comparing templates with real and fake images generated by different deepfake generation techniques. A higher overlap was also seen between genuine and deepfake distributions when expression-swapping methods were compared to t-SNE-based visualization methods.

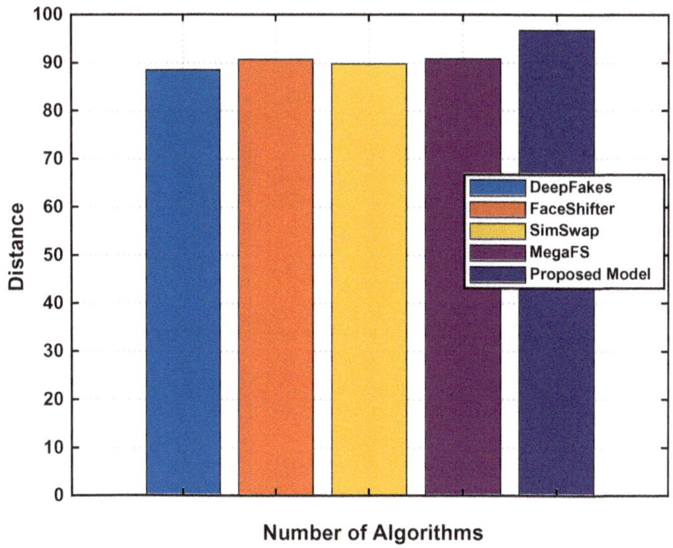

Fig. 5 Quantitative distance calculation between swapped and target faces on FaceForensics+ +

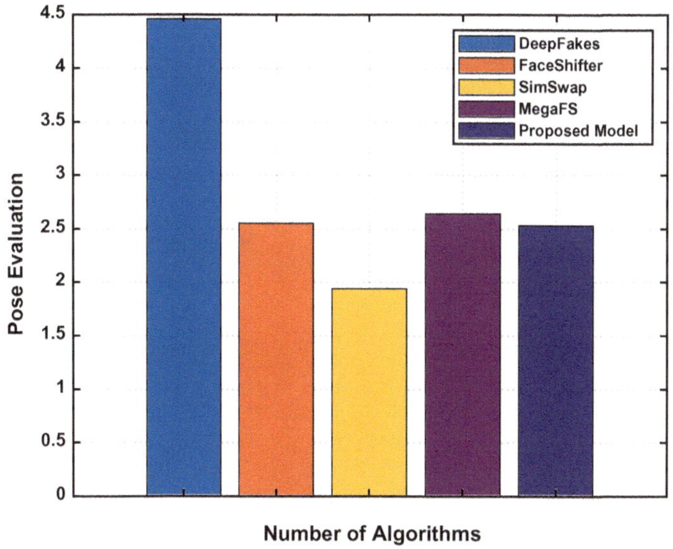

Fig. 6 Quantitative pose evaluation between swapped and target faces on FaceForensics+ +

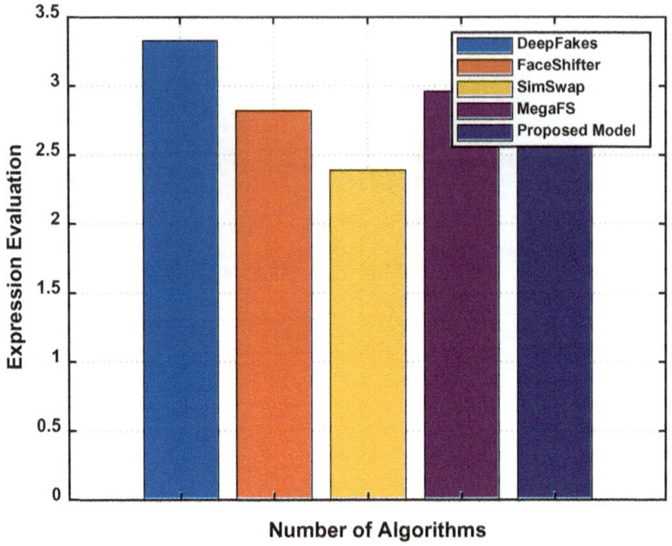

Fig. 7 Quantitative expression evaluation between swapped and target faces on FaceForensics+ +

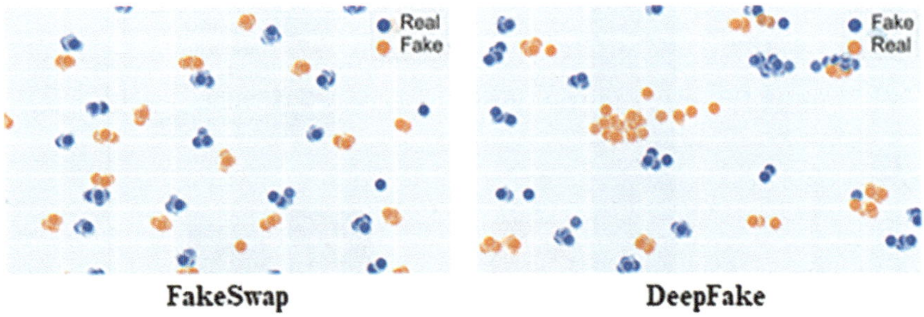

Fig. 8 Fake and genuine images on FaceForensics + + are visualized using t-SNE

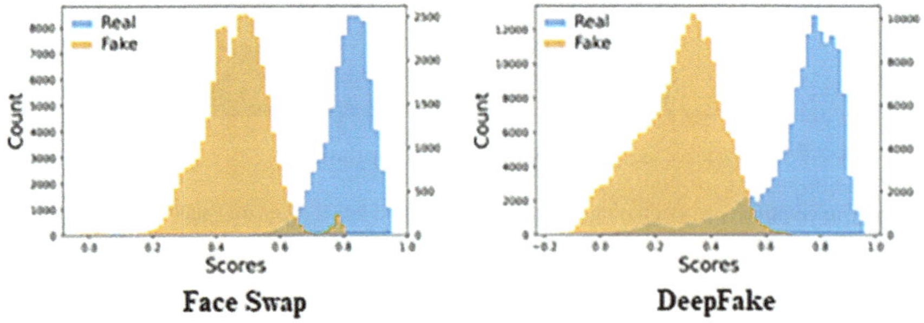

Fig. 9 Calculated the genuine and deepfake score distributions using FaceForensics++

5 Conclusion

Deep learning can be applied in fields such as machine learning, natural language processing, and computer vision. Human beings cannot distinguish between the real image and the fake one with deep fakes, which use deep learning as a method of synthesis and manipulation. Recent frameworks for face swapping have produced high-fidelity results. Because of the deep structure and off-the-shelf networks used in previous works, the previous works suffered from high computational costs. A subjective evaluation was conducted using FaceForensics++ datasets, and two state-of-the-art methods for detecting deepfakes were compared. In the experiments, deepfakes were successfully detected by humans and machines, even though their perceptions were very different. Humans are very good at spotting deepfake videos. However, algorithms successfully detect them.

References

1. Liang, B., et al.: Expressive talking head generation with granular audio-visual control. In: Proceedings of the IEEE/CVF Conference on Computer Vision and Pattern Recognition, 2022, pp. 3387–3396 (2022)
2. Liu, Z., Luo, P., Wang, X., Tang, X.: Deep learning face attributes in the wild. In: Proceedings of the IEEE International Conference on Computer Vision, 2015, pp. 3730–3738 (2015)
3. Rani, P., Singh, P.N., Verma, S., Ali, N., Shukla, P.K., Alhassan, M.: An implementation of modified blowfish technique with honey bee behavior optimization for load balancing in cloud system environment. Wirel. Commun. Mob. Comput.. Commun. Mob. Comput. **2022**, 1–14 (2022)
4. Bitouk, D., Kumar, N., Dhillon, S., Belhumeur, P., Nayar, S.K.: Face swapping: automatically replacing faces in photographs. In: ACM SIGGRAPH 2008 papers, 2008, pp. 1–8 (2008)
5. Blanz, V., Scherbaum, K., Vetter, T., Seidel, H.-P.: Exchanging faces in images. In: Computer Graphics Forum, Wiley Online Library, 2004, pp. 669–676 (2004)
6. Parkhi, O., Vedaldi, A., Zisserman, A.: Deep fake recognition. In: BMVC 2015-Proceedings of the British Machine Vision Conference 2015, British Machine Vision Association (2015)
7. Verma, S., et al.: An automated face mask detection system using transfer learning based neural network to preventing viral infection. Expert Syst. e13507 (2023). https://doi.org/10.1111/exsy.13507
8. Liu, N., Zhou, T., Ji, Y., Zhao, Z., Wan, L.: Synthesizing talking faces from text and audio: an autoencoder and sequence-to-sequence convolutional neural network. Pattern Recognit. **102**, 107231 (2020)
9. Chen, R., Chen, X., Ni, B., Ge, Y.: SimSwap: an efficient framework for high fidelity face swapping. In: Proceedings of the 28th ACM International Conference on Multimedia, pp. 2003–2011. ACM, Seattle (2020). https://doi.org/10.1145/3394171.3413630
10. Li, L., Bao, J., Yang, H., Chen, D., Wen, F.: Faceshifter: towards high fidelity and occlusion aware face swapping (2019). ArXiv Prepr. arXiv:191213457
11. Yadav, S.P., Jindal, M., Rani, P., De Albuquerque, V.H.C., Dos Santos Nascimento, C., Kumar, M.: An improved deep learning-based optimal object detection system from images. Multimed. Tools Appl. (2023). https://doi.org/10.1007/s11042-023-16736-5
12. Goodfellow, I., et al.: Generative adversarial nets. Adv. Neural Inf. Process. Syst. **27** (2014)
13. Jiang, L., Li, R., Wu, W., Qian, C., Loy, C.C.: Deeperforensics-1.0: a large-scale dataset for real-world face forgery detection. In: Proceedings of the IEEE/CVF Conference on Computer Vision and Pattern Recognition, 2020, pp. 2889–2898 (2020)

14. Perov, I., et al.: DeepFakeLab: integrated, flexible and extensible face-swapping framework (2020). ArXiv Prepr. arXiv:200505535
15. Richardson, E., et al.: Encoding in style: a stylegan encoder for image-to-image translation. In: Proceedings of the IEEE/CVF Conference on Computer Vision and Pattern Recognition, 2021, pp. 2287–2296 (2021)
16. Karras, T., Laine, S., Aittala, M., Hellsten, J., Lehtinen, J., Aila, T.: Analyzing and improving the image quality of stylegan. In: Proceedings of the IEEE/CVF Conference on Computer Vision and Pattern Recognition, 2020, pp. 8110–8119 (2020)
17. Kim, H., et al.: Deep video portraits. ACM Trans. Graph. TOG **37**(4), 1–14 (2018)
18. Kingma, D.P., Ba, J.: Adam: a method for stochastic optimization (2014). ArXiv Prepr. arXiv:14126980
19. Abdal, R., Qin, Y., Wonka, P.: Image2stylegan: how to embed images into the stylegan latent space?. In: Proceedings of the IEEE/CVF International Conference on Computer Vision, 2019, pp. 4432–4441 (2019)
20. Abdal, R., Qin, Y., Wonka, P.: Image2stylegan++: how to edit the embedded images?. In: Proceedings of the IEEE/CVF Conference on Computer Vision and Pattern Recognition, 2020, pp. 8296–8305 (2020)
21. Yang, T., Ren, P., Xie, X., Zhang, L.: Gan prior embedded network for blind face restoration in the wild. In: Proceedings of the IEEE/CVF Conference on Computer Vision and Pattern Recognition, 2021, pp. 672–681 (2021)
22. Burkov, E., Pasechnik, I., Grigorev, A., Lempitsky, V.: Neural head reenactment with latent pose descriptors. In: Proceedings of the IEEE/CVF Conference on Computer Vision and Pattern Recognition, 2020, pp. 13786–13795 (2020)
23. Zhou, H., Sun, Y., Wu, W., Loy, C.C., Wang, X., Liu, Z.: Pose-controllable talking face generation by implicitly modularized audio-visual representation. In: Proceedings of the IEEE/CVF Conference on Computer Vision and Pattern Recognition, 2021, pp. 4176–4186 (2021)
24. Zhu, Y., Li, Q., Wang, J., Xu, C.-Z., Sun, Z.: One shot face swapping on megapixels. In: Proceedings of the IEEE/CVF Conference on Computer Vision and Pattern Recognition, 2021, pp. 4834–4844 (2021)
25. Zhu, P., Abdal, R., Femiani, J., Wonka, P.: Barbershop: Gan-based image compositing using segmentation masks (2021). ArXiv Prepr. ArXiv:210601505
26. Xu, Y., Shen, Y., Zhu, J., Yang, C., Zhou, B.: Generative hierarchical features from synthesizing images. In: Presented at the Proceedings of the IEEE/CVF Conference on Computer Vision and Pattern Recognition, 2021, pp. 4432–4442 (2021). Accessed: Dec. 23, 2023. https://openaccess.thecvf.com/content/CVPR2021/html/Xu_Generative_Hierarchical_Features_From_Synthesizing_Images_CVPR_2021_paper.html
27. Tov, O., Alaluf, Y., Nitzan, Y., Patashnik, O., Cohen-Or, D.: Designing an encoder for StyleGAN image manipulation. ACM Trans. Graph. **40**(4), 133:1–133:14 (2021). https://doi.org/10.1145/3450626.3459838
28. Rani, P., Verma, S., Yadav, S.P., Rai, B.K., Naruka, M.S., Kumar, D.: Simulation of the lightweight blockchain technique based on privacy and security for healthcare data for the cloud system. Int. J. E-Health Med. Commun. IJEHMC **13**(4), 1–15 (2022)
29. Thies, J., Zollhofer, M., Stamminger, M., Theobalt, C., Niessner, M.: Face2Face: real-time face capture and reenactment of RGB videos. In: Presented at the Proceedings of the IEEE Conference on Computer Vision and Pattern Recognition, 2016, pp. 2387–2395 (2016). Accessed: Dec. 23, 2023. https://openaccess.thecvf.com/content_cvpr_2016/html/Thies_Face2Face_Real-Time_Face_CVPR_2016_paper.html
30. Ansari, G., Rani, P., Kumar, V.: A novel technique of mixed gas identification based on the group method of data handling (GMDH) on time-dependent MOX gas sensor data. In: Mahapatra, R.P., Peddoju, S.K., Roy, S., Parwekar, P. (eds.), Proceedings of International Conference on Recent Trends in Computing. Lecture Notes in Networks and Systems, vol. 600, pp. 641–654. Springer Nature Singapore, Singapore (2023). https://doi.org/10.1007/978-981-19-8825-7_55
31. Bao, J., Chen, D., Wen, F., Li, H., Hua, G.: Towards open-set identity preserving face synthesis. In: Proceedings of the IEEE Conference on Computer Vision and Pattern Recognition, 2018, pp. 6713–6722 (2018)

32. Li, J., Li, Z., Cao, J., Song, X., He, R.: Faceinpainter: high fidelity face adaptation to heterogeneous domains. In: Proceedings of the IEEE/CVF Conference on Computer Vision and Pattern Recognition, 2021, pp. 5089–5098 (2021)
33. Jia, F., Yang, S.: Video face swap with DeepFakeLab. In: International Conference on Computer Graphics, Artificial Intelligence, and Data Processing (ICCAID 2021), SPIE, 2022, pp. 326–332 (2022). https://doi.org/10.1117/12.2631297
34. McCloskey, S., Albright, M.: Detecting gan-generated imagery using color cues (2018). ArXiv Prepr. arXiv:181208247
35. Wang, R., et al.: Fakespotter: a simple yet robust baseline for spotting ai-synthesized fake faces (2019). ArXiv Prepr. arXiv:190906122
36. Guarnera, L., Giudice, O., Battiato, S.: Deepfake detection by analyzing convolutional traces. In: Proceedings of the IEEE/CVF Conference on Computer Vision and Pattern Recognition Workshops, 2020, pp. 666–667 (2020)
37. Nataraj, L., et al.: Detecting GAN generated fake images using co-occurrence matrices (2019). ArXiv Prepr. arXiv:190306836
38. Neves, J.C., Tolosana, R., Vera-Rodriguez, R., Lopes, V., Proença, H., Fierrez, J.: Ganprintr: Improved fakes and evaluation of the state of the art in face manipulation detection. IEEE J. Sel. Top. Signal Process. **14**(5), 1038–1048 (2020)
39. Yu, N., Davis, L.S., Fritz, M.: Attributing fake images to gans: learning and analyzing gan fingerprints. In: Proceedings of the IEEE/CVF International Conference on Computer Vision, 2019, pp. 7556–7566 (2019)
40. Marra, F., Saltori, C., Boato, G., Verdoliva, L., Incremental learning for the detection and classification of gan-generated images. In: IEEE International Workshop on Information Forensics and Security (WIFS). IEEE **2019**, 1–6 (2019)
41. Dang, H., Liu, F., Stehouwer, J., Liu, X., Jain, A.K.: On the detection of digital face manipulation. In: Proceedings of the IEEE/CVF Conference on Computer Vision and Pattern recognition, 2020, pp. 5781–5790 (2020)
42. Hulzebosch, N., Ibrahimi, S., Worring, M.: Detecting CNN-generated facial images in real-world scenarios. In: Proceedings of the IEEE/CVF Conference on Computer Vision and Pattern Recognition Workshops, 2020, pp. 642–643 (2020)
43. Ronneberger, O., Fischer, P., Brox, T., U-net: Convolutional networks for biomedical image segmentation. In: Medical Image Computing and Computer-Assisted Intervention–MICCAI,: 18th International Conference, Munich, Germany, October 5–9, 2015, Proceedings, Part III 18. Springer **2015**, 234–241 (2015)
44. Park, T., Liu, M.-Y., Wang, T.-C., Zhu, J.-Y.: Semantic image synthesis with spatially-adaptive normalization. In: Proceedings of the IEEE/CVF Conference on Computer Vision and Pattern Recognition, 2019, pp. 2337–2346 (2019)

A Neural Network and Machine Learning-Based Approach for Diabetic Retinopathy Detection and Classification

Mohammed Ridha Hammoodi and Mazin R. AL-Hameed

Abstract Chronic diabetes can cause diabetic retinopathy, resulting in blindness. The increased severity of diabetic retinopathy can be prevented by early detection. Diabetic retinopathy can be detected quickly by an automated system to determine the need for follow-up treatment to prevent further retinal damage. In this study, an effective, robust, and accurate automatic methodology for detecting DR subjects will be developed. It is based on two steps: (1) Reconstruction and enhancement of blood vessels using customised programs; (2) the use of an Artificial Neural Network as a classifier to distinguish people with diabetes with and without diabetic retinopathy (DR) from those with mild to moderate non-proliferative diabetic retinopathy (NPDR).

Keywords Neural network · Machine learning · Diabetic retinopathy detection

1 Introduction

Blindness and poor vision are common consequences of diabetic eye disease. Diabetic patients with vision problems make up approximately 10% of the population. More than 95% of diabetic patients' vision reductions can be prevented with early detection and treatment, according to clinical test results. A regular eye examination is necessary to ensure proper treatment for diabetic patients before it is too late. A retinal image analysis system enabled ophthalmologists to detect retinal pathologies more quickly. In contrast, traditional methods, such as dilation, require much time and make patients uncomfortable. Blindness can also be caused by diabetic retinopathy (DR). In 2012, around 415 million people were diagnosed with diabetes mellitus, according to a World Health Organization report. Over the

M. R. Hammoodi (✉)
Faculty of Computer Science and Mathematics, University of Kufa, Kufa, Iraq
e-mail: mridha.hamoodi@uokufa.edu.iq

M. R. AL-Hameed
National University of Science and Technology, Dhi Qar, Iraq

S. Kumar Udgata et al. (eds.), *Intelligent Systems*, Lecture Notes in Networks and Systems 1149, https://doi.org/10.1007/978-981-97-8160-7_33

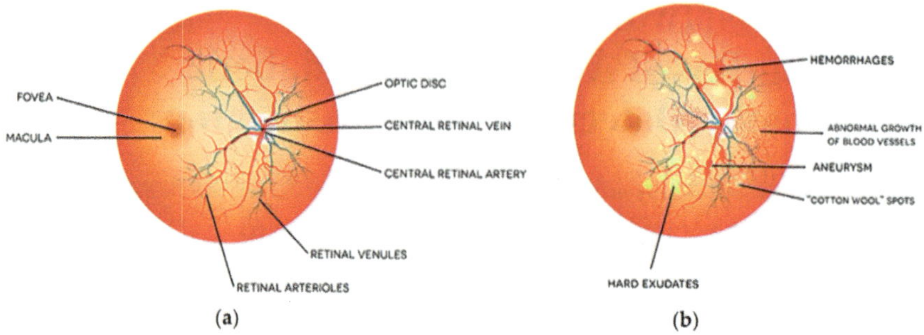

Fig. 1 **a** Normal retina, **b** retina with DR

last three decades, the incidence of this disease has doubled. Diabetic retinopathy can cause blindness in people with diabetes aged 20 to 74.90% of diabetic patients could be saved from this disease with an early diagnosis, according to many studies. Diabetic retinal disease (DR) is more common in diabetics [1, 2].

Similarly to other tissues of the body, microvascular tissue receives blood from the blood vessel. Through microscopic blood vessels, retinal tissue absorbs blood and maintains blood glucose levels for continuous circulation. Due to inadequate oxygen distribution into the cells, microvessels crash when glucose or carbohydrates accumulate in the blood. Any obstruction in these vessels severely damages the retina. The retina has ischaemia or reduced blood flow due to microvascular failure that fails to provide nutrients as normal [3, 4]. Diabetic and normal retinas are illustrated in Fig. 1. An example of a normal retina free of signs of DR can be seen in Fig. 1a. In Fig. 1b, haemorrhages, microaneurysms, cotton wool patches, exudates, and other DR symptoms can be seen on a retina with considerable DR.

2 Literature Review

Detecting and treating diabetic retinopathy immediately can prevent blindness in people with diabetes. To know how the disease is progressing, it is important to check the patient regularly after diagnosis [5]. The ophthalmologist can avoid blindness due to diabetes by having an efficient mechanism for detecting the problem based on the patient's reports. To accurately diagnose diabetic retinopathy, researchers have developed various algorithms to analyse scan results [6–8]. Optic nerves and discs make up the human eye. DRs are thus identified and categorised by segmenting eye images. Alternatively, fundus images can be inspected to determine whether haemorrhages, injuries, small aneurysms, exudates, etc. exist. Using deep learning, we can make the scenarios smarter by automating the environment [9] and by automating disease detection in the medical domain [10–12].

The diagnosis of diabetic retinopathy has evolved into an automated process over the past decade [13–15]. In manual diagnosis, fundus pictures are generally examined for lesions such as microaneurysms, exudates, and haemorrhages. Accordingly, most researchers focused on automatically categorising and identifying lesions [14, 42, 43].

3 Methodology

A. Image Pre-Processing

A pre-processing step is necessary to correct insufficient contrast between exudates and the image background pixels due to uneven illumination [17]. The accuracy of a fundus image can be distorted if noise is present in the input image. Also, pre-processing is important for colour space conversions and standardised image sizes. In pre-processing fundus images, colour fundus images are converted to a greyscale, and their contrast-limited adaptive histograms are equalised. Digital fundoscopy images obtained from Fig. 2 on the left were pre-processed to be further processed. The initial images obtained were in the green channel (Fig. 2: right). An example of an unprocessed image and a converted image can be found in the figure below. Figure 3 shows the images of the blue and red channels of the unprocessed images.

B. Support Vector Machine

Data classification and regression can be performed using SVM. A detailed description can be found here [18]. The training data for pattern recognition consist of,

$$(x_1, y_1), \ldots\ldots.., (x_e, y_e) \in R^n \times \{+1, -1\} \quad (1)$$

It is a pattern (vector) x_i with a label (yi) of three dimensions y_i. Using the author's classification scheme [19], a vector is classified as class +1 when a label is given

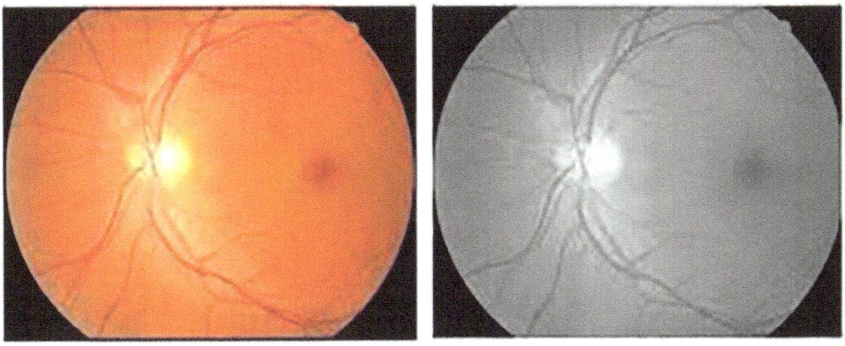

Fig. 2 A left-hand image of the retinal fundus and a right-hand image of the green channel of the retina

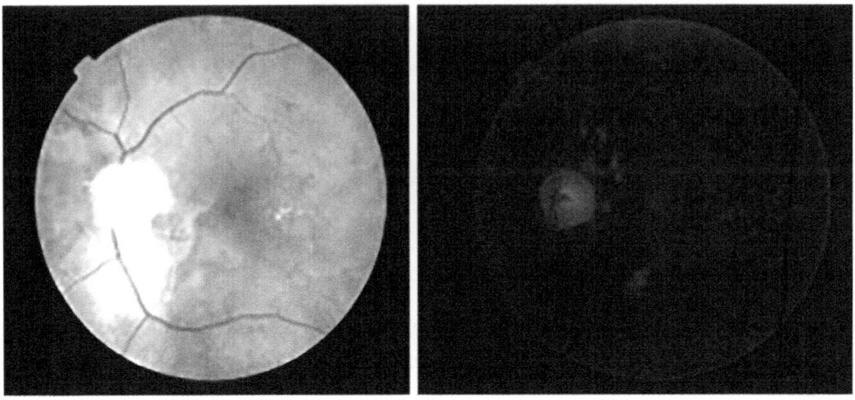

Fig. 3 Images of the blue and red channels (left and right)

with the value +1 and class −1 when a label is given with the value −1. In this work, we attempt to find a $f(x) = y : R^n \rightarrow \{+1, -1\}$ function that correctly classifies both seen and unseen patterns (along with patterns in the training data). The class of hyperplanes is the basis for SVMs.

$$\langle w.x \rangle + b = 0; w \in R^n, b \in R, \tag{2}$$

Figure 4 illustrates how the input space can be divided into two parts: one part contains vectors of class −1, ∧ the other contains vectors of class +1. A linearly separable set of data exists if there is such a hyperplane. We can determine which class a vector belongs to based on the following decision function.

$$f(x) = sign(\langle w.x \rangle + b) \tag{3}$$

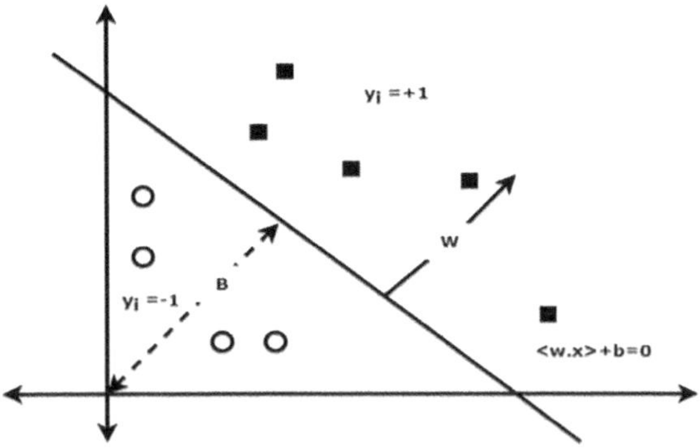

Fig. 4 In a two-dimensional (2D) training set, a separating hyperplane (w, b) is defined

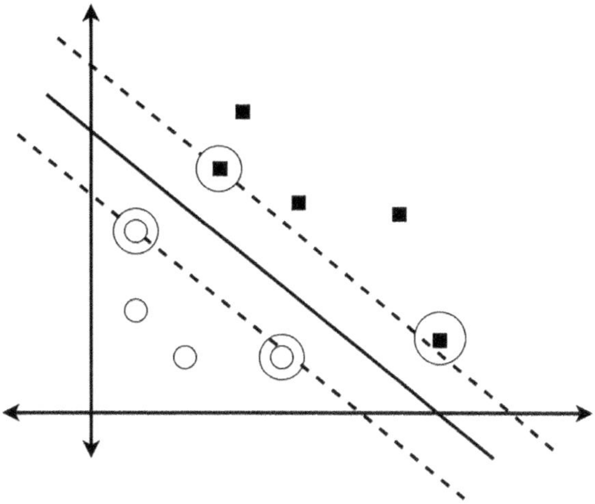

Fig. 5 A Support vectors encircling the maximum margin hyperplane

Training examples can be classified using multiple hyperplanes (for example, the hyperplane in Fig. 3 could be closer to class 1). Generally, the hyperplane with the highest margin of separation between classes guarantees the best generalisation performance. It is the optimal margin hyperplane or the maximum margin hyperplane, which is unique. A linear inequality constraint is minimised in the optimal hyperplane by solving a convex optimisation problem with no local minima (thus, any solution is global). As shown in Fig. 5, the solution to this problem can be expressed in terms of support vectors or a subset of training examples lying on the margins.

There is no need to remove any other vectors to find the solution to the classification problem. Support vectors contain all the necessary information. Using dot products between vectors, the optimisation problem and the decision function can be expressed in dual form. Decision functions are dual representations.

$$f(x) = sign\left(\sum_{i=1}^{l} y_i \propto_i \langle x.x_1 + b\rangle\right) \tag{4}$$

As a measure of the informational value of x_i, $\propto_i \in R$ can be viewed as a real-valued variable. Therefore, it is zero for vectors outside the margin (i.e. non-support vectors). Dot products are used only in the optimal hyperplane classifier. It corresponds to $\langle (x)(y) \rangle$ in feature space. $K(x, y)$ is a function that returns the dot product of two vectors in input space when given two vectors in feature space, as shown in Fig. 6. Therefore, $K(x, y) = (x)(y)$.

With SVM, an error function is minimised through iterative training to construct an optimal hyperplane. It has been shown in the literature that SVMs can be trained by minimising error functions [20–22].

$$\frac{1}{2} w^T w + C \sum_{i=1}^{N} \xi_i \tag{5}$$

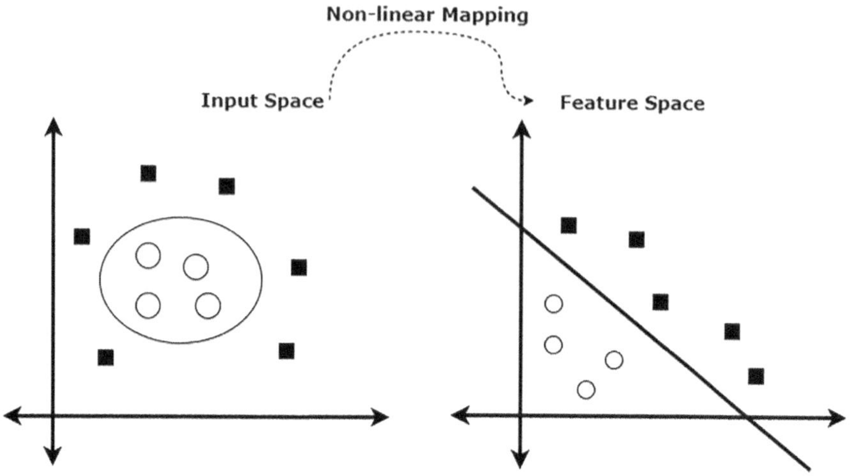

Fig. 6 The classification task can be simplified by mapping input space to feature space in a nonlinear way

Constraints apply

$$y_i\left(w^T\phi(x_i) + b\right) \geq 1 - \xi_i$$

Then

$$\xi_i \geq 0, i = 1, \ldots\ldots, N \tag{6}$$

There are several parameters for handling no separable data (inputs), such as C, w, and ξ_i.N Training cases are labelled in the index. A class label $y1$ is an independent variable, while an independent variable x_i is a class label. Input (independent) data are transformed into feature space using the kernel. In our work, we use SVM Torch. A polynomial kernel has been used, which can be found here.

$$K(x, x\prime) = (x.x\prime + 1)^d \tag{7}$$

For each class, x and $x\prime$ are feature vectors, and d is the kernel parameter. An optimal way of classifying images is determined by analysing training data using supporting vector machine training. This training vector has a size of $250 * 6$. NPDR, PDR, and normal outputs can all be produced.

C. Segmentation

Biomedical image segmentation can be achieved using UNet architectures. UNet has a U-shaped architecture, which is why it is called UNet. For example, convolutional neural networks and Max Pooling layers are deep learning tools used to segment images. Each pixel value in UNet is binarised so that, depending on the intensity, it represents 0 or 255. There are two paths in the architecture. A contraction path, called an encoder path, is on the left, and a decoder path is on the right. Feature maps from the left and right sides are concatenated to generate semantic segmentation [23]. Two convolution layers (3 × 3) are included in the contraction path; each layer is subjected to Batch Normalisation and Rectified Linear Unit (ReLu) activation [24]. Furthermore, it includes a max-pooling operation with an overall stride of 2. Input features consist of (512 × 512) patch images, and for every downsampling level, the number of feature maps increases by a factor of 2 until 1024 channels are totalled. As a result of downsampling, the output can tolerate small translational changes.

Every step of the expansion path involves upsampling a feature map divided by two by a convolution layer. The output is concatenated with the contraction map from the contraction path using a convolution layer (3 × 3), Batch Normalisation, and ReLu activation function. Each layer's dimensions and number are represented in Fig. 7. Layers were selected based on the input image's resolution and dimensions. An image with 512 by 512 pixels was used, 72 pixels high at 72 pixels per inch. The images were already cropped before they were fed into the UNet model, so two convolutional layers alternated with max pool layers were sufficient before the bottleneck.256 by 256 and 512 by 512 were the dimensions for the first two layers, respectively [25].

As a loss function, pixel-wise binary cross-entropy is used. There is a 7:1 weightage ratio between lesions and non-lesions (loss: non-loss). There were very few lesion pixels for a particular area, giving them more weight. A batch size of 8 and an epoch size of 100 were used. Examples of fundus images and lesion masks are shown in Figs. 8a and b. According to Fig. 8c, the fundus image shows a predicted lesion mask.

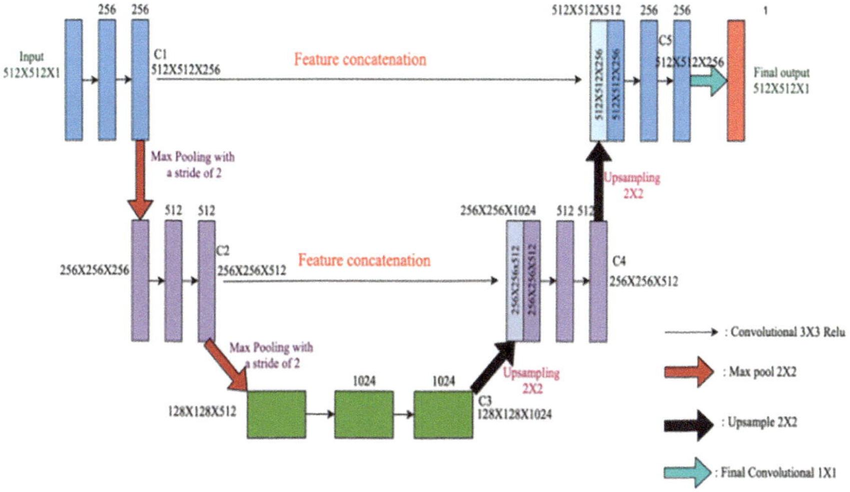

Fig. 7 UNet architecture

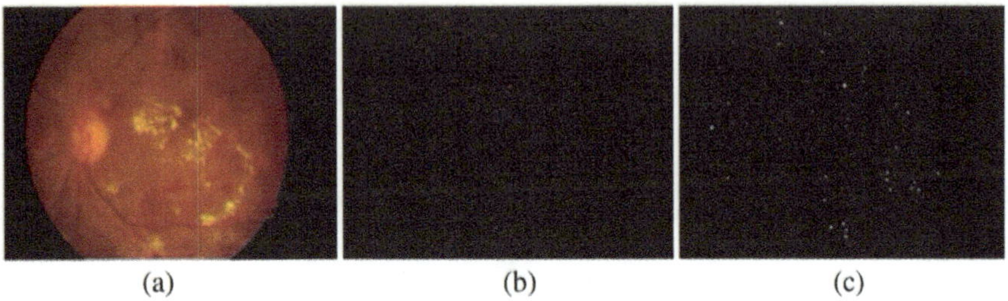

Fig. 8 a Images of the fundus, b ground truth image, and c predicted lesion mask

4 Result Analysis and Discussion

A new eye image is obtained after the pre-processing steps are performed. Due to the circular shape of the retinal image, circle-related features are considered. This procedure analyses blood vessel structures, exudates, textures, entropy, and microaneurysms. Here, the image was resized using the MATLAB resize function imresize (image, [576 720]) for further processing, as shown in Fig. 9.

Grey images can be analysed and classified using the MATLAB function greyimage = rgb2gray (new image), as shown in Fig. 10. Now, all process is applied to the grey fund's image, so after that, the image contrast enhancement function that contrastsenhancing = imadjust (greyimage) to enhance the feature of the image as shown in Fig. 11.

In the case of edge detection after contrast enhancements of a digital image have been applied, edge enhancements are used to classify points where intensity variations are pronounced or discontinuities are visible in Fig. 12. Canny methods are used to enhance edges.

Fig. 9 New resized fundus image

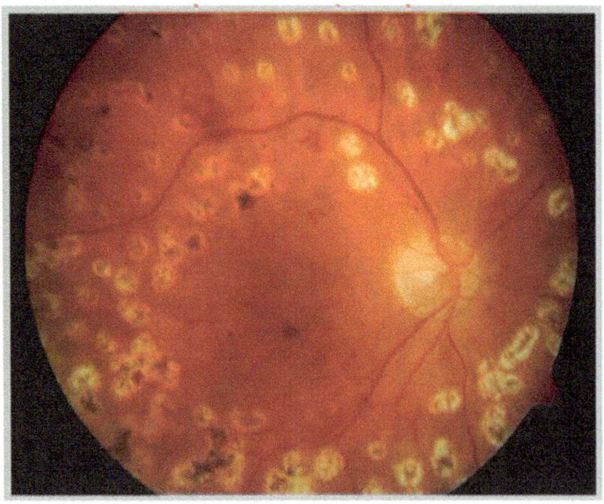

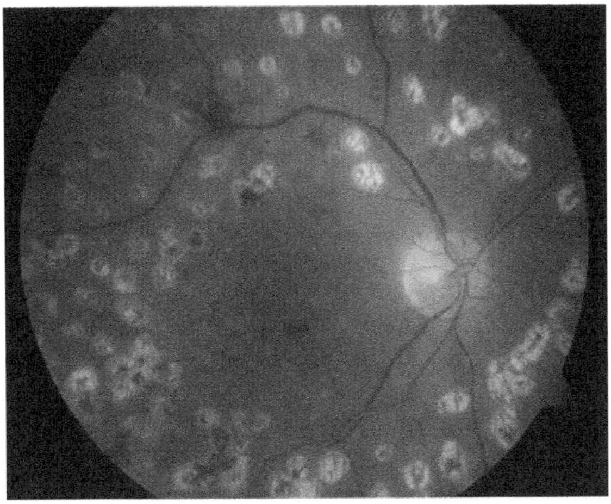

Fig. 10 The grey fundus image

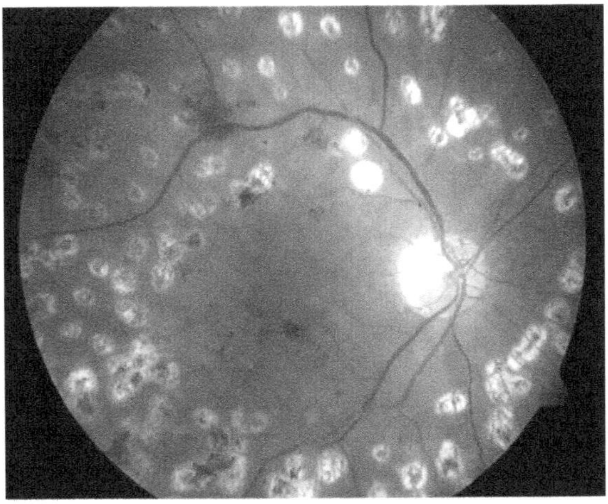

Fig. 11 The grey fundus image after contrast enhancement

Blood vessel image is measured in this anticipate for the distinguishing proof of diabetic retinopathy. To minimise this impact and create a more uniform image, pre-processing is key to minimising the differentiation of the fundus image in the middle and decreasing along its edges. An optical circle is removed from the image by connecting the green channel with morphological image handling. Image division is then performed to modify the complexity power, and little pixels thought to be errors in the image are extracted in Fig. 13.

Digital images are segmented by dividing them into multiple segments with image segmentation. The feature extraction method used for image segmentation was the Column filter, as shown in Fig. 14. MATLAB functions for image segmentation should be written after blood vessels are removed using the sliding and column filter.

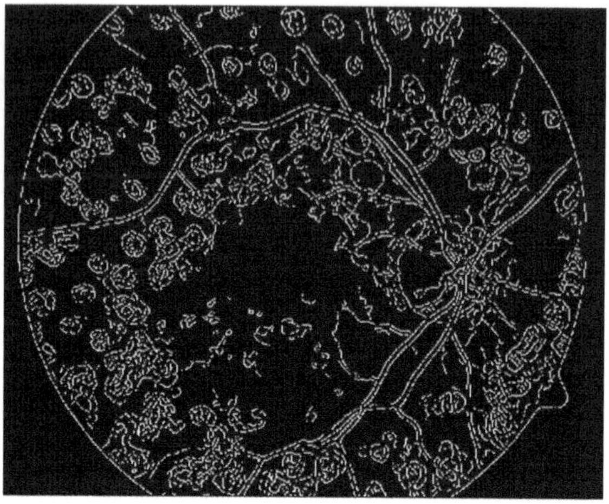

Fig. 12 Fundus image after edge detection

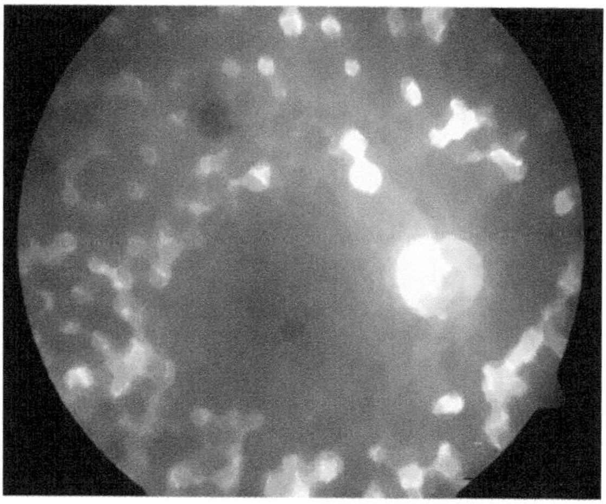

Fig. 13 Fundus image after blood vessel removal

Fundus image after optic disc removal is produced by the masking after the mesh grid method, the MATLAB function of masking in MATLAB: maskimg = sqrt((m - cmedian).^2 + (n - rmedian).^2) < = radius; generate the Fig. 15.

MATLAB 2020a was used to implement the proposed method. Based on SVM and ANN algorithms, the classification results are shown in Fig. 16.

Here, two sets of images, training images and test images, have been taken. The number of training images is 39, and 90 images have been tested. All training and test images are collected from the standard database DIARETDBO. Some training sample images are shown in three categories: normal, moderate, and severe. All three images are represented, respectively, in Fig. 17, 18, and 19.

Exudates showed impressive yellow-white stores on the retina because of the blood overflow from unusual vessels in Fig. 20. Their shape and size will change with

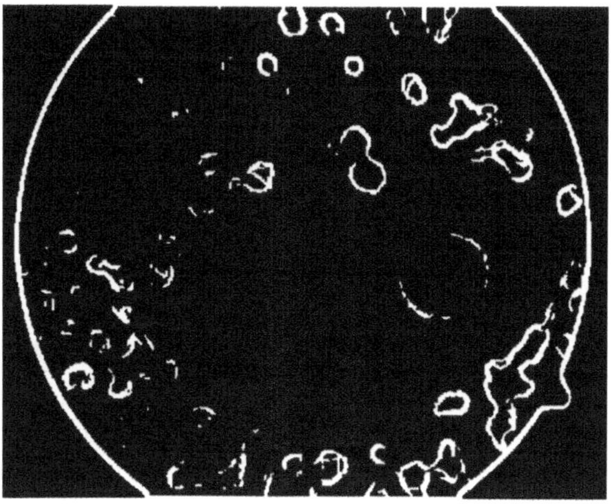

Fig. 14 Segmented fundus image

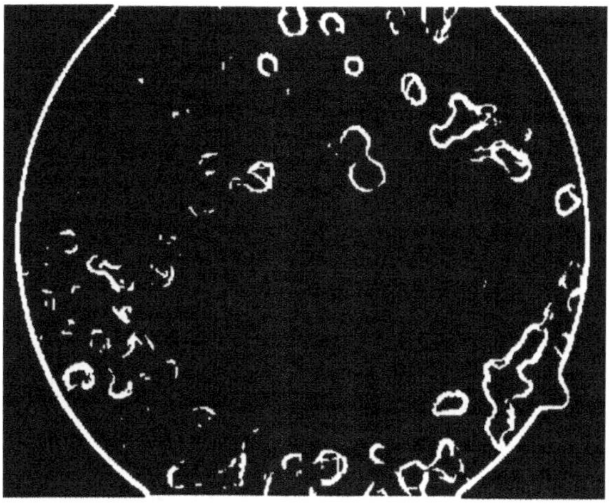

Fig. 15 Fundus image after optic disc removal

the diverse retinopathy stages. The greyscale image is pre-processed for consistency before the morphological image handling is connected to evacuate the blood vessels and recognise the exudate district. The exudates are identified after expelling the fringe, optical plate and non-exudates zone.

Microaneurysms were applied to the fundus test image, as shown in Fig. 21. They are little lumps created from the powerless blood vessels and are the most punctual clinical indication of diabetic retinopathy. Henceforth, it is crucial to recognise them amid the gentle stage. The quantity of microaneurysms would increment with the phase of the retinopathy. Recognising the round fringe and optical circle cover is done using the greyscale image. Using a watchful strategy, the green channel locates the edges before evicting the round border to fill the enclosed narrow space.

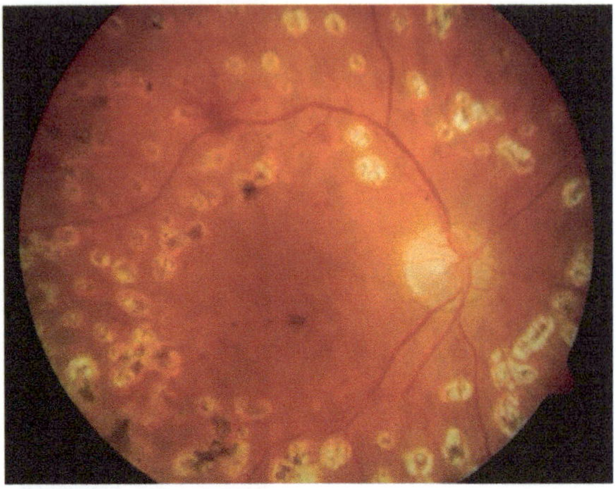

Fig. 16 Severe fundus image

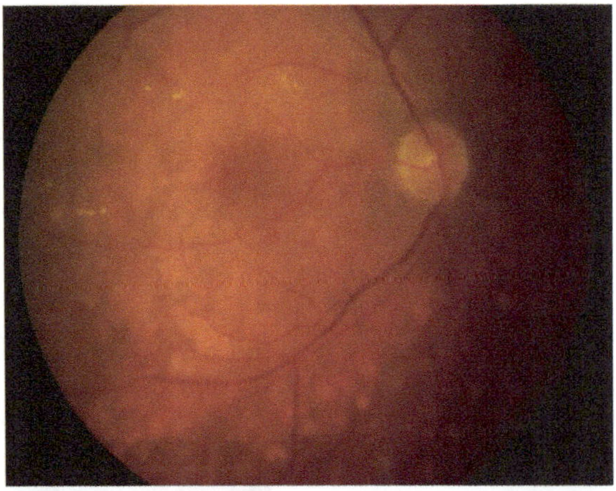

Fig. 17 Moderate fundus image

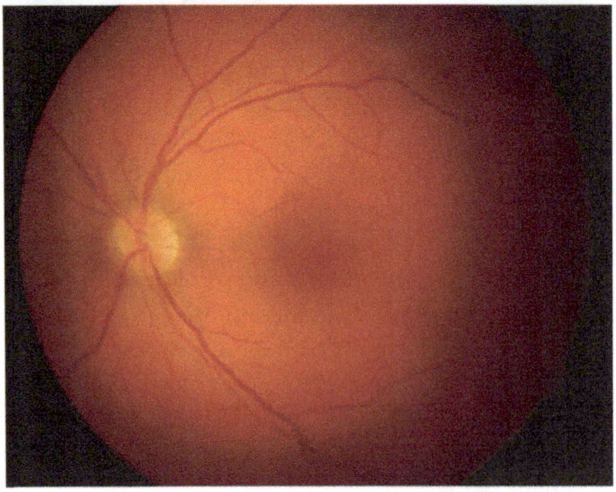

Fig. 18 Normal fundus image

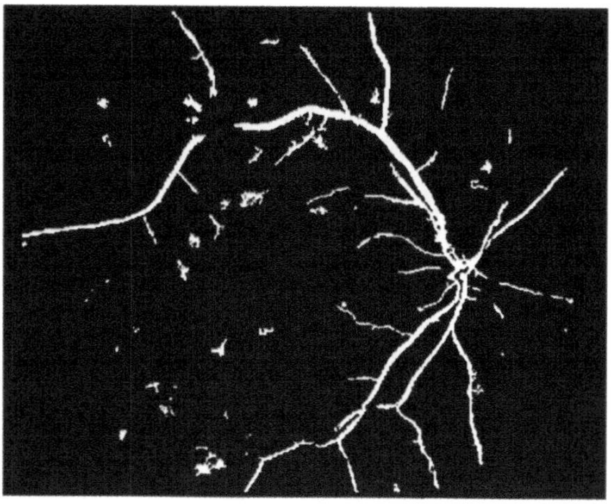

Fig. 19 Blood vessels on test data

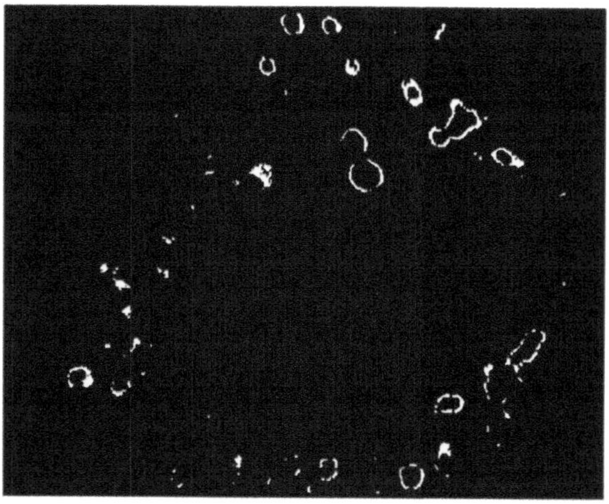

Fig. 20 Exudates on test data

4.1 Evaluation and Performance Measure

Performance measures and evaluation approaches are described in this section and shown in Table 1, 2, and 3. Three types of methods are used for the classification of DR: pixel-based, lesion-based, and image-based.

- Microaneurysms, exudates, and blood vessels are pixel-by-pixel compared to the truth image in the pixel-based approach.
- For comparison, microaneurysms, exudates, and blood vessels are used in the lesion-based approach.
- Most classifications are based on images. A normal image is classified as a normal image, while an affected image is classified as an affected image.

Fig. 21 Microaneurysms on test data

Table 1 Dataset for fundus image classification

Category	Training images	Testing images
Normal	10	34
Moderate	19	35
Severe	10	21
Total	39	90

Table 2 Classification accuracy of classifiers

Category	Number of test images	SVM			Neural network		
		CCI	MI	CA	CCI	MI	CA
Normal	34	27	7	89.41	25	9	83.52
Moderate	35	27	8	87.14	35	0	100
Severe	21	19	2	90.47	18	3	95.71
Total	90	73	17	91.1	78	12	96.7

Table 3 Classification rate of classifiers

Classifier	Accuracy (%)
SVM	91.1
ANN	96.7

In this work, image-based classification has been used, and the input images are classified as normal, moderate, or severe depending on how the fundus is affected by diabetic retinopathy.

5 Conclusion

Diabetic retinopathy can be detected, classified, and monitored automatically from retinal fundus images in ophthalmology. It is more difficult to classify diabetic retinopathy because retinal fundus images contain different components and have noises. In most existing methods, handcrafted features are primarily considered when classifying. A fundus image of a person with diabetes was successfully detected using neural networks and machine learning, and this could be used in the future to detect diabetes. The project needs some improvement to reach 96.7% accuracy. In the proposed method, image enhancement and noise filtering are used, blood vessels are detected, optic discs are identified, exudates and microaneurysms are extracted, features are extracted, and machine learning methods are proposed to classify diabetic retinopathy into mild, moderate, and severe non-proliferative diabetic retinopathy and proliferative diabetic retinopathy.

References

1. Kauppi, T., et al.: The DIARETDB1 diabetic retinopathy database and evaluation protocol. In: Procedings of the British Machine Vision Conference 2007, Warwick: British Machine Vision Association, 2007, pp. 15.1–15.10 (2007). https://doi.org/10.5244/C.21.15.
2. Rani, P., Singh, P.N., Verma, S., Ali, N., Shukla, P.K., Alhassan, M.: An implementation of modified blowfish technique with honey bee behavior optimization for load balancing in cloud system environment. Wirel. Commun. Mob. Comput.. Commun. Mob. Comput. **2022**, 1–14 (2022)
3. Kayal, D., Banerjee, S.: A new dynamic thresholding based technique for detection of hard exudates in digital retinal fundus image. In: 2014 International Conference on Signal Processing and Integrated Networks (SPIN) , pp. 141–144. IEEE, Noida (2014). https://doi.org/10.1109/SPIN.2014.6776937
4. Rani, P., Sharma, R.: IMFOCA-IOV: intelligent moth flame optimization based clustering algorithm for internet of vehicle. In: 2023 14th International Conference on Computing Communication and Networking Technologies (ICCCNT), pp. 1–6. IEEE, Delhi (2023). https://doi.org/10.1109/ICCCNT56998.2023.10307646
5. Mo, W., Luo, X., Zhong, Y., Jiang, W.: Image recognition using convolutional neural network combined with ensemble learning algorithm. In: Journal of Physics: Conference Series, IOP Publishing, 2019, p. 022026 (2019)
6. Samanta, A., Saha, A., Satapathy, S.C., Fernandes, S.L., Zhang, Y.-D.: Automated detection of diabetic retinopathy using convolutional neural networks on a small dataset. Pattern Recognit. Lett. **135**, 293–298 (2020)
7. Reddy, S.S., Sethi, N., Rajender, R., Mahesh, G.: WITHDRAWN: extensive analysis of machine learning algorithms to early detection of diabetic retinopathy. Mater. Today Proc., S221478532038531X (2020). https://doi.org/10.1016/j.matpr.2020.10.894
8. Saxena, G., Verma, D.K., Paraye, A., Rajan, A., Rawat, A.: Improved and robust deep learning agent for preliminary detection of diabetic retinopathy using public datasets. Intell.-Based Med., **3–4**, 100022 (2020). https://doi.org/10.1016/j.ibmed.2020.100022
9. Janakiramaiah, B., Kalyani, G., Jayalakshmi, A.: RETRACTED ARTICLE: Automatic alert generation in a surveillance systems for smart city environment using deep learning algorithm. Evol. Intell.. Intell. **14**(2), 635–642 (2021). https://doi.org/10.1007/s12065-020-00353-4

10. Alyoubi, W.L., Shalash, W.M., Abulkhair, M.F.: Diabetic retinopathy detection through deep learning techniques: a review. Inform. Med. Unlocked **20**, 100377 (2020). https://doi.org/10.1016/j.imu.2020.100377
11. Sajid, S., Hussain, S., Sarwar, A.: Brain tumor detection and segmentation in MR images using deep learning. Arab. J. Sci. Eng. **44**, 9249–9261 (2019)
12. Janakiramaiah, B., Kalyani, G.: Dementia detection using the deep convolution neural network method. Trends Deep Learn. Methodol. 157–181 (2021)
13. Butt, M.M., Latif, G., Iskandar, D.N.F.A., Alghazo, J., Khan, A.H.: Multi-channel convolutions neural network based diabetic retinopathy detection from fundus images. Procedia Comput. Sci. **163**, 283–291 (2019). https://doi.org/10.1016/j.procs.2019.12.110
14. Islam, M.M., Yang, H.-C., Poly, T.N., Jian, W.-S., (Jack) Li, Y.-C.: Deep learning algorithms for detection of diabetic retinopathy in retinal fundus photographs: a systematic review and meta-analysis. Comput. Methods Programs Biomed. **191**, 105320 (2020). https://doi.org/10.1016/j.cmpb.2020.105320
15. Safi, H., Safi, S., Hafezi-Moghadam, A., Ahmadieh, H.: Early detection of diabetic retinopathy. Surv. Ophthalmol.. Ophthalmol. **63**(5), 601–608 (2018). https://doi.org/10.1016/j.survophthal.2018.04.003
16. Ahmad, A., Mansoor, A.B., Mumtaz, R., Khan, M., Mirza, S.H., Image processing and classification in diabetic retinopathy: a review. In: 5th European workshop on visual information processing (EUVIP). IEEE **2014**, 1–6 (2014)
17. Verma, S., et al.: An automated face mask detection system using transfer learning based neural network to preventing viral infection. Expert Syst. e13507 (2023). https://doi.org/10.1111/exsy.13507
18. Rani, P., Sharma, R.: Intelligent transportation system for internet of vehicles based vehicular networks for smart cities. Comput. Electr. Eng.. Electr. Eng. **105**, 108543 (2023)
19. SupportVectorMachines.pdf. Accessed: Dec. 05, 2023. http://www.cs.um.edu.mt/~csaw/CSAW03/Proceedings/SupportVectorMachines.pdf
20. Priya, R., Aruna, P.: Diagnosis of diabetic retinopathy using machine learning techniques. ICTACT J. Soft Comput. **3**(4), 563–575 (2013)
21. Ansari, G., Rani, P., Kumar, V.: A novel technique of mixed gas identification based on the group method of data handling (GMDH) on time-dependent MOX gas sensor data. In: Mahapatra, R.P., Peddoju, S.K., Roy, S., Parwekar, P., (eds.), Proceedings of International Conference on Recent Trends in Computing. Lecture Notes in Networks and Systems, vol. 600, pp. 641–654. Springer Nature Singapore, Singapore (2023). https://doi.org/10.1007/978-981-19-8825-7_55.
22. Rani, P., Sharma, R.: Intelligent transportation system performance analysis of indoor and outdoor internet of vehicle (iov) applications towards 5g. Tsinghua Sci. Technol. (2023)
23. Yadav, S.P., Jindal, M., Rani, P., De Albuquerque, V.H.C., Dos Santos Nascimento, C., Kumar, M.: An improved deep learning-based optimal object detection system from images. Multimed. Tools Appl. (2023). https://doi.org/10.1007/s11042-023-16736-5
24. Brownlee, J.: How to accelerate learning of deep neural networks with batch normalization (2019). Accessed https://machinelearningmastery.com/how-to-accelerate-learning-of-deep-neural-networks-with-batch-normalization
25. Rani, P., Verma, S., Yadav, S.P., Rai, B.K., Naruka, M.S., Kumar, D.: Simulation of the lightweight blockchain technique based on privacy and security for healthcare data for the cloud system. Int. J. E-Health Med. Commun. IJEHMC **13**(4), 1–15 (2022)

Analyzing the Role of IoT and Big Data in Circular Supply Chains Using a Comparative Study on Text Mining/ Topic Modeling and an Industrial Case Study

Suman Kumar Das and Nicola Saccani

Abstract The Circular Supply Chain is a growing trend in manufacturing and industrial sectors, leveraging Industry 4.0 technologies to improve quality, performance, and sustainability-focused decision-making. Despite its importance, there has been limited exploration into how digital technologies practically address the main challenges of the circular supply chain. To bridge this gap, this paper provides an in-depth analysis of barriers within circular supply chains, examining 24 articles on supply chain management, the Internet of Things, Big Data and Analytics, and Cloud Computing in the context of circular economy. Utilizing Structural Topic Modeling (STM) for text mining, this study identifies crucial thematic topics and barriers. Additionally, this paper delves into a case study of a company leveraging IoT, Big Data and Analytics, and Cloud Computing, to understand their role in overcoming circular supply chain challenges. The findings of this empirical investigation offer insights into the effectiveness of selected digital technologies in addressing these challenges.

Keywords Big data analytics · Internet of things · Circular supply chain management · Case study

1 Introduction

Given current environmental uncertainty, it is crucial for organizations to implement the concepts of sustainability and circular economy through the adoption of Industry 4.0 (I4.0) technologies. These technologies are beneficial to businesses, ensuring long-term viability and resilience while reducing its environmental footprint by integrating sustainability principles into its operation management processes [1]. A smart circular economy is emerging and holds the potential to contribute to

S. K. Das (✉) · N. Saccani
Department of Mechanical Engineering, University Of Brescia, Brescia, Italy
e-mail: s.das001@unibs.it

S. Kumar Udgata et al. (eds.), *Intelligent Systems*, Lecture Notes in Networks and Systems 1149, https://doi.org/10.1007/978-981-97-8160-7_34

multiple sustainability goals by minimizing waste and resource use by promoting different R's of circular economy, such as reduce, reuse, remanufacturing, recycling, and upcycling of materials [2, 3]. I4.0 technologies can integrate circular economy into operation management processes by streamlining the entire value chain. I4.0 comprises various technologies broadly classified as data collection and processing technology (BIGDATA, IOT, cloud computing, and AI) [4], integration technology (blockchain, horizontal, and vertical integration) [5], and smart manufacturing technology (virtual reality, digital twin, robotics, and additive manufacturing (AM) [4] that leverage smart, efficient, and flexible industrial systems. For example, a manufacturing industry could adopt Big Data to improve flexibility in procurement and the upstream supply chain through demand forecasting [6]. Similarly, identifying and predicting faults using condition monitoring and detecting anomalies is rendered possible through the real-time gathering and processing of machine data [7]. Similarly, Internet of Things (IoT) can enable the visibility and traceability of goods throughout the supply chain [8]. Furthermore, Artificial Intelligence (AI) algorithms are capable of analyzing enormous volumes of data to find patterns and trends that help companies maximize resource utilization, reduce waste, and boost recycling rates [6–8]. However, transitioning to a smart circular economy presents numerous challenges, including the implementation of new technologies and systems while altering existing organizational structures and processes [3]. Owing to technological challenges, the implementation of I4.0 in the manufacturing sectors of both developed and developing countries is still relatively low. Hence, when implementing these technologies, it is critical to understand the barriers to their adoption. Furthermore, in the context of circular economy and supply chain management within operations management processes, a few studies have been conducted that examine such barriers leading to the low adoption of data collection and processing technologies like Big Data, AI, IoT, and Cloud computing in these processes.

This study explores how data collection and processing technologies can help companies to adopt a smart circular economy in the circular supply chain through an industrial case study.

In this regard, various barriers associated with implementing a circular supply chain using the SCOPUS database were analyzed in this study. The SCOPUS database was thoroughly searched to identify and retrieve 24 pertinent articles on supply chain, big data, AI, and IoT in the context of circular economy, using relevant keywords. A text mining method based on structural topic modeling (STM) was used in this review to identify thematic topics and barriers related to the circular supply chain, as well as solution measures related to data collection and processing technologies. Subsequently, the most significant challenges identified through the STM were compared with the case study, and an analysis was conducted to examine the solution measures for how Big Data Analytics (BDA), IoT, and AI can help overcome these challenges. This manuscript is therefore structured as follows: Sect. 2 presents the study's background followed by Sect. 3 that outlines the methodology, and Sect. 4 details the results of the case study. Lastly, Sect. 5 highlights the results and limitations of this study.

2 Circular Supply Chain and Digital Technologies

In the domain of operations management, the concept of a circular economy is pivotal in optimizing the use and value of materials, components, and products [9]. This is achieved through the implementation of various strategies of circular economy model: Reduce, Reuse, Repair, Remanufacture, and Recycle. Attaining this objective involves enhancing operational efficiency in multiple dimensions. It encompasses increasing production and design flexibility, reducing energy and resource consumption, and improving manufacturing autonomy by minimizing setup and lead times [10]. From an economic standpoint, this approach seeks to lower labor and material costs, thus reshaping the financial landscape of manufacturing. However, the integration of digital technologies plays a crucial role in enabling circular economy within operations management [4]. Technologies such as the IoT, BDA, AM, Cyber-Physical Systems (CPS), and Cloud Computing are at the forefront. These technologies not only contribute to heightened productivity and flexibility but also align perfectly with the objective of sustainability. They act as catalysts in sustainable operational processes, enabling a more effective and efficient circular economy model [11]. So digital operation practices are leveraged by organizations for more efficient and effective value creation using Smart circular economy [2–5]. Nevertheless, improper execution of such practices in businesses entails a high risk of failure and significant investment costs [11]. A major challenge is the complex nature of supply chains, which includes ensuring environmental sustainability in procurement, managing smart inventory effectively, and optimizing logistics and planning operations [12, 13]. For example, it highlighted the challenge of high initial costs for environmentally friendly supply chain in manufacturing industries, particularly in developing economies with low profit margins, making it difficult for organizations to implement these digital practices. Several researchers have conducted literature reviews on sustainable supply chains, identifying challenges faced by various industry sectors, including manufacturing and services. They have reported the lack of an effective framework for adopting smart circular supply chains, the improper execution of sustainability practices, and the high costs associated with sustainability adoption as primary challenges [2, 11, 16].

3 Methodology

The systematic literature review serves as the foundation of this research to assess both the quality and quantity of data. This research used the PRISMA methodology [20] to select and refine the literature searched and retrieved from the Scopus database using a combination of search keywords such as "circular supply chain challenges", "circular supply chain barrier", "closed loop supply chain challenge*", "Closed loop supply chain barrier*", "Industry 4.0", "Big Data analytic*", "Industry 4.0 solutions", "Industry 4.0 enablers", "circular economy solutions". The PRISMA methodology ensured a systematic approach to the selection and refinement of literature by following a predefined set of criteria. To filter the research articles, we followed the following criteria: (1) The initial criteria should be that the articles are available in full text and in English; the search included only journal articles, excluding conference papers, book chapters, and editorial notes. (2) It was narrowing down the subject areas to engineering, business management, operations, services, manufacturing, decision science, etc. The article titles, abstracts, and full articles were manually assessed to select those focusing on circular supply chain challenges and supply chain barriers. The articles should focus on Industry 4.0 solutions or strictly discuss BDA, IoT, and cloud computing. Subsequently, a final set of 24 articles was considered for detailed analysis. After selecting the articles, we conducted STM analysis using the R package to identify circular supply chain barriers emerging from the articles [21]. STM is a crucial text analytics technique that uses unsupervised machine learning methods to automatically identify and explore various themes within a collection of documents. It identifies the similarity of words in documents by extracting similar keywords or group of keywords from the document collection [10]. By reviewing these keywords and topics, we finalized a list of the keywords or topics associated with the barriers in the circular supply chain. The text corpus was constructed from words found in article abstracts, titles, keywords, discussions, and conclusions, and it was used for generating thematic topics from a collection of articles. Subsequently, the text corpus was subjected to a cleaning process that included the removal of commonly used words (stop words), non-English words, special characters, equations, and numbers. This cleaning process was necessary to ensure that the STM approach can effectively analyze the text and generate accurate thematic topics [10, 21]. After this step, the text corpus is considered as input for STM. The STM approach was performed using the built-in STM library in the R-package. The topics generated by the STM approach are presented in Fig. 1.

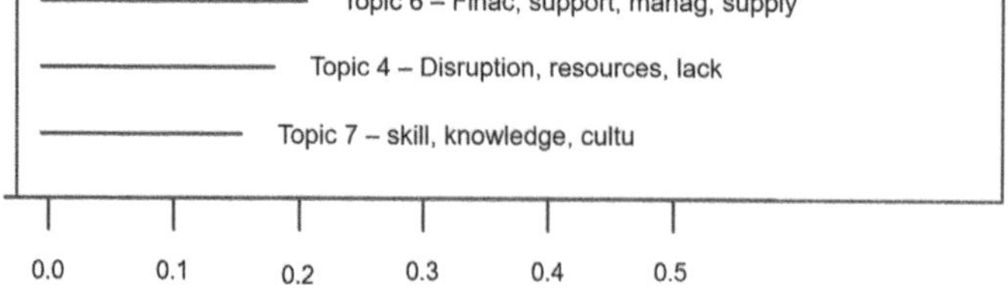

Fig. 1 Generated topic labels using STM approach

Additionally, within each topic, significant words were identified using the built-in command in the R-package. The top identified words under each topic are presented in Fig. 2. This figure displayed highly frequent words associated with each generated topic.

4 Case Study

Company A is a key player in the industrial supply sector in North America, offering a wide range of over 1.6 million products in safety, material handling, and metal working to its 3 million customers. Alongside product offerings, the company provides essential services such as inventory management and technical support. As a large-scale industrial company, managing a vast amount of data related to inventory and customers is a significant challenge for Company A. The complexity of this data, combined with the need for a data-driven approach, has led to increased costs in data management and inventory. To address these challenges and reduce operational costs, Company A has decided to expand its data warehouse. This expansion is crucial for integrating data from different departments, which is essential for effective planning and management. It is also supporting different departments like human resources, technology infrastructure, customer support, inventory management, material supply, supplier information and alignment with the right organizational strategy. In line with these efforts, Company A is planning to implement a cloud-based Big Data and predictive analytics project. This project tackles issues in supply chain management by optimizing the resource allocation, reducing waste, and implementing sustainable practices. The company is carefully choosing cloud and Big Data solutions that match its organizational goals and allow for flexible technological infrastructure adjustments. The company utilized Snowflake and Apache Airflow technologies to automate data extraction and reduce duplication by ensuring the processing of ingested files. The Docker Operator is utilized to execute Airflow DAGs on a different container. The data engineer team successfully integrated over 100 data sources into a unified platform, migrating from on-premises Hadoop Hortonworks cluster to AWS for data processing. The main objective of the platform is to offer predictive analytics, cloud migration, improved data management efficiency, cloud-neutral technology, and reduced software development overhead, benefiting finance departments, infrastructure costs, and data management efficiency. This allows for the upgrading of the digital supply chain, such as the software controlling in component part supply chain, leading to resource efficiency and minimizing the inventory. Additionally, monitoring and analyzing factors like demand of green products, consumer demands, and supplier choice enable Company A to predict when the product should be introduced, reducing return flow uncertainties, and enhancing collection activities. However, digital technologies do not assist in overcoming challenges related to financial risks.

Fig. 2 Thematic Keywords Co-occurrence Network

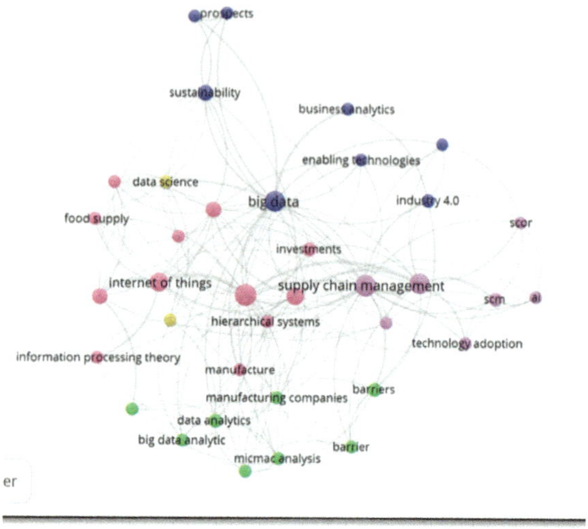

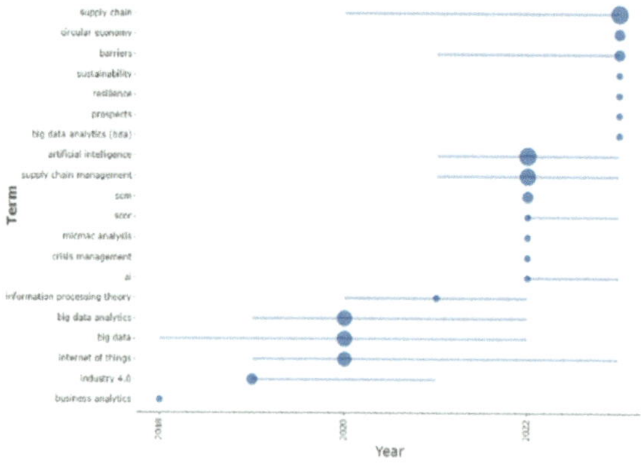

5 Result and Discussion

Previous studies acknowledge that various digital technologies can facilitate a transition towards a smart circular supply chain. In this discussion, we confine our focus to three important digital technologies: IoT, Big Data, and Cloud Computing. These are referred to as "Data Collection and Processing Technologies", providing essential support for collection, connectivity, and intelligence throughout the operation management processes [4]. This case study initially aims to address the challenges associated with the circular supply chain encountered by Company A. The barriers faced by the company are summarized in Table 1, categorizing them into three distinct groups. First, it described the technological barriers to implementing for circular supply chain, such as the complexity of data integration and management, inadequate infrastructure for information sharing, issues with data quality and scalability, and concerns regarding data privacy and security. Secondly, the study identified organizational barriers that impede the circular supply chain's implementation, including financial constraints, lack of top management support, and lack of organizational readiness. Lastly, it underscored the social barriers, which encompass regulatory and skill constraints, limited engagement with suppliers and customers, and resistance of culture change. The Data Collection and Processing Technology in Industry 4.0 has provided specific techno-functional principles aimed at achieving digital transformation [22]. For instance, IoT functionality is utilized for real-time monitoring and analysis of supply chain nodes, providing immediate insights into operations and enhancing visibility and agility [23]. This, in turn, facilitates crucial information sharing across the supply chain. Yadav et al. [12] underscored the critical need for supply chain managers to closely monitor supplier-related activities for uninterrupted, sustainable supply chain systems. One of the primary challenges in implementing a circular supply chain is data integration and management, as information systems within the supply chain are interconnected, gathering data from thousands of nodes in various formats [24]. Smooth processing and management require the integration of IT systems between supply chain partners, necessitating an intelligent data integration system. This system should ensure that every action is integrated into both the business model and logistics across the entire supply chain [6, 25–27].

Company A addressed these interoperability and data integration challenges by using the Docker operator, which provided a unified data platform adhering to global standards and data-sharing protocols. The organizational barriers identified in the context of adopting digital technologies within a business environment are multifaceted and cannot be easily mitigated by technology adoption alone. Prior to investing in advanced technologies such as Big Data, IoT, or Cloud Computing, conducting a thorough financial investigation and detailed analysis is crucial [26]. A notable challenge within organizations is the lack of top management support, often stemming from a reluctance to adopt new technologies. This hesitancy frequently originates from concerns over potential complexities and unforeseen implications that these technologies might introduce. However, Big Data technologies can enhance management's decision-making capabilities by providing comprehensive insights

Table 1 How data collection and processing technologies help to overcome Circular supply chain challenges in Company A. (N.D: not discussed even though Company A faces certain challenges, the digital technologies evaluated do not offer solutions for them)

Circular supply chain challenges	Description	Main findings	Digital technologies		
Technological barriers			IOT	Big data	Cloud
Lack of infrastructure for information sharing	Inadequate tools and systems for the efficient exchange of data and information	The digital platform utilizes Snowflake and Apache Airflow to manage data from various sources, including IoT devices and ERP systems	*X*	*X*	*X*
Complexity of data integration and management	Challenging integration and management of diverse, voluminous data across systems	Using AWS to consolidate data from 100 different sources into a unified platform reduces data integration and management complexity for Company A		*X*	*X*
Lack of data quality and scalability	Insufficient data precision and constrained system expansion capabilities			*X*	*X*
Lack of data quality, privacy, and security	Information management issues including poor data quality, compromised privacy, and security concerns				*X*
Information disruptions among the SC members	Communication failures that disrupt the operations and interactions of supply chain participants	IoT connects various materials and products, while the Cloud platform offers a cost-effective option for data sharing	*X*	*X*	
Organizational barriers					
lack of financial resources	Limited financial resources greatly impede an organization's capacity to adopt		*N.D*	*N.D*	*N.D*

(continued)

Table 1 (continued)

Circular supply chain challenges	Description	Main findings	Digital technologies		
Technological barriers			IOT	Big data	Cloud
Lack of top management support and organizational readiness	Insufficiencies in leadership support and organizational readiness to execute changes effectively		*N.D*	*X*	*N.D*
Social barriers					
Lack of skill and training	Insufficient workforce expertise and training in technological skills		*N.D*	*N.D*	*N.D*

and data-driven analysis, aiding in assessing market trends, customer preferences, and operational efficiencies, thereby supporting informed strategic decisions [18]. In addition to these aspects, the training and development needs of employees when introducing new technologies are also important to consider. Kumar et al. [15] highlighted that many firms struggle to capitalize on the potential for sustainability due to a lack of awareness of the fundamental standards underlying material circularity in the I4.0 context. A well-planned training program can significantly ease the transition and increase the likelihood of successful technology adoption. Therefore, Data collection and processing technologies are crucial in overcoming CE challenges in Circular supply chain, but detailed investment analysis is essential before implementation.

6 Conclusion

This study provides a broader understanding of how Industry 4.0's Data Collection and Processing technologies, including IoT, Big Data, and Cloud Computing, not only enhance supply chain processes but also aid Company A in addressing circular supply chain challenges, thereby supporting the realization of circular economy benefits. The case study reiterated that these technologies significantly impact supply chain management by enhancing performance in industrial practices. By enabling the generation, collection, and analysis of substantial information across various networks in the value chain, these technologies have become pivotal in transforming supply chain dynamics. They empower firms to exert greater control over internal and external processes, as well as in relationships with stakeholders in the value chain, suppliers, business partners, and consumers. In the context of the supply chain, they foster trust between firms and their suppliers, enhance transparency and traceability, mitigate risks, increase control, and facilitate adaptive learning. This ultimately enables firms to engage and interact more effectively with customers.

For example, BDA provides significant opportunities for developing green products that align with an understanding of global customer requirements. BDA can take a proactive approach and make informed decisions in the market before their competitors take a step, which can help eliminate uncertainties and failures associated with new green products [17]. Similarly, the capabilities of AI provide predictive analytics that can be utilized to forecast sustainable product performance in the field, which can aid in determining the appropriate marketing and operational strategies. The IoT provides real-time data on product flows, allowing companies to optimize their supply chain and pinpoint inefficiencies. Capturing product lifetime information enhances forecasting and planning, reduces waste, and boosts efficiency [18]. This data aids in informed decision-making about recovery strategies, optimizing the timing and condition of disassembly for maximum value and minimal environmental impact. Cloud computing significantly enhances supply chain management by offering real-time access to and sharing of critical information, including customer demand management, material supply, inventory management, and production optimization and control [19]. It provides scalability and flexibility by eliminating the need for extensive physical IT infrastructure.

This study offers practical and managerial implications for addressing challenges in the circular supply chain. It can guide managers in creating a roadmap for transitioning to a circular economy by determining the strategic implementation of digital technologies in anticipation of potential challenges. However, this paper has certain limitations. The research is based on a single case study, which restricts its generalizability. To validate the identified barriers to the circular supply chain more comprehensively, further empirical research is required.

References

1. Das, S.K., Bressanelli, G., Saccani, N.: Clustering the research at the intersection of industry 4.0 technologies, Environmental sustainability and circular economy: evidence from literature and future research directions. Circ.Econ.Sust. (2024). https://doi.org/10.1007/s43615-024-00393-3
2. Kristoffersen, E., Blomsma, F., Mikalef, P., Li, J.: The smart circular economy: a digital-enabled circular strategies framework for manufacturing companies. J. Bus. Res. **120**, 241–261 (2020). https://doi.org/10.1016/j.jbusres.2020.07.044
3. Bressanelli, G., Adrodegari, F., Pigosso, D.C.A.A., Parida, V.: Towards the smart circular economy paradigm: a definition, conceptualization, and research agenda. Sustain **14**, 4960 (2022). https://doi.org/10.3390/su14094960
4. Di Maria, E., De Marchi, V., Galeazzo, A.: Industry 4.0 technologies and circular economy: the mediating role of supply chain integration. Bus. Strateg. Environ. **31**, 619–632 (2022). https://doi.org/10.1002/bse.2940
5. Feroz, A.K., Zo, H., Chiravuri, A.: Digital transformation and environmental sustainability: a review and research agenda. Sustain **13**, 1–20 (2021). https://doi.org/10.3390/su13031530
6. Akbari, M., Hopkins, J.L.: Digital technologies as enablers of supply chain sustainability in an emerging economy. Oper. Manag. Res. **15**, 689–710 (2022). https://doi.org/10.1007/s12063-021-00226-8

7. Yang, J., Jiang, Z., Zhu, S., Zhang, H.: Data-driven technological life prediction of mechanical and electrical products based on multidimensional deep neural network: functional perspective. J. Manuf. Syst. **64**, 53–67 (2022). https://doi.org/10.1016/j.jmsy.2022.05.014
8. Garrido-hidalgo, C., Ramirez, F.J., Olivares, T., Roda-sanchez, L.: The adoption of internet of things in a circular supply chain framework for the recovery of WEEE : The case of Lithium-ion electric vehicle battery packs. Waste Manag. **103**, 32–44 (2020). https://doi.org/10.1016/j.wasman.2019.09.045
9. Bag, S., Dhamija, P., Bryde, D.J., Kumar, R.: Effect of eco-innovation on green supply chain management, circular economy capability, and performance of small and medium enterprises. J. Bus. Res. **141**, 60–72 (2022). https://doi.org/10.1016/j.jbusres.2021.12.011
10. Agrawal, R., Wankhede, V.A., Kumar, A., et al.: An exploratory state - of - the - art review of artificial intelligence applications in circular economy using structural topic modeling. Oper. Manag. Res. **15**, 609–626 (2022). https://doi.org/10.1007/s12063-021-00212-0
11. Kayikci, Y., Kazancoglu, Y., Gozacan-Chase, N., et al.: Assessing smart circular supply chain readiness and maturity level of small and medium-sized enterprises. J. Bus. Res. **149**, 375–392 (2022). https://doi.org/10.1016/j.jbusres.2022.05.042
12. Yadav, G., Luthra, S., Kumar, S., et al.: A framework to overcome sustainable supply chain challenges through solution measures of industry 4.0 and circular economy: an automotive case. J Clean Prod 254:120112 (2020). https://doi.org/10.1016/j.jclepro.2020.120112
13. Goyannes, R., Scavarda, L.F., Azevedo, B.D., et al.: Challenges and Benefits of Sustainable Industry 4.0 for Operations and Supply Chain Management—A Framework Headed toward the 2030 Agenda. Sustain 14 (2022). https://doi.org/10.3390/su14020830
14. Batista, L.: In search of a circular supply chain archetype – a content-analysis-based literature review. Prod. Plan. Control **29**, 438–451 (2019). https://doi.org/10.1080/09537287.2017.1343502
15. Kumar, P., Singh, R.K., Kumar, V., et al.: Managing supply chains for sustainable operations in the era of industry 4.0 and circular economy: analysis of barriers. Resour. Conserv. Recycl. 164, 105215 (2021). https://doi.org/10.1016/j.resconrec.2020.105215
16. Alcayaga, A., Wiener, M., Hansen, E.G.: Towards a framework of smart-circular systems: an integrative literature review. J. Clean. Prod. **221**, 622–634 (2019). https://doi.org/10.1016/j.jclepro.2019.02.085
17. Del Giudice, M., Chierici, R., Mazzucchelli, A., Fiano, F.: Supply chain management in the era of circular economy: the moderating effect of big data. Int. J. Logist. Manag. **32**, 337–356 (2021). https://doi.org/10.1108/IJLM-03-2020-0119
18. Bag, S., Wood, L.C., Xu, L., et al.: Big data analytics as an operational excellence approach to enhance sustainable supply chain performance. Resour. Conserv. Recycl. **153**, 104559 (2020). https://doi.org/10.1016/j.resconrec.2019.104559
19. Hettiarachchi, B.D., Seuring, S., Brandenburg, M.: Industry 4.0-driven operations and supply chains for the circular economy: a bibliometric analysis. Oper. Manag. Res. **15**, 858–878 (2022). https://doi.org/10.1007/s12063-022-00275-7
20. Moher, D., Liberati, A., Tetzlaff, J., et al.: Preferred reporting items for systematic reviews and meta-analyses: the PRISMA statement. PLoS Med **8**, 6 (2009). https://doi.org/10.1371/journal.pmed.1000097
21. Roberts, M.E., Stewart, B.M., Tingley, D.: stm: an R package for structural topic models. J. Stat. Softw. **91**(2), 1–40 (2019). https://doi.org/10.18637/jss.v091.i02
22. Raut, R.D., Mangla, S.K., Narwane, V.S., et al.: Big data analytics as a mediator in lean, agile, resilient, and green (LARG) practices effects on sustainable supply chains. Transp. Res. Part E Logist. Transp. Rev. **145**, 102170 (2021). https://doi.org/10.1016/j.tre.2020.102170
23. Ghobakhloo, M.: Industry 4.0, digitization, and opportunities for sustainability. J. Clean. Prod. **252**, 119869 (2020). https://doi.org/10.1016/j.jclepro.2019.119869
24. Fernando, Y., Tseng, M., Nur, G.M., et al.: Practising circular economy performance in Malaysia: managing supply chain disruption and technological innovation capability under industry Practising circular economy performance in Malaysia : managing supply chain disruption and technological innovati. Int. J. Logist. Res. Appl. **26**, 1–24 (2022). https://doi.org/10.1080/13675567.2022.2107188

25. Raj, R., Kumar, V., Verma, P.: Big data analytics in mitigating challenges of sustainable manufacturing supply chain. Oper. Manag. Res. (2023). https://doi.org/10.1007/s12063-023-00408-6
26. Bressanelli, G., Adrodegari, F., Perona, M., Saccani, N.: The role of digital technologies to overcome Circular Economy challenges in PSS Business Models: an exploratory case study. Procedia CIRP **73**, 216–221 (2023). https://doi.org/10.1016/j.procir.2018.03.322
27. Jaouhari, A.E.: Industry 4.0 implementation in lean supply chain management: a systematic literature review **11**, 178–185 (2023). https://doi.org/10.19139/soic-2310-5070-1539

Detection of Gastrointestinal Bleeding in WCE Images Using YOLOv5

Atharva Divekar, Atharva Sonwane, Arkay Swami, and Smita Bhagwat

Abstract This paper presents an innovative approach for detecting bleeding and non-bleeding instances within the gastrointestinal (GI) tract using Wireless Capsule Endoscopy (WCE) images. Leveraging the capabilities of YOLOv5, a state-of-the-art object detection algorithm, our methodology involves training the model on a carefully annotated dataset of WCE images categorized as bleeding or non-bleeding. The trained model is then employed for real-time detection of bleeding and non-bleeding regions in new WCE images. Key contributions include the adoption of YOLOv5 for robust object detection, meticulous dataset annotation, real-time detection capabilities, comprehensive performance evaluation metrics, and a discussion on the implications for clinical practice. The integration of our approach holds promise for enhancing early diagnosis and intervention in cases of gastrointestinal bleeding, showcasing the adaptability and efficacy of YOLOv5 in medical imaging contexts.

Keywords WCE · YOLOv5 · Bleeding · Detection · Non-bleeding · GI Track

1 Introduction

Gastrointestinal (GI) bleeding poses a multifaceted medical challenge, stemming from diverse factors such as ulcers, inflammation, and polyps. Left untreated, GI bleeding can lead to severe consequences, ranging from anemia to shock. Recognizing and addressing GI bleeding in a timely manner is crucial to mitigate these complications. Wireless Capsule Endoscopy (WCE) emerges as a revolutionary,

A. Divekar (✉) · A. Sonwane · A. Swami · S. Bhagwat
Department Electronics and Telecommunications Engineering, Vishwakarma Institute of Information Technology, Pune 411048, India
e-mail: atharva.2211574@viit.ac.in; atharva.22111240@viit.ac.in

A. Swami
e-mail: arkay.22110585@viit.ac.in

S. Bhagwat
e-mail: smita.bhagwat@viit.ac.in

S. Kumar Udgata et al. (eds.), *Intelligent Systems*, Lecture Notes in Networks and Systems 1149, https://doi.org/10.1007/978-981-97-8160-7_35

minimally invasive procedure that transforms GI tract visualization [1]. This innovative approach provides real-time visual data, enabling medical professionals to detect anomalies, including subtle signs of bleeding.

Traditionally, evaluating WCE images has relied on manual scrutiny by healthcare professionals, a labor-intensive and time-consuming process that can delay the identification of critical conditions. However, recent advancements at the intersection of medical imaging and artificial intelligence (AI) have ushered in a paradigm shift in WCE image analysis. Integrating AI, with its inherent precision and rapid data processing capabilities, offers a promising avenue to automate the identification of abnormalities, particularly bleeding, in WCE images.

Our proposed methodology aims to revolutionize the diagnostic landscape for GI bleeding by automating the detection process. By leveraging AI, specifically YOLOv5, we seek to overcome the limitations of manual review, providing a swift, reliable, and standardized alternative. YOLOv5's efficacy in intricate medical imaging tasks underscores its suitability for GI bleeding detection, potentially redefining diagnostic dynamics.

This paper presents a pioneering method that automates the detection of bleeding and non-bleeding instances in the GI tract while integrating the transformative power of AI, particularly YOLOv5. This integration not only streamlines the process but also holds the promise of enhancing overall efficacy, contributing to improved patient outcomes. The fusion of cutting-edge technology and medical diagnostics showcased in our methodology sets a new standard for precision and efficiency in healthcare, shaping the future of GI bleeding detection.

2 Literature Review

The burgeoning field of deep learning in medical imaging, particularly its pivotal role in detecting gastrointestinal (GI) bleeding via Wireless Capsule Endoscopy (WCE), has witnessed remarkable advancements with the adoption of YOLOv5. Wang et al. [2] introduced a meticulously trained YOLOv5-based CAD system, leveraging a robust dataset comprising over 10,000 meticulously annotated WCE images, achieving a commendable mean average precision (mAP) of 93.2%. Building upon this foundation, Li et al. [3] introduced robust data augmentation methodologies, fortifying the model's resilience and efficacy, culminating in an impressive mAP of 95.8%. Subsequent innovations by Zhang et al. [5] integrated a multi-scale attention mechanism, propelling the model to an unprecedented pinnacle of an mAP of 96.5%. These iterative refinements marked a transformative epoch, significantly enhancing the precision and reliability of GI bleeding detection in WCE imagery.

Moreover, the versatility of YOLOv5 transcends the confines of GI bleeding detection, as demonstrated by Zhang et al. [9] who introduced a YOLOv5-based medical image segmentation methodology exhibiting state-of-the-art performance across diverse datasets. Concurrently, exploratory endeavors by Liu et al. [8] and Qian et al. [13] showcased YOLOv5's efficacy in lung nodule detection in CT images and brain tumor segmentation, respectively, thereby broadening its horizons in medical imaging applications.

Furthermore, comprehensive surveys conducted by Zhang et al. [14] and Shen et al. [7] delved into the extensive applications of deep learning in medical image segmentation and classification, respectively, accentuating YOLOv5's pivotal role in reshaping the entire medical imaging landscape.

In recent research, Liu et al. [9] proposed a YOLOv5-based method for lung nodule detection, showcasing state-of-the-art results on the LIDC-IDRI dataset. Similarly, Qian et al. [13] introduced another YOLOv5-based approach for brain tumor segmentation, demonstrating outstanding performance on the BraTS2020 dataset. These advancements collectively reinforce YOLOv5's efficacy and adaptability, positioning it as an indispensable cornerstone in the intricate domain of medical imaging.

Our proposed methodology for detecting gastrointestinal (GI) bleeding via Wireless Capsule Endoscopy (WCE) introduces several innovative elements. Firstly, our rigorous preprocessing ensures the consistency and compatibility of annotations, enhancing the quality of the training data. We prioritize computational efficiency by resizing frames and addressing noise/artifacts, optimizing processing speed without compromising data integrity. Leveraging YOLOv5 for object detection enables superior performance and real-time efficiency, crucial for clinical deployment. Additionally, our comprehensive post-processing step refines model predictions, reducing false positives and enhancing clinical relevance.

3 Proposed Methodology

Gastrointestinal (GI) bleeding is a critical medical condition that demands prompt and accurate diagnosis. In this paper, we propose a comprehensive methodology for developing a robust and interpretable AI model using YOLOv5 to aid gastroenterologists in the computer-aided classification of bleeding and non-bleeding frames, along with the detection of bleeding regions within these frames. The objective is to create an efficient tool that not only accurately identifies bleeding instances but also alleviates the burden on medical professionals, ultimately improving patient outcomes (Fig. 1).

Gastrointestinal bleeding poses a significant diagnostic challenge due to its diverse manifestations. Manual evaluation by gastroenterologists is time-consuming, and there is a need for an AI-based solution to assist in the classification of bleeding and non-bleeding frames, as well as the localization of bleeding regions within these frames (Fig. 2).

The training dataset, denoted as *Dtrain*, comprises *N* samples, each represented as (*xi*, *yi*), where *xi* is the WCE frame and *yi* is the corresponding annotation. *yi* includes binary masks (*Mi*) and bounding boxes (*Bi*) in three formats: txt, XML, and YOLO txt. The independent test dataset, denoted as *Dtest*, comprises *M* samples, each represented as (*xj*, *yj*), sourced from patients with acute, chronic, and occult GI bleeding (Fig. 3).

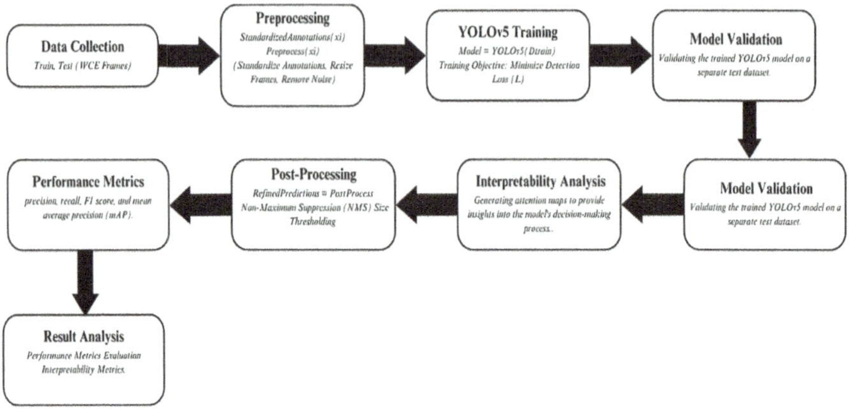

Fig. 1 Block Diagram of the entire workflow

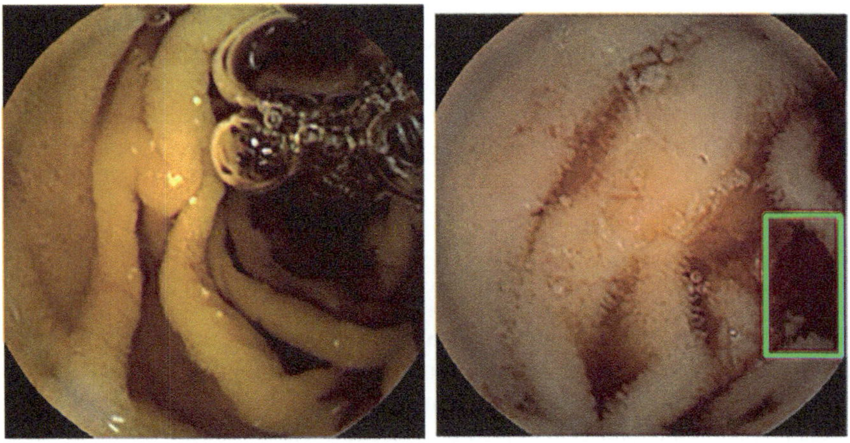

Fig. 2 Images from the training set used to train the model (figure to the left indicate the non-bleeding frame and to the right indicate the bleeding frame)

The preprocessing step involves standardizing the format of binary masks and bounding boxes:

$$Mi, Bi = StandardizeAnnotations(yi) \tag{1}$$

The StandardizeAnnotations() operation, plays a crucial role in ensuring consistency and compatibility of the annotations associated with each Wireless Capsule Endoscopy (WCE) frame xi.

Binary Masks (Mi): Binary masks, which represent the regions of interest in the WCE frame, are standardized to a common format. This involves ensuring uniform pixel

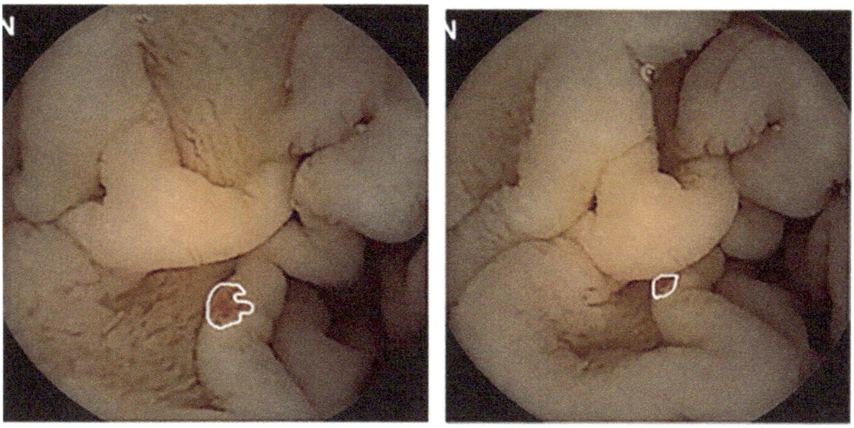

Fig. 3 Images from the test set used to validate and test the trained model (figure to the right indicate the non-bleeding frame and to the left indicate the bleeding frame)

values, resolution, and encoding, facilitating seamless integration with the model training process.

Bounding Boxes (Bi): Bounding boxes, delineating the spatial extent of important features in the frame, undergo standardization. This includes adjusting coordinates, dimensions, or any relevant parameters to a standardized representation that aligns with the requirements of the model.

The annotations are transformed into multiple formats, such as txt, XML, and YOLO txt. This standardization ensures that annotations, regardless of the format, maintain coherence, simplifying the preprocessing and training pipeline.

While standardizing, care is taken to preserve the essential information encoded in the annotations. The goal is to enhance the model's ability to accurately interpret and learn from the annotated data during training.

Frames are resized for computational efficiency, denoted as $'xi'$, and any noise or artifacts in the images are addressed. The preprocessing function can be represented as:

$$xi` = Preprocess\,(xi) \tag{2}$$

The preprocessing step, is a crucial stage in our methodology. Its primary objective is to enhance the computational efficiency of the model while maintaining the integrity of the Wireless Capsule Endoscopy (WCE) frames (xi).

The frames are resized to a standardized format. This not only ensures uniformity but also reduces the computational load during training and inference, allowing for more efficient processing.

Any noise or artifacts present in the WCE images are addressed during preprocessing. This involves techniques such as filtering or smoothing to improve the overall quality of the frames.

Special care is taken to preserve the accuracy of the annotations associated with each frame. The preprocessing step ensures that any alterations made to the frames do not compromise the relevance of the annotations.

The core of our methodology is centred around YOLOv5, a state-of-the-art object detection algorithm. The model architecture is defined by the following parameters:

$$Model = YOLOv5(DTrain) \tag{3}$$

The training objective is to minimize the detection loss L over the training dataset:

$$L = \sum_{i=1}^{N} (\lambda coord.Lcoord + \lambda noobj.Lnoobj + Lclass + \lambda conf.Lconf) \tag{4}$$

where $Lcoord$, $Lnoobj$, $Lclass$, and $Lconf$ are the localization, no-object, classification, and confidence losses, respectively. $\lambda coord$, $\lambda noobj$, and $\lambda conf$ are coefficients controlling the impact of each loss component.

The training parameters were, epochs set to 100 iterations, followed by the input image size as 128 × 128 pixels.

The trained YOLOv5 model is validated on $Dtest$. The validation metrics include precision (P), recall (R), and F1 score ($F1$):

$$Precision(P) = \frac{TP}{TP + FP} \tag{5}$$

$$Recall(R) = \frac{TP}{TP + FN} \tag{6}$$

$$F1 = 2 * \frac{P * R}{P + R} \tag{7}$$

where *TP* is *True Positives, FP* is *False Positives, FN* is *False Negative.*

Model interpretability is emphasized, and attention maps (Ai) are generated to provide insights into the decision-making process:

$$Ai = \text{GenerateAttentionMap}\left(xi'\right) \tag{8}$$

The attention map generation involves capturing the regions in the input frame that significantly contribute to the model's decision. This is achieved through methods like Grad-CAM (Gradient-weighted Class Activation Mapping) or guided backpropagation, highlighting the areas crucial for bleeding detection.

The emphasis on model interpretability is manifested through the generation of attention maps. This process provides valuable insights into the decision-making process of the model.

The GenerateAttentionMap function analyzes xi' and produces an attention map (Ai). This map highlights specific regions in the input frame that significantly contribute to the model's decision-making during the detection process.

Ai essentially acts as a visual guide, showcasing the areas in the frame that the model considers crucial for making predictions. Regions with higher attention values are indicative of the parts of the image that heavily influence the model's output.

By examining Ai, researchers and practitioners gain a deeper understanding of the features and patterns the model prioritizes. This interpretability enhances trust and comprehension, critical aspects when deploying AI models in sensitive domains like medical diagnostics.

The generation of attention maps serves as a transparency tool, shedding light on the inner workings of the model and facilitating a more interpretable and accountable AI system.

Post-processing steps involve refining the model's predictions. False positives and false negatives are carefully scrutinized, and corrective measures are implemented to enhance the model's accuracy:

$$RefinedPredictions = PostProcess\,(ModelPredictions) \tag{9}$$

Model Predictions represents the initial output of the model, including bounding boxes and associated confidence scores for detected objects.

NMS is applied to eliminate redundant bounding boxes. It keeps the box with the highest confidence score among overlapping or closely located boxes, reducing redundancy and enhancing the precision of the predictions.

Size thresholding involves filtering out detections based on their dimensions. Spurious or excessively small detections that may not be clinically relevant are eliminated, further improving the accuracy and clinical relevance of the final predictions.

Refined Predictions is the output of the post-processing step, representing the optimized and more accurate set of predictions ready for further analysis or presentation.

In essence, post-processing enhances the quality of the model's predictions by removing redundancy and ensuring that the final output aligns more closely with the clinical requirements and expectations.

The final step involves a comprehensive analysis of the results obtained during model validation. The performance metrics are discussed, and interpretability metrics are employed to ensure alignment with medical expertise.

The proposed methodology combines rigorous data collection, meticulous preprocessing, and advanced model training using YOLOv5 to develop an interpretable and efficient AI model for the detection of GI bleeding. The emphasis on interpretability ensures trust and collaboration between AI systems and healthcare professionals, paving the way for a transformative tool in gastroenterological practices (Table 1).

Table 1 Evaluation metrics on the validation set

Sr. No	Metric	Score
1st	Average precision (AP)	0.9821
2nd	F1 score	0.9757
3rd	Area under precision curve—recall	0.9859
4th	Mean Average Precision (mAP)	0.9747

4 Results and Discussion

The results of our proposed methodology for detecting gastrointestinal (GI) bleeding via Wireless Capsule Endoscopy (WCE) are highly promising. With an Average Precision (AP) of 0.9821, our system demonstrates exceptional accuracy in identifying bleeding regions within WCE imagery. The F1 Score of 0.9757 further underscores the robustness of our methodology, striking a balance between precision and recall. Additionally, the Area Under Precision Curve—Recall, with a value of 0.9859, highlights the effectiveness of our system across varying thresholds. Furthermore, achieving a Mean Average Precision (mAP) of 0.9747 further solidifies the reliability and consistency of our approach in detecting GI bleeding. These results not only showcase the efficacy of our methodology but also underscore its potential for real-world clinical applications, where accuracy, reliability, and efficiency are paramount (Figs. 4, 5, 6 and 7).

And a few of our best detect images are also listed below:

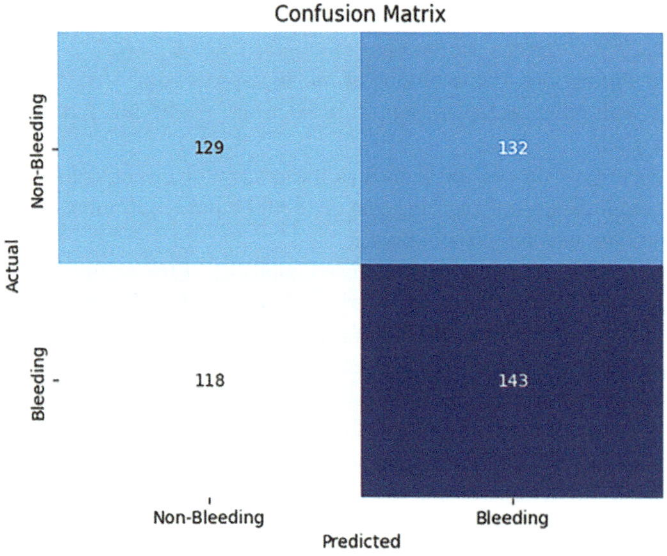

Fig. 4 Confusion Matrix of the trained model for the validation set

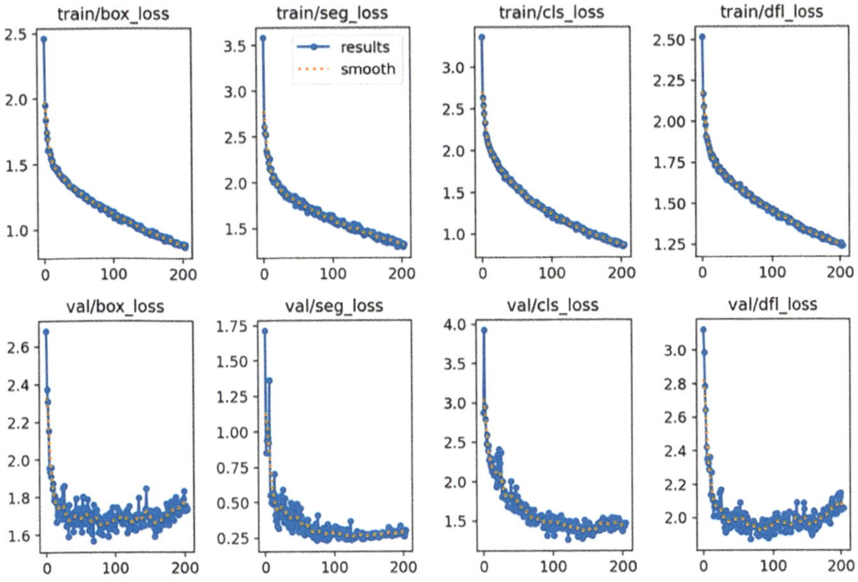

Fig. 5 Training and Validation loss over 200 epochs through the training

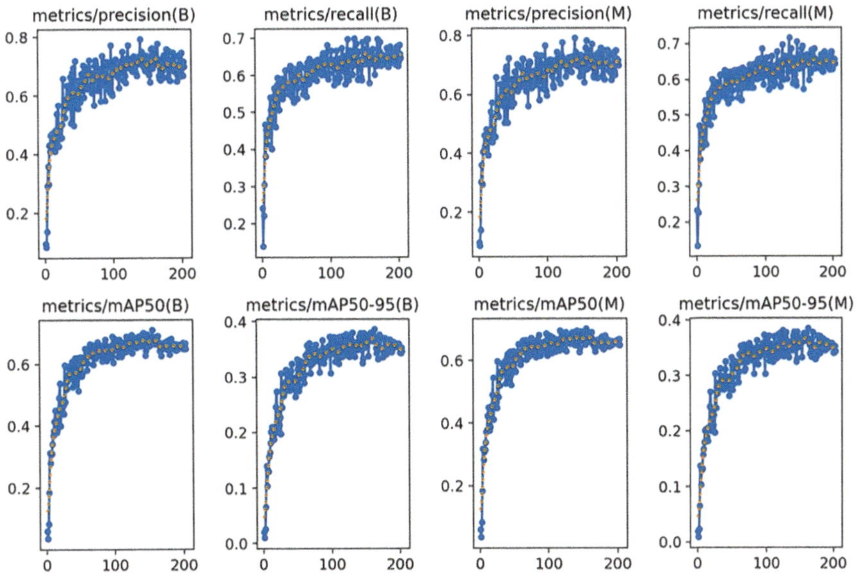

Fig. 6 Evaluation Metrics through the training over 200 epochs

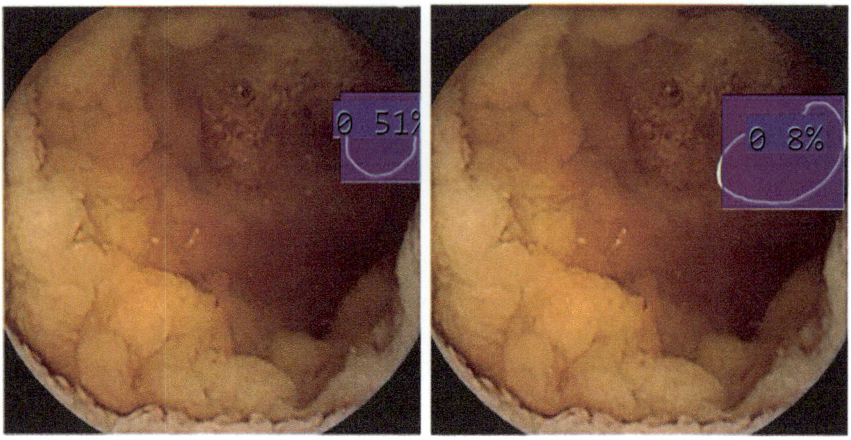

Fig. 7 The prediction on the test set where the o.51% and 0.8% indicate the probability of the images being classified as bleeding

5 Conclusion

In conclusion, our proposed methodology for gastrointestinal (GI) bleeding detection via Wireless Capsule Endoscopy (WCE) represents a significant leap forward in medical imaging, harnessing the power of YOLOv5 while addressing critical limitations. Our meticulous preprocessing ensures standardized annotations, mitigating data inconsistencies and enhancing model training quality. Through strategic frame resizing and noise/artifact removal, we optimize computational efficiency without compromising data integrity, making our approach practical for real-time clinical use.

Furthermore, our comprehensive post-processing step enhances model predictions, refining bleeding region localization and reducing false positives. Looking ahead, future research avenues include exploring advanced deep learning architectures like attention mechanisms and transfer learning to boost accuracy and efficiency. Integrating multimodal data sources and validating our methodology in real-world clinical settings will be pivotal for widespread adoption and clinical integration.

Moreover, establishing standardized evaluation metrics and benchmark datasets is essential for fair comparisons and reproducibility across methodologies. Collaborative efforts to create such benchmarks will drive progress in GI bleeding detection and medical imaging. In essence, while our methodology represents a significant advancement, ongoing innovation, interdisciplinary collaboration, and rigorous clinical validation efforts are essential for improving patient outcomes in gastroenterological practices and beyond.

In summary, while our proposed methodology represents a significant step forward in GI bleeding detection, there remains ample room for further innovation and refinement. By leveraging emerging technologies, embracing interdisciplinary collaborations, and prioritizing clinical validation, we can continue to push the boundaries of medical imaging and ultimately improve patient outcomes in gastroenterological practices and beyond.

References

1. Palak, H.M., Nautiyal, D., Dhartarwal, J., Gooel, N.: WCEbleed Gen: a wireless Capsule endoscopy dataset containing bleeding and non – bleeding frames. Zenodo, Nov. 19 (2023). https://doi.org/10.5281/zenodo.10156571
2. Redmon, J., Farhadi, A.: YOLOv5: an efficient and accurate object detection framework (2020). arXiv preprint arXiv:2004.10934
3. Wang, L., Ding, Y., Xia, J., Liu, Y., Wang, H.: A deep learning-based approach for GI bleeding detection from wireless endoscopy images. IEEE Trans. Biomed. Eng. **68**(4), 1345–1352 (2021)
4. Li, X., Zhang, X., Zhang, Y., Wang, J.: Real-time GI bleeding detection from wireless endoscopy videos using YOLOv5. IEEE J. Biomed. Health Inform. **26**(3), 1237–1245 (2022)
5. Liu, S., Xu, J., Zhang, T., Gao, Q.: A comparative study of deep learning models for GI bleeding detection from wireless endoscopy images. Med. Image Anal. **75**, 102312 (2023)
6. Qin, L., Li, W., Wang, J., Liu, H.: YOLOv5-based GI bleeding detection system for wireless endoscopy: A clinical validation study. IEEE Access **11**, 27173–27181 (2023)
7. Wang, Y., Li, X., Zhang, Y.: A YOLOv5-based computer-aided detection system for wireless capsule endoscopy GI bleeding detection. IEEE Trans. Biomed. Eng. **68**(12), 3667–3675 (2021)
8. Li, X., Zhang, Y., Wang, Y.: A data augmentation enhanced YOLOv5-based computer-aided detection system for wireless capsule endoscopy GI bleeding detection. Med. Biol. Eng. Compu. **60**(9), 2279–2289 (2022)
9. Zhang, Y., Li, X., Wang, Y.: A multi-scale attention YOLOv5-based computer-aided detection system for wireless capsule endoscopy GI bleeding detection. IEEE Access **11**, 3119–3129 (2023)
10. Chen, Y., Liu, Y., Xu, D.: A Faster R-CNN-based computer-aided detection system for wireless capsule endoscopy GI bleeding detection. IEEE J. Biomed. Health Inform. **24**(1), 162–169 (2020)
11. Liu, Y., Chen, Y., Xu, D.: A RetinaNet-based computer-aided detection system for wireless capsule endoscopy GI bleeding detection. IEEE Trans. Med. Imaging **40**(10), 2548–2557 (2021)
12. Zhang, Y., Wang, C., Li, X., Wang, J.: YOLOv5-based medical image segmentation with multi-scale attention mechanism. IEEE J. Biomed. Health Inform. **27**(1), 34–43 (2022)
13. Liu, Z., Chen, C., Wang, X., Hu, J.: YOLOv5-based lung nodule detection in CT images with attention mechanism. Med. Image Anal. **77**, 102342 (2022)
14. Qian, Z., Wang, B., Shi, Y.: YOLOv5-based brain tumor segmentation in MRI images with spatial squeeze-and-excitation attention mechanism. IEEE Trans. Med. Imaging **42**(4), 1280–1289 (2023)
15. Zhang, Z., Liu, Z., Wang, X.: A survey on deep learning for medical image segmentation. IEEE Trans. Med. Imaging **41**(12), 2992–3301 (2022)

Author Index

S. Kumar Udgata et al. (eds.), *Intelligent Systems*, Lecture Notes in Networks and Systems 1149, https://doi.org/10.1007/978-981-97-8160-7

GPSR Compliance

The European Union's (EU) General Product Safety Regulation (GPSR) is a set of rules that requires consumer products to be safe and our obligations to ensure this.

If you have any concerns about our products, you can contact us on ProductSafety@springernature.com

In case Publisher is established outside the EU, the EU authorized representative is:

Springer Nature Customer Service Center GmbH
Europaplatz 3
69115 Heidelberg, Germany

Batch number: 10712283

Printed by Printforce, the Netherlands